# 175
# Advances in Biochemical Engineering/Biotechnology

## Aims and Scope

This book series reviews current trends in modern biotechnology and biochemical engineering. Its aim is to cover all aspects of these interdisciplinary disciplines, where knowledge, methods and expertise are required from chemistry, biochemistry, microbiology, molecular biology, chemical engineering and computer science.

Volumes are organized topically and provide a comprehensive discussion of developments in the field over the past 3–5 years. The series also discusses new discoveries and applications. Special volumes are dedicated to selected topics which focus on new biotechnological products and new processes for their synthesis and purification.

In general, volumes are edited by well-known guest editors. The series editor and publisher will, however, always be pleased to receive suggestions and supplementary information. Manuscripts are accepted in English.

In references, Advances in Biochemical Engineering/Biotechnology is abbreviated as *Adv. Biochem. Engin./Biotechnol.* and cited as a journal.

More information about this series at http://www.springer.com/series/10

Erdmann Rapp • Udo Reichl
Editors

# Advances in Glycobiotechnology

With contributions by

K. Alagesan · M. J. Betenbaugh · M. Butler · J. F. Buyel · S. Cajic · N. Callewaert · Ö. Can · M. M. J. Cox · M. P. DeLisa · C. De Wachter · L. Elling · R. Fischer · S. L. Flitsch · H. Frohnmeyer · J. Gottschalk · G. Goyette-Desjardins · K. M. Heffner · D. B. Hizal · M. Hoffmann · T. Holland · R. Jefferis · D. Kolarich · D. Laaf · B. Lepenies · A. Malik · J. E. R. Martinez · D. C. Mills · T. D. Moeller · L. A. Palomares · A. Pralow · O. T. Ramírez · E. Rapp · U. Reichl · T. Rexer · M. Sack · S. Schillberg · K. Schön · P. H. Seeberger · I. K. Srivastava · E. Stoger · B. Thomas · R. M. Twyman · L. Van Landuyt · D. Varón Silva · Q. Wang · K. B. Weyant · L. E. Yates

Springer

*Editors*
Erdmann Rapp
Bioprocess Engineering
Max Planck Institute for Dynamics
of Complex Technical Systems
Magdeburg, Germany

glyXera GmbH
Magdeburg, Germany

Udo Reichl
Bioprocess Engineering
Max Planck Institute for Dynamics
of Complex Technical Systems
Magdeburg, Germany

Otto-von-Guericke University Magdeburg
Magdeburg, Germany

ISSN 0724-6145 ISSN 1616-8542 (electronic)
Advances in Biochemical Engineering/Biotechnology
ISBN 978-3-030-69589-7 ISBN 978-3-030-69590-3 (eBook)
https://doi.org/10.1007/978-3-030-69590-3



This Springer imprint is published by the registered company Springer Nature Switzerland AG.
The registered company address is: Gewerbestrasse 11, 6330 Cham, Switzerland

# Preface

From the different languages on the molecular level of life, only the genome has become a subject of wider public interest, while the proteome and even more the glycome remain reserved for specialists.

Biopharmaceuticals – also known as biologicals or biologics – include purified and recombinantly expressed therapeutic proteins like monoclonal antibodies, erythropoietin, insulin, growth and coagulation factors, hormones, interferons, and others. In addition, vaccines mainly developed and applied as inactivated or attenuated intact pathogens or as purified or recombinantly expressed major antigens play a major role. Recombinant human insulin was among the first substances to be approved for therapeutic purpose, and currently, there are nearly 300 biopharmaceutical products which have been approved and are available in the market. The global biopharmaceutical market size was about $240 billion in 2019 and is expected to grow at a compound annual growth rate (CAGR) of 14.20% during the forecasting period (2020–2027). Most biopharmaceuticals are glycoproteins, which are comprised of proteins and glycans – complex carbohydrates consisting of glycosidically linked monosaccharides. Out of the top 10 drugs by global sales in 2019, seven were recombinant glycoproteins. In terms of value, monoclonal antibodies represent the largest market share with vaccines being second.

Glycosylation is not only important for the correct folding of glycoproteins, it has a large impact on the mechanisms of action of biopharmaceuticals, their pharmacokinetics and their pharmacodynamics. Moreover, it plays a crucial role in various biological processes such as cell proliferation, cell–cell recognition, pathogen–host interaction, and immune responses. For recombinant therapeutic glycoproteins, glycosylation is therefore classified by the authorities as a critical quality attribute (CQA). However, in contrast to proteins that are defined by the nucleotide sequence of genes, glycans are synthesized from a complex matrix of enzymes without a predefined template. This results in large heterogeneity in biomanufacturing glycoproteins with consequent high variability in their properties. Therefore, in glycobiotechnology, the engineering and analysis of glycosylation is of critical interest, and significant efforts have been made to improve glycoengineering and glycoanalytical toolboxes.

In order to present the complex field of glycobiotechnology and its latest developments in full breadth, our book "Advances in Glycobiotechnology" will address the following topics:

- The impact of the expression system on glycosylation
- The influence of culture conditions on glycosylation
- Synthetic and biosynthetic glycoengineering
- Glycoengineering via cell line design
- Technologies and methods for glycoanalysis
- Challenges in the industrial production of therapeutics and vaccines

Written by selected experts in the field, and divided into 14 chapters, this book will provide a wide coverage on the state of the art in analytics, pharmaceutical process technologies, and medical applications in glycobiotechnology.

Magdeburg, Germany

Erdmann Rapp
Udo Reichl

# Contents

# Animal Cell Expression Systems

**M. Butler and U. Reichl**

## Contents



The original version of this chapter was revised: An erratum to this chapter can be found at DOI 10.1007/10_2018_60.

A correction to this chapter can be found at DOI 10.1007/10_2020_121.

M. Butler
National Institute of Bioprocessing Research & Training (NIBRT), Dublin 4, Ireland
e-mail: michael.butler@nibrt.ie

U. Reichl (✉)
Lehrstuhl für Bioprozesstechnik, Otto-von-Guericke-Universität Magdeburg, Magdeburg, Germany
e-mail: udo.reichl@ovgu.de

**Abstract** The glycan profile of therapeutic recombinant proteins such as monoclonal antibodies is a critical quality attribute, which affects the efficacy of the final product. The cellular glycosylation process during protein expression is dependent upon a number of factors such as the availability of substrates in the media, the intracellular content of nucleotide sugars, and the enzyme repertoire of the host cells. In order to control the variability of glycosylation it is important to understand the critical process parameters and their acceptable range of values to enable reproducible production of proteins with a predetermined glycan profile providing the desired biological function or therapeutic effect. The depletion of critical nutrients such as glucose or galactose, which may occur toward the end of a culture process, can lead to truncated glycans. Terminal galactosylation and sialyation are particularly variable but may be controlled by the presence of some key media components. Ammonia accumulation, pH, and dissolved oxygen levels are also known to be key bioprocess parameters that affect the glycosylation of recombinant proteins. Specific enzyme inhibitors can be added to the media to drive the formation of selected and predetermined glycan profiles. Various attempts have been made to predict the glycan profiles of cellular expressed proteins and have led to metabolic models based upon knowledge of metabolic flux and the kinetics of individual glycosylation reactions.

In contrast to single recombinant proteins, the glycan profiles of viral vaccines are far more complex and difficult to predict. The example of influenza A virus shows that hemagglutinin, the major antigenic determinant, has three to nine *N*-glycans, which may influence the antigenicity and efficacy of the vaccine. Glycosylation of the influenza A virus has been largely unmonitored in the past as production has been from eggs, where glycan profiles of antigens are difficult if not impossible to control. Over the past decade, however, there have been various commercial influenza vaccines made available from cell technology using animal host cells. Analysis of glycosylation control shows that the type of host cell has the greatest influence on the final analyzed glycan profile. Other factors such as the virus strain, the cultivation system, or various process parameters have been shown to have only a minor effect on the glycosylation pattern. We predict that the analysis of glycan profiles in viral vaccines will become increasingly important in the development and consistent manufacturing of safe and potent vaccines.

**Graphical Abstract**

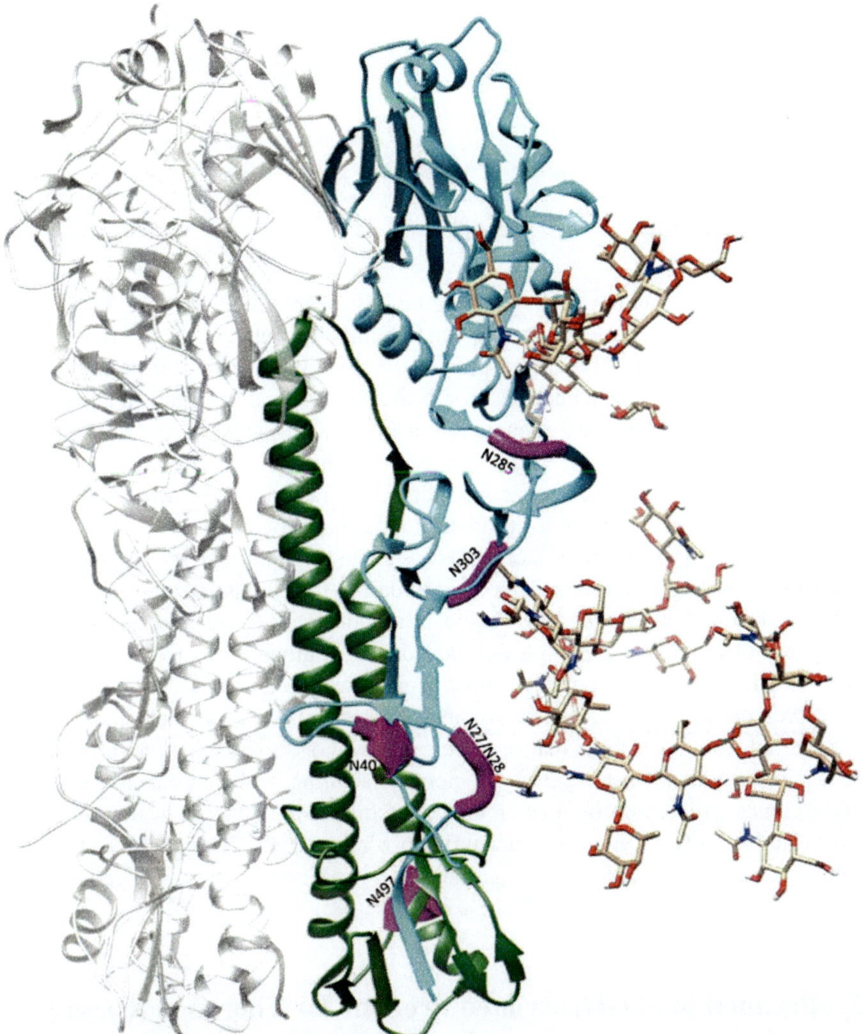


**Keywords** ADCC, Ammonia, Castanospermine, Galactosylation, Glycosylation, Hemagglutinin, Influenza, MDCK cells, Oxygen, Process conditions, QbD, Sialylation, Swainosine, Vero cells

## Abbreviations

| | |
|---|---|
| GDP | Guanosine diphosphate |
| Glc | Glucose |
| GlcNAc | *N*-Acetylglucosamine |

| | |
|---|---|
| HexNAc | *N*-Acetyl hexosamine |
| IgG | Immunglobulin G |
| Man | Mannose |
| ManNAc | *N*-Acetylmannosamine |
| UDP | Uridine diphosphate |

# 1 Introduction

The global biopharma market is growing rapidly, with a recent report showing an annual growth rate of 8.6% and an expectation that the global market will reach US $ 291 billion by 2021 [1]. Current products include more than 200 protein-based therapeutics that have been approved for the treatment of unmet medical needs such as cancer, multiple sclerosis, rheumatoid arthritis, and many other serious human conditions [2]. In terms of value, monoclonal antibodies represent the largest market share, with vaccines being second. In contrast, the worldwide production volume of vaccines far exceeds all other biopharmaceuticals.

The availability of biopharmaceuticals has come about through the ability to grow animal cells in large-scale bioreactors and to design these cells for the production of specific glycoprotein targets. Most biopharmaceuticals are glycoproteins, which are composed of proteins and complex carbohydrates. Although the protein structure is determined by the sequence of nucleotides in the gene, the carbohydrates or glycans are far more complex because of their branched structures and the fact that they are synthesized from a complex matrix of enzymes without a predefined template. This results in considerable heterogeneity in biomanufacturing, with consequent variability in the clinical efficacy of the final products. We first address the production of monoclonal antibodies and then the manufacturing of viral antigens. The focus is on Chinese hamster ovary (CHO) cells and various other cell lines used for virus replication.

# 2 Production of Glycosylated Proteins for Therapeutic Use

Glycosylation of proteins is now identified as crucial quality attribute (CQA) that is essential for the effective clinical function of the final drug product. The glycosylation profile is unique to each glycoprotein in terms of attachment of a glycan at a specific peptide site (macroheterogeneity) and structural variation at each site (microheterogeneity). Variability in the profile can affect physicochemical properties, including solubility, thermal stability [3], protease resistance [4], and aggregation [5, 6], and result in clinical variation in serum half-life [7], immunogenicity [8, 9], and therapeutic efficacy [10]. Unlike proteins and nucleic acids, the polysaccharide structure of a glycan is not governed by a template. This means that the variability is affected by the portfolio of enzyme activities in the producer cell line, the availability of precursors, and the environmental conditions of the bioprocess.

## 2.1 Process Variables

There are several aspects of process control associated with the glycosylation of a protein destined as a therapeutic product. First, it is essential to maintain batch-to-batch consistency so that the structure and function of the final product does not vary. For this, it is important to identify the process parameters that cause variability and determine the range of values that are permissible in limiting product variability. This is the intrinsic method of quality by design (QbD), a procedure recommended by the Food and Drug Administration (FDA) to understand how the variables of a manufacturing process influence product quality.

Second, it is important to determine the range of process variables that produce a glycoprotein with a predetermined glycan profile that will provide the desired biological function or therapeutic effect. This second level is not so easy because it requires both an understanding of the structure-to-function relationship of a single glycoform and the ability to produce a restricted glycoform product consistently during biomanufacture. An example is the discovery that a nonfucosylated antibody can elicit significantly higher clinical efficacy than its fucosylated counterpart through enhancement of the antibody-dependent cellular cytotoxicity (ADCC). Rituxan (rituximab) was developed by Biogen and Genentech as a humanized anti-CD20 monoclonal antibody (Mab) and since 1997 has been used as a highly effective treatment for chronic lymphocytic leukemia. However, more recently, a nonfucosylated form of rituximab was obtained by gene manipulation of the producer cell to manufacture Ganzya (obinutuzumab). This nonfucosylated form of the antibody has a 35-fold enhanced ADCC compared with rituximab [11].

## 2.2 Culture Media: The Contribution of Nutrients to Glycosylation

The composition of the complex medium required to grow animal cells is a key factor in ensuring consistent recombinant protein production from a producer cell. Media formulations can contain 60–100 components, which change in concentration during a batch culture [12]. The gradual depletion of nutrients during the course of culture certainly results in time-dependent effects on glycosylation. Glucose and glutamine are key nutrients utilized for energy metabolism during cell growth. However, there are also precursors for glycosylation and several studies suggest that when the concentration of either of these compounds reaches a critically low level in the medium then the glycosylation process is compromised in preference to primary energy metabolism.

The control of microheterogeneity by nutrient feeding is crucial in producing consistent biopharmaceuticals and in avoiding significant batch-to-batch product variation and diminished therapeutic efficacy. However, each cell line and clone may have specific metabolic characteristics that can affect protein glycosylation [13]. Accordingly, to ensure consistent product quality, metabolic analysis of culture parameters and high-throughput glycan analytics are necessary in order to monitor factors that affect glycosylation.

It was recognized some time ago that underglycosylation and abnormal truncated glycans could result from glucose depletion in the medium [14]. Glucose starvation is attributed to a shortage of glucose-derived precursors and results in an intracellular depleted state, giving rise to a higher proportion of high mannose glycans [15]. Curling et al. showed a reduction in glycan site occupancy of gamma-interferon toward the end of a batch culture of CHO cells [16]. In a follow-up experiment using a different mode of culture but with the same cells, Hayter et al. detected a high proportion of nonglycosylated gamma-interferon from a glucose-limited chemostat; however, normal levels of glycosylation were restored by pulsed additions of glucose [17]. Furthermore, it was shown that there is a critical concentration of glucose (<0.5 mM) below which reduced site-occupancy of *N*-glycans is observed in IgG produced from mouse myeloma cells [18].

Liu et al. showed that depletion of glucose resulted in synthesis of a reduced size glycan in the lipid-linked oligosaccharide precursor, from the 14-oligomer dolichol-GlcNAc$_2$-Man$_9$-Glc$_3$ structure to the shorter dolichol-GlcNAc$_2$-Man$_5$, correlating with a reduction in glycan occupancy on a Mab [19]. This effect also correlated with a reduced concentration of intracellular nucleotide sugars, GDP-sugars. and UDP-hexosamines [20]. It was also shown that the time of cell exposure to glucose-depleted medium was tightly correlated with reduced galactosylation of the fraction of Mabs that were glycosylated [19]. This finding is pertinent to the operation of fed-batch cultures, which operate with cycles of nutrient feeding. These periodic cycles may include times of depleted nutrients, which might not affect cell growth but could increase glycan heterogeneity.

Intracellular nucleotide sugars are the immediate precursors of protein glycosylation in the endoplasmic reticulum (ER) and Golgi apparatus. So, it is not unexpected that low or depleted levels of glucose or glutamine in the medium can result in decreased intracellular concentrations of these precursors, which in turn affects glycosylation with enhanced macroheterogeneity (glycan occupancy) and microheterogeneity (variable glycan structures) [20–23]

In several studies, elevated intracellular levels of UDP-HexNAc resulted in higher antennarity of the glycan structures of several proteins [24–27]. This has been attributed to higher ammonia levels, although this does not have to be the case, as shown with cells adapted to glutamine-free medium but still showing a correlation between elevated UDP-HexNAc and glycan antennarity [28].

Media supplementation with nucleotide precursors such as glucosamine and uridine for UDP-GlcNAc synthesis [29, 30], uridine and galactose for UDP-Gal synthesis [31, 32], galactose, glucosamine, or *N*-acetylmannosamine (ManNAc) [33] have been successful in increasing the nucleotide sugar availability and promoting specific glycosylation targets. However, it is well recognized that the ratios of sugar nucleotides are also important. For example, it has been shown that elevated levels of UDP-HexNAc impair cytidine monophosphate (CMP)-acetylneuraminic acid transport into the Golgi apparatus, thus reducing sialylation [34]. Furthermore, enhanced sialylation has been shown in gamma-interferon production by supplementation with ManNAc [35].

However, other factors such as enzymes for nucleotide sugar biosynthesis or transporters may be limiting in some cell lines [36]. Thus, enhanced sialylation was improved by overexpression of a CMP-sialic acid transporter [23] and supplementation of the culture with galactose, glucosamine, and ManNAc [33]. In human embryonic kidney (HEK293) cells, the hexosamine biosynthetic pathway and the *N*-acetylglucosamine (GlcNAc) transferases, which control glycan branching, may influence the uptake of glutamine and essential amino acids under low nutrient conditions and allow increased cell growth [37].

## 2.3 Galactosylation

Terminal galactosylation of glycans of recombinant antibodies exhibits significant variability, depending on the state of the medium. Because of the sequential nature of the transferase enzymes, the addition of galactose to a glycan chain is a prerequisite for terminal sialylation. Feeding cultures with galactose can ensure high levels of terminal galactosylation, as shown in the production of a number of antibodies [38]. Galactose feeding was shown to increase uridine diphosphate (UDP)-galactose pool in the cell up to 20-fold compared with control levels and corresponded to a concentration of 7 fmol/$10^5$ viable cells. However, in a separate study, Clark et al. showed that the sialic acid content of a glycoprotein is not increased by galactose feeding [39]. They attributed this to enhanced intracellular sialidase activity in the galactose-fed cultures that increased the potential for desialylation. Kildegaard et al. determined the effect of eight independent supplements on the glycoprofile of an immunoglobulin produced from CHO cells in fed-batch cultures. This work showed that supplementing the medium with galactose consistently enhanced galactosylation, whereas addition of GlcNAc or mannose caused a small but significant decrease [40]. Addition of up to 40 mM galactose to culture media resulted in enhanced galactosylation and sialylation of a recombinant fragment crystallizable region (Fc)-fusion protein, with minimal effect on culture performance apart from a reduction in glucose uptake [41].

Specific glycosyltransferase reactions could be enhanced by the availability of substrates and specific cofactors. A cocktail of supplements comprising uridine, manganese, and galactose (UMG) was found to stimulate the galactosylation process [32], which is often measured by the galactosylation index (GI = 0–1) [42]. Figure 1 shows how the availability of glucose to the cells directly correlates with the extent of galacatosylation and sialylation of the synthesized antibody [19]. Although commercially available Mabs are associated with a relatively low GI level ($<0.35$) [43], use of UMG supplement enhanced the galactosylation of a chimeric human–llama Mab to an even higher level than shown in Fig. 1, giving GI values up to 0.83 [19]. Galactosyltransferase requires manganese for activity and manganese addition alone can increase galactosylation in late day cultures [44]. It has been shown that the individual components of the UMG cocktail can be altered through a statistical design-of-experiment (DOE) to control the galactosylation of a protein to the desired level [31].

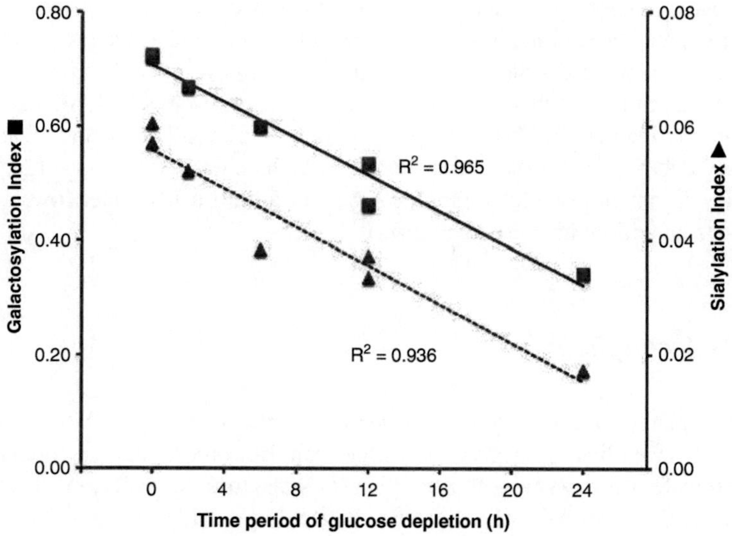


**Fig. 1** Change in galactosylation and sialylation indices with exposure of cells to glucose-depleted media for various times [19]

## 2.4 *Sialylation*

The addition of neuraminic acid to the terminal end of a glycan is known as sialylation or "capping." The two predominant forms of neuraminic acid are *N*-acetylneuraminic acid (NANA) and *N*-glycolylneuraminic acid (NGNA). Sialylated human glycoproteins contain almost exclusively NANA. Although NGNA can be synthesized by other mammals, including mice, it may be immunogenic to humans. The sialylation of CHO-secreted products gives predominantly NANA, whereas glycoproteins from murine producer cells may have both forms of neuraminic acid, giving rise to the possibility of immunogenicity [45, 46]. Glycosylation from CHO cells is often described as "human-like," but the linkage of NANA to the adjacent galactose is predominantly α2-6 in humans and exclusively α2-3 in hamster [47]. The extent of protein sialylation can vary considerably in bioprocesses, but its control is important in order to maintain consistent clinical efficacy of the final product. The residence time of a therapeutic glycoprotein in the blood stream is highly dependent on the extent of sialylation. The bioprocess parameters that influence sialylation were investigated by Lewis et al., who found that lower sialylation levels during large-scale production could be attributed to low levels of dissolved oxygen (DO) [48]. They attributed this relationship to a lower flux through the hexosamine pathway, resulting in a reduced intracellular level of NANA, as the key precursor of sialylation. It is interesting to note that this effect appears to be independent of the effect of reduced DO levels on galactosylation, observed earlier in a murine hybridoma [49].

Enhanced sialylation of a glycoprotein can be attained by cell engineering through transfection and expression of a sialyltransferase. Most of these studies involved use of the α2-6 sialyltransferase to enable the human-like linkage [47, 50, 51]. The protein structure can also influence the extent of sialylation, with only modest levels (<10%) reported for Mabs because of the restricted space within the protein pocket for glycosylation. In this case, greater accessibility of the enzyme can be enabled by removal of an aromatic group, which occurs in the F243A (replacement of phenylalanine 243 by alanine) mutation in an antibody, allowing increased sialylation to >30% [52].

## 2.5 *Ammonia*

Glutamine is an effective substrate for cell growth because of its rapid transfer through the cell membrane and incorporation into tricarboxylic acid cycle intermediates. However, this process is associated with the accumulation of ammonia, which eventually becomes inhibitory to cell growth [53], an effect shown to be greater at high pH values [54].

Accumulated ammonia also exerts an effect on glycosylation by decreasing terminal sialylation [27, 55, 56]. There are two possible mechanisms to explain this effect. First, an increase in intracellular UDP-GlcNAc levels occurs through incorporation of ammonia into glucosamine. This enhanced UDP-GlcNAc competes with the transport of the sialic acid precursor CMP-NANA into the Golgi apparatus, which is an essential step prior to sialylation [24]. The second possible mechanism for reduced sialylation in the presence of ammonia is that the pH value of the Golgi apparatus increases, shifting away from the optimal pH for the sialyltransferase enzyme [57]. Glycosylation-related gene expression in non-immunoglobulin-secreting murine myeloma (NS0) cells was shown to be insensitive to moderate increases in ammonia, which suggests that the mechanistic effects of ammonia on glycosylation are probably metabolic and not at the transcriptional level [58].

It was shown some time ago that the substitution of glutamine in the medium by less ammoniagenic substrates could be effective in reducing the accumulation of ammonia, as shown in several cell lines, but to the detriment of cell growth [59]. More recently, enhanced sialylation was shown for an Fc-fusion protein expressed in CHO cells by replacing glutamine in the medium with α-ketoglutarate [60]. This effect was attributed to the lowering of metabolically produced ammonia by 75%, but at the expense of a longer lag phase and reduced cell growth.

## 2.6 *pH Value*

Cells are normally cultured at neutral pH, with various reports showing that optimal growth and/or cell productivity occurs at a pH range of 7.0–7.8 [61–63]. Glycan site occupancy decreases at lower (<6.9) and or higher (>8.2) pH values, a phenomenon that can be explained by the effect of adverse external pH conditions on the internal pH of the Golgi apparatus, resulting in reduced activity of glycosylation enzymes [62]. Some reports have also indicated specific changes in glycan microheterogeneity caused by changes in pH. The galactosylation of an antibody produced from human cells decreased with increasing pH value in the range of 6.8–7.6 [63]. This effect was confirmed by Aghamohseni et al. for Mab production from CHO cells [61]. In this report, a shifted pH strategy is described, in which cells are grown at a normal pH of 7.8 to maximum cell density, followed by a shift to pH 6.8 during a stationary phase. This strategy had the benefit of increasing galactosylation and sialylation while enabling maximum cell growth and antibody productivity.

Zalai et al. reported a link between high specific productivity (Qp) and the formation of high mannose glycan structures [64]. They also showed in their system that the high mannose structures were more likely to occur at a lower pH value (6.9) than the control (pH 7.2). A relationship between premature glycan structures, including high mannose glycans, has also been shown for higher Qp produced under hypothermic conditions [65].

## 2.7 *Oxygen*

The DO level is a key parameter for the intracellular metabolism of producer cells in bioprocesses [66, 67]. A high rate of oxygen consumption occurs during aerobic metabolism and may decrease during depletion of nutrients in the medium [64]. A change in the glycoform profile of a recombinant protein may well result from such metabolic changes [68]. Control of the DO set-point in a bioreactor is important for minimizing the possibility of an altered metabolism affecting a change in the glycan profile. The terminal galactosylation of an antibody has been directly related to the DO level, with a gradual decrease in the digalactosylated (G2) form from 30% at the higher oxygen level to 12% at low DO [49]. The mechanism for the effect of DO is unclear, but it is probably caused by a change in cellular metabolism rather than a direct effect through the activity of the galactosyltransferase enzyme [69].

## 2.8 *Use of Inhibitors to Control Glycosylation*

A number of highly specific alkaloid-based inhibitors of glycosylation can be added to the growth medium to cause predetermined changes in the glycan profile. Kifunensine blocks the removal of mannose by inhibiting mannosidase I, resulting in a high mannose structure [70, 71]. Because the mannosidase reaction occurs prior to fucosylation in the glycosylation process, this can lead to a nonfucosylated oligomannose glycan structure when an antibody is expressed. This structure demonstrates the ADCC normally associated with nonfucosylated antibodies [72].

Other inhibitors include Castanospermine (Cas), which is a glucosidase inhibitor preventing the removal of terminal glucose sugars from the high mannose glycan attached to protein in the ER [73]. A third example is Swainsonine (Swa), which is a mannosidase II inhibitor. Mannosidase II is the second mannosidase in the glycosylation pathway and removes mannose from one arm of the glycan structure; its inhibition results in the formation of hybrid glycan structures [74]. A typical glycan profile resulting from the addition of these inhibitors to antibody-secreting CHO cells is shown in Fig. 2.

## *2.9 Predictive Metabolic Models of Glycosylation*

The unpredictability of glycosylation is a major problem in the bioindustry. This applies to the unpredictability of the relationship between clinical efficacy and the glycoprofile structure, as well as the unpredictability of process control of glycosylation [76].

Metabolic modeling analysis could improve the understanding of how shifts in nutrients affect key cellular metabolites in glycan synthesis and glycosylation outcomes, as well as cell growth and productivity during process development [77, 78]. However, this is a difficult task because glycosylation is such a complex system, affected by many aspects of cellular metabolism, some of which may yet be unknown. Furthermore, because glycosylation is highly variable between cell types and clones [13], new parameters must be established for each one.

The glycosylation process occurs inside the Golgi apparatus of the cell and is a sequential attachment/detachment of nucleotide sugars to the backbone of the protein to form a complex glycan. The series of successive reactions are catalyzed in the Golgi apparatus by a small number of enzymes, which can be classified into two main groups: (a) exoglycosidases that act on one substrate and detach a mannose from the oligosaccharide chain and (b) glycotransferases (GTs) that act on two substrates and attach a particular monosacharaide to the chain. There are reported to be over 250 mammalian GT enzymes, which can be classified into groups based on the type of monosaccharide they act on [79]. Due to the sequential nature of the glycosylation process, GTs are distributed along the Golgi apparatus [79, 80]. Thus, glycan structure can be affected significantly by the localization of the enzymes.

A relatively simple model based on the probability of reactions that transition glycans from one structure to the next was developed based on random walking (a Markov chain model) and analysis of pre-existing glycan profiles [81]. This model does not require kinetic data, but relies on an in silico flux balance analysis and glycosylation as a stochastic process. The model predicts the effect of a specific enzyme deletion, but does not take into account changes in enzyme activity that lead to variations in kinetics or variable access to substrates.

Metabolic flux analysis in continuous culture can be useful in understanding the effects of altering key nutrients (glucose and glutamine) on the glycan profile. Hossler et al. [80] assumed 341 glycans and addressed this complex network of reactions with a relationship matrix [80] and the development of vizualization software, GlycoVis [82]. Two metabolic models, a dynamic model based on flux analysis and the GlycoVis software model, were used to study and visualize the relationships between glutamine, glucose, pH value, ammonia, and glycosylation in

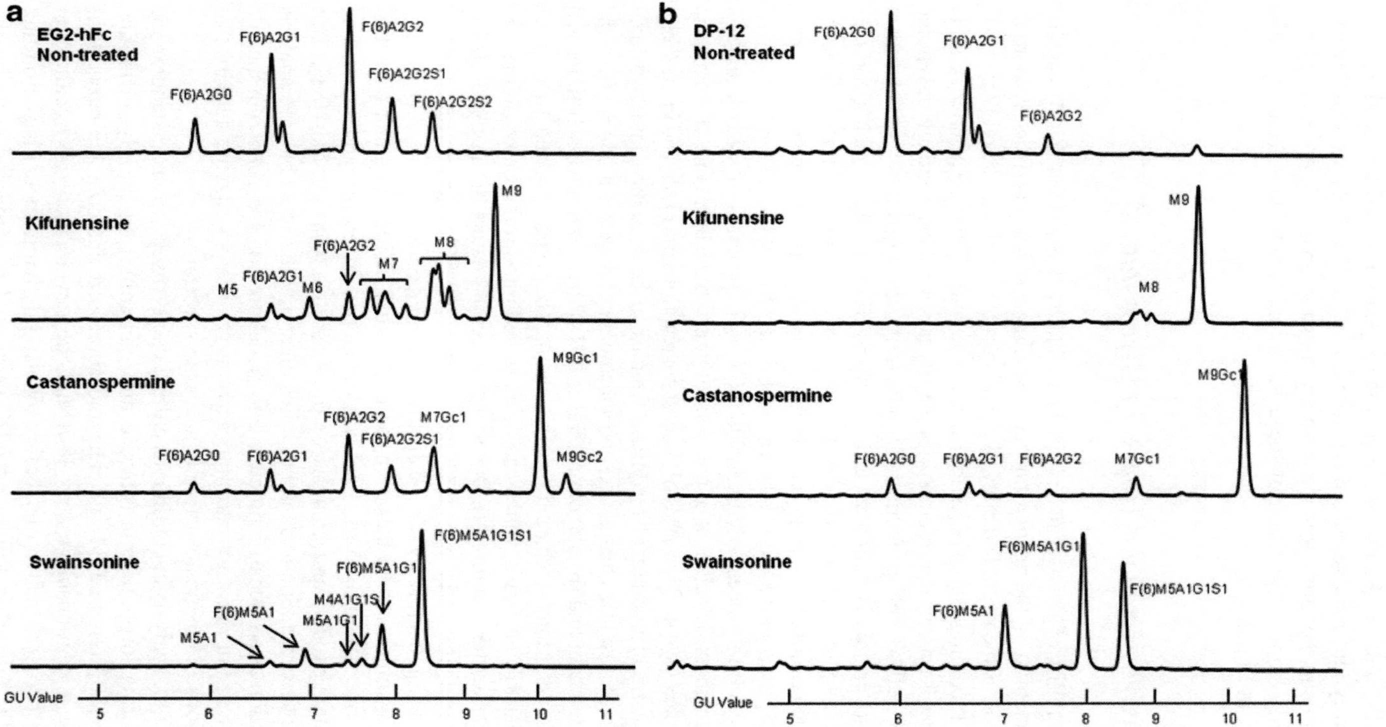


**Fig. 2** Glycan profiles of two monoclonal antibodies, (**a**) CHO EG2-hFc and (**b**) CHO DP-12, expressed by CHO cells grown in media supplemented with inhibitors of glycosylation. Both cell lines were cultured in the presence of the inhibitors Kifunensine (Kif), Castanospermine (Cas), and Swainsonine (Swa). Glycans were isolated from purified Mabs, labeled with 2-aminobenzamide, and separated by a HILIC/normal phase column standardized using a dextran ladder [75]

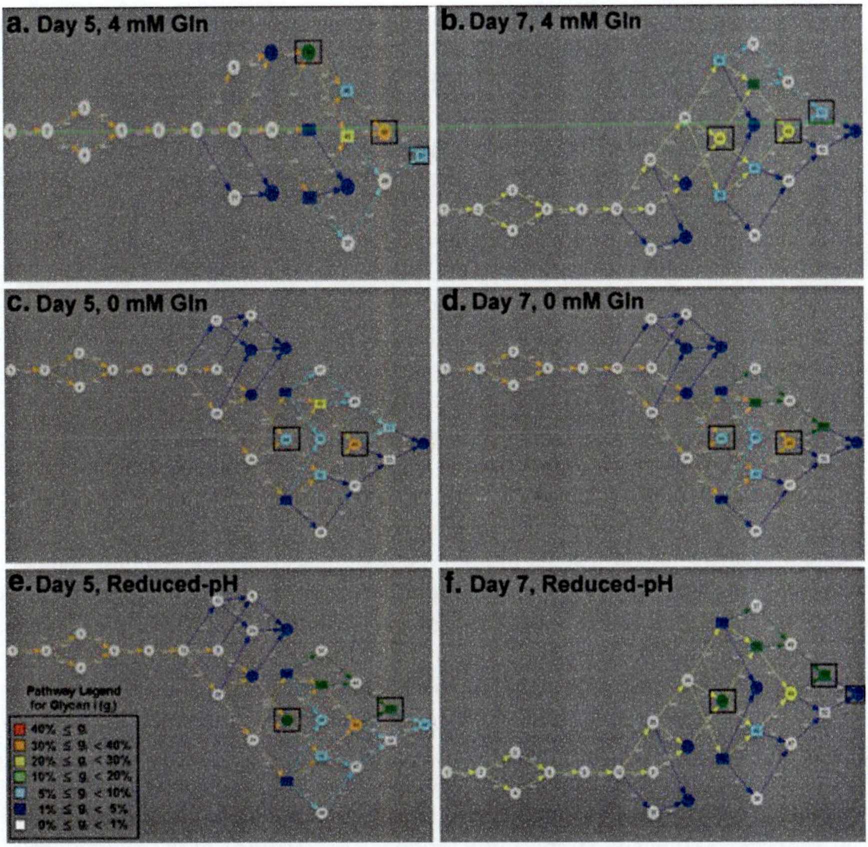


**Fig. 3** Glycan distribution network for Mab-producing CHO cells grown under different conditions: (**a**) 4 mM glucose at day 5, (**b**) 4 mM glutamine (Gln) at day 7, (**c**) 0 mM Gln at day 5, (**d**) 0 mM Gln at day 7, (**e**) reduced pH at day 5, and (**f**) reduced pH at day 7. The representation is based on the GlycoVis program [61]

batch cultures [61]. Reducing glutamine levels can lower glucose consumption along with cell yield, but increase galactosylation and sialylation. Expression of this data using the GlycoVis software is shown in Fig. 3. Extracellular ammonia concentration was correlated with UDP-GlcNAc synthesis, and pH value with inhibition of sialylation. In another similar study, low glutamine conditions reduced sialylation and antennarity of human chorionic gonadotropin and correlated with reduced UDP-GlcNAc [83].

Data on the reaction kinetics of specific glycosylation enzymes based upon values reported in the literature were used to produce one of the first detailed kinetic models [84]. This was extended by Krambeck et al. to embrace the kinetics of 19 glycosylation enzymes into a reaction network to predict glycan profiles analyzed by mass spectrometry [85, 86]. An application of this type of kinetic analysis was prediction of galactosyl transferase (GalT) IV as a major control point for glycosylation branching. Computer simulation and, subsequently, experimental data showed that the downregulation of GalT increased the tri- and tetra-antennary glycan structures of human chorionic gonadotropin expressed in CHO cells [87].

Nucleotide sugars are the direct precursors of glycosylation in the ER and Golgi apparatus and their abundance or depletion determines the extent of glycosylation. They form a matrix of interrelated reactions in the cytoplasm, which is supplied by extracellular nutrients. Two models have been proposed to describe the effect of extracellular conditions on the glycoprofiles for CHO cell cultures [88] and for a murine hybridoma [89]. Both of these models propose to connect the extracellular environment described by a metabolic flux model to the glycosylation processes occurring in the Golgi apparatus through nucleotide sugars. In general, both reported dynamic models are of very large dimensions. For example, the model of Ohadi et al. involves 10 mass balances for extracellular species, 8 differential equation for nucleotides and nucleotide sugars, 104 differential equations describing the component balances of individual glycans, and 4 nucleotide sugars serving as substrates for the glycosylation reactions.

The in silico metabolic model established by Jedrzejewski et al. also has a framework of reactions of nucleotides and nucleotide sugars in the cytoplasm, including kinetic measurements into the pool from extracellular culture components and out of the pool through nine transport rate equations of the nucleotide sugars feeding into the ER and Golgi apparatus (thus feeding the glycosylation reactions). This matrix of nucleotide sugar reactions was developed in an attempt to predict the pathway of glycosylation for different levels of extracellular metabolites [89]. The model gives good predictive results compared with experimental data generated from a murine hybridoma.

The analysis of nucleotide sugar metabolism was extended by del Val et al., who took into account the stoichiometric requirements of host cell proteins, glycolipids, and secreted recombinant proteins [90]. Following an analysis of the requirements for glycosylation, it was concluded that the consumption rates of nucleotide sugars toward cellular and recombinant proteins were of the same order of magnitude. However, the partition of requirements between these demands depends on the relative values of cell-specific productivity and growth rate, both of which can change during the time course of a culture. Undoubtedly, this type of mathematical framework can lead to an understanding of the specific requirements of nucleotide sugars and allow the development of rational feeding strategies.

There is considerable value in the development of robust mathematical models that can be predictive in terms of the effect of changes of media components on the glycosylation profile of the final product. Such models are aided by multivariate data analysis to indicate critical bioprocess parameters associated with alterations in glycan profiles [91]. Many of the existing models are product-specific and/or cell line-specific. A robust model that could be extended to multiple culture systems would be valuable for bioprocess control so that culture conditions could be altered predictably to enable the formation of a predetermined glycan profile. This would also offer the possibility of reverse analysis, which would be applicable to the diagnosis of clinical conditions such as congenital disorders of glycosylation. Here, the objective would be to pinpoint the precise metabolic defect associated with the abnormal glycan profile obtained from a patient sample.

## 3 Production of Cell Culture-Derived Viral Antigens

Immunization is one of the most powerful and cost-effective ways to prevent disease and save millions of lives. Vaccination can significantly reduce virus spreading and, therefore, the emergence of endemics and pandemics. This includes new challenges such as SARS-coronavirus or HIV/AIDS, viruses that evolve very quickly (e.g., influenza A virus) or spread to a new population (e.g., Ebola virus, West Nile virus), and reemerging diseases (e.g., drug-resistant tuberculosis). The complete list of bacterial and viral vaccines licensed for human use in the USA currently comprises more than 160 entries [92]. Vaccines can contain live attenuated viruses that do not cause disease, inactivated organisms or viruses, inactivated toxins, or segments of pathogens (subunit and conjugate vaccines). In addition, nucleic acid vaccines, often called "third generation" vaccines, are currently being developed. These are relatively inexpensive, easy to produce, and offer the chance to reduce lead times in vaccine manufacturing in the case of pandemics [93]. This section focuses on cultivation processes for production of viruses and viral antigens using cells of higher animals and mammals (i.e., from avian, canine, ape, and human origin).

Historically, viruses were propagated only on whole organisms such as mice, rats, rabbits, foxes, or monkeys. With the establishment of viral tissue culture methods in the 1950s, the first generation of inactivated polio vaccines (Salk) and live attenuated oral polio vaccines (Sabin) became available; polio is now close to being eradicated worldwide. During the same period, egg-based systems for virus production were established that are still in use for viruses such as influenza, measles-mumps-rubella (MMR), rabies, and yellow fever vaccines. The advent of modern cell culture techniques enabled the manufacture of today's large range of biological pharmaceutical products at the industrial scale, including viral vaccines for human and veterinary use [94]. Early attempts to use primary cell lines for research and vaccine production date back to the 1920s and 1930s. Today, primary cell lines, diploid cells, and spontaneously transformed continuous cell lines are routinely used for production of viral vaccines [95–99]. In addition, designed continuous cell lines such as PER.C6 [100, 101], AGE1.CR [102], and EB66 [103–105] are considered as substrates for virus propagation. For example, licensed vaccines against measles and mumps are still produced in chicken embryo fibroblasts [106], diploid cell lines are used for rabies vaccine production [107], and Vero cell cultures have been established for inactivated polio vaccine production [108]. In addition, following a WHO recommendation in 1995 [109] the first generation of cell culture-derived seasonal human influenza vaccines was approved in 2007 for a manufacturing process using a MDCK suspension cell line [110].

Like the recombinant therapeutic proteins produced in hybridoma cells and CHO cells, vaccines are highly complex products. In contrast to chemically synthesized low molecular weight drugs, biologics cannot be fully characterized by existing analytical methods. For establishment of cell culture-derived viral vaccine production processes according to current good manufacturing practice (cGMP), the main focus is on potency and safety of the final products. Quality control includes characterization of source materials (virus strains, cells, media), monitoring of cultivation conditions (cell concentrations, virus yields, process parameters), and analysis of in-process samples and testing of the samples before release of the final product. For example, for control of a cell culture-derived monovalent influenza virus pool (whole virus, inactivated), the following tests need to be performed: (1) effective inactivation, (2) antigen concentration (hemagglutinin (HA) content), (3) presence of neuraminidase (viral surface protein that enables virus release after replication), (4) identity (antigenic specificity), (5) extraneous agents, (6) purity (contaminating protein, DNA), and (7) contamination with chemicals used in production (e.g., detergents, organic solvents, inactivant) [111]. Interestingly, and in stark contrast to regulations for production of therapeutic proteins discussed at the beginning of this chapter (e.g. [112]), except for the potency of the final product, physicochemical and specific immunological and biological properties of the antigen are not considered for viral vaccines. This concerns, in particular, the *N*-linked glycosylation of whole virus particles and viral antigens (split and subunit vaccines) using attenuated (live vaccines) or nonattenuated (inactivated vaccines) virus strains. However, and in line with the guidelines for production of therapeutic proteins, the glycosylation patterns, purity, amino acid sequence, and molecular size of recombinant proteins are considered for quality control in the manufacture of vaccines involving the expression of virus surface proteins in recombinant bacteria, yeast, animal cells, or plants [113].

For monoclonals and other recombinant proteins, it is well known that glycosylation has a significant impact on the pharmacokinetics of theses product and modulates several of their immunogenic properties. Accordingly, tests and acceptance criteria need to be defined for relevant glycosylation structures of therapeutic proteins and protein drug products to comply with cGMP guidelines and to realize process analytical technology (PAT) and QbD objectives. For cell culture-derived viral antigens, however, questions regarding the impact of glycosylation on the immunogenicity of a vaccine are so far only addressed in research [114–117], and no specific requirements exist regarding glycoprofiling of viral antigens for the quality control of conventional viral vaccines. Nevertheless, it is widely accepted that the presence or absence of host cell-derived carbohydrates can modulate the antigenicity of antigens by either preventing the binding of neutralizing antibodies or masking epitopes that are recognized by $CD4^+$ T cells that help other lymphocytes to lyse virus-infected cells [118]. In the case of HIV, for example, the high density of HIV-1 envelope glycosylation is considered an evolving "glycan shield" mechanism, whereby specific changes in glycan packing prevent neutralizing antibody binding but not receptor binding [119]. Similar findings were reported for other viral antigens, for instance the HA of influenza A virus, or the envelope glycoproteins E1 and E2 of hepatitis C [120–124]. However, the impact of glycosylation should be carefully evaluated because other findings suggest that changes in the glycosylation pattern of antigen domains do not necessarily influence the immunogenicity of vaccines. For a DNA vaccine encoding the HA of avian H5N1 influenza viruses, for example, modifications of the influenza virus HA1 domain had little impact on the antibody response in a mouse model [125]. Obviously, differences in the glycosylation of viral antigens as a result of host cell selection or modification of glycan structure and composition caused by changes in process condition have to be carefully evaluated to assess their impact on potency. The *N*-linked glycosylation of viral antigens is involved in various other crucial functions such as entry into host cells, proteolytic processing, protein trafficking, and virus release that can have a significant impact on the establishment and optimization of cell culture-derived vaccine production.

There are several reasons for the lack of monitoring of large-scale vaccine production regarding the glycosylation of virus particles (whole virus vaccines) and viral proteins (split and subunit vaccines). First, during the early years of vaccine development, the focus in vaccine development was on potency and safety aspects. In addition, knowledge of carbohydrate structures, their composition, and the impact of glycosylation of viral antigens on immunogenicity was very limited. Second, licensed viral vaccines (whole virus presentations, split and subunit vaccines) significantly exceed other biologicals (i.e., recombinant proteins) in complexity regarding the number of antigens involved in the immune response and the high number of glycan structures involved (i.e., whole virus preparations). Third, it is not clear which specific glycosylation patterns/glycan structures of viral antigens are required for the generation of high potency vaccines. Fourth, besides macro- and microheterogeneity of glycosylated antigens caused by host cell processes, the high mutation rate of many viruses and the complex and dynamic distribution of variants (viral quasispecies) increase the overall heterogeneity of vaccine preparations. Fifth, until very recently there were only limited options for detailed characterization of how antigens shape the human antibody repertoire, which is crucial to our understanding of B-cell immunity and the targeted design of effective immunogens [114, 126, 127].

Over the last few years, protein glycosylation analysis has seen significant methodological progress with improvements in mass spectrometry (MS)-based platforms [121, 128, 129] and the establishment of methods using capillary gel electrophoresis for high-throughput analysis of glycosylation patterns of viral antigens and native virions with high resolution [130, 131]. In addition, there have been significant advances in clinical glycoproteomics and in high-throughput antibody repertoire sequencing using large-scale computational structural modeling and analysis [127, 132, 133]. Based on these analytical tools, it will be possible to make significant progress in the characterization of adaptive responses following vaccination. Such progress will broaden our understanding of the role of antigen glycosylation on infectivity and the modulation of immunogenicity at an unprecedented scale, which opens exciting perspectives for the development of viral vaccines with improved potency and safety profiles.

Recent advances in high-throughput glycan profiling [128, 130, 131] using multiplexed capillary gel electrophoresis with laser-induced fluorescence detection (xCGE-LIF) have enabled the characterization of virus seeds, monitoring of virus replication in animal cell culture regarding the number of infectious and noninfectious virus particles produced, and detailed study of the impact of cultivation conditions on the status and changes in antigen glycosylation patterns over the course of virus replication. These affect not only the selection of virus strains and the host cells used for vaccine manufacture, but also the choice of media, adherent or suspension growth, process parameters (temperature, pH, DO), and inactivation conditions.

### *3.1 Impact of Cultivation Conditions on Glycosylation of Hemagglutinin of Influenza A Virus*

In contrast to the processes established for production of recombinant proteins, there is very limited information available on the impact of host cells, virus strains, and process parameters on the glycosylation of virus particles and viral antigens in animal cell culture-derived vaccine production. The rest of this section gives some examples of upstream processing of influenza A virus using various animal cell lines.

### *3.2 Influenza A Virus Hemagglutinin*

Influenza A virus expresses two membrane-bound surface glycoproteins, hemagglutinin (HA) and neuraminidase (NA). Both proteins carry *N*-linked glycosylations, which can contain a mixture of high-mannose, complex, or hybrid-type oligosaccharides. As a result of the action of viral NA during the budding process, the complex-type oligosaccharides of the released virus particles lack sialic acid but the glycans can be sulfated as a further structural modification [124].

HA is a homotrimeric integral membrane glycoprotein of rod-like shape that protrudes approximately 13.5 nm from the viral surface. During virus replication, HA monomers are transcribed and translated, and then undergo complex post-translational modifications (glycosylation, phosphorylation, sulfation, acylation, etc.) in the ER and Golgi apparatus. After export to the cell surface, the HA monomer is cleaved from HA0 into a globular head (HA1) and a long helical chain anchored in the membrane (HA2); they are linked by a disulfide bond. Typically, three to nine *N*-linked glycans are attached to the intact HA protein backbone (Fig. 4).

HA is the major antigenic determinant of commercial influenza vaccines, and modifications of the glycosylation sites of the globular head of the HA1 influence not only receptor binding and fusion activity but also antigenicity, virulence, and the immune evasion of influenza viruses [94, 124, 134]. Recently, broadly protective vaccine candidates targeting the conserved HA2 stalk domain have been identified and efforts are being made by several research groups toward the development of universal influenza vaccines and therapeutic monoclonal antibodies [92, 135–137].

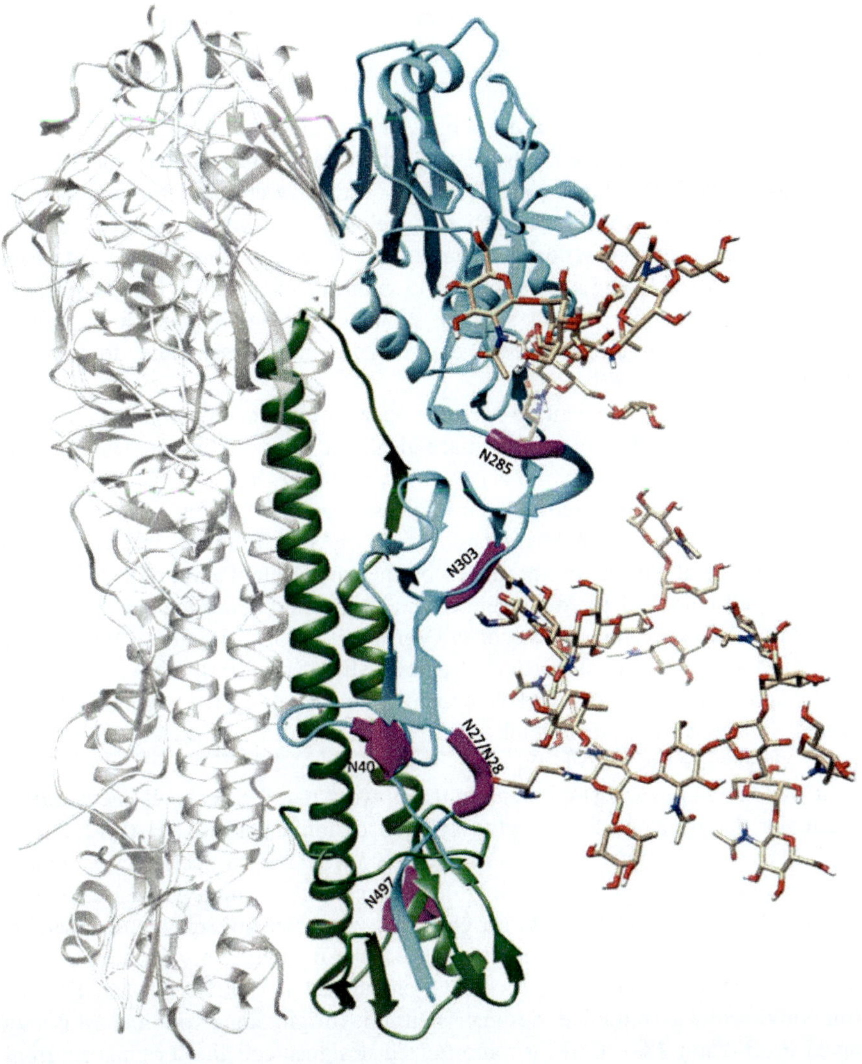


**Fig. 4** HA with *N*-glycans attached. Homotrimeric HA (PDB ID: 1ru7) and attached *N*-glycans (LinucsID 1893, retrieved from http://www.glycosciences.de) generated using Chimera (version 1.10.2). The six hypothetical *N*-glycosylation sites of the HA monomer (UniProtKB: P03452) are highlighted in *magenta*. Five *N*-glycosylation sites (N27/N28, N40, N285, and N303) are located within the HA1 domain (*green*). The HA2 domain (*cyan*) harbors site number six (N497). Complex tri-antennary *N*-glycan structures attached to N27/N28, N285, and N303 represent a possible glycome of the HA monomer. *N*-Glycosylation of the remaining HA monomers (*greyed out*) has been omitted to facilitate visualization (Source: Alexander Pralow, Max Planck Institute for Dynamics of Complex Technical Systems, Magdeburg, Germany)

## 3.3 Impact of Host Cells on the Hemagglutinin Glycosylation Pattern

Egg-based human influenza vaccines have been available for more than 70 years and the majority of vaccine doses are still produced in embryonated chicken eggs [138]. Since the 1990s, however, cell culture-derived veterinary (equine influenza) and human influenza vaccines have been developed to overcome problems associated with egg-based production. This has led to a reduction in lead times and easier logistics compared to egg supply, propagation of influenza strains that are difficult to isolate in eggs (i.e., avian strains), no need for egg-adapted high growth reassortants, and an aseptic closed environment in upstream and downstream processing [95, 139, 140].

In 2001, the first cell culture-derived seasonal human influenza vaccine was licensed (Influvac® TC, Solvay Pharmaceuticals Inc.) but discontinued after acquisition by Abbott Laboratories [141]. A few years later, in 2007, the cell culture-derived (trivalent, inactivated) human influenza vaccine (Optaflu®, Novartis) was approved by the European Medicines Agency (EMA), followed in 2012 by US FDA licensing of Flucelvax® produced by the same manufacturer [110, 142]. In addition, cell culture-derived pandemic whole virion influenza vaccines are propagated in African green monkey kidney (Vero) cells (H5N1, Baxter AG) [108] and duck embryonic stem cells (EB66®, Valneva SE, GSK) [105], and various other production systems are under development (Per.C6, etc.) [103]. A recombinant protein influenza vaccine, Flublok® ($SF^{+}$ insect cells, Protein Sciences Corporation), was licensed in 2013 [143].

It was shown in early studies and in recent work that the host cell line is one of the major determinants for the glycosylation of influenza virus HA [121, 144–146]. In particular, newer studies performed for a wide range of cell lines, including egg-derived virus, demonstrated clear differences in the complexity and composition of HA *N*-glycosylation. As an example, *N*-glycan fingerprints obtained by xCGE-LIF (one peak corresponds to at least one distinct *N*-glycan structure) of influenza A virus (H1N1, Puerto Rico/8/34) propagated in MDCK, Vero, AGE1.CR® (immortalized and modified designer cell line originating from the Muscovy duck, ProBioGen AG), CAP® (immortalized designer cell line originating from primary human amniocytes, Cevec Pharmaceuticals GmbH), and embryonated eggs (IDT Biologika GmbH) are shown in Fig. 5 [146].

As expected, the *N*-glycan fingerprints show strict host cell specificity. HA *N*-glycan fingerprints clearly differ, as reported previously by Schwarzer et al. [146], who also showed by exoglycosidase digestions that all *N*-glycan structures attached to MDCK cell-derived HA are of the complex type with either terminal α- or β-galactose, whereas most *N*-glycan structures of Vero cell-derived HA are of the complex type with terminal β-galactose (with a few other structures of the high mannose type). These results correspond to studies by An et al. [121], who

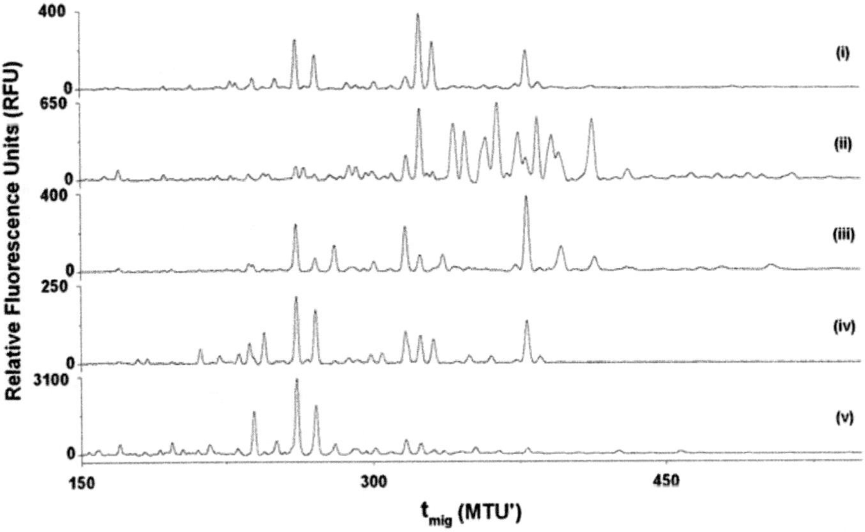


**Fig. 5** Impact of host cell line on the HA *N*-glycosylation pattern. Overlay of *N*-glycan fingerprints. Relative fluorescence units (RFU) are plotted over the migration time ($t_{mig}$) in normalized migration time units (MTU′). Influenza A virus (H1N1, Puerto Rico/8/34) was produced in adherent Vero cells (*i*), adherent MDCK cells (*ii*), human-derived CAP® suspension cells (*iii*), duck-derived AGE1.CR.pIX suspension cells (*iv*), and embryonated chicken eggs (*v*)

characterized H5N1 HA expressed in three insect cell lines, a human cell (HEK283), and embryonated chicken eggs using MS-based analytics. In particular, the authors reported quantitative and qualitative differences in the overall *N*-glycosylation patterns and structures produced by different cell types, and addressed the identification of $\alpha$1,3-fucosylated structures in the core region of *N*-glycans on HA produced by High Five™ cells, which may be allergenic in humans. In a more general experimental setup, species-specific differences in *N*-glycosylation were also shown by Raju et al. [147] for peptide-*N*-glycosidase-F (PNGase F)-treated IgG from 13 different animal species using MALDI-TOF-MS. Furthermore, intracellular glycosylation processing is affected by the enzyme repertoire of the host cell, transit time in the Golgi apparatus, and the availability of intracellular sugar nucleotide donors [148, 149].

A major determinant of HA glycosylation is the selection of the host cell line used in vaccine manufacturing. However, various changes in cultivation conditions to optimize virus production can also have a significant impact on HA glycosylation patterns. In particular, the adaptation of adherent cell lines to growth in suspension using a serum-free medium drastically altered not only the proteome of MDCK cells [150] but also their HA *N*-glycan fingerprints (Fig. 6). The total number of different *N*-glycan structures was reduced, and the *N*-glycans expressed show a

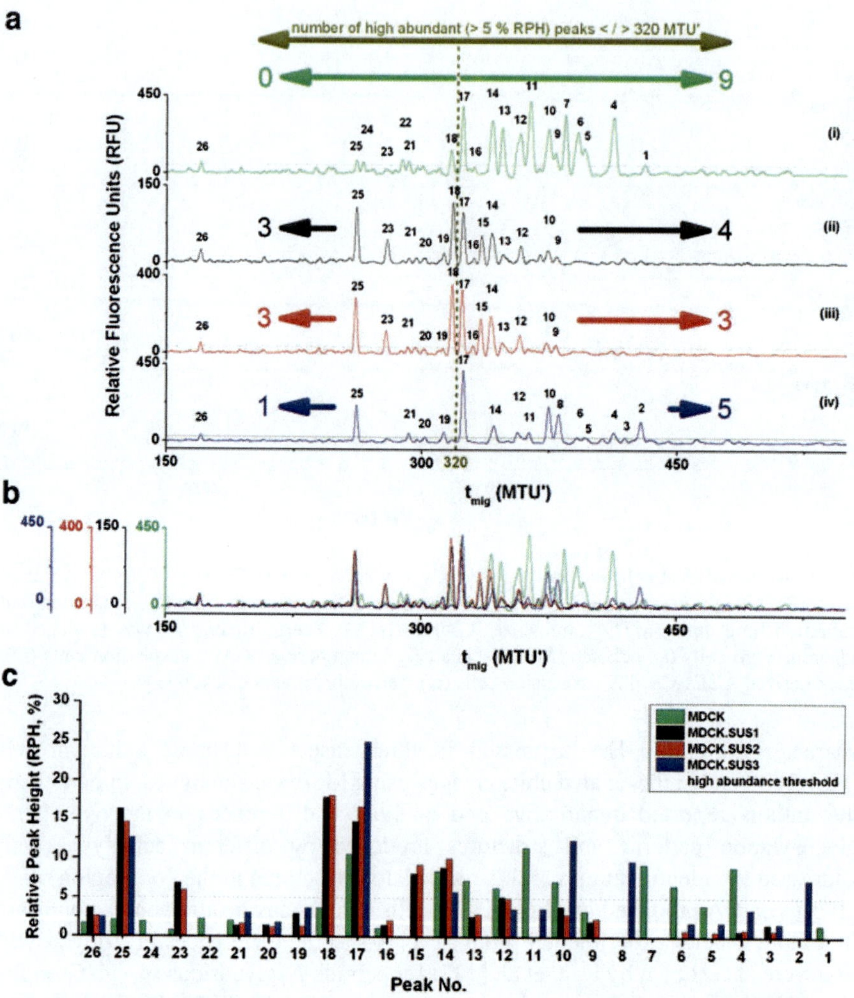


**Fig. 6** Impact of host cell adaptation to serum-free suspension growth on the HA *N*-glycosylation pattern of influenza A virus (H1N1, Puerto Rico/8/34). (**a**) HA *N*-glycan fingerprints. Relative fluorescence units (RFU) are plotted over the migration time ($t_{mig}$) in normalized migration time units (MTU′). All peaks exceeding the 10× baseline noise threshold of at least one fingerprint are annotated. Serum-requiring adherent MDCK cell line (*i*), MDCK cell line adapted to serum-free suspension growth (*ii*; MDCK.SUS1), and further adapted MDCK.SUS1 cell line to better growth characteristics (*iii*; MDCK.SUS2) [151]. Biological duplicate of the first adaptation step (*iv*: MDCK.SUS3). The number of high abundant peaks (RPH >5%) with migration times below or above 320 MTU′ is indicated. (**b**) Overlay of all four *N*-glycosylation fingerprints. (**c**) Relative *N*-glycan structure abundance (RPH) as percentage of the total peak height (TPH, sum of all annotated peaks). Peaks <5% RPH (*dashed lines*) are defined as low abundance

tendency toward smaller structures [152]. Interestingly, without adaptation to growth in suspension, the switch from serum-containing to serum-free growth of adherent MDCK cells caused no significant changes in the total number of HA-associated *N*-glycan peaks but only their relative abundance (not shown). This suggests that adaptation of host cells to a new medium or changes in medium composition to optimize cell growth or process yields have only a slight impact on the host cell's glycosylation machinery.

### *3.4 Impact of Virus Strains on the Hemagglutinin Glycosylation Pattern*

Besides the impact of host cells, the HA glycosylation patterns are influenced by the specific virus type and subtype used for the generation of seed virus. Typically, candidate vaccine viruses matching those recommended for inclusion in seasonal and pandemic vaccines are produced in fertilized eggs and distributed by WHO Collaborating Centers for Reference and Research. In addition, genetic reassortment of influenza A viruses is performed to improve yields and robustness of vaccine production processes [153–155]. Identity testing and sequence analysis is performed to confirm their similarity to the reference strains, but comparison of the glycosylation patterns of the field strains with the virus strains used for vaccine production and evaluation of the impact of differences on immunogenicity of vaccines are not required. In addition, it is well known that egg propagation can affect antigenicity [156, 157] and egg-derived high growth reassortants do not necessarily result in high yield cell culture processes. Therefore, the use of cell-only passaged virus instead of one that has been egg-derived might be favorable and should be considered for cell culture-derived vaccine production [158, 159].

In contrast to the host cell, the selection of virus strain has only a modest influence on the *N*-glycan fingerprint. A comparison of various MDCK cell-derived influenza A subtypes [H1N1, Puerto Rico/8/34, reassortant California/07/2009 (pandemic) and H3N2, reassortant Uruguay/716/2007 (H3N2) × PR8/34 (H1N1), reassortant Victoria/210/2009 (H3N2) × PR8/34 (H1N1)] showed that most peaks were present for all tested viruses (except for some low abundance peaks; Fig. 7a, b). Overall, the relative peak abundance varied with a maximum difference of 25.2% (peak 22, Fig. 7c) whereas strain-specific differences were less than 12.4% (H1N1 strains) and 9.8% (H3N2 strains).

The results suggest a closer relation between the H1N1 and H3N2 strains. Differences in *N*-glycan structure abundance (i.e., missing peaks) are probably a result of minor variations in the three-dimensional conformation of HA monomers of the four analyzed IVA strains, as already highlighted in 1997 by MirShekari et al. [160]. The authors demonstrated for Madin Darby bovine kidney cells that *N*-glycosylation is site-specific, and that glycans at the same site of the HA1 subunit are occupied by more or less conserved *N*-glycan structures depending on the

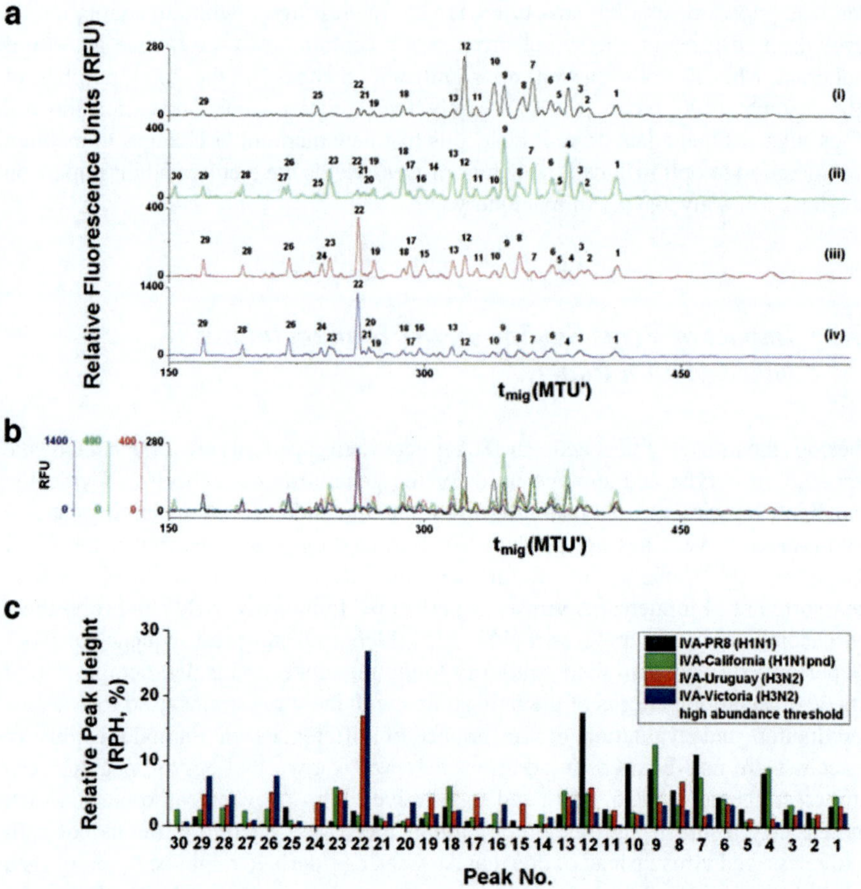


**Fig. 7** Impact of virus strain on MDCK cell-derived HA *N*-glycosylation patterns. (**a**) Overlay of HA *N*-glycan fingerprints. Relative fluorescence units (RFU) are plotted over the migration time ($t_{mig}$) in normalized migration time units (MTU′). Influenza A virus (IVA)-PR8 (*i*; H1N1, Puerto Rico/8/34), IVA-California (*ii*; H1N1, reassortant California/07/2009, pandemic), IVA-Uruguay [*iii*; reassortant Uruguay/716/2007 (H3N2) × PR8/34 (H1N1)], and IVA-Victoria [*iv*; reassortant Victoria/210/2009 (H3N2) × PR8/34 (H1N1)] were produced in MDCK cell culture. All peaks exceeding the 10× baseline noise threshold of at least one fingerprint are annotated. (**b**) Direct overlay of HA *N*-glycan fingerprints. (**c**) Relative *N*-glycan structure abundance (RPH) as percentage of the total peak height (TPH, sum of all annotated peaks). Peaks are defined as high abundance if RPH >5% (*dashed lines*)

specific site characteristic. In loop regions, bi-, tri- and tetra-antennary complex *N*-glycans are present. In contrast, the glycosylation site buried in the α-helix is mostly occupied by high mannose structures, indicating that these glycans are not easily accessible for glycosylation modulating enzymes. Specific glycosylation characteristics also influence glycan processing; for example, Harpaz and Schachter [161] demonstrated that the presence of bisecting GlcNAc inhibits GlcNAc transferases and, therefore, further glycan branching.

In addition, Roedig et al. [158] showed that extended passaging of influenza A virus (H1N1, Puerto Rico/8/34) in adherent MDCK cells had little influence on HA *N*-glycosylation [159]. Over a total number of 10 passages in roller bottles, the HA fingerprints of all passages featured the same 15 main peaks, with the maximum difference in the relative peak heights not exceeding 3.5% (not shown). For the same subtype, neither the harvest time point (24–96 h after infection) nor β-propiolactone inactivation (37°C, 24 h, final β-propiolactone concentration 3 mM) had a significant impact on HA *N*-glycosylation [152].

### 3.5 Impact of Cultivation Vessels and Process Parameters on the Hemagglutinin Glycosylation Pattern

In contrast to the production of recombinant proteins, relatively little is known about the impact of cultivation conditions on viral antigen glycosylation. For the HA of influenza A virus (H1N1, Puerto Rico/8/34) produced in adherent MDCK cells in serum-containing medium, changes in cultivation vessel (T75 flask, roller bottle, spinner vessel, stirred tank reactor), cultivation scale (50 mL–4.5 L working volume), cell concentration (standard batch versus high cell density), and temperature during virus replication (33–39°C) had no significant impact on the HA *N*-glycosylation pattern. At best, minor changes in the relative *N*-glycan structure abundances were identified [162]. As an example, HA glycosylation patterns for a wide range of cultivation vessel are shown in Fig. 8.

## 4 Conclusions

As more is understood about the glycan structures attached to glycoproteins, it has become increasingly obvious that they have a crucial role in the therapeutic effects of recombinant biopharmaceuticals. Notable examples are the role of sialylation on the residence time of erythropoietin in the blood stream, and the inflammatory properties of antibodies. Even small structural changes such as fucosylation can have a dramatic effect on receptor binding, which is crucial to targeted cancer treatment using therapeutic antibodies.

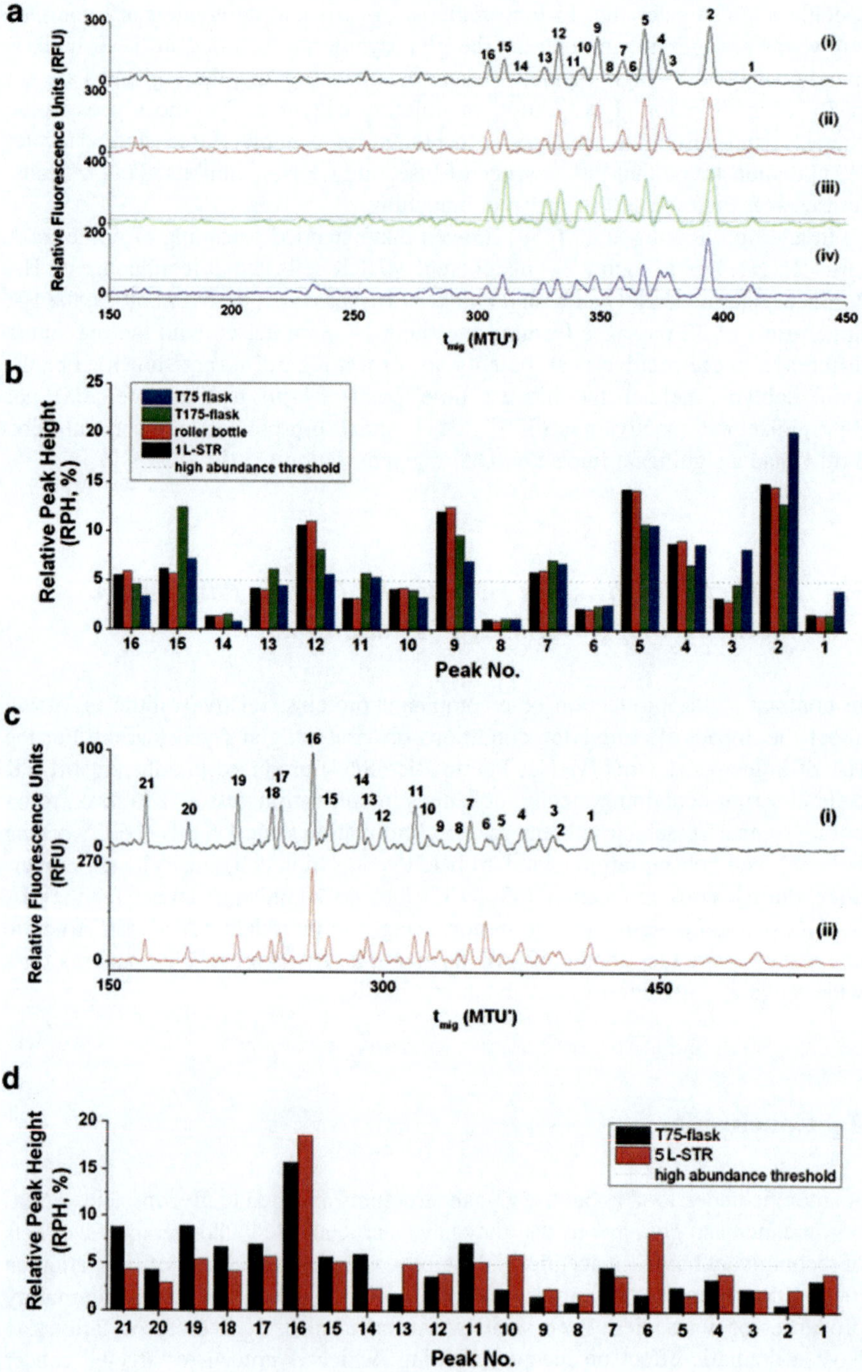


**Fig. 8** Impact of cultivation scale and vessel on the HA *N*-glycosylation pattern of two influenza A virus strains. (**a**, **b**) Puerto Rico/8/34 (H1N1) was produced in T75-flask (*i*), T175-flask (*ii*),

Precise control of glycosylation leading to a homogenous glycoform profile can be performed at the cellular level through metabolic engineering by the functional addition or removal of specific genes associated with glycosylation. However, the success of this approach is dependent upon the availability of precursors present in the culture media. A limited supply of substrates or cofactors in the medium limits full glycosylation, irrespective of the activity of enzymes.

Some of the bioprocess parameters explained in this chapter can influence glycosylation. To maintain a consistency of glycosylation it is important to identify the critical parameters and their optimal set points. Consistent glycosylation during a large-scale bioprocess requires sufficient information about the "design space" around these set points. This concept is consistent with the regulatory important QbD approach, which defines the limited boundaries of each identified bioprocess parameter. This leads to confidence in product consistency if there is a deviation from a set point but within the boundaries of the design space.

Compared with recombinant glycoproteins produced in animal cell culture, relatively little is known regarding the impact of host cells, virus strain/recombinant protein, and cultivation conditions on *N*-glycosylation of viral antigens. Based on the limited studies performed so far, it seems that selection of the expression system (host cell, adherent/suspension growth) has the largest impact on the glycan fingerprint. In addition, the virus strains and their passage history (egg-based versus animal cell-derived virus seeds) should be considered carefully. Interestingly, cultivation vessel, process parameters, choice of medium, harvest time point, and inactivation only seem to modulate peak abundance but not peak presence. Application of the powerful new analytical approaches developed over the last few years (i.e., MS-based platforms and xCGE-LIF), should be encouraged for monitoring the glycosylation status of viral antigens in vaccine production processes. In addition, the impact of changes in the *N*-glycosylation of viral antigens on cellular and humoral immunity of virus preparations should be carefully evaluated to enable the design of potent and safe vaccines using the optimal production platform.

---

**Fig. 8** (continued) roller bottle (*iii*) or 1 L stirred tank reactor (STR, *iv*) using MDCK cell culture. (**c, d**) The reassortant Uruguay/716/2007 (H3N2) × PR8/34 (H1N1) was produced in T75 flask (*i*) and 5 L STR (*ii*) using MDCK cell culture. (**a, c**) Overlay of HA *N*-glycan fingerprints, relative fluorescence units (RFU) are plotted over the migration time ($t_{mig}$) in normalized migration time units (MTU′). All peaks exceeding the 10 × baseline noise threshold of at least one fingerprint are annotated. (**b, d**) Relative peak height (RPH) in % of the total peak height (TPH, sum of all annotated peaks). Peaks are defined as high abundant if RPH >5% (*dashed lines*)

## References

1. Mordor Intelligence (2016) Global biopharmceuticals market growth, trends and forecasts (2016–2021). Mordor Intelligence, Hyderabad
2. Research and Markets (2013) Biopharmaceuticals – a global market overview. Research and Markets, Dublin
3. Zheng K, Bantog C, Bayer R (2011) The impact of glycosylation on monoclonal antibody conformation and stability. MAbs 3:568–576
4. Sareneva T, Pirhonen J, Cantell K, Julkunen I (1995) N-glycosylation of human interferon-gamma: glycans at Asn-25 are critical for protease resistance. Biochem J 308(Pt 1):9–14
5. Onitsuka M, Kawaguchi A, Asano R, Kumagai I, Honda K, Ohtake H, Omasa T (2013) Glycosylation analysis of an aggregated antibody produced by Chinese hamster ovary cells in bioreactor culture. J Biosci Bioeng 117:639–644
6. Rodriguez J, Spearman M, Huzel N, Butler M (2005) Enhanced production of monomeric interferon-beta by CHO cells through the control of culture conditions. Biotechnol Prog 21:22–30
7. Wright A, Morrison SL (1997) Effect of glycosylation on antibody function: implications for genetic engineering. Trends Biotechnol 15:26–32
8. Chung CH, Mirakhur B, Chan E, Le QT, Berlin J, Morse M, Murphy BA, Satinover SM, Hosen J, Mauro D, Slebos RJ, Zhou Q, Gold D, Hatley T, Hicklin DJ, Platts-Mills TA (2008) Cetuximab-induced anaphylaxis and IgE specific for galactose-alpha-1,3-galactose. N Engl J Med 358:1109–1117
9. Noguchi A, Mukuria CJ, Suzuki E, Naiki M (1995) Immunogenicity of N-glycolylneuraminic acid-containing carbohydrate chains of recombinant human erythropoietin expressed in Chinese hamster ovary cells. J Biochem 117:59–62
10. Sola RJ, Griebenow K (2010) Glycosylation of therapeutic proteins: an effective strategy to optimize efficacy. BioDrugs 24:9–21
11. Ratner M (2014) Genentech's glyco-engineered antibody to succeed Rituxan. Nat Biotechnol 32:6–7
12. Butler M (2015) Serum and protein free media. In: Al-Rubeai M (ed) Animal cell culture, Cell engineering, vol.9. Springer, Dordrecht, pp 223–236
13. van Berkel PH, Gerritsen J, Perdok G, Valbjorn J, Vink T, van de Winkel JG, Parren PW (2009) N-linked glycosylation is an important parameter for optimal selection of cell lines producing biopharmaceutical human IgG. Biotechnol Prog 25:244–251
14. Davidson SK, Hunt LA (1985) Sindbis virus glycoproteins are abnormally glycosylated in Chinese hamster ovary cells deprived of glucose. J Gen Virol 66(Pt 7):1457–1468
15. Rearick JI, Chapman A, Kornfeld S (1981) Glucose starvation alters lipid-linked oligosaccharide biosynthesis in Chinese hamster ovary cells. J Biol Chem 256:6255–6261
16. Curling EM, Hayter PM, Baines AJ, Bull AT, Gull K, Strange PG, Jenkins N (1990) Recombinant human interferon-gamma. Differences in glycosylation and proteolytic processing lead to heterogeneity in batch culture. Biochem J 272:333–337
17. Hayter PM, Curling EM, Baines AJ, Jenkins N, Salmon I, Strange PG, Tong JM, Bull AT (1992) Glucose-limited chemostat culture of Chinese hamster ovary cells producing recombinant human interferon-gamma. Biotechnol Bioeng 39:327–335
18. Tachibana H, Kim JY, Shirahata S (1997) Building high affinity human antibodies by altering the glycosylation on the light chain variable region in N-acetylglucosamine-supplemented hybridoma cultures. Cytotechnology 23:151–159
19. Liu B, Spearman M, Doering J, Lattova E, Perreault H, Butler M (2014) The availability of glucose to CHO cells affects the intracellular lipid-linked oligosaccharide distribution, site occupancy and the N-glycosylation profile of a monoclonal antibody. J Biotechnol 170:17–27
20. Villacres C, Tayi VS, Lattova E, Perreault H, Butler M (2015) Low glucose depletes glycan precursors, reduces site occupancy and galactosylation of a monoclonal antibody in CHO cell culture. Biotechnol J 10:1051–1066

21. Kochanowski N, Blanchard F, Cacan R, Chirat F, Guedon E, Marc A, Goergen JL (2008) Influence of intracellular nucleotide and nucleotide sugar contents on recombinant interferon-gamma glycosylation during batch and fed-batch cultures of CHO cells. Biotechnol Bioeng 100:721–733
22. Nyberg GB, Balcarcel RR, Follstad BD, Stephanopoulos G, Wang DI (1999) Metabolic effects on recombinant interferon-gamma glycosylation in continuous culture of Chinese hamster ovary cells. Biotechnol Bioeng 62:336–347
23. Wong DCF, Wong KTK, Goh LT, Heng CK, Yap MGS (2005) Impact of dynamic online fed-batch strategies on metabolism, productivity and N-glycosylation quality in CHO cell cultures. Biotechnol Bioeng 89:164–177
24. Grammatikos SI, Valley U, Nimtz M, Conradt HS, Wagner R (1998) Intracellular UDP-N-acetylhexosamine pool affects N-glycan complexity: a mechanism of ammonium action on protein glycosylation. Biotechnol Prog 14:410–419
25. Ryll T, Valley U, Wagner R (1994) Biochemistry of growth inhibition by ammonium ions in mammalian cells. Biotechnol Bioeng 44:184–193
26. Yang M, Butler M (2002) Effects of ammonia and glucosamine on the heterogeneity of erythropoietin glycoforms. Biotechnol Prog 18:129–318
27. Zanghi JA, Mendoza TP, Knop RH, Miller WM (1998) Ammonia inhibits neural cell adhesion molecule polysialylation in Chinese hamster ovary and small cell lung cancer cells. J Cell Physiol 177:248–263
28. Taschwer M, Hackl M, Hernandez Bort JA, Leitner C, Kumar N, Puc U, Grass J, Papst M, Kunert R, Altmann F, Borth N (2012) Growth, productivity and protein glycosylation in a CHO EpoFc producer cell line adapted to glutamine-free growth. J Biotechnol 157:295–303
29. Baker KN, Rendall MH, Hills AE, Hoare M, Freedman RB, James DC (2001) Metabolic control of recombinant protein N-glycan processing in NS0 and CHO cells. Biotechnol Bioeng 73:188–202
30. Hills AE, Patel A, Boyd P, James DC (2001) Metabolic control of recombinant monoclonal antibody N-glycosylation in GS-NS0 cells. Biotechnol Bioeng 75:239–251
31. Grainger RK, James DC (2013) CHO cell line specific prediction and control of recombinant monoclonal antibody N-glycosylation. Biotechnol Bioeng 110:2970–2983
32. Gramer MJ, Eckblad JJ, Donahue R, Brown J, Shultz C, Vickerman K, Priem P, van den Bremer ET, Gerritsen J, van Berkel PH (2011) Modulation of antibody galactosylation through feeding of uridine, manganese chloride, and galactose. Biotechnol Bioeng 108:1591–1602
33. Wong NS, Wati L, Nissom PM, Feng HT, Lee MM, Yap MG (2010) An investigation of intracellular glycosylation activities in CHO cells: effects of nucleotide sugar precursor feeding. Biotechnol Bioeng 107:321–336
34. Pels Rijcken WR, Overdijk B, van den Eijnden DH, Ferwerda W (1995) The effect of increasing nucleotide-sugar concentrations on the incorporation of sugars into glycoconjugates in rat hepatocytes. Biochem J 305(Pt 3):865–870
35. Gu X, Wang DI (1998) Improvement of interferon-gamma sialylation in Chinese hamster ovary cell culture by feeding of N-acetylmannosamine. Biotechnol Bioeng 58:642–648
36. Fan Y, del Val Jimenez I, Muller C, Wagtberg Sen J, Rasmussen SK, Kontoravdi C, Weilguny D, Andersen MR (2015) Amino acid and glucose metabolism in fed-batch CHO cell culture affects antibody production and glycosylation. Biotechnol Bioeng 112:521–535
37. Abdel Rahman AM, Ryczko M, Nakano M, Pawling J, Rodrigues T, Johswich A, Taniguchi N, Dennis JW (2015) Golgi N-glycan branching N-acetylglucosaminyltransferases I, V and VI promote nutrient uptake and metabolism. Glycobiology 25:225–240
38. Kiss R, Magreta P, Gray B, Stupp T, Kaneshiro S, Ryll T (1999) Control of glycosylation by environmental manipulations in CHO cell cultures: ll culture effects on the glycosylation. American Chemical Society, Division of Biochemical Technology, Anaheim

39. Clark KJ, Griffiths J, Bailey KM, Harcum SW (2005) Gene-expression profiles for five key glycosylation genes for galactose-fed CHO cells expressing recombinant IL-4/13 cytokine trap. Biotechnol Bioeng 90:568–577
40. Kildegaard HF, Fan Y, Sen JW, Larsen B, Andersen MR (2016) Glycoprofiling effects of media additives on IgG produced by CHO cells in fed-batch bioreactors. Biotechnol Bioeng 113:359–366
41. Liu J, Wang J, Fan L, Chen X, Hu D, Deng X, Poon HF, Wang H, Liu X, Tan WS (2015) Galactose supplementation enhance sialylation of recombinant Fc-fusion protein in CHO cell: an insight into the role of galactosylation in sialylation. World J Microbiol Biotechnol 31:1147–1156
42. Majid FA, Butler M, Al-Rubeai M (2007) Glycosylation of an immunoglobulin produced from a murine hybridoma cell line: the effect of culture mode and the anti-apoptotic gene, bcl-2. Biotechnol Bioeng 97:156–169
43. Raju S (2003) Glycosylation variations with expression systems and their impact on bioactivity of therapeutic immunoglobulins. Bioproc Int 1:44–53
44. Crowell CK, Grampp GE, Rogers GN, Miller J, Scheinman RI (2007) Amino acid and manganese supplementation modulates the glycosylation state of erythropoietin in a CHO culture system. Biotechnol Bioeng 96:538–549
45. Brooks SA (2004) Appropriate glycosylation of recombinant proteins for human use: implications of choice of expression system. Mol Biotechnol 28:241–255
46. Ghaderi D, Zhang M, Hurtado-Ziola N, Varki A (2012) Production platforms for biotherapeutic glycoproteins. Occurrence, impact, and challenges of non-human sialylation. Biotechnol Genet Eng Rev 28:147–175
47. Lin N, Mascarenhas J, Sealover NR, George HJ, Brooks J, Kayser KJ, Gau B, Yasa I, Azadi P, Archer-Hartmann S (2015) Chinese hamster ovary (CHO) host cell engineering to increase sialylation of recombinant therapeutic proteins by modulating sialyltransferase expression. Biotechnol Prog 31:334–346
48. Lewis AM, Croughan WD, Aranibar N, Lee AG, Warrack B, Abu-Absi NR, Patel R, Drew B, Borys MC, Reily MD, Li ZJ (2016) Understanding and controlling sialylation in a CHO Fc-fusion process. PLoS One 11:e0157111
49. Kunkel JP, Jan DC, Jamieson JC, Butler M (1998) Dissolved oxygen concentration in serum-free continuous culture affects N-linked glycosylation of a monoclonal antibody. J Biotechnol 62:55–71
50. Raymond C, Robotham A, Spearman M, Butler M, Kelly J, Durocher Y (2015) Production of alpha2,6-sialylated IgG1 in CHO cells. MAbs 7:571–583
51. Yin B, Gao Y, Chung CY, Yang S, Blake E, Stuczynski MC, Tang J, Kildegaard HF, Andersen MR, Zhang H, Betenbaugh MJ (2015) Glycoengineering of Chinese hamster ovary cells for enhanced erythropoietin N-glycan branching and sialylation. Biotechnol Bioeng 112:2343–2351
52. Mimura Y, Kelly RM, Unwin L, Albrecht S, Jefferis R, Goodall M, Mizukami Y, Mimura-Kimura Y, Matsumoto T, Ueoka H, Rudd PM (2016) Enhanced sialylation of a human chimeric IgG1 variant produced in human and rodent cell lines. J Immunol Methods 428:30–36
53. Butler M, Spier RE (1984) The effects of glutamine utilisation and ammonia production on the growth of BHK cells in microcarrier cultures. J Biotechnol 1:187–196
54. Doyle C, Butler M (1990) The effect of pH on the toxicity of ammonia to a murine hybridoma. J Biotechnol 15:91–100
55. Andersen DC, Goochee CF (1994) The effect of cell-culture conditions on the oligosaccharide structures of secreted glycoproteins. Curr Opin Biotechnol 5:546–549
56. Yang M, Butler M (2000) Effects of ammonia on CHO cell growth, erythropoietin production, and glycosylation. Biotechnol Bioeng 68:370–380

57. Valley U, Nimtz M, Conradt HS, Wagner R (1999) Incorporation of ammonium into intracellular UDP-activated N-acetylhexosamines and into carbohydrate structures in glycoproteins. Biotechnol Bioeng 64:401–417
58. Brodsky AN, Caldwell M, Bae S, Harcum SW (2014) Glycosylation-related genes in NS0 cells are insensitive to moderately elevated ammonium concentrations. J Biotechnol 187:78–86
59. McDermott RH, Butler M (1993) Uptake of glutamate, not glutamine synthetase, regulates adaptation of mammalian cells to glutamine-free medium. J Cell Sci 104(Pt 1):51–58
60. Ha TK, Lee GM (2014) Effect of glutamine substitution by TCA cycle intermediates on the production and sialylation of Fc-fusion protein in Chinese hamster ovary cell culture. J Biotechnol 180:23–29
61. Aghamohseni H, Ohadi K, Spearman M, KRAHN N, Moo-Young M, Scharer JM, Butler M, Budman HM (2014) Effects of nutrient levels and average culture pH on the glycosylation pattern of camelid-humanized monoclonal antibody. J Biotechnol 186:98–109
62. Borys MC, Linzer DI, Papoutsakis ET (1993) Culture pH affects expression rates and glycosylation of recombinant mouse placental lactogen proteins by Chinese hamster ovary (CHO) cells. Biotechnology 11:720–724
63. Seo JS, Kim YJ, Cho JM, Baek E, Lee GM (2013) Effect of culture pH on recombinant antibody production by a new human cell line, F2N78, grown in suspension at 33.0° C and 37.0° C. Appl Microbiol Biotechnol 97:5283–5291
64. Zalai D, Hever H, Lovasz K, Molnar D, Wechselberger P, Hofer A, Parta L, Putics A, Herwig C (2016) A control strategy to investigate the relationship between specific productivity and high-mannose glycoforms in CHO cells. Appl Microbiol Biotechnol 100:7011–7024
65. Sou SN, Sellick C, Lee K, Mason A, Kyriakopoulos S, Polizzi KM, Kontoravdi C (2015) How does mild hypothermia affect monoclonal antibody glycosylation? Biotechnol Bioeng 112:1165–1176
66. Heidemann R, Lutkemeyer D, Buntemeyer H, Lehmann J (1998) Effects of dissolved oxygen levels and the role of extra- and intracellular amino acid concentrations upon the metabolism of mammalian cell lines during batch and continuous cultures. Cytotechnology 26:185–197
67. Jan DC, Petch DA, Huzel N, Butler M (1997) The effect of dissolved oxygen on the metabolic profile of a murine hybridoma grown in serum-free medium in continuous culture. Biotechnol Bioeng 54:153–164
68. Chotigeat W, Watanapokasin Y, Mahler S, Gray PP (1994) Role of environmental conditions on the expression levels, glycoform pattern and levels of sialyltransferase for hFSH produced by recombinant CHO cells. Cytotechnology 15:217–221
69. Kunkel JP, Yan WY, Butler M, Jamieson JC (2003) Decreased monoclonal IgG1 galactosylation at reduced dissolved oxygen concentration is not a result of lowered galactosyltransferase activity in vitro. Glycobiology 13:875
70. Elbein AD, Tropea JE, Mitchell M, Kaushal GP (1990) Kifunensine, a potent inhibitor of the glycoprotein processing mannosidase I. J Biol Chem 265:15599–15605
71. Weng S, Spiro RG (1996) Endoplasmic reticulum kifunensine-resistant alpha-mannosidase is enzymatically and immunologically related to the cytosolic alpha-mannosidase. Arch Biochem Biophys 325:113–123
72. Zhou Q, Shankara S, Roy A, Qiu H, Estes S, Mcvie-Wylie A, Culm-Merdek K, Park A, Pan C, Edmunds T (2008) Development of a simple and rapid method for producing non-fucosylated oligomannose containing antibodies with increased effector function. Biotechnol Bioeng 99:652–665
73. van Leeuwen JE, Kearse KP (1997) Reglucosylation of N-linked glycans is critical for calnexin assembly with T cell receptor (TCR) alpha proteins but not TCRbeta proteins. J Biol Chem 272:4179–4186
74. Crispin M, Harvey DJ, Chang VT, Yu C, Aricescu AR, Jones EY, Davis SJ, Dwek RA, Rudd PM (2006) Inhibition of hybrid- and complex-type glycosylation reveals the presence of the GlcNAc transferase I-independent fucosylation pathway. Glycobiology 16:748–756

75. Krahn N, Spearman M, Meier M, Dorion-Thibaudeau J, McDougall M, Patel TR, de Crescenzo G, Durocher Y, Stetefeld J, Butler M (2017) Inhibition of glycosylation on a camelid antibody uniquely affects its FcgammaRI binding activity. Eur J Pharm Sci 96:428–439
76. Walsh G (2010) Post-translational modifications of protein biopharmaceuticals. Drug Discov Today 15:773–780
77. Gerdtzen ZP (2012) Modeling metabolic networks for mammalian cell systems: general considerations, modeling strategies, and available tools. Adv Biochem Eng Biotechnol 127:71–108
78. Sellick CA, Croxford AS, Maqsood AR, Stephens GM, Westerhoff HV, Goodacre R, Dickson AJ (2015) Metabolite profiling of CHO cells: molecular reflections of bioprocessing effectiveness. Biotechnol J 10:1434–1445
79. Stanley P (2011) Golgi glycosylation. Cold Spring Harb Perspect Biol 3:a005199
80. Hossler P, Mulukutla BC, Hu WS (2007) Systems analysis of N-glycan processing in mammalian cells. PLoS One 2:e713
81. Spahn PN, Hansen AH, Hansen HG, Arnsdorf J, Kildegaard HF, Lewis NE (2016) A Markov chain model for N-linked protein glycosylation--towards a low-parameter tool for model-driven glycoengineering. Metab Eng 33:52–66
82. Hossler P, Goh LT, Lee MM, Hu WS (2006) GlycoVis: visualizing glycan distribution in the protein N-glycosylation pathway in mammalian cells. Biotechnol Bioeng 95:946–960
83. Burleigh SC, van de Laar T, Stroop CJ, van Grunsven WM, O'Donoghue N, Rudd PM, Davey GP (2011) Synergizing metabolic flux analysis and nucleotide sugar metabolism to understand the control of glycosylation of recombinant protein in CHO cells. BMC Biotechnol 11:95
84. Umana P, Bailey JE (1997) A mathematical model of N-linked glycoform biosynthesis. Biotechnol Bioeng 55:890–908
85. Krambeck FJ, Bennun SV, Narang S, Choi S, Yarema KJ, Betenbaugh MJ (2009) A mathematical model to derive N-glycan structures and cellular enzyme activities from mass spectrometric data. Glycobiology 19:1163–1175
86. Krambeck FJ, Betenbaugh MJ (2005) A mathematical model of N-linked glycosylation. Biotechnol Bioeng 92:711–728
87. McDonald AG, Hayes JM, Bezak T, Gluchowska SA, Cosgrave EF, Struwe WB, Stroop CJ, Kok H, van de Laar T, Rudd PM, Tipton KF, Davey GP (2014) Galactosyltransferase 4 is a major control point for glycan branching in N-linked glycosylation. J Cell Sci 127:5014–5026
88. Ohadi K, Aghamohseni H, Gädke J, Moo-Young M, Legge R, Scharer J, Budman H (2013) Novel dynamic model to predict the glycosylation pattern of monoclonal antibodies from extracellular cell culture conditions. 12th IFAC symposium on computer applications in biotechnology. IFAC Proc Vol 46(31): 30–35
89. Jedrzejewski PM, del Val IJ, Constantinou A, Dell A, Haslam SM, Polizzi KM, Kontoravdi C (2014) Towards controlling the glycoform: a model framework linking extracellular metabolites to antibody glycosylation. Int J Mol Sci 15:4492–4522
90. del Val IJ, Polizzi KM, Kontoravdi C (2016) A theoretical estimate for nucleotide sugar demand towards Chinese Hamster ovary cellular glycosylation. Sci Rep 6:28547
91. Rathore AS, Kumar Singh S, Pathak M, Read EK, Brorson KA, Agarabi CD, Khan M (2015) Fermentanomics: relating quality attributes of a monoclonal antibody to cell culture process variables and raw materials using multivariate data analysis. Biotechnol Prog 31:1586–1599
92. Cho A, Wrammert J (2016) Implications of broadly neutralizing antibodies in the development of a universal influenza vaccine. Curr Opin Virol 17:110–115
93. Steel J (2015) A paradigm shift in vaccine production for pandemic influenza. Ann Transl Med 3:165
94. Kim JI, Park MS (2012) N-linked glycosylation in the hemagglutinin of influenza A viruses. Yonsei Med J 53:886–893

95. Genzel Y, Reichl U (2009) Continuous cell lines as a production system for influenza vaccines. Expert Rev Vaccines 8:1681–1692
96. Genzel Y, Rodig J, Rapp E, Reichl U (2014) Vaccine production: upstream processing with adherent or suspension cell lines. In: Portner R (ed) Animal cell biotechnology: methods and protocols3rd edn. Humana, Totowa
97. Jordan I, Sandig V (2014) Matrix and backstage: cellular substrates for viral vaccines. Virus 6:1672–1700
98. Rodrigues AF, Soares HR, Guerreiro MR, Alves PM, Coroadinha AS (2015) Viral vaccines and their manufacturing cell substrates: new trends and designs in modern vaccinology. Biotechnol J 10:1329–1344
99. Zahoor MA, Khurshid M, Qureshi R, Naz A, Shahid M (2016) Cell culture-based viral vaccines: current status and future prospects. Futur Virol 11:549–562
100. Pau MG, Ophorst C, Koldijk MH, Schouten G, Mehtali M, Uytdehaag F (2001) The human cell line PER.C6 provides a new manufacturing system for the production of influenza vaccines. Vaccine 19:2716–2721
101. Sanders BP, Oakes Ide L, van Hoek V, Liu Y, Marissen W, Minor PD, Wimmer E, Schuitemaker H, Custers JH, Macadam A, Cello J, Edo-Matas D (2015) Production of high titer attenuated poliovirus strains on the serum-free PER.C6((R)) cell culture platform for the generation of safe and affordable next generation IPV. Vaccine 33:6611–6616
102. Jordan I, Vos A, Beilfuss S, Neubert A, Breul S, Sandig V (2009) An avian cell line designed for production of highly attenuated viruses. Vaccine 27:748–756
103. Genzel Y (2015) Designing cell lines for viral vaccine production: where do we stand? Biotechnol J 10:728–740
104. Guehenneux F, Moreau K, Esnault M, Mehtali M (2008) Duck embryonic derived stem cell lines for the production of viral vaccines. Patent WO/2008/129058 A1
105. Schuind A, Segall N, Drame M, Innis BL (2015) Immunogenicity and safety of an EB66 cell-culture-derived influenza A/Indonesia/5/2005(H5N1) AS03-adjuvanted vaccine: a phase 1 randomized trial. J Infect Dis 212:531–541
106. Betakova T, Svetlikova D, Gocnik M (2013) Overview of measles and mumps vaccine: origin, present, and future of vaccine production. Acta Virol 57:91–96
107. Jacobs JP, Jones CM, Baille JP (1970) Characteristics of a human diploid cell designated MRC-5. Nature 227:168–170
108. Barrett PN, Mundt W, Kistner O, Howard MK (2009) Vero cell platform in vaccine production: moving towards cell culture-based viral vaccines. Expert Rev Vaccines 8:607–618
109. Cox N, Hannoun C, Hay A, Kaverin NV, Kilbourne ED, Webster R, Levandowski RA, Schild G, Wood J, Ghendon Y, Grachev V, Griffiths E, Martinez LJ (1995) Cell-culture as a substrate for the production of influenza vaccines – memorandum from a WHO meeting. Bull World Health Organ 73:431–435
110. Doroshenko A, Halperin SA (2009) Trivalent MDCK cell culture-derived influenza vaccine Optaflu (Novartis vaccines). Expert Rev Vaccines 8:679–688
111. WHO (2005) Recommendations for the production and control of influenza vaccine (inactivated). WHO Technical Report Series, No. 927, Annex 3. World Health Organization, Geneva. http://www.who.int/biologicals/publications/trs/areas/vaccines/influenza/ANNEX%203%20InfluenzaP99-134.pdf?ua=1. http://www.who.int/biologicals/publications/trs/areas/vaccines/influenza/en/
112. EMEA (2008) Guideline on development, production, characterisation and specifications for monoclonal antibodies and related products. Committee For Medicinal Products For Human Use (CHMP), European Medicines Agency, London. http://www.ema.europa.eu/docs/en_GB/document_library/Scientific_guideline/2009/09/WC500003074.pdf. http://www.ema.europa.eu/ema/index.jsp?curl=pages/regulation/general/general_content_000867.jsp&mid=WC0b01ac058002956b

113. WHO (2011) Guidelines on regulatory preparedness for human pandemic influenza vaccines WHO Technical Report Series, No. 963. Annex 2. World Health Organization, Geneva. http://www.who.int/biologicals/vaccines/Annex_2_WHO_TRS_963-3.pdf
114. Hutter J, Rodig J, Hoper D, Reichl U, Seeberger PH, Rapp E, Lepenies B (2012) Glycosylation matters: role of viral hemagglutinin N-glycosylation in immunogenicity. Immunology 137:760–761
115. Lancaster C, Pristatsky P, Hoang VM, Casimiro DR, Schwartz RM, Rustandi R, Ha S (2016) Characterization of N-glycosylation profiles from mammalian and insect cell derived chikungunya VLP. J Chromatogr B Analyt Technol Biomed Life Sci 2016:218–223
116. Lin SC, Jan JT, Dionne B, Butler M, Huang MH, Wu CY, Wong CH, Wu SC (2013) Different immunity elicited by recombinant H5N1 hemagglutinin proteins containing pauci-mannose, high-mannose, or complex type N-glycans. PLoS One 8:e66719
117. Liu WC, Lin YL, Spearman M, Cheng PY, Butler M, Wu SC (2016) Influenza virus hemagglutinin glycoproteins with different N-glycan patterns activate dendritic cells in vitro. J Virol 90:6085–6096
118. Wolfert MA, Boons GJ (2013) Adaptive immune activation: glycosylation does matter. Nat Chem Biol 9:776–784
119. Wei XP, Decker JM, Wang SY, Hui HX, Kappes JC, Wu XY, Salazar-Gonzalez JF, Salazar MG, Kilby JM, Saag MS, Komarova NL, Nowak MA, Hahn BH, Kwong PD, Shaw GM (2003) Antibody neutralization and escape by HIV-1. Nature 422:307–312. (Erratum, Nature 423: 197)
120. Abe Y, Takashita E, Sugawara K, Matsuzaki Y, Muraki Y, Hongo S (2004) Effect of the addition of oligosaccharides on the biological activities and antigenicity of influenza A/H3N2 virus hemagglutinin. J Virol 78:9605–9611
121. An Y, Rininger JA, Jarvis DL, Jing X, Ye Z, Aumiller JJ, Eichelberger M, Cipollo JF (2013) Comparative glycomics analysis of influenza Hemagglutinin (H5N1) produced in vaccine relevant cell platforms. J Proteome Res 12:3707–3720
122. Helle F, Duverlie G, Dubuisson J (2011) The hepatitis C virus glycan shield and evasion of the humoral immune response. Viruses 3:1909–1932
123. Ren Y, Min Y-Q, Liu M, Chi L, Zhao P, Zhang X-L (2016) N-glycosylation-mutated HCV envelope glycoprotein complex enhances antigen-presenting activity and cellular and neutralizing antibody responses. Biochim Biophys Acta-Gen Subj 1860:1764–1775
124. Tate MD, Job ER, Deng YM, Gunalan V, Maurer-Stroh S, Reading PC (2014) Playing hide and seek: how glycosylation of the influenza virus hemagglutinin can modulate the immune response to infection. Virus 6:1294–1316
125. Bright RA, Ross TM, Subbarao K, Robinson HL, Katz JM (2003) Impact of glycosylation on the immunogenicity of a DNA-based influenza H5 HA vaccine. Virology 308:270–278
126. Cao R, Mejias A, Ramilo O (2016) Systems immunology: beyond antibody titers. J Infect 72 (Suppl):S115–S118
127. Dekosky BJ, Kojima T, Rodin A, Charab W, Ippolito GC, Ellington AD, Georgiou G (2015) In-depth determination and analysis of the human paired heavy- and light-chain antibody repertoire. Nat Med 21:86–91
128. Hennig R, Rapp E, Kottler R, Cajic S, Borowiak M, Reichl U (2015) N-Glycosylation fingerprinting of viral glycoproteins by xCGE-LIF. Methods Mol Biol 1331:123–143
129. Huffman JE, Pucic-Bakovic M, Klaric L, Hennig R, Selman MHJ, Vuckovic F, Novokmet M, Kristic J, Borowiak M, Muth T, Polasek O, Razdorov G, Gornik O, Plomp R, Theodoratou E, Wright AF, Rudan I, Hayward C, Campbell H, Deelder AM, Reichl U, Aulchenko YS, Rapp E, Wuhrer M, Lauc G (2014) Comparative Performance of four methods for high-throughput glycosylation analysis of immunoglobulin G in genetic and epidemiological research. Mol Cell Proteomics 13:1598–1610
130. Behne A, Muth T, Borowiak M, Reichl U, Rapp E (2013) glyXalign: high-throughput migration time alignment preprocessing of electrophoretic data retrieved via multiplexed

capillary gel electrophoresis with laser-induced fluorescence detection-based glycoprofiling. Electrophoresis 34:2311–2315

131. Schwarzer J, Rapp E, Reichl U (2008) N-glycan analysis by CGE-LIF: profiling influenza A virus hemagglutinin N-glycosylation during vaccine production. Electrophoresis 29:4203–4214
132. McDaniel JR, Dekosky BJ, Tanno H, Ellington AD, Georgiou G (2016) Ultra-high-throughput sequencing of the immune receptor repertoire from millions of lymphocytes. Nat Protoc 11:429–442
133. Plomp R, Bondt A, de Haan N, Rombouts Y, Wuhrer M (2016) Recent advances in clinical glycoproteomics of immunoglobulins (Igs). Mol Cell Proteomics 15:2217–2228
134. Sun X, Jayaraman A, Maniprasad P, Raman R, Houser KV, Pappas C, Zeng H, Sasisekharan R, Katz JM, Tumpey TM (2013) N-linked glycosylation of the hemagglutinin protein influences virulence and antigenicity of the 1918 pandemic and seasonal H1N1 influenza A viruses. J Virol 87:8756–8766
135. Berlanda Scorza F, Tsvetnitsky V, Donnelly JJ (2016) Universal influenza vaccines: Shifting to better vaccines. Vaccine 34:2926–2933
136. Eggink D, Goff PH, Palese P (2014) Guiding the immune response against influenza virus hemagglutinin toward the conserved stalk domain by hyperglycosylation of the globular head domain. J Virol 88:699–704
137. Krammer F (2016) Novel universal influenza virus vaccine approaches. Curr Opin Virol 17:95–103
138. Weir JP, Gruber MF (2016) An overview of the regulation of influenza vaccines in the United States. Influenza Other Respir Viruses 10:354–360
139. Milian E, Kamen AA (2015) Current and emerging cell culture manufacturing technologies for influenza vaccines. Biomed Res Int 2015:11
140. Noh JY, Kim WJ (2013) Influenza vaccines: unmet needs and recent developments. Infect Chemother 45:375–386
141. Brands R, Visser J, Medema J, Palache AM, van Scharrenburg GJ (1999) Influvac: a safe Madin Darby canine kidney (MDCK) cell culture-based influenza vaccine. Dev Biol Stand 98:93–111
142. Manini I, Domnich A, Amicizia D, Rossi S, Pozzi T, Gasparini R, Panatto D, Montomoli E (2015) Flucelvax (Optaflu) for seasonal influenza. Expert Rev Vaccines 14:789–804
143. Cox MMJ, Hollister JR (2009) FluBlok, a next generation influenza vaccine manufactured in insect cells. Biologicals 37:182–189
144. Mochalova L, Gambaryan A, Romanova J, Tuzikov A, Chinarev A, Katinger D, Katinger H, Egorov A, Bovin N (2003) Receptor-binding properties of modern human influenza viruses primarily isolated in Vero and MDCK cells and chicken embryonated eggs. Virology 313:473–480
145. Romanova J, Katinger D, Ferko B, Voglauer R, Mochalova L, Bovin N, Lim W, Katinger H, Egorov A (2003) Distinct host range of influenza h3n2 virus isolates in vero and mdck cells is determined by cell specific glycosylation pattern. Virology 307:90–97
146. Schwarzer J, Rapp E, Hennig R, Genzel Y, Jordan I, Sandig V, Reichl U (2009) Glycan analysis in cell culture-based influenza vaccine production: influence of host cell line and virus strain on the glycosylation pattern of viral hemagglutinin. Vaccine 27:4325–4336
147. Raju TS, Briggs JB, Chamow SM, Winkler ME, Jones AJS (2001) Glycoengineering of therapeutic glycoproteins: in vitro galactosylation and sialylation of glycoproteins with terminal N-acetylglucosamine and galactose residues. Biochemistry 40:8868–8876
148. Butler M (2006) Optimisation of the cellular metabolism of glycosylation for recombinant proteins produced by Mammalian cell systems. Cytotechnology 50:57–76
149. Rudd PM, Dwek RA (1997) Glycosylation: heterogeneity and the 3D structure of proteins. Crit Rev Biochem Mol Biol 32:1–100

150. Kluge S, Benndorf D, Genzel Y, Scharfenberg K, Rapp E, Reichl U (2015) Monitoring changes in proteome during stepwise adaptation of a MDCK cell line from adherence to growth in suspension. Vaccine 33:4269–4280
151. Lohr V, Genzel Y, Behrendt I, Scharfenberg K, Reichl U (2010) A new MDCK suspension line cultivated in a fully defined medium in stirred-tank and wave bioreactor. Vaccine 28:6256–6264
152. Roedig JV, Rapp E, Bohne J, Kampe M, Kaffka H, Bock A, Genzel Y, Reichl U (2013) Impact of cultivation conditions on N-glycosylation of influenza virus a hemagglutinin produced in MDCK cell culture. Biotechnol Bioeng 110:1691–1703
153. Gerdil C (2003) The annual production cycle for influenza vaccine. Vaccine 21:1776–1779
154. Murakami S, Horimoto T, Ito M, Takano R, Katsura H, Shimojima M, Kawaoka Y (2012) Enhanced growth of influenza vaccine seed viruses in vero cells mediated by broadening the optimal pH range for virus membrane fusion. J Virol 86:1405–1410
155. Ping J, Lopes TJ, Nidom CA, GHEDIN E, Macken CA, Fitch A, Imai M, Maher EA, Neumann G, Kawaoka Y (2015) Development of high-yield influenza A virus vaccine viruses. Nat Commun 6:8148
156. Aguilar-Yanez JM, Portillo-Lara R, Mendoza-Ochoa GI, Garcia-Echauri SA, Lopez-Pacheco F, Bulnes-Abundis D, Salgado-Gallegos J, Lara-Mayorga IM, Webb-Vargas Y, Leon-Angel FO, Rivero-Aranda RE, Oropeza-Almazan Y, Ruiz-Palacios GM, Zertuche-Guerra MI, Dubois RM, White SW, Schultz-Cherry S, Russell CJ, Alvarez MM (2010) An influenza A/H1N1/2009 hemagglutinin vaccine produced in *Escherichia coli*. PLoS One 5: e11694
157. Robertson JS (1993) Clinical influenza-virus and the embryonated Hen's egg. Rev Med Virol 3:97–106
158. Roedig JV, Rapp E, Hoper D, Genzel Y, Reichl U (2011) Impact of host cell line adaptation on quasispecies composition and glycosylation of influenza A virus hemagglutinin. PLoS One 6:10
159. Roedig JV, Rapp E, Hoper D, Genzel Y, Reichl U (2011) Impact of host cell line adaptation on quasispecies composition and glycosylation of influenza A virus hemagglutinin. PLoS One 6:e27989
160. Mirshekari SY, Ashford DA, Harvey DJ, Dwek RA, Schulze IT (1997) The glycosylation of the influenza A virus hemagglutinin by mammalian cells – a site-specific study. J Biol Chem 272:4027–4036
161. Harpaz N, Schachter H (1980) Control of glycoprotein-synthesis. 5. Processing of asparagine-linked oligosaccharides by one or more rat-liver golgi alpha-D-mannosidases dependent on the prior action of UDP-N-acetylglucosamine-alpha-D-mannoside beta-2-N-acetylglucosaminyltransferase-I. J Biol Chem 255:4894–4902
162. Roedig JV (2014) Impact of cultivation conditions on N-glycosylation of influenza A virus hemagglutinin, on quasispecies composition, and on immunogenicity of virus preparations. PhD thesis, Otto von Guericke University, Magdeburg, Germany

# Glycoengineering of Mammalian Expression Systems on a Cellular Level

**Kelley M. Heffner, Qiong Wang, Deniz Baycin Hizal, Özge Can, and Michael J. Betenbaugh**

## Contents



K. M. Heffner, Q. Wang, D. B. Hizal, and M. J. Betenbaugh (✉)
Department of Chemical and Biomolecular Engineering, Johns Hopkins University, Baltimore, MD, USA
e-mail: beten@jhu.edu

Ö. Can
Department of Medical Engineering, Acibadem Mehmet Ali Aydinlar University, Istanbul, Turkey

**Abstract** Mammalian expression systems such as Chinese hamster ovary (CHO), mouse myeloma (NS0), and human embryonic kidney (HEK) cells serve a critical role in the biotechnology industry as the production host of choice for recombinant protein therapeutics. Most of the recombinant biologics are glycoproteins that contain complex oligosaccharide or glycan attachments representing a principal component of product quality. Both *N*-glycans and *O*-glycans are present in these mammalian cells, but the engineering of *N*-linked glycosylation is of critical interest in industry and many efforts have been directed to improve this pathway. This is because altering the *N*-glycan composition can change the product quality of recombinant biotherapeutics in mammalian hosts. In addition, sialylation and fucosylation represent components of the glycosylation pathway that affect circulatory half-life and antibody-dependent cellular cytotoxicity, respectively. In this chapter, we first offer an overview of the glycosylation, sialylation, and fucosylation networks in mammalian cells, specifically CHO cells, which are extensively used in antibody production. Next, genetic engineering technologies used in CHO cells to modulate glycosylation pathways are described. We provide examples of their use in CHO cell engineering approaches to highlight these technologies further. Specifically, we describe efforts to overexpress glycosyltransferases and sialyltransfereases, and efforts to decrease sialidase cleavage and fucosylation. Finally, this chapter covers new strategies and future directions of CHO cell glycoengineering, such as the application of glycoproteomics, glycomics, and the integration of 'omics' approaches to identify, quantify, and characterize the glycosylated proteins in CHO cells.

**Graphical Abstract**

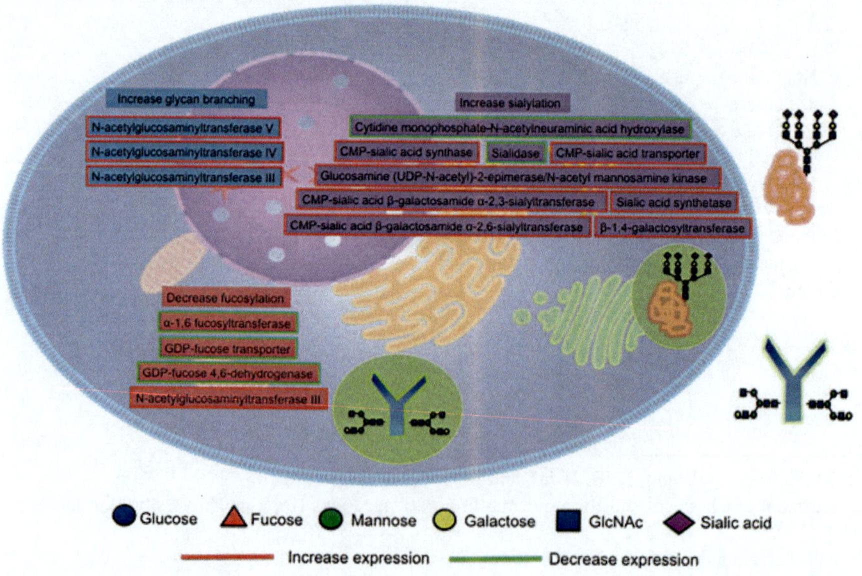


**Keywords** Chinese hamster ovary, CHO, CRISPR/Cas9, Fucosylation, Glycoengineering, Glycomics, Glycoproteomics, Mammalian expression systems, *N*-linked glycosylation, *O*-linked glycosylation, Sialylation, TALEN, ZFN

## Abbreviations

| | |
|---|---|
| ADCC | Antibody-dependent cellular cytotoxicity |
| Asn | Asparagine |
| BHK | Baby hamster kidney |
| CDC | Complement-dependent cytotoxicity |
| CHO | Chinese hamster ovary |
| CMP-SAT | cytidine 5′-monophosphate (CMP)-sialic acid transporter |
| CRISPR | Clustered regularly interspaced short palindromic repeats |
| Dol-P | Dolichol phosphate |
| EPO | Erythropoietin |
| ER | Endoplasmic reticulum |
| ESI-MS | Electrospray ionization mass spectrometry |
| Fc | Fragment crystallizable |
| FcγRIIIa | Fc gamma receptor IIIa |
| FUT8 | α-1,6-fucosyltransferase |
| FX | GDP-4-keto-6-D-deoxymannose epimerase/GDP-4-keto-6-L-galactose reductase |
| GFPP | GDP-fucose pyrophosphorylase |
| GFT | GDP-fucose transporter |
| GlcNAc | *N*-acetylglucosamine |
| GMD | GDP-fucose 4,6-dehydratase |
| GNE/MNK | Uridine diphosphate-*N*-acetyl glucosamine 2-epimerase/*N*-acetyl mannosamine kinase |
| GnT-1 or Mgat1 | *N*-acetylglucosaminyltransferase I |
| GnT-II or Mgat2 | Beta-1,2-*N*-acetylglucosaminyltransferase II |
| GnT-III or Mgat3 | Beta-1,4-*N*-acetylglucosaminyltransferase III |
| GnT-IV or Mgat 4 | Beta-1,2-*N*-acetylglucosaminyltransferase IV |
| GnT-V or Mgat 5 | Beta-1,2-*N*-acetylglucosaminyltransferase V |
| HEK | Human embryonic kidney |
| HNF1-alpha | Hepatocyte nuclear factor 1-alpha |
| HPLC | High-performance liquid chromatography |
| LacNAc | Acetyl lactosamine |
| mAb | Monoclonal antibody |
| MALDI-TOF | Matrix-assisted laser desorption/ionization time-of-flight |
| ManII | Alpha-mannosidase II |
| Neu5Gc | *N*-glycolylneuraminic acid |
| NK | Natural killer |
| OST | Oligosaccharyltransferase |
| RCA-I | *Ricinus communis* agglutinin I |

| | |
|---|---|
| Ser | Serine |
| shRNA | Short hairpin RNA |
| siRNA | Small interfering RNA |
| SPEG | Solid phase extraction of glycosylated peptides |
| TALEN | Transcription activator-like effector nuclease |
| Thr | Threonine |
| tPA | Tissue plasminogen activator |
| ZFN | Zinc finger nuclease |

# 1 Introduction

Both *N*-glycosylation and *O*-glycosylation pathways serve as key targets for mammalian cell engineering efforts. The efficiency and control of glycosylation in recombinant protein production is critical, because changes in protein glycosylation can affect circulatory half-life, bioactivity, and product quality. Improving the degree of glycosylation and sialylation can reduce clearance of the therapeutic product from the patient. Additionally, control of glycan antennarity helps to maintain lot-to-lot consistency during the drug production. Controlling fucosylation has important effects on therapeutic efficacy by regulating antibody-dependent cellular cytotoxicity (ADCC). Decreasing core fucosylation can improve antibody effector function and clinical efficacy. In summary, glycosylation control is crucial during the process of biotherapeutics development. This section introduces the Chinese hamster ovary (CHO) glycosylation pathway, as well as the importance of sialylation and fucosylation.

## 1.1 Glycosylation

Therapeutic glycoproteins include several classes, such as monoclonal antibodies (mAbs), immunoglobulin G fragment crystallizable domain (Fc)-fusion proteins (Fc-fusion proteins), enzymes, hormones, cytokines, growth factors, and hormones [1–3]. Overall, the biotechnology industry generates billions of dollars of sales from these glycoproteins [4]. The increasing demand for biotherapeutics for the treatment of cancer, autoimmune disorders, infectious diseases, genetic disorders, and metabolic disorders requires the development and precise control of glycotherapeutics production.

An overview of both *N*-glycosylation and *O*-glycosylation is shown in Fig. 1. During *N*-glycosylation, various carbohydrate chains are added to asparagine (Asn) [5] residues of proteins [5]. In contrast, *O*-glycosylation involves the addition of carbohydrate chains to serine (Ser) or threonine (Thr) [6]. While *N*-glycans are the most common modification in biotherapeutics such as mAbs, there are examples of therapeutic glycoproteins, such as erythropoietin (EPO) and etanercept, that also contain *O*-glycosylation [7]. Glycosylation is a critical post-translational modification found in most biotherapeutics; interestingly, the cellular process generates

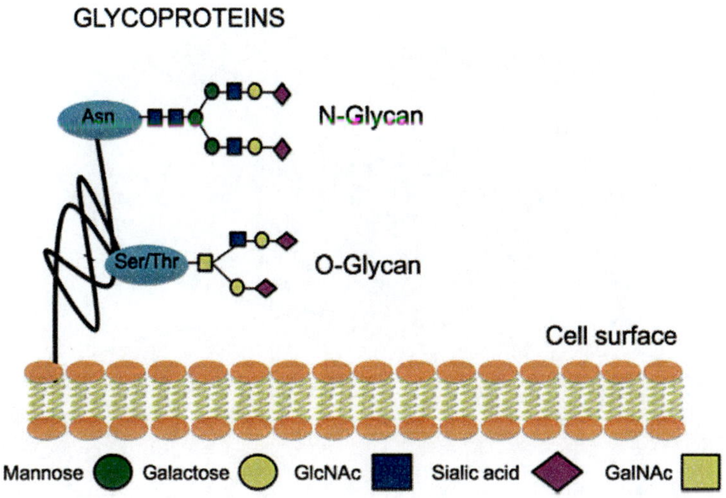


**Fig. 1** Examples of *N*- and *O*-linked glycosylation. *N*-linked glycosylation involves the asparagine (Asn) [5] residue, whereas *O*-linked glycans extend from serine (Ser) or threonine (Thr) residues. GlcNAc: *N*-acetylglucosamine, GalNAc: *N*-acetylgalactosamine

structural diversity that includes a number of different structures even for a single protein from one organism. The variety of glycoforms expands dramatically when the protein is produced in another host cell or species even under different reactor conditions. Most importantly, the pattern of glycosylation can play a major role in modulating a number of product quality characteristics [8].

A prerequisite for *N*-glycosylation is the requirement that *N*-glycans be linked to the Asn of the Asn-X-Ser/Thr consensus sequence, where X represents any amino acid except for proline [9]. A similar consensus sequence for *O*-linked glycosylation has not been identified [8]. As proteins are processed through the endoplasmic reticulum (ER) and Golgi apparatus prior to secretion, a number of enzymes can act to shorten or extend the *N*-glycan chain, as shown in Fig. 2. Since the enzymes do not act on every protein that traverses a particular compartment, the stochastic nature of the interactions creates heterogeneity, owing to the variability in glycosylation site occupancy and the diversity of glycoforms that are formed during passage through the secretory apparatus. In addition, there is continuous interplay between enzymes and oligosaccharide substrates. Since more than one enzyme can act on a glycan substrate, a wide arsenal of glycoproteins can be generated [10, 11].

The complex *N*-linked glycosylation reaction network shown in Fig. 2 involves glycosidases and glycosyltransferases that catalyze enzymatic modifications in different cellular compartments. First, the biosynthesis of mammalian *N*-glycans begins with the transfer of *N*-acetylglucosamine-1-phosphate (GlcNAc-P) from uridine diphosphate-*N*-acetylglucosamine (UDP-GlcNAc) to the dolichol phosphate (Dol-P) lipid carrier to generate dolichol pyrophosphate *N*-acetylglucosamine (Dol-P-P-GlcNAc) at the cytoplasmic face of the ER membrane [12]. Next, sugars are sequentially added to Dol-P-P-GlcNAc to form an oligosaccharide precursor

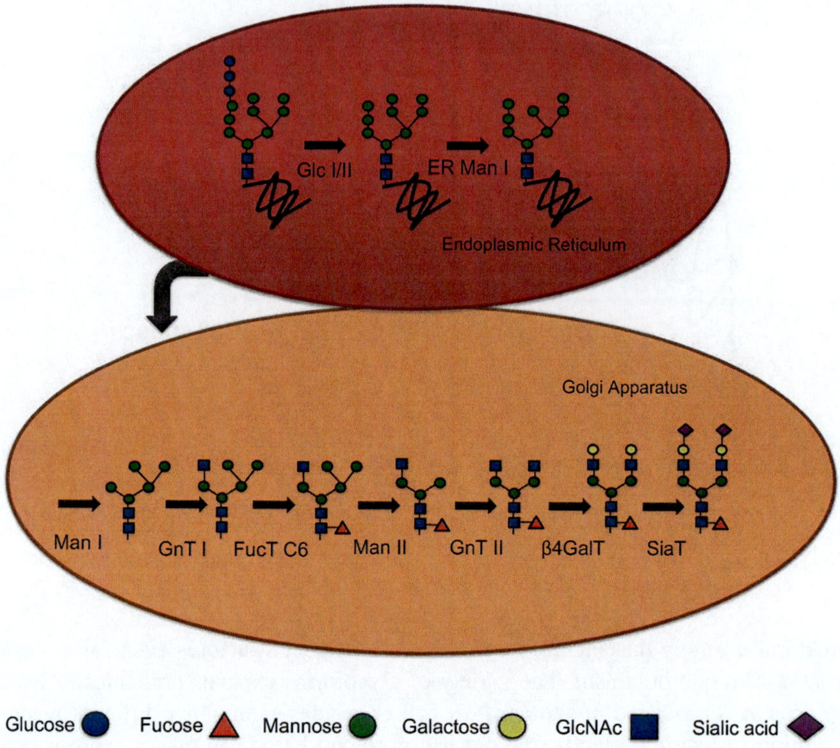


**Fig. 2** Overview of *N*-linked glycosylation generating biantennary sialylated glycans. During *N*-linked glycosylation, various enzymes extend and trim the glycoprotein as it passes from the endoplasmic reticulum to the Golgi apparatus. Abbreviations: Glc I/II glucosidase I/II, ER Man I endoplasmic reticulum mannosidase I, Man I mannosidase I, GnT I *N*-acetylglucosaminyltransferase I, FucT C6 α(1,6)-fucosyltransferase, Man II mannosidase II, GnT II *N*-acetylglucosaminyltransferase II, β4GalT β-1,4-galactosyltransferase, SiaT sialyltransferase

known as Dol-P-P-$GlcNAc_2Man_5$ [12]. The structure is then flipped to the ER side and further extended to generate $Glc_3Man_9GlcNAc_2$-P-P-Dol. Oligosaccharyltransferase (OST) identifies the consensus sequence (Asn-X-Ser/Thr) in the nascent polypeptide and transfers $Glc_3Man_9GlcNAc_2$ from the dolichol-linked donor to the side chain amide of Asn, while releasing the Dol-P-P during the process [13]. The glucose residues on the precursor are then sequentially trimmed by ER alpha-glucosidase I and ER alpha-glucosidase II to form a monoglucosylated glycan. This intermediate plays a role in the ER lectin chaperones calnexin/calreticulin-associated glycoprotein folding control cycle [14]. When the precursor is folded, it is next trimmed to yield $Man_8GlcNAc_2$-protein before exiting the ER; this step is catalyzed by ER alpha-mannosidase I. The $Man_8GlcNAc_2$ glycoform is translocated into the cis-Golgi apparatus, where it is trimmed again, to $Man_5GlcNAc_2$, a key intermediate along the pathway to form hybrid and complex *N*-glycans, and sometimes found as a final glycan product; this step is catalyzed by Golgi alpha-mannosidases I.

In order to generate hybrid and complex *N*-glycans, *N*-acetylglucosaminyltransferase I (GnT-1 or Mgat1) is required to add GlcNAc to $Man_5GlcNAc_2$ in the medial Golgi apparatus [12]. *N*-Glycans are trimmed by Golgi alpha-mannosidase II, which removes two mannoses from $GlcNAcMan_5GlcNAc_2$ to generate GlcNAc-$Man_3GlcNAc_2$. Hybrid *N*-glycans result from the incomplete removal of mannose residues, which occurs when a structure such as $GlcNAcMan_5GlcNAc_2$ undergoes no further extension or trimming and the structure ends up with one or two terminal Man residues. In an alternative process, GlcNAc can be added to the innermost Man group by the enzyme beta-1,4-*N*-acetylglucosaminyltransferase III (GnT-III or Mgat3) in the medial Golgi apparatus, a process which generates bisecting GlcNAc structures that alter the capacity for other downstream enzymes to act on the glycan structure.

The precursor for all multi-antennary complex *N*-glycans is $GlcNAc_2Man_3GlcNAc_2$, which is generated by the action of beta-1,2-*N*-acetylglucosaminyltransferase II (GnT-II or Mgat2) that adds GlcNAc to the $GlcNAcMan_3GlcNAc_2$ structure. Tri-antennary and tetra-antennary branches are created through the addition of GlcNAc at the alpha-(1,3)-mannose site by *N*-acetylglucosaminyltransferase IV (GnT-IV or Mgat 4) and at the alpha-(1,6)-mannose site by *N*-acetylglucosaminyltransferase V (GnT-V or Mgat 5).

There can be further modifications, such as fucosylation, branch extension, and sialylation, which generate even more complex glycans. Fucosylation occurs in the trans Golgi apparatus with the addition of core alpha-(1,6)-fucose to the GlcNAc adjacent to Asn of the *N*-glycan by alpha-(1,6)-fucosyltransferase. Branch extension involves the addition of a beta-linked galactose residue to GlcNAc, which yields Gal-beta-1-4GlcNAc, also known as acetyl lactosamine (LacNAc). For sialylation, terminal Gal residues can be acted upon by alpha-(2,3)- or alpha-(2,6)-sialyltransferases that add sialic acid residues to the glycan [12].

One reason for the widespread use of CHO cell lines in biotechnology is their capacity to produce complex glycans that are compatible with the human immune system [1, 15]. Alternative mammalian cell lines can also produce biopharmaceuticals, but their use is not as widespread in industry because of their potential for immunogenicity and difficulty in manufacturing scale-up; examples include baby hamster kidney (BHK), murine myeloma and hybridoma cell lines (NS0 and Sp2/0), and human host cell lines, such as human embryonic kidney (HEK-293) and human retinal cells (PER.C6) [1, 2, 16].

When glycans are generated outside of human hosts, it is critical to avoid the production of non-human glycans, such as terminal Gal-alpha-1,3-Gal linkages (alpha-Gal) and *N*-glycolylneuraminic acid (Neu5Gc) residues, which may result in adverse immunogenic reactions if given to humans with a sensitivity to these residues [1, 17]. Mouse cells such as NS0 have an alpha-1,3-galactosyltransferase enzyme that produces glycans containing the alpha-Gal linkage [18]. The second potential immunogenic reaction from Neu5Gc is common in all non-primate mammalian cells, owing to the presence of the enzyme *N*-acetylneuraminic acid hydroxylase, which converts cytidine monophosphate (CMP)-*N*-acetylneuraminic acid (Neu5Ac) to CMP-Neu5Gc in all mammals other than old-world primates

[1, 19]. Humans exhibit a circulating polyclonal anti-Neu5Gc antibody response, so it is desirable to avoid Neu5Gc in biotherapeutics production [1, 17]. In contrast to the alpha-Gal epitope, Neu5Gc can be metabolically incorporated into glycoforms during cell culture from metabolites in cell culture media. Mouse myeloma cells (NS0 and Sp2/0) thus exhibit the highest potential for immunogenicity because they express higher levels of alpha-Gal and Neu5Gc than CHO cells, which can be an issue if biotherapeutics with these modifications are provided to patients at large doses or for long periods [19–21]. These subtle differences in glycosylation processing are one of the principal reasons why CHO cells are preferred for bioproduction.

Aside from immunogenic epitopes, glycosylation patterns in CHO cells and humans often differ in other ways too [22]. One reason is that CHO cells lack bisecting GlcNAc residues because they typically do not express GnT-III; the resulting difference may affect the efficacy of the glycotherapeutics [23]. Human cells contain GnT III and can produce glycans with bisecting GlcNAc; in comparison, NS0 and SP2/0 cells are able to generate only a portion of glycans with bisecting GlcNAc residues [24].

Overall, the glycoform profiles on glycoproteins can vary widely depending on the cell lines, growth, and bioreactor conditions such as pH, temperature, media, and feeding strategies. The interplay of various glycosylation enzymes is responsible for the great diversity of glycoproteins. Some examples relevant to glycotherapeutics are shown in Fig. 3. Specifically, the degree of antennarity varies across glycoproteins. Glycosylation in biotherapeutics directly affects product quality because it plays a role in solubility, stability, protease resistance [25], aggregation [1, 2], serum half-life [26], immunogenicity [8], efficacy [27, 28], and ligand binding [29].

These impacts of glycosylation highlight the need for glycoengineering in order to yield glycotherapeutics with consistent and desirable glycoform profiles. In the next section of the chapter, we examine targets and review genetic engineering

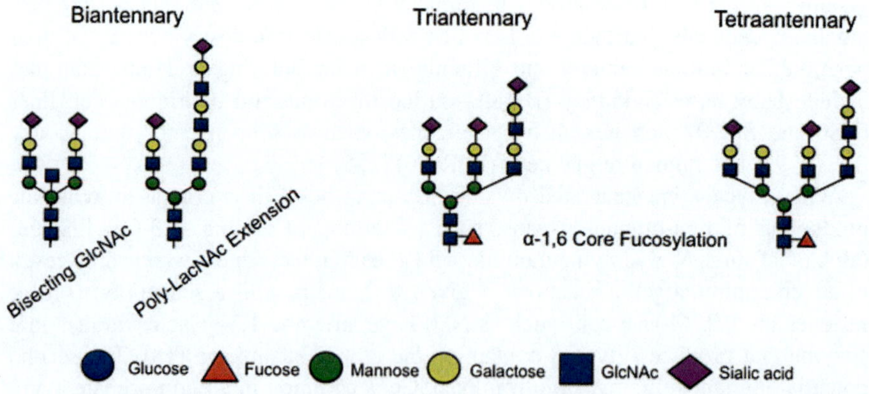


**Fig. 3** Examples of *N*-glycans with different antennarities. The differences between bi-antennary, tri-antennary, and tetra-antennary glycoforms are shown. These correspond to 2, 3, and 4 branches, respectively. Poly-*N*-acetyllactosamine (Poly-Lac)

approaches to control glycosylation, such as increasing the expression of glycosylation and sialylation enzymes, or reducing the expression of sialidase cleavage and fucosylation enzymes. Recently, multiple genes have been modified simultaneously, and new strategies such as zinc finger nucleases (ZFNs), transcription activator-like effector nucleases (TALENs), and clustered regularly interspaced short palindromic repeats (CRISPR)/CRISPR-associated protein-9 (Cas9) offer new technologies for glycoengineering applications.

## 1.2 Sialylation

Sialic acid addition is the final step of the *N*-glycosylation pathway. While both Neu5Ac and Neu5Gc are present in CHO cells, humans lack Neu5Gc. Sialic acid is a negatively charged acidic 9-carbon sugar moiety that is typically attached by an alpha-glycosidic linkage on the C3- or C6-hydroxyl group of terminal galactose by alpha-2,3-sialyltransferases (ST3) or alpha-1,6-sialyltransferases (ST6) individually [30–33]. The sialic acid moiety may also on occasion be attached to the C8 position of sialic acid to form sialic acid multimers, but this is not typically observed for glycoprotein therapeutics [34]. As the terminal end cap, sialic acid is especially relevant for the half-life and bioactivity of recombinant glycoproteins. The removal of sialic acid by sialidase exposes the terminal galactose, and its cleavage often decreases circulatory half-life. Without sialic acid capping the glycoprotein, the galactose molecule is recognized by the hepatocyte asialoglycoprotein receptor and is cleared from the body [32, 35, 36]. For improving the circulatory half-life of recombinant therapeutics, preventing recognition by this receptor is desirable. Additionally, fully sialylated glycoproteins can increase the size and overall charge of the molecule. Therefore, it is often desirable to enhance or maximize sialylation in CHO cells to improve the production of recombinant therapeutics [14]. Of course, for the case of biosimilars, it may also be relevant to match the sialylation profile of the innovator molecule. If the host cell line of the innovator and biosimilar are different from each other, matching the sialylation profile can be even more difficult. Therefore, both sialyltransferases and sialidases are targets for genetic engineering, because they affect opposing processes.

## 1.3 Fucosylation

Fucosylation, or the addition of fucose to glycoforms, occurs through both de-novo and salvage pathways. In the first pathway, D-glucose uptake into the cytoplasm generates guanosine diphosphate (GDP)-mannose. The enzymes GDP-mannose 4,6-dehydratase and GDP-keto-6-deoxymannose 3,5-epimerase, 4-reductase convert GDP-mannose into GDP-fucose [37]. In contrast, the salvage pathway utilizes L-fucose from extracellular and lysosomal sources. Fucokinase phosphorylates

L-fucose into L-fucose-1-phosphate, and GDP-fucose pyrophosphorylase (GFPP) converts L-fucose-1-phosphate to GDP-fucose. GDP-fucose is subsequently transported to the Golgi apparatus by the GDP-fucose transporter and fucose is added onto the glycan chains of maturing glycoproteins by fucosyltransferases. Thus, enzymes are important in fucosylation and they are also important targets for cell engineering.

This step of glycosylation is critically important for antibody fragment crystallizable (Fc) receptor-mediated activity, which can strongly affect ADCC activity. During ADCC, an antibody first binds to a cell-surface antigen and then recruits the immune effector cells to destroy the target cells, such as cancer cells carrying antigens recognized by antibodies. The Fc gamma receptor IIIa (FcγRIIIa) on natural killer (NK) cells binds to the Fc region of the antibody, which region has a major antibody effector function in the immune system. This binding results in lysis and apoptosis of the targeted cell by NK-cell-mediated killing. A decrease of fucosylation at Asn297 in the antibody Fc domain significantly increased antibodies' binding affinity to FcγRIIIa and, further, improved ADCC potency [38, 39, 40].

## 2 Technologies for Glycoengineering Through Gene Knocking Down, In, and Out

Strategies for CHO glycoengineering include the knockdown or knockout of enzymes such as sialidase or fucosyltransferase, along with the overexpression or knocking-in of glycosylation enzymes, such as glycosyltransferases and sialyltransferases. Genetic engineering approaches, including small interfering RNA (siRNA), short hairpin RNA (shRNA), ZFN, TALEN, and CRISPR/Cas9, aim to modify gene expression [37, 41–51], while other methods can amplify the expression of a target gene, such as by overexpression and knockin. Both siRNA and shRNA have extensive use in decreasing gene expression, thus playing a role in both reduced fucosylation and sialidase cleavage. Table 1 compares the current

**Table 1** Comparison of genetic engineering technologies

| Technology | Established | Design | Specificity | On-target | Target site | Reference |
|---|---|---|---|---|---|---|
| Zinc finger nucleases (ZFN) | 2003 | Hard | Low | Low-medium | 18–36 bp | [46, 47, 51] |
| Transcription activator-like effector nucleases (TALEN) | 2011 | Easy | Medium-high | Medium-high | 24–38 bp | [48, 51] |
| Clustered regularly interspersed short palindromic repeats (CRISPR)/CRISPR-associated protein-9 (Cas9) | 2013 | Easy | High | High | >22 bp | [49–51] |

technologies for ZFN, TALEN, and CRISPR/Cas9, all of which can be used to modify expression in glycosylation pathways.

As an example, fucosylation is often controlled by gene knockdown and knockout strategies. The removal of core fucose can be highly advantageous for improving the therapeutic efficacy of mAbs. The core fucosylation is defined by the transferring of fucose from GDP-fucose to GlcNAc in an $\alpha$-1,6 linkage catalyzed by an $\alpha$-1,6-fucosyltransferase (encoded by $\alpha$-1,6-fucosyltransferase [FUT8]). In one study, overexpression of GnT-III was able to compete with native fucosyltransferase and produce a afucosylated antibody [52]. The results in that study indicated that GnT-III inhibited the core FUT8, increasing the production of a bisected afucosylated antibody with enhanced ADCC activity [52]. Coexpression of GnT-III with Golgi alpha-mannosidase II (ManII) resulted in more complex oligosaccharides compared with the expression of GnT-III alone [52]. The overall results indicate the importance of decreased fucosyltransferase activity for improving ADCC. A number of strategies can be implemented to lower or silence fucosyltransferase activity, including siRNA, shRNA, ZFN, TALEN, and CRISPR/Cas9 [37, 41–51].

## 2.1 *siRNA*

siRNA can be used as a transient or a stable method to suppress specific gene expression using RNA interference. Two siRNA sequences were found that reduced the expression of FUT8 in CHO DG44 cells to 20% of the level in parental controls [45]. The decrease in mRNA expression corresponded to a 40% fucosylated antibody with 100 times the ADCC of that for control cells [45]. Additionally, clone stability was demonstrated, as the ability to produce antibody with decreased fucosylation continued over repeated passages and fed-batch culture [45]. Interestingly, FUT8 knockdown was more effective in the exponential phase than in the stationary phase of culture [45]. In summary, this siRNA approach did not completely knockout FUT8 expression, but the decreased expression resulted in decreased fucosylation and enhanced ADCC.

In another study, a CHO cell line, also with FUT8 knocked down using siRNA, was created and compared with two lectin-mutated defucosylation cell lines—an endogenous GDP-fucose 4,6-dehydratase (GMD)-deficient cell line (Lec13) and an endogenous GnT-1-deficient cell line (Lec1) [41, 53]. These lectin-mutated cell lines produced afucosylated antibody, but over culture time, the percentage of fucosylated antibody increased [41]. In contrast, in this study, the FUT8 siRNA cell line produced completely afucosylated antibody throughout cell culture [41]. Subsequent scaling of the experiment to bioreactors with pH and dissolved oxygen control yielded similar results, including afucosylated antibody and enhanced ADCC, from the FUT8 siRNA cells [41]. Thus, siRNA is an important tool for controlling fucosylation at different scales in bioprocess development.

Moreover, three key enzymes in the fucosylation pathways in CHO cells have been identified, and knockdown of these key enzymes—FUT8, GDP-fucose transporter (GFT), and GDP-fucose 4,6-dehydratase (GMD)—using separate siRNA vectors, has also been achieved to study the effect on fucosylation of recombinant glycoproteins. Both the FUT8 and GMD siRNA cell lines were separately found to produce afucosylated antibodies [37]. In contrast, knockdown of 98% GFT expression at the mRNA level yielded only 40% reduction of the Fc fucosylated oligosaccharide [37]. After it was demonstrated that GMD inhibition with siRNA removed intracellular GDP-fucose and yielded afucosylated antibodies, it was shown that GMD-KO CHO DG44 cells produced fucosylated antibodies upon medium supplementation of L-fucose during culture [37]. Cell culture samples were obtained and the level of UDP-glucose and the oligosaccharide profiles were determined with high-performance liquid chromatography (HPLC) and matrix-assisted laser desorption/ionization time-of-flight (MALDI-TOF) mass spectrometry (MS), respectively [37]. These findings highlight that the degree of fucosylation can be controlled through both cell engineering and media manipulation.

In related experiments, knockdown of FUT8, GFT, or GMD resulted in low levels of fucosylated antibody [43]. Furthermore, the combined knockdown of FUT8 and GMD, using siRNA, synergistically improved the fraction of afucosylated antibody [43]. From these results, a tandem expression vector was designed to simultaneously knockdown FUT8 and GMD expression in CHO cells [43]. This strategy produced completely afucosylated antibody at constant levels during passaging and adaptation to serum-free medium for 2 months [43]. This result highlights that combined knockdown of fucosylation enzymes can result in antibodies that are completely devoid of fucosylation.

## 2.2 *shRNA*

One disadvantage of siRNA is the quick degradation of the RNA in the cell. shRNA is more stable because, following transfection, the shRNA becomes an active double strand. Using an alternative strategy,CHO DG44 cells transfected with FUT8 shRNA showed less than 5% FUT8 mRNA expression, which resulted in the production of 12% fucosylated antibody and enhanced ADCC compared with that in the parental cells [42]. Glycoform profiles were determined by electrospray ionization mass spectrometry (ESI-MS) [42]. One benefit of shRNA over siRNA technology is the extended efficacy of the former. Stability studies demonstrated that FUT8 knockdown was maintained for over 4 weeks [42]. After prolonged culture, the mRNA expression of FUT8 and the percentage of fucosylated antibody remained consistently low [42]. Thus, it is possible to achieve significant reduction in antibody fucosylation using either siRNA or shRNA.

### 2.3 Random Mutagenesis and Homologous Recombination Knockout Selection

Originally, the knockout of genes such as FUT8 required the screening of numerous clones to find one in which the gene had been randomly mutated. Sequential homologous recombination was used to knockout both alleles of FUT8 [44]. Gene targeting by homologous recombination is a useful strategy to genetically modify any chosen allele in a predetermined way without affecting any other locus in the genome [54]. This strategy produced completely afucosylated antibodies, with the growth and viability of the cell culture being similar to that in the parental controls [44]. Assays to determine binding activity, ADCC, and complement-dependent cytotoxicity (CDC) revealed no effect on the binding activity or CDC of the FUT8−/− knockout, whereas the ADCC was increased 100-fold over that of a commercial antibody, Rituxan (rituximab; Genentech), without the FUT8 knockout [44]. The FUT8−/− knockout showed significantly stronger binding to FcγRIIIa than the parental FUT8+/+ antibodies [44]. Additionally, knockout of one or both alleles of FUT8 was compared and it was found that a hemizygous FUT8+/− knockout did not reduce fucosylation completely [44]. Thus, knockout of both FUT8 alleles can be used as a strategy to produce completely afucosylated antibody therapeutics from CHO cells.

Mutants can also be used to understand glycosylation and identify new targets for intervention. Treating CHO cells with the cytotoxic lectin *Ricinus communis* agglutinin I (RCA-I), which is specific for terminal beta-1,4-linked galactose [55], was designed to select mutants with defects in the *N*-glycosylation pathway upstream of galactose addition. Surprisingly, RCA-I-resistant CHO mutants contained mutations in the *N*-acetylglucosaminyltransferase I (GnT-I) gene similar to those in the Lec1 mutant [56]. Possibly, RCA-I may not be specific for terminal beta-1,4-linked galactose, and may bind other glycan structures, except for $Man_5GlcNAc_2$ [57]. Without functional GnT-I, cells fail to transfer GlcNAc to $Man_5GlcNAc_2$. By restoring functional GnT-I in these mutants, the sialic acid content of recombinant proteins in transient expression and stably transfected clones increased [56]. While the molecular mechanism for this phenomenon remains unknown [58], recombinant EPO generated in the RCA-I-restored mutant cell line with GnT-1 exhibited an increase in sialylation of 30% over the control [59]. In addition, the percentage of tri- and tetra-antennary glycans on EPO produced by the GnT I-restored CHO-GnT I-deficient cells increased, as measured by MS [59].

### 2.4 ZFNs

An alternative to random mutagenesis is to apply ZFN technology. Zinc fingers are transcription factors that recognize three to four bases of a sequence and can be used to target a specific sequence. ZFNs contain the zinc finger domain and FokI endonuclease domain, which must dimerize for activity that ensures specificity [60]. In one of the initial applications, ZFNs were designed to eliminate FUT8 function [46]. The benefit of this technique is the applicability of the created ZFNs to any CHO cell line [46]. The technology allows for targeting point mutations with in-frame, short deletions [46]. Zinc finger-transfected cells had growth, antibody productivity, and glycosylation patterns similar to those in the parental controls; however, the antibodies produced were completely afucosylated [46].

In another experiment, ZFNs were used to generate CHO cell lines deficient in mannosyl (alpha-1,3-)-glycoprotein beta-1,2-*N*-acetylglucosaminyltransferase (GnT-I) [47]. This resulted in the production of glycoproteins with high Man5 species [47]. Using ZFNs reduced off-target effects and maintained the same growth and productivity as that in the parental cell line, thus demonstrating process robustness and potential for scale-up [47]. An application of this knockdown is to generate mannose receptor targeted biologics. ZFNs offer an alternative to generating knockouts and may have important applications in future cell engineering strategies to control glycosylation.

## 2.5 TALENs

Another novel technology for genetic engineering is TALENs. This technology is more flexible than ZFN technology, because TALENs are dimeric transcription factor nucleases—composed of 33–35 amino acid modules—that can each target a single nucleotide [60]. Many companies offer custom design of TALENs, thus reducing the cost of TALENs compared with that of ZFNs. In recent research, knockout of CHO FUT8 via the simultaneous TALEN-mediated integration of an antibody cassette was demonstrated, and this process produced afucosylated antibodies [61]. Another novel technology applied TALEN and precise integration into target chromosome (PITCh) vector-mediated integration of long gene cassettes in CHO cells [48]. Results showed over 9 kb whole plasmid integration and over 7 kb backbone-free integration at the defined genomic locus, and the production of a recombinant single-chain Fv (variable region)-Fc(constant region) protein [48]. The method demonstrated the applicability of TALENs for high-throughput knockin of large DNA into CHO cells. Thus, TALENs can serve as a beneficial tool for biotechnology applications, such as improving the generation of high-producing cell lines with desirable glycosylation.

## 2.6 CRISPR/Cas9

Finally, CRISPR/Cas9 represents one of the newest and most rapidly expanding methods for genome engineering in CHO cells. First, Cas9 generates a double-strand DNA break at a site determined by the guide RNA; the system is different from those of ZFNs or TALENs because it relies on bacterial adaptive immunity [60]. Multiple guide RNAs can be used to facilitate simultaneous mutations, and the system can be applied to activator or repressor domains to control gene expression [60]. One potential disadvantage, however, is the shorter length of the guide RNA compared with ZFNs and TALENs, which could result in off-target effects [60]. The first published results of CRISPR/Cas9 genome editing in CHO cells demonstrated the successful disruption of C1GALT1 specific chaperone 1 (COSMC) and FUT8 [49]. The single guide RNAs generated an indel frequency of 47.3% in COSMC and 99.7% in FUT8 (with lectin selection) [49]. In addition, the bioinformatics tool CRISPy was established to identify the single guide RNA sequences in the CHO genome [49]. In related research, CRISPR/Cas9 was used to simultaneously disrupt FUT8, BCL2 antagonist/killer, and BCL2 associated X in CHO cells [50]. Single cell sorting revealed that, among 97 clones, there were 34 triple-, 23 double-, and four single-disrupted cell lines [50]. The triple-disrupted clones were confirmed to have removal of BAK and BAX, as well as decreased fucosylation [50]. Additionally, the disrupted cell lines were more resistant to apoptosis than the parental cells [50]. Further, instead of targeting the FUT8 gene, the knockout of key enzymes in fucosylation pathways provided alternatives to suppress fucosylation, such as knockout GDP-D-mannose-4,6-dehydratase (GMD) and GDP-4-keto-6-D-deoxymannose epimerase/GDP-4-keto-6-L-galactose reductase (FX), which are involved in the de-novo synthesis of GDP-fucose [62]. Disruption of both alleles of the FX gene via CRISPR/Cas9 led to the expression of an antibody with fully afucosylated glycan profiles [62]. CRISPR/Cas9 can thus serve as a useful tool for glycoengineering, because of its potential to affect multiple genes involved in glycosylation, sialylation, and fucosylation. These successes highlight the applicability of CRISPR/Cas9 for genome editing.

Recently, the combination of ZFNs, TALENs, and CRISPR/Cas9 was used for CHO glycoengineering to inactivate the GDP-fucose transporter and improve ADCC [51]. Mass spectrometry was used to identify that the EPO-Fc and anti-Her2 antibody produced in the modified cell lines lacked core fucosylation [51]. Removal of the core fucose did not affect cell growth or productivity as compared with these properties of the parental cell lines [51]. This experiment shows that genome editing techniques are applicable to CHO glycoengineering and can provide results that aid bioprocess development.

## 3 CHO Glycoengineering

CHO glycoengineering efforts aim to alter glycosylation steps by either increasing or decreasing specific glycan attachment, including terminal sialylation, or, alternatively, by reducing the cleavage of sialic acid by sialidase. The previous section highlighted the efforts made to reduce core fucosylation by gene knockdown and knockout. Glycoengineering strategies are described in this section, including the overexpression of glycosyltransferases, overexpression of galactosyltransferases, overexpression of sialyltransferases, and reduction in sialidase cleavage. These modifications also include efforts to alter terminal sialylation by overexpression or, alternatively, by reducing the cleavage of sialic acid by sialidase.

### *3.1 Overexpression of GnT Genes*

During *N*-glycosylation, various monosaccharides are added to the oligosaccharide chains of glycoproteins. One strategy to improve glycosylation was through the overexpression of rat GnT-III in CHO DG44 cells producing recombinant antibody [63]. Glycan analysis by HPLC revealed that most glycoproteins displayed bisecting GlcNAc residues [63]. This resulted in a 10- to 20-fold improvement in ADCC, as determined by the increased affinity of the antibody to Fcγ receptor III (FcγRIII), without affecting cell growth or antibody productivity [63]. Similarly, increased expression of GnT-III increased the bisecting GlcNAc residues [64–66]. Improving the proportion of glycans that have GlcNAc residues has positive effects on therapeutic efficacy. Additionally, it has been shown that GnT-III competes with beta-1,4-galactosyltrasferase; as bisecting GlcNAc residues increase, there is a concomitant decrease in the complexity of the glycans [64–66]. Even more importantly, GnT-III will compete with the core FUT8 enzyme, which leads to decreased fucosylation, while increasing bisecting GlcNAc residues. Ultimately, the reduction in fucosylation may be the primary reason for the increased ADCC observed in the CHO cells that overexpress GnT-III. Recently, GnT-III was coexpressed with fucosyltransferase 7 in order to optimize glycoengineering by localizing the glycosyltransferase in the Golgi machinery [67]. The approach was able to control the *N*-glycans with defined structural motifs; the addition of bisecting GlcNAc, as measured by HPLC and MS, resulted in an increased ADCC for the therapeutic agent cetuximab [67].

GnT-IV and GnT-V are involved in multiantennary glycan formation [64–66]. Overexpression of branching genes can increase complexity, as well as increasing sialylation acceptor sites. Shown in Fig. 3 are examples of bi-, tri-, and tetra-antennary structures. The structures include complex-type *N*-glycans with GlcNAc that can be extended to contain the disaccharide Gal-beta-1,4-GlcNAc, sometimes capped by a terminal sialic acid. The formation of tri- and tetra-antennary *N*-glycans is controlled by the enzymatic actions of GnT-IV and GnT-V. Cell proliferation, cell-surface signaling [23], cancer metastasis, regulation of T-cell activation [68], and the rate of therapeutics clearance by the kidneys are all affected by the actions of GnT-IV and GnT-V [69]. In one study, only a small fraction of glycoproteins produced in a CHO cell line contained GlcNAc beta-1-6 branching controlled by GnT-V [66]. This suggested genetic engineering approaches targeting GnTs might serve to improve the production of recombinant therapeutics. Overexpression of GnT-IV or GnT-V individually was found to increase the antennarity of the glycoform profile, as determined by an increase in reactivity with *Datura stramonium* agglutinin [70] lectin blot [64–66].

In order to control the multi-antennary glycoforms of recombinant proteins, the overexpression of GnT-IV and GnT-V was used in CHO cells producing human interferon (IFN)-gamma and EPO [66, 71]. In both cases, tri- and tetra-antennary sugar chains comprised more than 50% of the total sugar chains [71]. At the same time, this resulted in higher levels of poly LacNAc [66, 71]. In another study, mouse ST3 and/or rat ST6 were incorporated into CHO cell lines stably transfected with GnT-V that were producing IFN-gamma [65]. Results showed that over 60% of the glycoforms were sialylated with alpha-2,3- and alpha-2,6-linkages [65].

Recently, a combined approach was used to increase both branching and sialylation in CHO-K1 cells producing EPO [71]. Both GnT-IV and GnT-V, as well as human alpha-2,6-sialyltransferase (ST6Gal1) were incorporated in the CHO-K1 cells, resulting in a pool of 92% *N*-glycans with tri- and tetra-antennarity [71]. This also improved sialylation, as measured by an increase of 45% in tetra-sialylation [71]. The approach showed that combining the genetic integration of complementary genes could significantly enhance glycosylation branching complexity, as well as enhancing overall improvements in sialylation.

*O*-linked glycosylation can also be modified through cell engineering approaches. Although studies of *O*-glycosylation are limited, there are important biological applications of *O*-glycans. During *O*-glycosylation, various carbohydrate chains are added to the serine or threonine residues of proteins. Cell engineering strategies have attempted to control *O*-glycosylation by altering GnT activity. In one experiment, the core 2 beta1-6GlcNAc transferase (C2GnT) was overexpressed in CHO DG44 cells [72]. The increase in enzyme activity was hypothesized to play a role in T-cell activation and immunodeficiency [72]. In another study, the combined overexpression of C2GnT and the knockdown of CMP-sialic acid: Gal-beta-1,3-GalNAc-alpha-2,3-sialyltransferase (ST3Gal1) was evaluated in CHO-K1 cells [73]. ST3Gal1 inhibition was predicted to redirect *O*-glycosylation toward the production of tetrasaccharide structures important for cell-cell interaction [73]. This experiment suggests that cell engineering can be used to simultaneously upregulate and downregulate competing enzymes involved in glycosylation.

Recently, extended C1 beta-3 GnT-III, C2 beta-3 GnT-I, and C3 beta-3 beta-1,4-*N*-acetylglucosaminyltransferase VI were transiently transfected into CHO cells and the resulting *O*-glycome was mapped by MS [39]. This transfection experiment resulted in extended core 1 and core 3 *O*-glycans, as well as the increased expression of core 2 *O*-glycans [39]. Overall, these results suggest that cell engineering can be applied to *O*-glycosylation in order to control the branching of glycans. This will aid bioprocess developments to generate mucin-type recombinant proteins.

### 3.2 Overexpression of Sialyltransferase and Galactosyltransferase Genes

Glycoengineering by increasing the expression of sialyltransferase enzymes has been an effective strategy to control sialylation; these enzymes add the sialic acid (Neu5Ac) residue to the terminal galactose. There are six beta-galactoside alpha 2,3-sialyltransferases (ST3GAL1-6) and two beta-galactoside alpha-2,6-sialyltransferases (ST6GAL1-2) that generate terminal sialic acids in mammalian cells. Whereas human glycoproteins contain both alpha-2,3- and alpha-2,6-linked sialic acid, CHO cells natively contain almost exclusively alpha-2,3-linked sialic acid on their glycoproteins. This means that efforts to generate more human-like glycoforms can be implemented in CHO cells. As stated above, normally, CHO cells produce almost exclusively alpha-2,3-linked sialic acid, whereas in humans, glycoproteins represent a pool of alpha-2,3- and alpha-2,6-linked sialic acid. Rat alpha-2,6-sialyltransferase was transfected into CHO cells producing tissue plasminogen activator (tPA) and it was observed that competing glycosyltransferases yielded glycoproteins with different sialic acid linkages [74]. Thus, recombinant proteins with a mixture of alpha-2,3- and alpha-2,6- sialic acid can be generated; this mixture is similar to the pool of sialylated proteins in humans.

A combination of ST3GAL3, ST3GAL4, and ST3GAL6 knockdown using siRNA has revealed that all three enzymes are involved in alpha-2,3-sialylation in CHO cells [33]. Of these enzymes, ST3GAL4 was the most critical for glycoprotein alpha-2,3-sialylation [33]. In contrast, in humans, ST6GAL1 prefers the Gal beta-1-4GlcNAc disaccharide sequence linked to a protein, whereas ST6GAL2 prefers free disaccharide Gal beta-1-4GlcNAc substrate [75].

Lee et al. found that competition between endogenous alpha-2,3-sialyltransferase and heterologous alpha-2,6-sialyltransferase yielded glycoproteins with alpha-2,3- and alpha-2,6- linkages in CHO cells [76]. As the expression of alpha-2,6-sialyltransferase increased, enzymatic assays revealed only a slight increase in total sialyltransferase activity in transfected cells; of this activity, 50% was correlated to alpha-2,6-sialyltrasferase [76]. Furthermore, the transfected cells attached alpha-2,6-sialic acid to 20% of terminal galactose [76]. Another group found that the expression of human alpha-2,6-sialyltransferase in CHO cells [77] resulted in an increased percentage of tri- (by 8%) and tetra- (by 16%) sialylated recombinant thyroid-stimulating hormone [77]. The increase in more fully sialylated protein did not affect hydrophobicity or bioactivity [77]. This research indicates the potential for more human-like sialylation of recombinant therapeutics.

Another means to increase sialylation is to make more sites available for adding sialic acid. This can be achieved by overexpressing human $\beta$ 1,4-galactosyltransferase in CHO cells in order to reduce the oligosaccharides terminating with GlcNAc [68]. In one study, the overexpression of human $\beta$ 1,4-galactosyltransferase in CHO cells significantly reduced oligosaccharides terminating with GlcNAc compared with results in controls [78].

The engineering of galactosyltransferases is frequently used in combination with sialyltransferase engineering. In order to increase the level of sialylation, the enzyme alpha-2,3-sialyltransferase was overexpressed along with beta-1,4-galactosyltransferase in CHO cells [78]. Similarly, the coexpression of alpha-2,6-sialyltransferase with beta-1,4-galactosyltransferase effectively increased sialic acid content [79]. Results indicated that the overexpression of galactosyltransferase improved the homogeneity of glycoforms, while the overexpression of sialyltransferase improved the sialylation of recombinant protein to 90% compared with the level in parental cells [78]. The effect of increased sialylation was verified in rat models, where it was shown that recombinant proteins with increased sialic acid had increased circulation time [78]. Matrix-assisted laser desorption/ionization time-of-flight (MALDI-TOF) MS was used to detect charged and neutral oligosaccharides in negative and positive ion modes, respectively [78]. Despite overexpression, glycoproteins that were not fully sialylated were detected, a finding which may be attributed to sialidase cleavage or steric hindrance [78]. In another study, Jeong et al. studied the effect of the overexpression of beta-1,4-galactosyltransferase and alpha-2,3-sialyltransferase in CHO cells producing recombinant EPO [80]. The coexpression of galactosyltransferase and sialyltransferase resulted in an increase in sialic acid content, from 6.7 to 8.2 mol of sialic acid per mole of EPO, and an increase in trisialylated glycans from 17.3% to 35.5% compared with levels in parental cells [80]. At the same time, cell growth, metabolism, and protein productivity were not affected [80]. There was virtually no change in tetrasialylated glycans, suggesting possible steric hindrance in attaching a fourth CMP-sialic acid to the trisialylated glycans or sialyltransferases having branch specificity [80]. This result highlights the importance of both galactosyltransferase and sialyltransferase in producing homogenous, sialylated glycoproteins. Both enzymes are important for maintaining the lot-to-lot consistency of glycoprotein therapeutics, which is required for consistent manufacturing and drug efficacy.

The studies cited above [78, 79, 80] showed the effect of sialyltransferase and galactosetransferase expression on the sialylation of glycoproteins. However, transporting CMP-sialic acid to the Golgi apparatus is a potential bottleneck, owing to the levels or activity of the CMP-sialic acid transporter (CMP-SAT) that transports CMP-Neu5Ac into the Golgi apparatus. Overexpression of CMP-SAT alone resulted in a 4–16% increase in the site sialylation of IFN-gamma [81]. Following these findings, combinatorial efforts have sought to engineer multiple genes in the pathway in order to improve sialic acid content in the intracellular pool and improve the transport of sialic acid substrates in the Golgi apparatus.

Another approach is to implement methods that increase the levels of the sialylation substrate, CMP-Neu5Ac (or CMP-sialic acid). In order to enhance both activities, human alpha-2,3-sialyltransferase and CMP-sialic acid synthase were simultaneously overexpressed in CHO cells producing recombinant EPO [82]. This resulted in increased sialylation; however, the increase was attributed to alpha-2,3-sialyltransferase alone [82]. Coexpression was found to increase the pool of intracellular CMP-sialic acid, suggesting that a bottleneck to sialylation is the transport of sialic acid into the Golgi apparatus [82]. Following this finding, alpha-

2,3-sialyltransferase, CMP-sialic acid synthase, and CMP-sialic acid transporter were simultaneously overexpressed and there was an additional increase in tri- and tetra- sialylated glycans concomitant with a decrease in monosialylated glycans [82]. This result highlights how genetic engineering strategies can be used to overcome pathway bottlenecks in both the generation of glycosylation substrates and the transfer of the substrate to the oligosaccharide target, by using a transferase enzyme in order to maximize the sialylation of recombinant proteins.

In another experiment, the enzymes uridine diphosphate-*N*-acetyl glucosamine 2-epimerase/*N*-acetyl mannosamine kinase (GNE/MNK), CMP-sialic acid transporter, and alpha-2,3-sialyltransferase were simultaneously introduced in CHO cells producing recombinant EPO [83]. GNE/MNK initiates sialic acid biosynthesis; a mutant variant of the enzyme was used in order to eliminate feedback control by the end product of the pathway. Subsequently, CMP-sialic acid transporter sequesters CMP-sialic acid into the Golgi apparatus, where sialyltransferase then adds sialic acid to the maturing glycoprotein. Results indicated that the sialic acid content of recombinant EPO increased by 43% compared with that in parental cells; additionally, there was a 32% increase in tetrasialylated EPO and declines of 50% in both monosialylated and asialylated EPO [83]. This study provides further evidence that the combined simultaneous transfection of multiple enzymes in the sialic acid biosynthetic and transfer pathways can have a significant impact on overall product sialylation.

### *3.3 Inhibition of Sialidase Activity*

Sialidases are exoglycosidases that catalyze the hydrolytic removal of sialic acid from sialoglycoconjugates (glycoproteins, polysaccharides, gangliosides) [84]. Four sialidases (Neu 1–4) have been identified in human, mouse, rat, and CHO cells, and their activity is localized to different subcellular compartments: Neu1 is located in the lysosome, Neu2 is located in the cytoplasm, Neu3 is located in the plasma membrane, and Neu4 is also located in the lysosome [84–86]. Thus, sialidase function varies as a result of the different substrate specificities and subcellular locations [32]. Sialidase cleavage occurs in cell culture as viability decreases, and this cleavage leads to the desialylation of recombinant glycoproteins [32, 85].

In order to decrease sialidase activity, a CHO cell line was developed that expressed sialidase antisense RNA [87]. Sialidase activity in this cell line was reduced by 40%, compared with the control culture; this reduction corresponded to an increase in sialic acid content ranging from 20 to 37% [87]. Over the culture duration, sialidase concentration increased in both the control and antisense cultures [87]. However, the sialidase level in the antisense culture remained 40% lower than that in the control cells [87]. Another important finding was the consistent viability between the control and antisense cultures, which suggests that sialidase antisense RNA is a useful strategy for reducing sialidase cleavage [87]. The finding that sialidase antisense RNA expression was not completely knocked out indicated the likelihood that no severe effects on growth or metabolism had occurred. This result highlights the capacity to manipulate sialidase levels as a means to maintain sialylated glycoforms for recombinant protein production.

In another experiment, Neu2 knockdown was used to decrease cytosolic sialidase activity in CHO cells producing IFN-gamma [85]. After siRNA sequences were compared, the most active sequence was transfected into CHO cells, resulting in a reduction in sialidase activity of 60% relative to control cells [85]. Cell culture glycan samples were analyzed by MS and it was found that reducing sialidase cleavage did not affect the glycan site distribution [85]. This finding is important for the development of the cell culture process so that consistent batches of recombinant protein are produced. For one clone with decreased sialidase cleavage, there was no change in the percentages of asialoglycans, monosialylated glycans, or disialylated glycans [85]. However, over the duration of the control cell culture, there was a decrease in disialylated glycans, with a concomitant increase in asialoglycans and monosialylated glycans [85]. RNAi knockdown of Neu2 did increase sialic acid content, but only when cells were in the death phase [85, 87, 88]. Although sialic acid content does not always increase during the growth phases with sialidase RNAi, it is possible to maintain consistent glycoforms and prevent the desialylation of glycoforms in later culture stages with this strategy.

In another study, siRNA and shRNA were used to knockdown Neu1 and Neu3 sialidase genes [89]. Reduced expression of Neu3 resulted in a 98% reduction in Neu3 sialidase activity in CHO cells, corresponding to increases in sialic acid content of 33% and 26% for samples from the cell stationary phase and death phase, respectively [89]. Interestingly, application of the siRNA technique to knockdown Neu3 (located in the plasma membrane) individually resulted in negligible sialidase activity, whereas knockdown of Neu2 (located in the cytoplasm) individually only reduced sialidase activity to 40% of the control level [32]. Unlike Neu2 knockdown effects that acted exclusively in the death phase, protein sialylation was increased throughout cell culture by Neu3 knockdown, suggesting different mechanisms of sialylation control by Neu2 and Neu3, respectively [32].

In summary, many strategies, involving both the upregulation and downregulation of enzymes involved in the glycosylation pathways, can be manipulated to control cellular glycosylation. Some of the approaches described in this chapter are highlighted in Fig. 4, which shows the effect of specific gene overexpression or the knockdown of enzymes involved in various glycan processing steps, including sialylation and fucosylation. In the next section, we introduce the importance of a systems biology approach to understand glycosylation and to elucidate glycan compositions that can be used to drive genetic engineering strategies in the future.

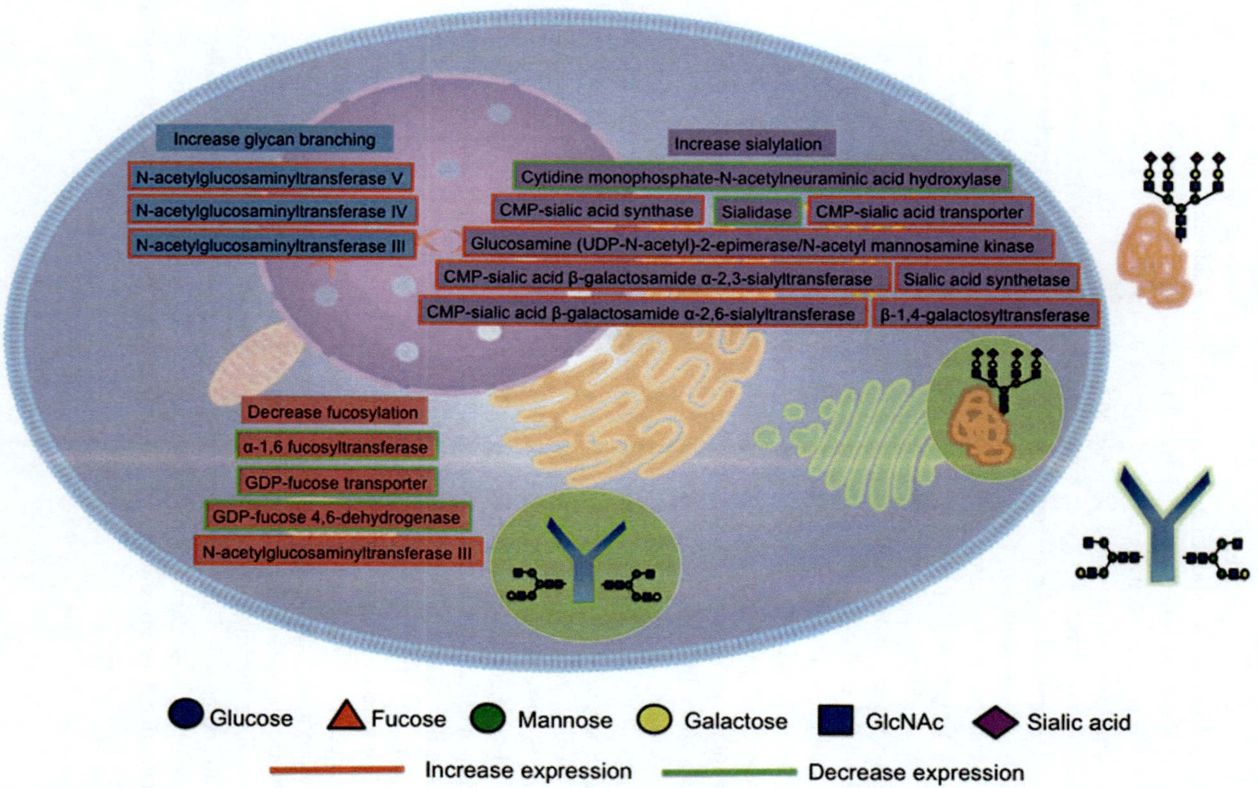


**Fig. 4** Examples of different glycoengineering targets. Strategies for glycoengineering include efforts to increase branching, increase sialylation and galactosylation, and decrease fucosylation. Enzyme targets are shown for different strategies and are outlined in red if expression is typically increased or green if expression is typically decreased. UDP uridine diphosphate,. . . Cytidine monophosphate (CMP)

# 4 CHO Glycoproteomics and Combined 'Omics'

Since glycan patterns are exposed on cell surfaces, they are ready targets for high-throughput technologies such as glycoproteomics [90, 91]. Indeed, the development of sophisticated analytical techniques [92–95] and data analysis tools [96–100] provides increasing opportunities to utilize high-throughput screening for glycans as disease markers and for the structural classification of therapeutic proteins. Glycogene microarrays, lectin chips, and RNA sequencing tools are widely used to analyze the whole glycogenome and the changes in glycosylation enzymes, as shown in Fig. 5. In addition to these tools, recent advances in MS)enable qualitative and quantitative analyses of glycans, glycosites, glycopeptides, and intact glycoproteins [101].

## 4.1 *Glycoproteomics*

Glycoproteomics, a field that evaluates glycosylated proteins and their glycosylation sites [102], involves glycoprotein enrichment of the samples followed by sophisticated proteomics methods, advanced MS techniques, and powerful bioinformatics tools. Label-free quantification [103], stable isotope labeling (SILAC) [104], isobaric tag for relative and absolute quantitation (iTRAQ) [105], and tandem mass tags (TMT) [106] are some of the methods that can be used to interpret the differential

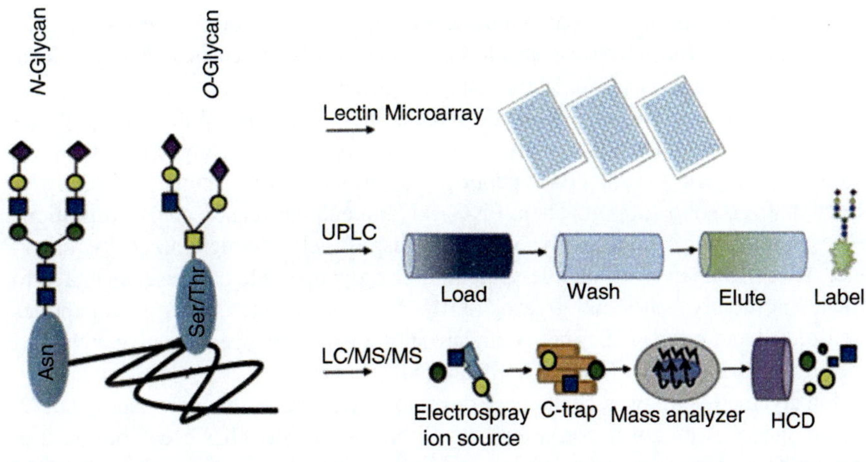


**Fig. 5** Methods for glycan analysis. New advances in sample preparation and analysis enable the identification and quantification of glycoproteins with high accuracy and reproducibility. Examples include lectin microarrays, ultra-performance liquid chromatography (UPLC), and liquid chromatography tandem mass spectrometry (LC/MS/MS). Abbreviations: Asn asparagine, Ser serine, Thr threonine, HCD higher-energy collisional dissociation

expression of glycoproteins between samples, such as different clones or changing process conditions.

Solid phase extraction of glycosylated peptides (SPEG) enables the identification of *N*-linked glycoproteins using hydrazide chemistry. In this method, a protein mixture is equilibrated with a hydrazide resin, which binds to the carbohydrate moieties on the glycoproteins. Then, polypeptides are oxidized and enzymatically removed by peptide-*N*-glycosidase F for liquid chromatography tandem mass spectrometry (LC-MS) analysis [107]. A previous CHO proteome analysis used a label-free approach to identify 6,164 total proteins and glycoproteins [108]. Of these, the SPEG method revealed that at least 1,292 proteins were *N*-glycosylated [108]. In recent years, more developments have been made to improve the identification and quantification of glycoproteins. Glycan quantification using isobaric tags, such as aminoxyTMT and iART, is difficult owing to their tertiary amine structure [109]. A novel MS-based technology, called quaternary amine-containing isobaric tag for glycan (QUANTITY), was recently developed to improve the complete labeling of glycans and increase reporter ion intensity upon second stage of mass (MS2) fragmentation [109]. The QUANTITY labeling approach has been coupled with solid-phase immobilization techniques for the glycomic comparison of CHO cells engineered with glycosyltransferases [109]. Samples are first denatured and immobilized on AminoLink resin (Thermo Fisher Scientific). To stabilize sialic acid groups, *p*-toluidine can be used with a carbodiimide coupling reagent, and then PNGaseF releases *N*-glycans from the solid support. Next, the aldehyde group of the GlcNAc at the reducing end of the glycans from each sample can be labeled with QUANTITY, followed by an analysis with liquid chromatography tandem mass spectrometry (LC/MS/MS). A global proteomics analysis can also be conducted by performing on-bead digestion [109].

Site-specific glycan occupancy and alterations in glycoproteins are also significantly important for bioprocess development. Previously, glycosites, glycopeptides, and glycans were studied separately owing to difficulties with simultaneous analysis. Solid-phase extraction of *N*-linked glycans and glycosite-containing peptides (NGAG) can simultaneously analyze glycans, glycosites, and glycopeptides from complex samples [110]. First, peptides are immobilized using an aldehyde-functionalized solid support. Then, PNGaseF and endoproteinase Asp-N digestions release the *N*-glycans and *N*-glycopeptides, respectively, through enzymatic cleavage. After MS analysis, a sample-specific intact glycopeptide database is created to document the glycosites and glycans [110]. At the same time, intact glycopeptides are isolated and run by MS. The spectra are subsequently mapped to a glycosylation-specific database using GPQuest software [110, 111].

Finally, methods have recently been developed to improve our understanding of *O*-glycosylation. A microwave-assisted beta-elimination method has been optimized to analyze *O*-glycans from cells, tissues, serum, and formalin-fixed paraffin-embedded tissues [112]. In summary, the use of 'omics' has expanded our ability to elucidate glycan structures, glycosites, and glycopeptide composition in order to understand the glycoproteome and glycoform profiles from CHO cell cultures. These efforts seek to identify deficiencies in glycosylation profiles that may be overcome through genetic engineering intervention.

## 4.2 *Combined 'Omics'*

More recent efforts have combined glycoproteomics with other 'omics' technologies for the validation and improved understanding of glycosylation. In one approach, genome-wide association studies were combined with high-throughput HPLC analysis of plasma proteins from 2,705 individuals, to reveal polymorphisms in FUT6 and FUT8, as well as those in hepatocyte nuclear factor 1-alpha (HNF1-alpha) [113]. The analysis was extended to 3,533 individuals to identify polymorphisms in MGAT5 and B3GAT1 and the protein pump SLC9A9 [114]. Another study combined epigenomics with proteomics to show that global changes in the DNA methylation of ovarian cancer epithelial cells could affect glycans by reducing core fucosylation, increasing branching, and increasing sialylation [115]. Altered expression of fucose biosynthetic genes and increased expression of MGAT5 were found to modify the branching and sialylation of secreted glycans [116]. These studies demonstrate how epigenomics and glycan structural analysis can be combined to study the effects of genes and pathways in human glycosylation that may also be important for CHO glycosylation processing.

In another approach, pathway mapping was used to correlate transcriptional regulation and glycan expression [117]. Increased polysialylation and alpha-Gal termination were observed in differentiated cell types, whereas alpha-Gal capped glycans were more abundant in extra-embryonic endodermal cells [117]. Another integration study mapped microRNA (miRNA) regulators onto glycan biosynthetic pathways by the introduction of glycomics data. Lectin microarrays were used to mimic miRNAs, enabling miRNA regulators of high mannose, fucose, and beta-GalNAc networks to be determined [6].

Finally, *N*-glycan and glycogene expression during the epithelial-to-mesenchymal transition was studied using a systems glycobiology approach [118]. Fucosylation and bisecting GlcNAc glycans were significantly decreased during the transition, whereas levels of high mannose type *N*-glycans were increased [118]. In this way, the integration of 'omics' tools has led to the improved understanding of how glycogene expression is controlled at genomic, transcriptomic, proteomic, and epigenomic levels.

# 5 Conclusions and Outlook

This review has highlighted the role of glycosylation as a critical quality attribute in the *production of biotherapeutics, and more importantly it has highlighted how these glycans can be manipulated* in CHO expression systems through cell engineering, as summarized in Table 2. Mammalian cell lines such as CHO can produce

**Table 2** Summary of glycoengineering efforts in Chinese hamster ovary (CHO) cells

| Target | Result | Reference |
|---|---|---|
| B4GALT1 | Expression increases galactose sites and sialic acid content without affecting growth, metabolism, or protein productivity | [68, 78–80] |
| CMP-*N*-acetylneuraminic acid hydroxylase | Knockdown decreases the Neu5Gc content | [120] |
| CMP-sialic acid synthase | Overexpression increases CMP-sialic acid pool | [82] |
| CMP-sialic acid transporter | Expression increases sialic acid content and tetrasialylated glycoforms and decreases monosialylated and asialylated glycoforms | [81–83] |
| Core 1 *β*3 GnT-III | Expression of extended core 1 and core 3 *O*-glycans is increased, and there is increased expression of core 2 *O*-glycans | [59] |
| Core 2 *β*1-6GlcNAc transferase | Overexpression increases GlcNAc transfer | [72, 73] |
| Core 2 *β*3 GnT-I | Expression of extended core 1 and core 3 *O*-glycans, as well as increased expression of core 2 *O*-glycans | [59] |
| Core 3 *β*3 GnT-VI | Expression of extended core 1 and core 3 *O*-glycans, as well as increased expression of core 2 *O*-glycans | [59] |
| FUT8 | Knockdown increases the percentage of afucosylated antibodies and ADCC activity | [37, 41–46, 49, 50, 61] |
| GDP-fucose 4,6-dehydrogenase | Knockdown increases the percentage of afucosylated antibodies | [37, 43, 62] |
| GDP-fucose transporter | Knockdown increases the percentage of afucosylated antibodies | [37, 43, 62] |
| GNE/MNK | Expression increases sialic acid content and tetrasialylated glycoforms and decreases asialylated glycoforms | [83] |
| GnT-III | Knockout eliminates bisecting GlcNAc and overexpression inhibits core *α*-1,6-fucosylation | [52, 63–67] |
| GnT-IV | Overexpression increases tri- and tetra-antennary sugar chains | [66, 71] |
| GnT-V | Overexpression increases tri- and tetra-antennary sugar chains | [66, 71] |
| Sialidase | Knockdown reduces sialidase cleavage and increases sialylation without affecting viability | [85, 87, 89] |
| *α*-2,3-Sialyltransferase | Expression increases sialic acid content and trisialylated glycoforms without affecting growth, metabolism, or protein productivity | [33, 73, 78, 80, 82, 83] |
| *α*-2,6-Sialyltransferase | Expression increases 2,6 sialic acid linkages | [74–77, 79, 121] |
| *α*-Mannosidase II | Expression increases complex glycans with increased ADCC activity | [52] |

*CMP* cytidine monophosphate, *Neu5Gc* *N*-glycolylneuraminic acid, *GnT-I* Beta-1,4-*N*-acetylglucosaminyltransferase I, *GnT-VI* Beta-1,4-*N*-acetylglucosaminyltransferase VI, *GDP* guanosine diphosphate, *ADCC* antibody-dependent cellular cytotoxicity, *GNE/MNK* uridine diphosphate-*N*-acetyl glucosamine 2-epimerase/*N*-acetyl 75 mannosamine kinase valuable recombinant proteins that can be accepted by humans as therapeutics. However, subtle differences exist between glycosylation in humans and other mammals, and understanding these differences requires knowledge of the physiological characteristics of each cell type. Efforts to exert control over protein glycosylation in CHO cells have been made by maximizing terminal sialylation through the overexpression of *N*-acetylglucosaminyltransferases, the overexpression of galactosyltransferases, the overexpression of sialyltransferases, the inhibition of sialidases, and the manipulation of CMP-sialic acid pathways. Equally important have been approaches to limit fucosylation through the overexpression of inhibiting *N*-acetylglucosaminyltransferases such as GnTIII, suppressing fucosyltransferase activity, and blocking the generation of the GDP-fucose substrate. The increasing use of advanced technologies such as ZFN, TALEN, and more recently CRISPR/Cas9, will greatly facilitate efforts to insert precise modifications of the glycosylation pathways into the CHO genome in future. Indeed, recent efforts have achieved comprehensive knockdown of multiple glycosyltransferases in order to control *N*-linked glycosylation in CHO cells [119]. This approach allows users to tailor the design of glycosylation for specific glycan profiles on recombinant glycoproteins. Furthermore, combinatorial glycoengineering approaches, including knockdowns, knockouts, knockins, and knockups, will be increasingly implemented to overcome multiple interacting pathway bottlenecks. These tools will enable highly refined and targeted modifications to be made to the processing capability of CHO cells in order to meet the need for flexible production capabilities, as well as meeting the need for the highly specified glycan targets required in biosimilar generation. Finally, the generation of 'omics' data sets is propelling a systems biology revolution to increase our understanding of CHO physiology and our capacity to modify glycans in different ways. Our ability to elucidate, characterize, quantify, and finally modify glycoproteins emerging from CHO, as well as the enzyme activities present in CHO, will facilitate the development of a superior CHO production platform that will yield consistent and desirable glycoforms in the future. In the coming decades, the emerging systems glycobiology integration of glycogenomics, glycoproteomics, glycomics, epiglycogenomics, and glycoinformatics, together with our ever-expanding toolkit for genome engineering, promises to accelerate our understanding of glycosylation in CHO and other mammalian cell lines, as well as increasing our capacity to control glycan processing more effectively.

## References


1. Ghaderi D et al (2012) Production platforms for biotherapeutic glycoproteins. Occurrence, impact, and challenges of non-human sialylation. Biotechnol Genet Eng Rev 28:147–175
2. Hossler P, Khattak SF, Li ZJ (2009) Optimal and consistent protein glycosylation in mammalian cell culture. Glycobiology 19(9):936–949
3. Lepenies B, Seeberger PH (2014) Simply better glycoproteins. Nat Biotechnol 32(5):443–445
4. Aggarwal RS (2014) What's fueling the biotech engine-2012 to 2013. Nat Biotechnol 32(1): 32–39

5. Jiménez D et al (2005) Contribution of N-linked glycans to the conformation and function of intercellular adhesion molecules (ICAMs). J Biol Chem 280(7):5854–5861
6. Agrawal P et al (2014) Mapping posttranscriptional regulation of the human glycome uncovers microRNA defining the glycocode. Proc Natl Acad Sci 111(11):4338–4343
7. Palomares LA, Estrada-Mondaca S, Ramirez OT (2004) Production of recombinant proteins: challenges and solutions. Methods Mol Biol 267:15–52
8. Walsh G, Jefferis R (2006) Post-translational modifications in the context of therapeutic proteins. Nat Biotechnol 24(10):1241–1252
9. Gavel Y, Vonheijne G (1990) Sequence differences between glycosylated and non-glycosylated Asn-X-Thr Ser acceptor sites – implications for protein engineering. Protein Eng 3(5):433–442
10. Stavenhagen K et al (2013) Quantitative mapping of glycoprotein micro-heterogeneity and macro-heterogeneity: an evaluation of mass spectrometry signal strengths using synthetic peptides and glycopeptides. J Mass Spectrom 48(6):i
11. Wang Q et al (2017) Glycoengineering of CHO cells to improve product quality. Methods Mol Biol 1603:25–44
12. Varki A, Schauer R (2009) Sialic acids. In: Varki A et al (eds) Essentials of glycobiology. Cold Spring Harbor, New York
13. Aebi M (2013) N-linked protein glycosylation in the ER. Biochim Biophys Acta 1833(11): 2430–2437
14. Aebi M et al (2010) N-glycan structures: recognition and processing in the ER. Trends Biochem Sci 35(2):74–82
15. Butler M, Meneses-Acosta A (2012) Recent advances in technology supporting biopharmaceutical production from mammalian cells. Appl Microbiol Biotechnol 96(4):885–894
16. Swiech K, Picanco-Castro V, Covas DT (2012) Human cells: new platform for recombinant therapeutic protein production. Protein Expr Purif 84(1):147–153
17. Padler-Karavani V, Varki A (2011) Potential impact of the non-human sialic acid N-glycolylneuraminic acid on transplant rejection risk. Xenotransplantation 18(1):1–5
18. Bosques CJ et al (2015) Chinese hamster ovary cells can produce galactose-alpha-1, 3-galactose antigens on proteins (vol 28, pg 1153, 2010). Nat Biotechnol 16(10):23849–23866
19. Muchmore EA et al (1989) Biosynthesis of N-glycolyneuraminic acid. The primary site of hydroxylation of N-acetylneuraminic acid is the cytosolic sugar nucleotide pool. J Biol Chem 264(34):20216–20223
20. Chung CH et al (2008) Cetuximab-induced anaphylaxis and IgE specific for galactose-alpha-1,3-galactose. N Engl J Med 358(11):1109–1117
21. Butler M, Spearman M (2014) The choice of mammalian cell host and possibilities for glycosylation engineering. Curr Opin Biotechnol 30:107–112
22. Croset A et al (2012) Differences in the glycosylation of recombinant proteins expressed in HEK and CHO cells. J Biotechnol 161(3):336–348
23. Zhao Y et al (2008) Branched N-glycans regulate the biological functions of integrins and cadherins. FEBS J 275(9):1939–1948
24. Raju TS, Jordan RE (2012) Galactosylation variations in marketed therapeutic antibodies. MAbs 4(3):385–391
25. Sareneva T et al (1995) N-glycosylation of human interferon-gamma – glycans at Asn-25 are critical for protease resistance. Biochem J 308:9–14
26. Wright A, Morrison SL (1997) Effect of glycosylation on antibody function: implications for genetic engineering. Trends Biotechnol 15(1):26–32
27. Spearman M, Butler M (2015) Glycosylation in cell culture. Anim Cell Cult 9:237–258
28. Sola RJ, Griebenow K (2010) Glycosylation of therapeutic proteins: an effective strategy to optimize efficacy. BioDrugs 24(1):9–21
29. Wright A et al (1991) Antibody variable region glycosylation: position effects on antigen binding and carbohydrate structure. EMBO J 10(10):2717–2723

30. Angata T, Varki A (2002) Chemical diversity in the sialic acids and related alpha-keto acids: an evolutionary perspective. Chem Rev 102(2):439–469
31. Harduin-Lepers A et al (2001) The human sialyltransferase family. Biochimie 83(8):727–737
32. Wang Q et al (2015) Strategies for engineering protein N-glycosylation pathways in mammalian cells. Methods Mol Biol 1321:287–305
33. Chung CY et al (2015) Assessment of the coordinated role of ST3GAL3, ST3GAL4 and ST3GAL6 on the alpha 2,3 sialylation linkage of mammalian glycoproteins. Biochem Biophys Res Commun 463(3):211–215
34. Stencel-Baerenwald JE et al (2014) The sweet spot: defining virus–sialic acid interactions. Nat Rev Microbiol 12:739
35. Ashwell G, Harford J (1982) Carbohydrate-specific receptors of the liver. Annu Rev Biochem 51:531–554
36. Cole ES et al (1993) In vivo clearance of tissue plasminogen-activator – the complex role of sites of glycosylation and level of sialylation. Fibrinolysis 7(1):15–22
37. Kanda Y et al (2007) Establishment of a GDP-mannose 4,6-dehydratase (GMD) knockout host cell line: a new strategy for generating completely non-fucosylated recombinant therapeutics. J Biotechnol 130(3):300–310
38. Shields RL et al (2002) Lack of fucose on human IgG1 N-linked oligosaccharide improves binding to human FcgammaRIII and antibody-dependent cellular toxicity. J Biol Chem 277: 26733–26740
39. Liu J et al (2015) O-glycan repertoires on a mucin-type reporter protein expressed in CHO cell pools transiently transfected with O-glycan core enzyme cDNAs. J Biotechnol 199:77–89
40. Chung S et al (2012) Quantitative evaluation of fucose reducing effects in a humanized antibody on Fcgamma receptor binding and antibody-dependent cell-mediated cytotoxicity activities. MAbs 4(3):326–340
41. Kanda Y et al (2007) Comparison of biological activity among nonfucosylated therapeutic IgG1 antibodies with three different N-linked Fc oligosaccharides: the high-mannose, hybrid, and complex types. Glycobiology 17(1):104–118
42. Beuger V et al (2009) Short-hairpin-RNA-mediated silencing of fucosyltransferase 8 in Chinese-hamster ovary cells for the production of antibodies with enhanced antibody immune effector function. Biotechnol Appl Biochem 53(Pt 1):31–37
43. Imai-Nishiya H et al (2007) Double knockdown of α1,6-fucosyltransferase (FUT8) and GDP-mannose 4,6-dehydratase (GMD) in antibody-producing cells: a new strategy for generating fully non-fucosylated therapeutic antibodies with enhanced ADCC. BMC Biotechnol 7(1):84
44. Yamane-Ohnuki N et al (2004) Establishment of FUT8 knockout Chinese hamster ovary cells: an ideal host cell line for producing completely defucosylated antibodies with enhanced antibody-dependent cellular cytotoxicity. Biotechnol Bioeng 87(5):614–622
45. Mori K et al (2004) Engineering Chinese hamster ovary cells to maximize effector function of produced antibodies using FUT8 siRNA. Biotechnol Bioeng 88(7):901–908
46. Malphettes L et al (2010) Highly efficient deletion of FUT8 in CHO cell lines using zinc-finger nucleases yields cells that produce completely nonfucosylated antibodies. Biotechnol Bioeng 106(5):774–783
47. Sealover NR et al (2013) Engineering Chinese hamster ovary (CHO) cells for producing recombinant proteins with simple glycoforms by zinc-finger nuclease (ZFN)—mediated gene knockout of mannosyl (alpha-1,3-)-glycoprotein beta-1,2-N-acetylglucosaminyltransferase (Mgat1). J Biotechnol 167(1):24–32
48. Sakuma T et al (2015) Homologous recombination-independent large gene cassette knock-in in CHO cells using TALEN and MMEJ-directed donor plasmids. Int J Mol Sci 16(10):23849
49. Ronda C et al (2014) Accelerating genome editing in CHO cells using CRISPR Cas9 and CRISPy, a web-based target finding tool. Biotechnol Bioeng 111(8):1604–1616
50. Grav LM et al (2015) One-step generation of triple knockout CHO cell lines using CRISPR/Cas9 and fluorescent enrichment. Biotechnol J 10(9):1446–1456

51. Chan KF et al (2016) Inactivation of GDP-fucose transporter gene (Slc35c1) in CHO cells by ZFNs, TALENs and CRISPR-Cas9 for production of fucose-free antibodies. Biotechnol J 11(3):399–414
52. Ferrara C et al (2006) Modulation of therapeutic antibody effector functions by glycosylation engineering: influence of Golgi enzyme localization domain and co-expression of heterologous beta1, 4-N-acetylglucosaminyltransferase III and Golgi alpha-mannosidase II. Biotechnol Bioeng 93(5):851–861
53. Patnaik SK, Stanley P (2006) Lectin-resistant CHO glycosylation mutants. Methods Enzymol 416:159–182
54. Tong C et al (2011) Generating gene knockout rats by homologous recombination in embryonic stem cells. Nat Protoc 6(6):827–844
55. Chan KF, Goh JSY, Song Z (2014) Improving sialylation of recombinant biologics for enhanced therapeutic efficacy. Pharm Bioprocess 2(5):363–366
56. Goh JS et al (2010) RCA-I-resistant CHO mutant cells have dysfunctional GnT I and expression of normal GnT I in these mutants enhances sialylation of recombinant erythropoietin. Metab Eng 12(4):360–368
57. Iskratsch T et al (2009) Specificity analysis of lectins and antibodies using remodeled glycoproteins. Anal Biochem 386(2):133–146
58. Goh JS et al (2014) Producing recombinant therapeutic glycoproteins with enhanced sialylation using CHO-gmt4 glycosylation mutant cells. Bioengineered 5(4):269–273
59. Goh JS et al (2014) Highly sialylated recombinant human erythropoietin production in large-scale perfusion bioreactor utilizing CHO-gmt4 (JW152) with restored GnT I function. Biotechnol J 9(1):100–109
60. Gaj T, Gersbach CA, Barbas CF (2013) ZFN, TALEN, and CRISPR/Cas-based methods for genome engineering. Trends Biotechnol 31(7):397–405
61. Cristea S et al (2013) In vivo cleavage of transgene donors promotes nuclease-mediated targeted integration. Biotechnol Bioeng 110(3):871–880
62. Louie S et al (2016) FX knockout CHO hosts can express desired ratios of fucosylated or afucosylated antibodies with high titers and comparable product quality. Biotechnol Bioeng 114(3):632–644
63. Davies J et al (2001) Expression of GnTIII in a recombinant anti-CD20 CHO production cell line: expression of antibodies with altered glycoforms leads to an increase in ADCC through higher affinity for FC gamma RIII. Biotechnol Bioeng 74(4):288–294
64. Fukuta K et al (2000) Control of bisecting GlcNAc addition to N-linked sugar chains. J Biol Chem 275(31):23456–23461
65. Fukuta K et al (2000) Genetic engineering of CHO cells producing human interferon-gamma by transfection of sialyltransferases. Glycoconj J 17(12):895–904
66. Fukuta K et al (2000) Remodeling of sugar chain structures of human interferon-gamma. Glycobiology 10(4):421–430
67. Reinl T et al (2013) Golgi engineering of CHO cells by targeted integration of glycosyltransferases leads to the expression of novel Asn-linked oligosaccharide structures at secretory glycoproteins. BMC Proc 7(6):P84
68. Demetriou M et al (2001) Negative regulation of T-cell activation and autoimmunity by Mgat5 N-glycosylation. Nature 409(6821):733–739
69. Misaizu T et al (1995) Role of antennary structure of N-linked sugar chains in renal handling of recombinant human erythropoietin. Blood 86(11):4097–4104
70. Park SS et al (2012) Effective correction of experimental errors in quantitative proteomics using stable isotope labeling by amino acids in cell culture (SILAC). J Proteome 75(12): 3720–3732
71. Yin B et al (2015) Glycoengineering of Chinese hamster ovary cells for enhanced erythropoietin N-glycan branching and sialylation. Biotechnol Bioeng 112(11):2343–2351
72. Bierhuizen MF, Fukuda M (1992) Expression cloning of cDNA encoding UDP-GlcNAc:Gal beta 1-3-GalNAc-R (GlcNAc to GalNAc) beta 1-6GlcNAc transferase by gene transfer into

CHO cells expressing polyoma large tumor antigen. Proc Natl Acad Sci U S A 89(19): 9326–9330
73. Prati EG et al (2000) Engineering of coordinated up- and down-regulation of two glycosyltransferases of the O-glycosylation pathway in Chinese hamster ovary (CHO) cells. Biotechnol Bioeng 68(3):239–244
74. Minch SL, Kallio PT, Bailey JE (1995) Tissue plasminogen activator coexpressed in Chinese hamster ovary cells with alpha(2,6)-sialyltransferase contains NeuAc alpha(2,6)Gal beta(1,4) Glc-N-AcR linkages. Biotechnol Prog 11(3):348–351
75. Krzewinski-Recchi MA et al (2003) Identification and functional expression of a second human beta-galactoside alpha2,6-sialyltransferase, ST6Gal II. Eur J Biochem 270(5):950–961
76. Lee EU, Roth J, Paulson JC (1989) Alteration of terminal glycosylation sequences on N-linked oligosaccharides of Chinese hamster ovary cells by expression of beta-galactoside alpha 2,6-sialyltransferase. J Biol Chem 264(23):13848–13855
77. Damiani R et al (2009) Stable expression of a human-like sialylated recombinant thyrotropin in a Chinese hamster ovary cell line expressing alpha2,6-sialyltransferase. Protein Expr Purif 67(1):7–14
78. Weikert S et al (1999) Engineering Chinese hamster ovary cells to maximize sialic acid content of recombinant glycoproteins. Nat Biotechnol 17(11):1116–1121
79. Raymond C et al (2015) Production of IgGs with a human-like sialylation in CHO cells. BMC Proc 9(Suppl 9):O3
80. Jeong YT et al (2008) Enhanced sialylation of recombinant erythropoietin in CHO cells by human glycosyltransferase expression. J Microbiol Biotechnol 18(12):1945–1952
81. Wong NS, Yap MG, Wang DI (2006) Enhancing recombinant glycoprotein sialylation through CMP-sialic acid transporter over expression in Chinese hamster ovary cells. Biotechnol Bioeng 93(5):1005–1016
82. Jeong YT et al (2009) Enhanced sialylation of recombinant erythropoietin in genetically engineered Chinese-hamster ovary cells. Biotechnol Appl Biochem 52(Pt 4):283–291
83. Son YD et al (2011) Enhanced sialylation of recombinant human erythropoietin in Chinese hamster ovary cells by combinatorial engineering of selected genes. Glycobiology 21(8): 1019–1028
84. Burg M, Muthing J (2001) Characterization of cytosolic sialidase from Chinese hamster ovary cells: part I: cloning and expression of soluble sialidase in Escherichia coli. Carbohydr Res 330(3):335–346
85. Ngantung FA et al (2006) RNA interference of sialidase improves glycoprotein sialic acid content consistency. Biotechnol Bioeng 95(1):106–119
86. Munzert E et al (1997) Production of recombinant human antithrombin III on 20-L bioreactor scale: correlation of supernatant neuraminidase activity, desialylation, and decrease of biological activity of recombinant glycoprotein. Biotechnol Bioeng 56(4):441–448
87. Ferrari J et al (1998) Chinese hamster ovary cells with constitutively expressed sialidase antisense RNA produce recombinant DNase in batch culture with increased sialic acid. Biotechnol Bioeng 60(5):589–595
88. Van Dyk DD et al (2003) Identification of cellular changes associated with increased production of human growth hormone in a recombinant Chinese hamster ovary cell line. Proteomics 3(2):147–156
89. Zhang M et al (2010) Enhancing glycoprotein sialylation by targeted gene silencing in mammalian cells. Biotechnol Bioeng 105(6):1094–1105
90. Tousi F, Hancock WS, Hincapie M (2011) Technologies and strategies for glycoproteomics and glycomics and their application to clinical biomarker research. Anal Methods 3(1):20–32
91. Zhang Y, Yin H, Lu H (2012) Recent progress in quantitative glycoproteomics. Glycoconj J 29(5–6):249–258
92. Ito S, Hayama K, Hirabayashi J (2009) Enrichment strategies for glycopeptides. Methods Mol Biol 534:195–203
93. Hua S, An HJ (2012) Glycoscience aids in biomarker discovery. BMB Rep 45(6):323–330

94. Furukawa J-i, Fujitani N, Shinohara Y (2013) Recent advances in cellular glycomic analyses. Biomol Ther 3(1):198
95. Bennun SV et al (2013) Integration of the transcriptome and glycome for identification of glycan cell signatures. PLoS Comput Biol 9(1):e1002813
96. Ranzinger R et al (2011) GlycomeDB—a unified database for carbohydrate structures. Nucleic Acids Res 39(Database issue):D373–D376
97. von der Lieth CW et al (2011) EUROCarbDB: an open-access platform for glycoinformatics. Glycobiology 21(4):493–502
98. Akune Y et al (2010) The RINGS resource for glycome informatics analysis and data mining on the web. OMICS 14(4):475–486
99. Lutteke T et al (2006) GLYCOSCIENCES.de: an Internet portal to support glycomics and glycobiology research. Glycobiology 16(5):71R–81R
100. Krambeck FJ et al (2009) A mathematical model to derive N-glycan structures and cellular enzyme activities from mass spectrometric data. Glycobiology 19(11):1163–1175
101. Yang S, Zhang H (2014) Glycomic analysis of glycans released from glycoproteins using chemical immobilization and mass spectrometry. Curr Protoc Chem Biol 6(3):191–208
102. Tian Y, Zhang H (2010) Glycoproteomics and clinical applications. Proteomics Clin Appl 4(2):124–132
103. Megger DA et al (2013) Label-free quantification in clinical proteomics. Biochim Biophys Acta 1834(8):1581–1590
104. Kashyap MK et al (2010) SILAC-based quantitative proteomic approach to identify potential biomarkers from the esophageal squamous cell carcinoma secretome. Cancer Biol Ther 10(8): 796–810
105. Tian Y, Bova GS, Zhang H (2011) Quantitative glycoproteomic analysis of optimal cutting temperature-embedded frozen tissues identifying glycoproteins associated with aggressive prostate cancer. Anal Chem 83(18):7013–7019
106. Raso C et al (2012) Characterization of breast cancer interstitial fluids by TmT labeling, LTQ-Orbitrap Velos mass spectrometry, and pathway analysis. J Proteome Res 11(6): 3199–3210
107. Zhang H et al (2003) Identification and quantification of N-linked glycoproteins using hydrazide chemistry, stable isotope labeling and mass spectrometry. Nat Biotechnol 21(6):660–666
108. Baycin-Hizal D et al (2012) Proteomic analysis of Chinese hamster ovary cells. J Proteome Res 11(11):5265–5276
109. Yang S et al (2015) QUANTITY: an isobaric tag for quantitative glycomics. Sci Rep 5:17585
110. Sun S et al (2016) Comprehensive analysis of protein glycosylation by solid-phase extraction of N-linked glycans and glycosite-containing peptides. Nat Biotechnol 34(1):84–88
111. Toghi Eshghi S et al (2015) GPQuest: a spectral library matching algorithm for site-specific assignment of tandem mass spectra to intact N-glycopeptides. Anal Chem 87(10):5181–5188
112. Furukawa J et al (2015) Quantitative O-glycomics by microwave-assisted beta-elimination in the presence of pyrazolone analogues. Anal Chem 87(15):7524–7528
113. Lauc G et al (2010) Genomics meets glycomics—the first GWAS study of human N-glycome identifies HNF1α as a master regulator of plasma protein fucosylation. PLoS Genet 6(12): e1001256
114. Huffman JE et al (2011) Polymorphisms in B3GAT1, SLC9A9 and MGAT5 are associated with variation within the human plasma N-glycome of 3533 European adults. Hum Mol Genet 20(24):5000–5011
115. Zoldos V et al (2012) Epigenetic silencing of HNF1A associates with changes in the composition of the human plasma N-glycome. Epigenetics 7(2):164–172
116. Saldova R et al (2011) 5-AZA-2′-deoxycytidine induced demethylation influences N-glycosylation of secreted glycoproteins in ovarian cancer. Epigenetics 6(11):1362–1372
117. Nairn AV et al (2012) Regulation of glycan structures in murine embryonic stem cells: combined transcript profiling of glycan-related genes and glycan structural analysis. J Biol Chem 287(45):37835–37856

118. Tan Z et al (2014) Altered N-glycan expression profile in epithelial-to-mesenchymal transition of NMuMG cells revealed by an integrated strategy using mass spectrometry and glycogene and lectin microarray analysis. J Proteome Res 13(6):2783–2795
119. Yang Z et al (2015) Engineered CHO cells for production of diverse, homogeneous glycoproteins. Nat Biotechnol 33(8):842–844
120. Chenu S et al (2003) Reduction of CMP-N-acetylneuraminic acid hydroxylase activity in engineered Chinese hamster ovary cells using an antisense-RNA strategy. Biochim Biophys Acta 1622(2):133–144
121. Zhang X, Lok SH, Kon OL (1998) Stable expression of human alpha-2,6-sialyltransferase in Chinese hamster ovary cells: functional consequences for human erythropoietin expression and bioactivity. Biochim Biophys Acta 1425(3):441–452

# Glycobiotechnology of the Insect Cell-Baculovirus Expression System

**Laura A. Palomares, Indresh K. Srivastava, Octavio T. Ramírez, and Manon M. J. Cox**

## Contents



**Abstract** The insect cell-baculovirus expression system technology (BEST) has a prominent role in producing recombinant proteins to be used as research and diagnostic reagents and vaccines. The glycosylation profile of proteins produced by the BEST is composed predominantly of terminal mannose glycans, and, in *Trichoplusia ni* cell lines, core α3 fucosylation, a profile different to that in mammals. Insects contain all the enzymatic activities needed for complex *N*- and *O*-glycosylation and sialylation, although few reports of complex glycosylation and sialylation by the BEST exist. The insect cell line and culture conditions determine the glycosylation profile of proteins produced by the BEST. The promoter used, dissolved oxygen tension, presence of sugar precursors, bovine serum or

The original version of this chapter was revised: The title of this chapter has been changed from "Glycobiotechnology of the Insect Cell-Baculovirus Expression System Technology" to "Glycobiotechnology of the Insect Cell-Baculovirus Expression System."

L. A. Palomares and O. T. Ramírez
Instituto de Biotecnología, Universidad Nacional Autónoma de México, Cuernavaca, Morelos, Mexico
e-mail: laura@ibt.unam.mx

I. K. Srivastava and M. M. J. Cox (✉)
Protein Sciences Corporation, A Sanofi Company, Meriden, CT, USA
e-mail: manoncox@nextwavebio.com

hemolymph, temperature, and the time of harvest all influence glycosylation, although more research is needed. The lack of activity of glycosylation enzymes possibly results from the transcription regulation and stress imposed by baculovirus infection. To solve this limitation, the glycosylation pathway of insect cells has been engineered to produce complex sialylated glycans and to eliminate α3 fucosylation, either by generating transgenic cell lines or by using baculovirus vectors. These strategies have been successful. Complex glycosylation, sialylation, and inhibition of α3 fucosylation have been achieved, although the majority of glycans still have terminal mannose residues. The implication of insect glycosylation in the proteins produced by the BEST is discussed.

**Graphical Abstract**

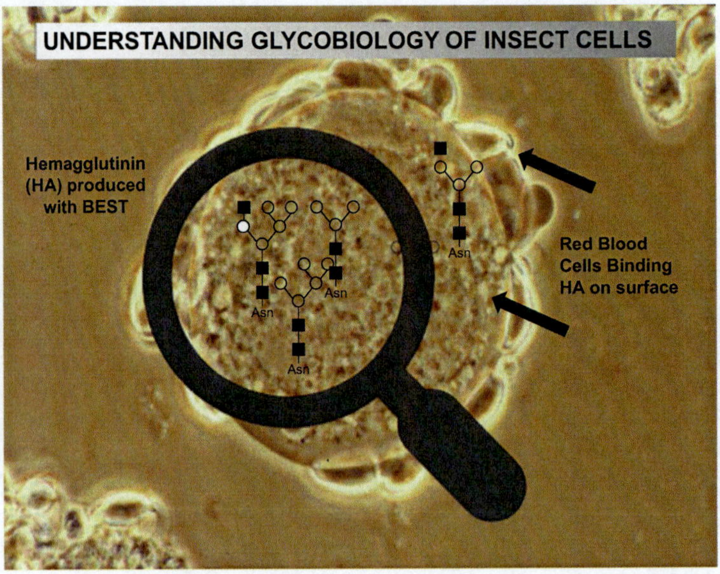


**Keywords** Baculovirus, Cell engineering, Glycobiotechnology, Glycosylation, Insect cells, Recombinant protein

## Abbreviations

| | |
|---|---|
| BEST | Baculovirus expression system technology |
| CHO | Chinese hamster ovary |
| CHST2 | Carbohydrate sulfotransferase 2 |
| CSAS | Sialic acid synthetase |
| DOT | Dissolved oxygen tension |
| eLH/CG | Recombinant equine luteinizing hormone/chorionic gonadotropin |

| | |
|---|---|
| ER | Endoplasmic reticulum |
| FBS | Fetal bovine serum |
| FDL | Fused lobes protein |
| Fuc | Fucose |
| FucT | Fucosyltransferase |
| FucT3 | α-1,3 Fucosyltransferase |
| GalT | β-(1→4)-Galactosyltransferase |
| Gal | Galactose |
| Gal3ST2 | Gal-3-*O*-sulfotransferase 2 |
| GalNAc | *N*-Acetylgalactosamine |
| Glc | Glucose |
| GlcNAc | *N*-Acetylglucosamine |
| GlcNAcase | β-*N*-Acetylglucosaminidase |
| GlcNAcT | GlcNAc transferase |
| hpi | Hours postinfection |
| IgG | Immunoglobulin |
| Man | Mannose |
| ManNAc | *N*-Acetylmannosamine |
| MP | Baculovirus basic protein promoter |
| ND | Not detected |
| Neu5Ac | *N*-Acetylneuraminic acid |
| NR | Not reported |
| PAP | Rat purple acid phosphatase |
| rHA | Recombinant influenza hemagglutinin |
| rLRE | Recombinant lutropin receptor ectodomain |
| RMD | Guanosine-5′-diphospho (GDP)4-dehydro-6-deoxy-D-mannose reductase |
| SAS | Sialic acid 9-phosphate synthase |
| SeAP | Secreted human alkaline phosphatase |
| SialT | Sialyltransferase |
| TPA | Tissue plasminogen activator |
| βhCG | β subunit of human chorionic gonadotropin |

## 1 Insect Glycobiology

The insect cell-baculovirus expression system technology (BEST) consists of the expression of a recombinant gene delivered to an insect cell culture by a recombinant baculovirus (reviewed by [1, 2]). The BEST is ideal for several applications, and especially for the production of complex proteins. The possibility of simultaneous expression of various proteins, the rapid and easy generation of new recombinant baculovirus, and the high productivity are useful benefits. With research, it is possible to evaluate different versions of a mutant protein in a eukaryotic context in a fast an efficient way (for example, [3, 4]). The BEST is a workhorse for production of recombinant proteins as research and diagnostics reagents, and is especially useful for the manufacture of virus-like particles and viral proteins in the market as vaccines [1, 5]. There is a very long list of virus-like particles (VLP) from a number of virus targets that have been produced in the BEST, as recombinant proteins that self-assemble efficiently as VLP [6]. The first recombinant human influenza seasonal vaccine, Flublok®, the human papillomavirus vaccine Cervarix ®, and the therapeutic cancer vaccine Provenge® were all produced using the BEST. In the case of influenza, the fast construction of new baculovirus vectors allows the fast and efficient strain change needed for opportune response to new influenza types, using a "plug and play" concept [5]. Several commercially available veterinary vaccines on the market are also produced using the BEST [5]. All its advantages give the BEST technology a huge potential.

An attribute of the BEST is the particular *N*-glycosylation profile of the produced proteins, which is different from that in proteins produced by mammalian cells. Mammalian cells produce glycoproteins with complex sialylated glycans, whereas the glycosylation of insect proteins mostly involves terminal mannose glycans. A comparison of the glycosylation profile of a model protein, secreted human alkaline phosphatase (SeAP) produced by CHO mammalian cells and by the commonly used insect cell line Tn5B1-4 (commercially known as High Five®), is shown in Fig. 1 (data from [7, 8]). Over 90% of glycans in SeAP produced by CHO cells were complex (without a terminal mannose residue), and more than 20% of glycans were sialylated [8]. In contrast, most of the glycans in SeAP produced by insect cells had terminal mannose residues and no sialylation was detected [7]. High mannose glycans, which are not processed in the medial Golgi, are more abundant in insect than in mammalian cells. Paucimannose glycans contain three or less mannose residues attached to the chitobiose core, and are the most abundant type of *N*-glycans in proteins produced by insects. The high abundance of terminal mannose glycans in insect proteins is different from the typical profiles of mammalian proteins. The consequences of these differences are discussed in Sect. 4. Table 1 lists some

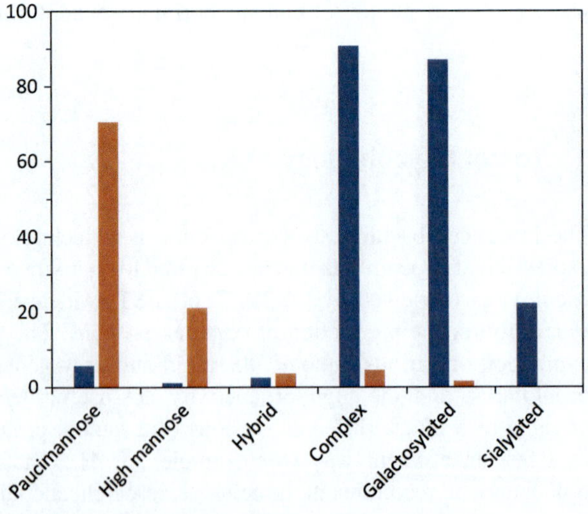


**Fig. 1** Comparison of the *N*-glycosylation profile of secreted human placental alkaline phosphatase (SeAP) produced by CHO mammalian (blue) and Tn5B1-4 insect (orange) cells. Complex glycans include galactosylated and sialylated glycans. Data from [78]

**Table 1** *N*-Glycosylation profiles of recombinant proteins expressed by the IC-BES

| Cell line | Recombinant protein | High mannose (%) | Paucimannose (%) | Hybrid[a] (%) | Complex (%) | α1,3Fuc (%) |
|---|---|---|---|---|---|---|
| DpN1[b] | SeAP | 24 | 44 | 6 | 26[c] | NR |
| *expres*SF+[d,e] | Influenza rHA | 34 | 56 | 10 | 0 | ND |
| High Five®[e,f] | Influenza rHA | 8 | 88 | 2 | Traces | 20 |
| High Five®[f,g] | Human transferrin | 25 | 54 | 7 | 0 | 56 |
| High Five®[f,h] | IgG2a | 0 | 35 | 30 | 35.6 | 18 |
| MBO503[i] | Human plasminogen | 35 | 2 | ND | 63[j] | NR |
| Sf9[d,k] | SeAP | 16 | 84 | ND | ND | NR |
| *SfSWT-7*[d,e,l] | Influenza HA | 48 | 10 | 23 | 18 | ND |
| Tn4h[m] | SeAP | 15 | 51 | 1 | 22[n] | NR |

*rHA* Recombinant hemagglutinin, *ND* Not detected, *NR* Not reported, *SeAP* Secreted human placental alkaline phosphatase, *IgG* Mouse immunoglobulin

[a]Hybrid glycans have a terminal mannose residue in one branch and a terminal glycan different to mannose in the other

[b]*Dannaus plexipus* cells, Palomares et al. [7]

[c]Includes 13% of sialylated glycans

[d]*Spodoptera frugiperda* cells

[e]An et al. [9]

[f]*Trichoplusia ni* cells

[g]Ailor et al. [9]

[h]Hsu et al. [10]

[i]Derived from *Mamestra brassicae*. Davidson and Castellino [11]

[j]Includes 33% of sialylated glycans

[k]Joshi et al. [12]

[l]Glycoengineered *expres*SF+ cells expressing mammalian *N*-acetylglucosaminyl-transferase II, β4-galactosylatransferase I, carbohydrate sulfotransferase 2 and galactose-3-*O*-sulfotransferase

[m]Derived from Tn5B1-4 cells. Joosten et al. [13]

[n]Includes 19.3% of sialylated glycans

glycosylation profiles reported for recombinant proteins produced by the BEST. Even when terminal mannose glycans are most common in proteins produced by insect cells, they can produce complex glycans at levels usually below 10% [7]. β-(1→4)-Galactosyltransferase (GalT) activity has been found and measured in Sf9 (barely detectable), High Five®, MBO503 (from *Mamestra brassicae*), and DpN1 (from *Dannaus plexipus*) cells [7, 14, 15], but the presence of galactosylated glycans produced by wild-type cells is seldom found. Wild-type Sf9 cells usually do not produce complex glycans and do not produce glycans with α1,3 fucose (Fuc). This is especially important, as discussed below. In contrast, α1,3 Fuc is found in glycoproteins produced by High Five® cells, which can also produce complex glycans under certain circumstances. Other non-conventional lepidopteran cell lines have an enhanced ability to produce complex glycans, even when they usually produce lower amounts of recombinant protein than Sf9 and High Five® cells. An important attribute that should not be overlooked is site occupancy. An et al. [16] found differences in site occupancy in recombinant influenza hemagglutinin (rHA) expressed in HEK293 or in insect cells, and Wang et al. [4] found that rat purple acid phosphatase (PAP) had two potential *N*-glycosylation sites occupied in the wild-type and Sf9 produced forms, although it was less extensively glycosylated when produced in CHO cells. Other reports have found that insect cells glycosylate in the same sites as in wild-type mammalian proteins.

The glycosylation profile in insect proteins is a result both of enzymatic activities present and of the availability of activated sugars for glycan formation. Table 2 summarizes the regulation of *N*-glycan diversity in insects by several enzymes, as reported by Walski et al. [17]. Fucose (Fuc), galactose (Gal), glucose (Glc), *N*-acetylglucosamine (NAcGlc), *N*-acetylgalactosamine (GalNAc), glucuronic acid, mannose (Man), xylose, and sialic acids have been found in insects [17], but it is their availability in activated form and the activity of glycosylation enzymes that determine the glycosylation profile of proteins. As in other organisms, insect development is severely affected by the disruption of glycosylation (Table 2). Figure 2 depicts the *N*-glycosylation pathway typical of insect cells used in the BEST. Processing in the Golgi is shown. High mannose glycans with five mannose residues are obtained in the *cis*-Golgi and are further processed by *N*-acetylglucosaminyl transferases and mannosidases. The branch point structure in insect *N*-glycosylation is indicated in the figure [18, 19]. The 2(GlcNAc)3(Man) α3 GlcNAc glycan can be processed by the GlcNAcTII and GalT to result in complex glycosylation. However, more often it is a substrate of GlcNAcases, which remove terminal GlcNAc residues and result in paucimannose forms, a reaction uncommon in mammalian cells. The extent of *N*-glycosylation in the BEST is a balance between GlcNAcase, GlcNAc transferase (GlcNAcT) I and II, and GalT activities in the Golgi, and the content of UDP-GlcNAc and UDP-Gal, activated sugar nucleotides. Paucimannosidic glycans are not only common in recombinant proteins produced by the BEST but are also major components in invertebrates [20]. The origin of such truncated glycans has been a subject of investigation of several groups. High hexosaminidase activity has been reported in insect cell cultures [7, 15, 21, 22]. The enzyme responsible for removal of the α3 branch GlcNAc was first discovered in *Drosophila* [23]. This *N*-acetylglucosaminidase (GlcNAcase) is encoded by the *fused lobes* gene, and an ortholog has been found in Sf9 insect cells, the most commonly used insect cell line, and in other invertebrate species [18, 19, 24, 25]. The fused lobes protein (FDL) is highly specific, although Dragosits et al. [26] found that, under extreme conditions, FDL can also trim other GlcNAc and Gal residues. Tomiya et al. [27] have reported a GlcNAcase that hydrolyzes in vitro terminal GlcNAc from the glycan core. It is possible that this enzyme also has a role in insect glycosylation, as removal of the α6 GlcNAc would require another enzyme. Nevertheless, it should be noted that the presence of α6 GlcNAcT in Sf9 cells is unlikely, as its activity has not been detected [28]. However, other GlcNAcase may have a role in other insect cell lines that can produce complex glycosylation (Table 1). Other exoglycosidase activities have also been detected in insect cells, such as those of sialidase and β-galactosidase [21].

Sialylation in insects and in recombinant proteins produced by the BEST has been studied by several groups. Synthesis and transfer of sialic acid had been considered as limited to the deuterstome lineage, but since 2002 it has been demonstrated that the protostome lineage is also capable of sialylation. A functional *N*-

**Table 2** Regulation of *N*-glycan diversity in insects

| Processing step | *N*-Glycans produced (examples) | Main enzymes involved | Regulation factors | Functions or processes affected by disruption |
|---|---|---|---|---|
| High mannose *N*-glycan trimming | 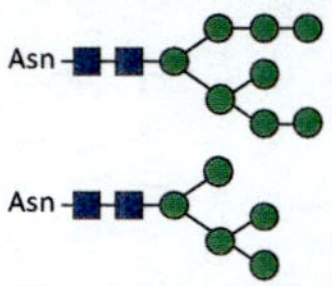 | ER class I α-mannosidase<br>Golgi class I α-mannosidase | Protein folding dynamics in ER<br>+ Transport of glycoprotein from ER to Golgi apparatus<br>− Steric hindrance caused by protein conformation | Glycoprotein quality control, impaired immune response<br>Wing, eye and nervous system development, immune response |
| Paucimannose glycan synthesis | 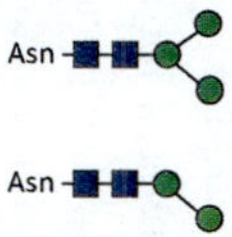 | GlcNAc transferase I, class II α-mannosidase, β-*N*-acetylhexosaminidase | + Transport of glycoprotein from ER to Golgi apparatus<br>+ Availability of $Man_5GlcNAc_2$ glycan | Eye and nervous system development, immune response, life span, fertility |
| Core fucosylation | 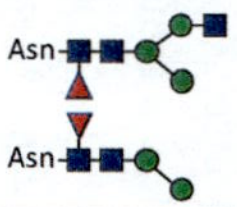 | α1,6-Fucosyltransferase<br>α1,3-Fucosyltransferase | + Terminal GlcNAc<br>− Removal of terminal GlcNac by β-*N*-acetylhexosaminidase | Wing development, immune response |
| Difucosylation | 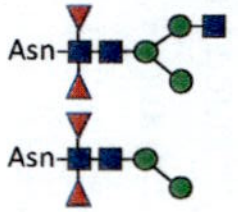 | α1,3-Fucosyltransferase | + Terminal GlcNAc<br>+Previous α1,6-fucosylation<br>− Substrate competition with *O*-fucosyltransferase<br>− Tissue-specific α1,3-fucosyltransferase expression | Nervous system development |
| Complex glycan synthesis | 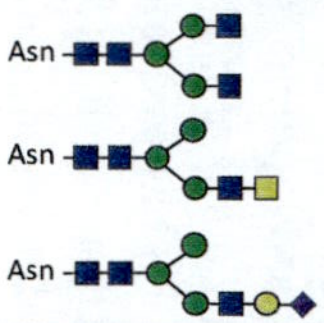 | GlcNAc transferase II, β1,4-Gal/GalNAc transferase, α2,6-sialyltransferase | − Temporality and spatial restricted expression of enzymes<br>− Removal of terminal GlcNAc by β-*N*-acetylhexosaminidase<br>− Low availability of substrate for sialylation (Neu5Ac) | Nervous system development, neuromuscular junctions |

Main steps of *N*-glycan production, their functions and elements involved in regulation of the process. + indicates factors that promote or are required for a given step. −indicates factors that inhibit a given step. Reprinted with permission from [17]. GlcNAc, blue square. Man, green circle. Fuc, red triangle. Gal, yellow circle. GalNAc, yellow square. Sialic acid, Purple diamond. ER, endoplasmic reticulum

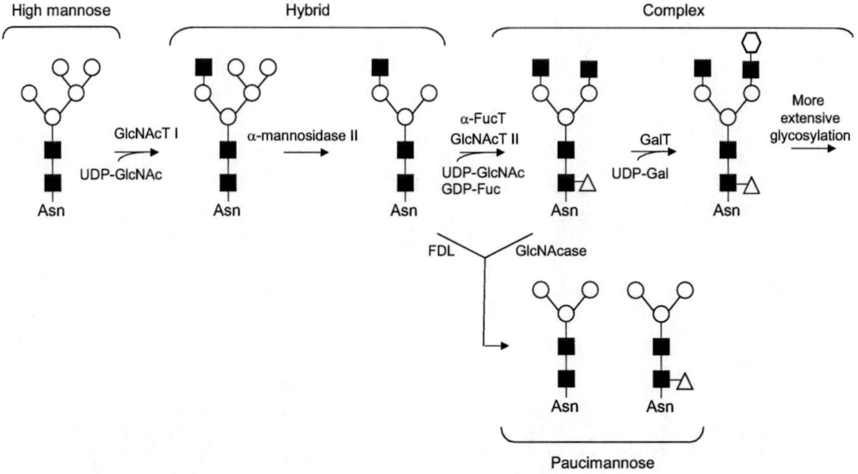


**Fig. 2** *N*-Glycosylation pathway in insect cells. High mannose glycans contain four or more mannose residues. They are only processed in the endoplasmic reticulum and cis Golgi. Hybrid glycans contain one terminal mannose residue. They accumulate because of low GlcNAcT I or II activities. The branch point structure for insect *N*-glycosylation is shown in a square [18]. FDL and a GlcNAcase activities result in the formation of paucimannose glycans, which only contain three or less mannose residues attached to the chitobiose glycan core. Square, GlcNAc. Circle, Man. Hexagon, Gal. Triangle, fucose

acetylneuraminic acid phosphate synthase, a CMP-sialic acid synthase, and a sialyltransferase have been found in *Drosophila* [29–31]. Sialylation has been observed in proteins of *Drosophila melanogaster* and *Philaenus spumarius* [32]. An active α-2,6-sialyltransferase (SialT) was found in the Lepidopteran *Bombyx mori* [33]. Sialylated glycans have been reported in proteins produced by Sf21 (from *Spodoptera frugiperda*), Tn4h (from *Trichoplusia ni*), DpN1 and MBO503 insect cells [7, 11, 12, 21, 34]. Watanabe et al. [22] observed sialylation in bovine interferon expressed by High Five® cells when β-*N*-acetylglucosaminidase (GlcNAcase) activity was inhibited. It is known that sialylation in insects, specifically in *Drosophila*, is a highly regulated process that occurs in specialized cells and at development stages. Such tight regulation can explain the few cases when sialylation of proteins produced by insects is reported. Supplementation of bovine serum has frequently been reported as a requirement for sialylation by insect cells. Hollister et al. [35] have demonstrated that insect cells can uptake from the culture medium and use sialylated *N*-glycans, *N*-acetylneuraminic acid (Neu5Ac), and GlcNAc to sialylate proteins when GalT and α-2,6-SialT genes are overexpressed, explaining the serum requirement. The presence of serum has introduced uncertainties regarding the possible presence of sialylated contaminants that copurify with the protein of interest [36, 37]. Hillar and Jarvis [37] believe that that is the case in reports of sialylation by Tn4h, Tn4s, and DpN1 cells ([7, 12, 13], among other reports), as they could not reproduce the results reported by other investigators. Because complete controls were included in the reports of sialylation of both cell lines, it is possible that the differences in the experiments performed by Hillar and

Jarvis, who used low sensitivity assays, can explain the different results obtained in their experiments.

Another peculiarity of insects is the presence of core α-1,3 linked fucose *N*-glycans and other modifications absent in mammalian proteins and that may cause hypersensitivity in patients with allergies [38]. α-1,3 Fucosyltransferase (FucT3) activity has been detected in *B. mori*, *Apis mellifera*, and *Drosophila* [39–41]. In insect cells commonly used for protein expression, core α-1,3 fucose has been found in proteins produced by High Five® *Trichoplusia ni* cells, but not in *Spodoptera frugiperda* Sf9 or *expres*SF+ cells [16, 38, 40]. Stanton et al. [42] recently reported the glycome of uninfected larvae from *Lymatria dispar*, *Trichoplusia ni*, and the High Five® cell line, and found that *N*-glycans in proteins from all three sources are decorated with sulfate, glucuronic acid, and phosphorylcholine, showing that insects and insect cells are capable of extensive glycan modification. They also found Lewis-like antenna fucosylated structures and *N*-acetylgalactosamine (GalNAc). An insect β4-*N*-acetylgalactosaminyltransferase has been detected and characterized [15, 43].

Insect cells are capable of *O*-glycosylation in the same sites as mammalian cells [44]. In the pseudorabies virus gp50, Sf9 cells produced protein with *O*-linked GalNAc and lower amounts of Galβ1-3GalNAc without sialic acid. The same protein produced by mammalian cells had higher amounts of Galβ1-3GalNAc and sialylation. Although the activity of UDP-GalNAc:polypeptide *N*-acetylgalactosaminyltransferase was comparable in Sf9, Vero and CHO cells, the Sf9 cells had a lower activity of UDP-Gal: *N*-GalNAc β1,3 galactosyltransferase. Similar results were found by Lopez et al. [45] in Sf9, High Five®, and SOC*Mb*-92-C6 (from *Mamestra brassicae*) insect cells. Gaunitz et al. [46] expressed a mucin-type protein in High Five® and Sf9 cells, which is different to those previously studied for *O*-glycosylation. They found *O*-glycans with glucuronic and galacturonic acids, sulfate, and phosphocholine. High Five® cells produced more extensively modified *O*-glycans than Sf9 cells.

## 2 Effect of Bioprocessing Conditions on the Glycosylation Profile of Proteins Produced by Insect Cells

Glycosylation demands high amounts of precursors and requires energy. It is therefore affected by nutrient availability and culture conditions [47]. In the case of the BEST, the recombinant gene is expressed most frequently under the polyhedrin (*polh*) promoter, which results in strong and very late expression. Cell growth and synthesis of cellular proteins is reduced after baculovirus infection [48, 49]. Van Die et al. [15] found that baculovirus infection decreases the activity of a glycosyltransferase to undetectable levels. Accordingly, different glycosylation profiles have been found in insect proteins in comparison with overexpressed recombinant proteins. Moreover, the increased metabolic activity of infected cells

**Table 3** Conditions that affect the *N*-glycosylation profile of proteins produced by insect cells

| Condition tested | Effect on recombinant protein *N*-glycosylation | Reference |
|---|---|---|
| Ammonia addition[a] | Addition of 40 mM ammonium sulfate (62 mM at the time of harvest) had no effect | [51] |
| Culture under simulated microgravity (HARV bioreactor)[a] | Sialylation of SeAP | [52] |
| Dissolved oxygen (DOT)[a] | High DOT (from 100 to 350% of air saturation) did not change the glycosylation profile of SeAP | [51] |
| Dissolved oxygen[a,b] | SeAP produced at 10% or 190% of air saturation had a higher abundance of high mannose glycans than cultures at 50% DOT | [53] |
| Fetal bovine serum (FBS) addition | Recombinant GST-SfManI was sialylated when SfB4GalT/ST6[c] cell cultures were supplemented with 10% FBS | [35] |
| FBS addition | Increased abundance of complex glycans in SeAP | [12] |
| Hemolymph addition[a] | c.a. 13% of sialylated glycans, reduction in SeAP yield | [13] |
| Inhibition of extracellular exoglycosidases[a] | No effect | [54] |
| Mannosamine addition[a,d] | Mannosamine addition (up to 20 mM) increased the abundance of GlcNAc terminal glycans | [55, 56] |
| Temperature[a] | Low temperature (20°C) increased the amount of terminal α(1,3)-mannose residues | [51] |
| Time of harvest[a] | A late time of harvest (120 hpi) increased the amount of mannosidase resistant glycans in SeAP | [51] |
| Use of a promoter earlier than *polh* | Increased protein concentration and increased sialylation | [57] |
| Use of *p10* promoter instead of *polh*[b] | Expression under the slightly earlier and weaker promoter *p10* resulted in rLRE secretion, complex glycosylation and sialylation | [58] |

*hpi* Hours postinfection, *SeAP* Human secreted alkaline phosphatase, *rLRE* Recombinant lutropin receptor ectodomain
[a]*Trichoplusia ni* cells
[b]Sf9 cells
[c]Sf9 cells engineered to express mammalian β-1,4-galactosyltransferase and a α2,6-sialyltransferase
[d]Sf21 cells

[50] and the energy needed for protein production and glycosylation can also affect the extent of modification. However, only a few papers have reported the effect of culture conditions in protein glycosylation, most of them from the 1990s and early 2000s. Results are summarized in Table 3. Few variables that can potentially affect insect glycosylation have been investigated, mostly because of the expected limited relevance of culture conditions on the formation of terminal mannose glycans, including the most abundant paucimannosidic forms. Ammonia concentrations well above those toxic to mammalian cells had no effect on glycosylation in the BEST [51]. Zhang et al. [53] observed a higher content of high mannose forms at low (10%) or high (190%) dissolved oxygen tension (DOT) in comparison with 50%

(with respect to air saturation), showing that in extreme conditions glycosylation was limited to processing in the endoplasmic reticulum (ER) and cis Golgi. Donaldson et al. [51] did not observe an impact on *N*-glycosylation at or above 100% DOT. Reduced productivity was observed under extreme conditions. It should be noted that Donaldson et al. [51] only used the semiquantitative fluorescence-assisted carbohydrate electrophoresis (FACE) method, whereas Zhang et al. [53] used capillary electrophoresis, a quantitative method that allows structure identification. Zhang et al. [53] controlled DOT in instrumented bioreactors, whereas Donaldson et al. [51] used spinners and manipulated DOT using an oxygen enriched environment, without control. Donaldson et al. [51] observed a more extensive glycan processing at low culture temperature (20°C). This contrasts with reports in mammalian cells, where a decrease in *N*-glycosylation was observed when cells were maintained at temperatures below 32°C [59]. An increase in mannosidase-resistant glycoforms was obtained at later culture times [51], and the addition of exoglycosidase inhibitors did not affect protein glycosylation [54]. Interestingly, Joshi et al. [52] observed sialylation of SeAP when Tn4h cells (*T. ni*) were cultured under simulated low gravity conditions. It can be hypothesized that a lower shear stress was present at low gravity, suggesting that shear stress can affect *N*-glycosylation. Aloi and Cherry [60] have observed sublethal effects of shear in insect cells, whereas Godoy-Silva et al. [61] observed changes on glycosylation at energy dissipation rates two orders of magnitude lower than lethal rates in CHO cell cultures. Fetal bovine serum addition increased the amount of complex glycans in Tn-4h cells (*T. ni*) [12], possibly because of its protective effect to shear stress [62], in addition to providing substrates.

Culture medium composition determines the glycosylation profile of proteins produced by the BEST. It has been shown that the pool of sugar nucleotides needed for complex glycosylation (UDP-GlcNAc, UDP-Gal) is similar in High Five®, Sf9, and mammalian cells [63], indicating that sugar nucleotides are not limiting complex glycosylation. Nevertheless, feeding of precursors of sugar nucleotides has been attempted to increase the extent of insect glycosylation [55, 56]. Interestingly, only the addition of mannosamine (ManN), a precursor of CMP-Neu5Ac (*N*-acetylneuraminic acid), increased the abundance of *N*-glycans with terminal GlcNAc. Estrada-Mondaca et al. [56] showed that ManN inhibits the activity of GlcNacase in vitro, explaining the observed results. It has been shown that supplementation of fetal bovine serum is needed to obtain sialylation in proteins produced by insect cells engineered to express mammalian β-1,4-galactosyltransferase and a α2,6-sialyltransferase, as insect cells can salvage sialic acid from the culture medium [35]. Hemolymph addition increased the abundance of sialylated glycans but lowered SeAP yield [13]. Even when it has not been evaluated, it can be expected that nutrient feeding in fed batch cultures alters the glycosylation profile, including site occupancy, of proteins produced by the BEST.

The strength and timing of baculovirus promoters driving recombinant gene expression affect the glycosylation profile of proteins. Baculoviruses are lytic viruses, and a more intact glycosylation machinery can be expected to exist at earlier infection times. Jarvis et al. [64] report that recombinant tissue plasminogen activator (TPA) was processed faster and more efficiently when expressed under the control of the immediate early *ie*1 promoter than when expressed under the very late *polh* promoter. Although Pajot-Augy et al. [58] attribute the more extensive processing and glycosylation of a recombinant lutropin receptor ectodomain to the use of the weaker *p10* promoter instead of *polh*, Sridhar et al. [57] report that it is the time of expression that results in an increased protein quality. In contrast with these reports, Toth et al. [65] did not find a difference in the quality of recombinant Western equine encephalitis virus glycoprotein produced under promoters with different timings and strengths, and suggests that effects are protein-dependent. The effects of other factors on glycosylation, such as nutrient feeding, other byproduct accumulation, etc., remain to be investigated. None of these papers have reported the effect of culture conditions on site occupancy. As strategies to increase complex glycosylation by insect cells are more widely used, it is likely that more research on the effects of culture conditions in protein glycosylation is performed.

## 3 Glycoengineering of Insect Cells

The glycosylation profile of proteins produced by the BEST impedes its use for production of glycoproteins that require mammalian glycosylation for their function. To overcome this problem, insect cells or the baculovirus vector have been engineered to produce mammalian enzymes. Table 4 summarizes progress toward obtaining mammalian glycosylation by the BEST. The first strategy used was to include glycosyltransferases into the baculovirus vector. Jarvis and Finn [66] first expressed a mammalian glycosyltransferase, GalT, in Sf9 cells, showing that the enzyme was active and that a baculovirus protein was galactosylated. This result confirmed the availability of UDP-Gal in Sf9 cells. In 2002, the Jarvis group reported the generation of a transgenic Sf9 cell line capable of sialylation when cultured in fetal bovine serum (FBS), called SfSWT-1, commercially available as the Mimic cell line through Thermo Scientific [28]. SfSWT-1 contains five mammalian glycosyltransferase genes (Table 4). The activities of GlcNAcT, GalT, and SialT in baculovirus-infected cells were confirmed. It was later reported that the α2,3 SialT had no activity in this cell line [67]. A decrease in paucimannose forms and the appearance of biantennary galactosylated forms were observed. Sialylation occurred only in one antenna. In 2003, Hollister et al. showed that sialylation could also be obtained with supplementation of other precursors in addition to FBS, Neu5Ac, or *N*-acetylmannosamine (ManNAc). It was found that Sf9 cells have a salvage pathway that allows the use of these precursors to sialylate proteins or as precursors for sialic acid synthesis. This shows that Sf9 cells produce negligible amounts of sialic acids and rely on taking up sialic acid or its precursors from the culture medium [35]. The presence of sialylated proteins upon addition of precursors for sialic acid synthesis shows that Sf9 cells have the required enzymatic machinery for sialic acid synthesis. Other groups that have used the Mimic cell line for recombinant protein expression have not found sialylation in the expressed proteins [77, 78].

**Table 4** Overview of the strategies used for glycosylation engineering in the IC-BES

| Gene and expression strategy | Results | References |
|---|---|---|
| Expression of bovine GalT under the *ie*1 promoter encoded in a baculovirus | Sf9 cells had GalT activity. Baculovirus gp64 was galactosylated | [66] |
| SfSWT-1 Mimic® cells (Thermo Scientific). Transgenic Sf9 cells expressing human GlcNAcT I, GlcNAcT II, bovine GalT, rat α2,6 SialT and mouse α2,3 SialT under the *ie*1 promoter | Biantennary, sialylated glycans in a recombinant protein produced with 10% bovine serum. Lower abundance of paucimannosidic forms than wild type cells. No α2,3 SialT activity. Slower cell growth | [28, 67] |
| Expression of human sialic acid 9-phosphate synthase (SAS) and UDP-GlcNAc 2-epimerase/ ManNAc kinase genes encoded in a baculovirus | Sialic acid generation by Sf9 cells | [68] |
| SfSWT-3 cells. SfSWT-1 cells expressing mouse SAS and sialic acid synthetase (CSAS) under the *ie*1 promoter | More extensive sialylation than SfSWT-1 in both antennae in serum free medium supplemented with ManNAc | [69] |
| Expression of human GlcNAcT II, GalT, and α2,6 SialT encoded in a baculovirus under the *polh* and *p10* promoters in Sf9 and Ea4 cells | Recombinant human antitrypsin produced by Ea4 cells sialylated, as determined by lectin blotting | [70] |
| Transgenic Sf21 cells expressing rat GlcNAcT III under the *ie*1 promoter | Cellular proteins and a recombinant human glycoprotein with bisecting GlcNAc | [71] |
| SfSWT-5. Transgenic Sf9 cells with inducible expression of GlcNAcT II, GalT, α2,6 SialT, α2,3 SialT, SAS and CSAS using *piggyBack* vectors inducible with doxycycline | No difference in cell growth or stability with or without induction. Cells stable for over 300 passages. No difference in recombinant glycoprotein yield. Sialylation detected by lectin blotting | [72] |
| SweetBac® (Geneva Biotech). Baculovirus encoding the *C. elegans* GlcNAcT II and bovine GalT | Complex galactosylated glycoforms not observed upon infection with wilt type baculovirus | [73] |
| SfSWT-21 cells. Transgenic Sf + cells expressing *E. coli* GlcNAc-6-P 2′ epimerase, mouse SAS, mouse CSAS, human Golgi CMP-sialic acid transporter, human GlcNAcT II, bovine GalT, rat α2,6 SialT under the *ie*1 promoter | Protein sialylation without ManNAc supplementation and without UDP-GlcNAc consumption without reduction of cell growth or yield | [18] |
| Short-hairpin RNA interference to stably silence expression of GlcNAcase | Reduced GlcNAcase activity | [74] |
| SfSWT-7 cells. *expres*SF+[a] cells cotransfected with dual *piggyBac* vectors encoding GlcNAcT II, GalT, carbohydrate sulfotransferase 2 (CHST2) and Gal-3-*O*-sulfotransferase 2 (Gal3ST2) | Engineered to produce biantennary, terminal Gal sulfated glycans. No evidence of sulfation observed. Eighteen percent of complex glycans. Reduced abundance of high mannose forms (Table 1) | [16] |

(continued)

**Table 4** (continued)

| Gene and expression strategy | Results | References |
|---|---|---|
| Expression of *Pseudomonas aeruginosa* guanosine-5′-diphospho (GDP)4-dehydro-6-deoxy-D-mannose reductase (RMD) encoded in a baculovirus under the *ie*1 promoter. Transgenic cell line expressing RMD | Consumption of the GDP-L-fucose precursor. Blocked α1,3 and α 1,6 fucosylation. Transgenic cells were unstable | [75] |
| Expression of RMD encoded in a baculovirus under the *gp64* promoter | Reduction of fucosylation in influenza hemagglutinin | [38] |
| Expression of human GlcNAcT II and GalT in *B. mori* pupae through bacmids with the actin A3 *B. mori* and *polh* promoters | Recombinant human IgG with terminal GlcNAc and Gal | [76] |

[a]Protein Sciences Corporation. USA

Kati et al. [76] obtained complex galactosylated glycans in a recombinant monoclonal antibody after using bacmids to express the GlcNAcT II and GalT in *B. mori* larvae. This strategy is useful for antibody production, as their mammalian *N*-glycosylation only contains terminal Gal complex glycans. Viswanathan et al. [68] infected Sf9 cells with baculovirus encoding for the human sialic acid 9-phosphate synthase (SAS) and UDP-GlcNAc 2-epimerase/ManNAc kinase genes. Infected Sf9 cells accumulated Neu5Ac without the need for precursor supplementation. The Jarvis group added the mouse SAS and sialic acid synthetase (CSAS) genes to the SfSWT-1 cells [69]. The resulting transgenic cell line, SfSWT-3, performed more extensive sialylation than SfSWT-1, resulting in bisialylated glycans. ManNAc addition was still needed in a serum-free medium. Interestingly, SfSWT-3 cells had growth kinetics similar to wild-type Sf9 cells, suggesting that the introduction of the two genes involved in sialic acid synthesis resulted in the elimination of the growth lag phase observed in sfSWT-1 cells [67]. In parallel, Chang et al. [70] inserted in a baculovirus the human GlcNAcT II, GalT, and α2,6 SialT genes under control of the very late *polh* and *p10* promoters and infected Sf9 and Ea4 (*Estigmena acrea*) cells. Only the recombinant glycoprotein expressed by Ea4 cells contained sialylated glycans. As these cells were cultured in serum-free medium, it is possible that Ea4 cells possess the ability to synthesize sialic acid. The presence of complex glycans in the absence of a mammalian GlcNAcT I demonstrates that overexpression of this enzyme is not needed for complex glycosylation, suggesting that it is active in wild-type insect cells. Okada et al. [71] introduced the GlcNAcT III gene into Sf21 cells and obtained both cellular proteins and a recombinant glycoprotein modified with bisected glycans. The use of transgenic cells has the advantages of having a unique host and that any baculovirus can be used for production of a glycoprotein of interest with complex glycosylation, although this approach provides limited flexibility and engineered cells may be unstable or have reduced growth or a high sensitivity to culture conditions. Interestingly, Aumiller et al. [72] used *piggyBac* vectors to construct a transgenic cell line with inducible expression of six mammalian glycosylation genes (SfSWT-5, Table 4). Sialylated proteins were obtained.

Stability of induced and non-induced cells was confirmed until more than 30 generations. No change in cell growth was observed. *piggyBac* vectors were used to modify the licensed *expres*SF+ insect cell line from Protein Sciences Corporation to introduce the GlcNAcT II, GalT, carbohydrate sulfotransferase 2 (CHST2), and Gal-3- *O*-sulfotransferase 2 (Gal3ST2) genes to obtain complex galactosylated sulfated glycans (SfSWT-7, [16]). Cells were infected with baculovirus coding for rHA. No evidence of sulfation was observed, even when expression of the related genes was detected. The relative abundance of each glycan type produced by SfSWT-7 cells is listed in Table 1. In this case, 18% of glycans were complex, but no detectable levels of complex glycans were observed in the Sf9 cell line.

As a solution to the problem of supplementation of sialic acid precursors to cultures, Geisler and Jarvis [18, 19] expressed the *Escherichia coli* GlcNAc-6-P 2′epimerase with other glycosylation proteins (SfSWT-21 cells, Table 4), and used its reverse reaction to obtain ManNAc-6P from Glc-6P, and then sialic acid, CMP-sialic acid, and sialylated proteins without the external addition of expensive ManNAc. All the effort on engineering insect cells has resulted in the production of biantennary structures that are still very simple compared to the glycosylation profile of many mammalian proteins. The advantage of having insect cells producing triantennary or tetranatennary glycans, in comparison to mammalian cells, remains to be evaluated.

Insect cells have been engineered to obtain complex sialylated glycans, but paucimannose glycans are still present in the produced proteins, albeit at a lower abundance than with wild-type insect cells. To prevent GlcNAcase cleavage, Kim et al. [74] used short-hairpin RNA interference to reduce the expression of a GlcNAcase. They observed a reduction in GlcNAcase activity but did not evaluate the glycosylation profile of proteins. Another undesirable activity in *T. ni* insect cells is the presence of α1,3Fuc. To prevent it, Palmberger et al. [38] and Mabashi-Asazuma et al. [75] expressed the *Pseudomonas aeruginosa* guanosine-5-′-diphospho (GDP)4-dehydro-6-deoxy-D-mannose reductase (RMD), which consumes GDP-4keto-6-deoxy-D-mannose, precursor of GDP-Fuc to form GDP-D-rhamnose, using baculovirus vectors. Mabashi-Asazuma et al. [75] constructed a transgenic insect cell line, but it was unstable. The reason for the instability of cells with depleted fucose is unknown. Walski et al. [17] report that fucosylation is needed for insect immune response and wing and nervous system development. The role of fucose in individual cells remains to be investigated. Although Mabashi-Asazuma did not observe fucosylation in a recombinant glycoprotein, Palmberger et al. [38] observed the absence of α1,3Fuc and a reduction in α1,6 fucosylation in rHA. The reason for this difference may lie in the two different promoters that were used for RMD expression. Palmberger et al. [38] used the *gp64* promoter, whereas Mabashi-Asazuma et al. [75] used the immediate early *ie*1 promoter. These promoters have different temporality, so it is possible that expression under the later *gp64* promoter was not efficient enough to impede fucosylation as expression under the *ie1* promoter [79].

Expression of glycosyltransferases can be achieved under a wide variety of conditions. The Jarvis group has mostly used the *ie*1 promoter, active in the immediate early phase of baculovirus infection and not requiring baculoviral factors for expression [64]. Thus, it is active in uninfected cells. The *ie*1 promoter is weaker than very late promoters, imposing a relatively low metabolic burden upon expression of several recombinant proteins. The metabolic burden imposed by overexpression of glycosylation enzymes has been perceived as a disadvantage of the transgenic cell line approach. Moreover, expression in uninfected cells when immediate early promoters are used can affect the physiology of the cells. The effect of expression of glycosylation enzymes in insect cells is not totally understood, and the performance of transgenic cells under industrial culture conditions has not been evaluated. Glycoengineering of insect cells is a very promising approach, but evaluation of transgenic cells under industrial manufacturing conditions is still needed.

## 4 Impact of Glycosylation on Proteins and Products Produced by Insect Cells

Even when one of the reasons for selecting the BEST for expressing a recombinant protein is its ability to perform glycosylation, few reports have systematically determined the role of insect glycosylation in the function of a recombinant protein. Some of these reports are listed in Table 5. In general, the lack or reduction of biological activity of proteins produced by the BEST is directly correlated with the absence of sialic acid. Sialic acid is negatively charged at physiological pH, and determines the distribution and half-life of proteins in the bloodstream [77]. Insect glycoproteins may therefore be active in vitro but lose all biological activity when tested in vivo [77]. Therefore, the main impact of insect glycosylation in proteins is the absence of in vivo activity caused by the lack of sialic acid.

Bantleon et al. [81] found that a recombinant IgE with high mannose and paucimannose glycosylation produced by Sf9 cells has the same immunoreactivity and binding to FcεRI as IgE with mammalian glycosylation. In some cases it is the absence of a glycan rather than an insect glycosylation profile that changes the activity of proteins [4]. It is believed that the particular glycosylation profile of proteins produced by insect cells would act as adjuvant in the case of vaccines, but there are no reports sustaining this. Lin et al. [78] found that non-sialylated glycosylation performed by Sf9 and sfSWT-1 cells induced higher anti rHA IgG titers than tetrasialylated rHA produced by CHO cells. However, lower rHA neutralizing IgG titers were obtained. Nevertheless, Dunkle et al. [82] have reported that Flublok®, a recombinant influenza vaccine (rHA) produced in the BEST, is more effective than an influenza vaccine produced in chicken eggs. The authors hypothesized that the higher efficacy of Flublok® results from its higher HA concentration or the presence of egg-derived mutations in the traditional comparator vaccine, not glycosylation. In any case, the insect *N*-glycosylation in rHA was not a disadvantage for the recombinant vaccine.

**Table 5** Effects of insect glycosylation on protein function

| Protein and characteristics | Observed effects | Reference |
|---|---|---|
| β subunit of human chorionic gonadotropin (βhCG) expressed in Sf9 cells under the baculovirus basic protein (MP) and *polh* promoters | βhCG produced under the MP promoter had higher bioactivity and sialic acid content than that produced under *polh* | [57] |
| Rat purple acid phosphatase (PAP) produced in CHO and Sf9 cells. PAP from rat and Sf9 cells was recognized by GNA lectin (terminal mannose) | Lower site occupancy reduces substrate affinity and catalytic activity | [4] |
| Equine eLH/CG expressed in sf9 and SfSWT-1 cells. Higher molecular weight of protein from SfSWT-1 cells. No sialylation detected in eLH/CG from both cell lines | No biological activity because of the absence of sialylation in eLH/CG from both cell lines | [77] |
| GP50 from *Taenia solium* expressed in Sf9 and High Five® cells | False positive reactivity with patient sera of GP50 produced by High Five® cells because of the presence of α1,3 linked fucose | [80] |
| rHA produced by Sf9, SfSWT-1, and CHO cells. SfSWT-1 cells did not produce sialylated glycans. rHA from CHO cells had up to tetrasialylated structures | Sf9 and SfSWT-1 rHA elicited higher anti HA IgG titers but lower neutralizing antibody titers than CHO rHA | [78] |
| Reduced fucosylation of recombinant influenza hemagglutinin produced by High Five® cells | Reduced binding of IgE from the sera of patients with allergy because of the absence of α1,3 fucose | [38] |
| Recombinant human IgE produced by Sf9 cells with paucimannosidic and high mannose *N*-glycosylation, whereas mammalian protein has complex *N*-glycosylation | No difference in immunoreactivity and FcεRI binding between insect and mammalian recombinant proteins | [81] |

Interestingly, it was found that α1,3 fucosylation produced by High Five® cells can result in cross reactivity of sera from patients, as the α1,3 fucose epitope is abundant in parasites [80]. This finding is especially important, as one of the most significant niches of the BEST is the production of proteins for diagnostics. Sf9 cells, which do not produce α1,3 fucose [16, 38, 40], would be a better cell line for production of proteins for diagnostics. In the same line, Palmberger et al. [38] observed that when α1,3 fucosylation was eliminated by RMD addition, binding from sera from allergic patients to rHA produced by High Five® cells was reduced.

## 5 Conclusions and Outlook

The BEST is a very popular system for recombinant protein expression, especially for complex proteins. To date, the particular glycosylation in proteins produced by insect cells has not limited its application and success because of its many advantages. The BEST is the system most used for production of the recombinant vaccines on the market for human or veterinary applications. The BEST is not ideal for expression of proteins that require sialylation or complex *N*-glycosylation for their biological function, as even when efforts on engineering insect cells for the production of complex sialylated glycans have been successful, mannose-terminal structures are still the most abundant. It is expected that the BEST will continue to be used extensively for research and diagnostics applications, that more manufacturers will benefit from the advantages of this technology, and that more products of the BEST will reach the market in future years.

**Acknowledgements** Research performed thanks to the support by UNAM-DGAPA-PAPIIT IT-200315. Technical assistance from Luis Alberto Diaz is acknowledged.

## References

1. Palomares LA, Realpe M, Ramírez OT (2015) An overview of cell culture engineering for the insect cell-baculovirus expression vector system (BEVS). In: Al-Rubeai M (ed) Animal cell culture. Cell engineering, vol 9. Cham, Springer International, pp 501–519
2. Cox MMJ, Hashimoto Y (2011) A fast track influenza virus vaccine produced in insect cells. J Invert Pathol 107:s31–s41
3. Orlova OV, Drutsa VL, Spirin PV, Popenko VI, Prasolov VS, Rubtsov PM, Kochetkov SN, Belzhelarskaya SN (2013) Role of N-linked glycans of HCV glycoprotein E1 in folding of structural proteins and formation of viral particles. Mol Biol 47:131–139
4. Wang Y, Norgård M, Andersson G (2005) N-Glycosylation influences the latency and catalytic properties of mammalian purple acid phosphatase. Arch Biochem Biophys 435: 147–156
5. Cox MMJ (2012) Recombinant protein vaccines produced in insect cells. Vaccine 30: 1759–1766
6. Palomares LA, Ramírez OT (2009) Challenges for the production of virus-like particles in insect cells: the case of rotavirus-like particles. Biochem Eng J 45:158–167
7. Palomares LA, Joosten CE, Hughes PR, Granados RR, Shuler ML (2003) Novel insect cell line capable of complex N-glycosylation and sialylation of recombinant proteins. Biotechnol Prog 19:185–192
8. Lipscomb ML, Palomares LA, Hernández V, Ramírez OT, Kompala DS (2005) Effect of production method and gene amplification on the glycosylation pattern of a secreted reporter protein in CHO cells. Biotechnol Prog 21:40–49
9. Ailor E, Takahashi N, Tsukamoto Y, Masuda K, Rahman BA, Jarvis DL, Lee YC, Betenbaugh MJ (2000) N-Glycan pattern of human transferrin produced in *Trichoplusia ni* insect cells: effects of mammalian galactosyltransferase. Glicobiology 10:837–847
10. Hsu TA, Takahashi N, Tsukamoto Y, Kato K, Shimada I, Masuda K, Whiteley E, Fan JQ, Lee YC, Betenbaugh MJ (1997) Differential N-glycan patterns of secreted and intracellular IgG produced by *Trichoplusia ni* cells. J Biol Chem 272:9062–9070

11. Davidson DJ, Castellino FJ (1991) Asparagine-linked oligosaccharide processing in Lepidopteran insect cells. Temporal dependence of the nature of the oligosaccharides assembled on asparagine-289 of recombinant human plasminogen produced in baculovirus vector infected *Spodoptera frugiperda* (IPLB-SF-21AE) cells. Biochemistry 30:6167–6174
12. Joshi L, Davis TR, Mattu TS, Rudd PM, Dwek RA, Shuler ML, Wood HA (2000) Influence of baculovirus-host cell interactions on complex N-linked glycosylation of a recombinant human protein. Biotechnol Prog 7:9–14
13. Joosten CE, Park TH, Shuler ML (2003) Effect of silkworm hemolymph on N-linked glycosylation in two *Trichoplusia ni* insect cell lines. Biotechnol Prog 83:695–705
14. Abdul-Rahman B, Ailor E, Jarvis D, Betenbaugh M, Lee YC (2002) β-(1→4)-Galactosyltransferase activity in native and engineered insect cells measured with time-resolved europium fluorescence. Carbohydr Res 337:2181–2186
15. van Die I, van Tetering A, Bakker H, van den Eijnden DH, Joziasse DH (1996) Glycosylation of lepidopteran insect cells: identification of a β1→4-N-acetylgalactosaminyltransferase involved in the synthesis of complex-type oligosaccharide chains. Glycobiology 6:157–164
16. An Y, Rininger JA, Jarvis DL, Jing X, Ye Z, Aumiller JJ, Eichelberger M, Cipollo JF (2013) Comparative glycomics analysis of influenza hemagglutinin (H5N1) produced in vaccine relevant cell platforms. J Proteome Res 12:3707–3720
17. Walski T, De Schutter K, Van Damme EJM, Smagghe G (2017) Diversity and function of protein glycosylation in insects. Insect Biochem Mol Biol 83:21–34
18. Geisler C, Jarvis DL (2012) Substrate specificities and intracellular distributions of three N-glycan processing enzymes functioning at a key branch point in the insect N-glycosylation pathway. J Biol Chem 287:7084–7097
19. Geisler C, Jarvis DL (2012) Innovative use if a bacterial enzyme involved in sialic acid degradation to initiate sialic acid biosynthesis in glycoengineered insect cells. Metab Eng 14: 642–652
20. Schachter H (2009) The functions of high mannose N-glycans in *Caenorhabditis elegans*. Trends Glycosci. Glycotechnol 119:131–148
21. Licari PJ, Jarvis DL, Bailey JE (1993) Insect cell hosts for baculovirus expression vectors contain endogenous exoglycosidase activity. Biotechnol Prog 9:146–152
22. Watanabe S, Kokuho T, Takakashi H, Takakashi M, Kubota T, Inumaru S (2002) Sialylation of N-glycans on the recombinant proteins expressed by a baculovirus-insect cell system under β-N-acetylglucosaminidase inhibition. J Biol Chem 277:5090–5093
23. Léonard R, Rendic D, Rabouille C, Wilson IBH, Préat T, Altmann F (2006) The *Drosophila fused lobes* gene encodes an N-acetylglucosaminidase involved in N-glycan processing. J Biol Chem 281:4867–4875
24. Altmann F, Schwihla H, Staudacher E, Glössl J, März L (1995) Insect cells contain an unusual, membrane bound β-N-acetylglucosaminidase probably involved in the processing of protein N-glycans. J Biol Chem 270:17344–17349
25. Geisler C, Aumiller JJ, Jarvis DL (2008) A *fused lobes* gene encodes the processing β-N-acetylglucosaminidase in Sf9 cells. J Biol Chem 283:11330–11339
26. Dragosits M, Yan S, Razzazi-Fazeli E, Wilson IBH, Rendic D (2015) Enzymatic properties and subtle differences in the substrate specificity of phylogenetically distinct invertebrate N-glycan processing hexosaminidases. Glycobiology 25:448–464
27. Tomiya N, Narang S, Park J, Abdul_Rahman B, Choi O, Singh S, Hiratake J, Sakata K, Betenbaugh MJ, Palter KB, Lee YC (2006) Purification, characterization and cloning of a *Spodoptera frugiperda* Sf9 β-N-acetylhexosaminidase that hydrolyzes terminal N-acetylglucosamine on the N-glycan core. J Biol Chem 281:19545–19560
28. Hollister J, Grabenhorst E, Nimtz M, Donradt H, Jarvis DL (2002) Engineering the protein N-glycosylation pathway in insect cells for production of biantennary, complex N-glycans. Biochemistry 41:15093–15104
29. Kim K, Lawrence SM, Park J, Pitts L, Vann WF, Betenbaugh MJ, Palter KB (2002) Expression of a functional *Drosophila melanogaster* N-acetylneuraminic acid (Neu5Ac) phosphate

synthase gene: evidence for endogenous sialic acid biosynthetic ability in insects. Glycobiology 12:73–83
30. Koles K, Irvine KD, Panin VM (2004) Functional characterization of *Drosophila* sialyltransferase. J Biol Chem 279:4346–4357
31. Viswanathan K, Tomiya N, Park J, Singh S, Lee YC, Palter K, Betenbaugh MJ (2006) Expression of a functional *Drosophila melanogaster* CMP-sialic acid synthetase. J Biol Chem 281: 15929–15940
32. Schauer R (2001) The occurrence and significance of sialic acids in insects. Trends Glycosci Glycotechnol 13:507–517
33. Kajiura H, Hamaguchi Y, Mizushima H, Misaki R, Fujiyama K (2015) Sialylation potentials of the silk worm, *Bombyx mori*; *B. mori* possesses an active α2,6-sialyltransferase. Glycobiology 25:1441–1453
34. Davidson DJ, Castellino FJ (1991) Structure of the asparagine-289-linked oligosaccharides assembled on recombinant human plasminogen expressed in a *Mamestra brassicae* cell line (IZD-MBO503). Biochemistry 30:6689–6696
35. Hollister J, Conradt H, Jarvis DL (2003) Evidence for a sialic acid salvaging pathway in lepidopteran insect cells. Glycobiology 13:487–495
36. Rendic D, Wilson IBH, Paschinger K (2008) The glycosylation capacity of insect cells. Croatica Chem Acta 81:7–21
37. Hillar A, Jarvis DL (2010) Re-visiting the endogenous capacity for recombinant glycoprotein sialylation by baculovirus-infected Tn-4h and DpN1 cells. Glycobiol 20:1323–1330
38. Palmberger D, Ashjaei K, Strell S, Hoffmann-Sommergruber K, Grabherr R (2014) Minimizing fucosylation in insect cell-derived glycoproteins reduces binding to IgE antibodies from sera of patients with allergy. Biotechnol J 9:1206–1214
39. Paschinger K, Staudacher E, Stemmer U, Fabini G, Wilson IBH (2005) Fucosyltransferase substrate specificity and the order of fucosylation in vertebrates. Glycobiology 15:463–474
40. Seismann H, Blank S, Braren I, Greunke K, Cifuentes L, Grunwald T, Bredehorst R, Ollert M, Spillner E (2010) Dissecting cross-reactivity in hymenoptera venom allergy by circumvention of α-1,3-core fucosylation. Mol Immunol 47:799–808
41. Minagawa S, Sekiguchi S, Nakaso Y, Tomita M, Takahisa M, Yasuda H (2015) Identification of core alpha 1,3-fucosyltransferase gene from silkworm: An insect popularly used to express mammalian proteins. J Insect Sci 15:110
42. Stanton R, Hykollary A, Eckmair B, Malzl D, Dragostis M, Palmberg D, Wang P, Wilson IBH, Paschinger K (2017) The underestimated N-glycomes of lepidopteran species. Biochim. Biophys Acta 1861:699–714
43. Vadaie N, Jarvis DL (2004) Molecular cloning and functional characterization of a lepidopteran insect β4-N-acetylgalactosaminyltransferase with broad substrate specificity, a functional role in glycoprotein biosynthesis and a potential functional role in glycolipid biosynthesis. J Biol Chem 279:33501–33518
44. Thomsen DR, Post LE, Elhammer AP (1990) Structure of O-glycosidically linked oligo-encephalitis virus glycoprotein saccharides synthesized by the insect cell line Sf9. J Cell Biochem 43:67–79
45. Lopez M, Tetaert D, Juliant S, Gazon M, Cerutti M, Verbert A, Delannoy P (1999) O-Glycosylation potential of lepidopteran insect cell lines. Biochim Biophys Acta 1427:49–61
46. Gaunitz S, Jin C, Nilsson A, Liu J, Karlsson NG, Holgersson J (2013) Mucin-type proteins produced in the *Trichoplusia ni* and *Spodoptera frugiperda* insect cell lines carry novel O-glycans with phosphocholine and sulfate substitutions. Glycobiology 23:778–796
47. Spearman M, Butler M (2015) Glycosylation in cell culture. In: Al-Rubeai M (ed) Animal cell culture, cell engineering 9. Springer International Publishing, Cham, pp 237–258
48. Ooi BG, Miller LK (1988) Regulation of host RNA levels during baculovirus infection. Virology 166:515–523
49. Palomares LA, Estrada-Mondaca S, Ramírez OT (2006) Principles and applications of the insect-cell baculovirus expression vector system. In: Ozturk SS, Hu WS (eds) Cell culture

technology for pharmaceutical and cell based therapies. Taylor and Francis, Boca Raton, pp 627–692
50. Palomares LA, López S, Ramírez OT (2004) Utilization of oxygen uptake rate to assess the role of glucose and glutamine in the metabolism of infected insect cell cultures. Biochem Eng J 19:87–93
51. Donaldson M, Wood HA, Kulakosky PC, Shuler ML (1999) Glycosylation of recombinant protein in the Tn5B1-4 insect cell line: influence of ammonia, time of harvest, temperature and dissolved oxygen. Biotechnol Bioeng 65:255–262
52. Joshi L, Shuler ML, Wood AH (2001) Production of a sialylated N-linked glycoprotein in insect cells. Biotechnol Prog 17:822–827
53. Zhang F, Saarinen MA, Itle LJ, Lang SC, Murhammer DW, Linhardt RJ (2002) The effect of dissolved oxygen (DO) concentration on the glycosylation of recombinant protein produced by the insect cell- baculovirus expression system. Biotechnol Bioeng 77:219–224
54. Joosten CE, Shuler ML (2003) Production of sialylated N-linked glycoprotein in insect cells: role of glycosidases and effect of harvest time on glycosylation. Biotechnol Prog 19:193–201
55. Donaldson M, Wood HA, Kulakosky PC, Shuler ML (1999) Use of mannosamine supplementation for inducing the addition of outer arm N-acetylglucosamine onto N-linked oligosaccharides of recombinant proteins in insect cells. Biotechnol Prog 15:168–173
56. Estrada-Mondaca S, Delgado-Bustos LA, Ramírez OT (2005) Mannosamine supplementation extends the N-acetylglucosaminylation of recombinant human secreted alkaline phosphatase produced in *Trichoplusia ni* (cabbage looper) insect cell cultures. Biotechnol Appl Biochem 42:25–34
57. Sridhar P, Panda AK, Pal R, Talwar GP, Hasnain SE (1993) Temporal nature of the promoter and not relative strength determines the expression of an extensively processed protein in a baculovirus system. FEBS Lett 315:282–286
58. Pajot-Augy E, Bozon V, Remy JJ, Couture L, Salesse R (1999) Critical relationship between glycosylation of recombinant lutropin receptor ectodomain and its secretion from baculovirus-infected insect cells. Eur J Biochem 260:635–648
59. Ahn WS, Jeon JJ, Jeong YR, Lee SJ, Yoon SW (2008) Effect of culture temperature on erythropoietin production and glycosylation in a perfusion culture of recombinant CHO cells. Biotechnol Bioeng 101:1234–1244
60. Aloi LA, Cherry RS (1994) Intracellular calcium response of Sf-9 insect cells exposed to intense fluid forces. J Biotechnol 33:21–31
61. Godoy-Silva R, Chalmers JJ, Casnocha SA, Bass LA, Ma N (2009) Physiological responses CHO cells to repetitive hydrodynamic stress. Biotechnol Bioeng 103:1103–1117
62. Ramirez OT, Mutharasan R (1992) Effect of serum on the plasma membrane fluidity of hybridomas: an insight into its shear protective mechanism. Biotechnol Prog 8:40–50
63. Tomiya N, Ailor E, Lawrence SM, Betenbaugh MJ, Lee YC (2001) Determination of nucleotides and sugar nucleotides involved in protein glycosylation by high-performance anion-exchange chromatography: sugar nucleotide contents in cultured insect cells and mammalian cells. Anal Biochem 293:129–137
64. Jarvis DL, Weinkauf C, Guarino LA (1996) Immediate-early baculovirus vectors for foreign gene expression in transformed or infected insect cells. Protein Expr Purif 8:191–203
65. Toth AM, Geisler C, Aumiller JJ, Jarvis DL (2011) Factors affecting recombinant Western equine encephalitis virus glycoprotein production in the baculovirus system. Protein Expr Purif 80:274–282
66. Jarvis DL, Finn EE (1996) Modifying the insect cell N-glycosylation pathway with immediate early baculovirus expression vectors. Nat Biotechnol 14:1288–1292
67. Geisler C, Jarvis D (2009) Insect cell glycosylation patterns in the context of biopharmaceuticals. In: Walsh G (ed) Post-translational modification of protein biopharmaceuticals. Wiley-VCH Weinheim, Weinheim, pp 165–191

68. Viswanathan K, Lawrence S, Hinderlich S, Yarema KJ, Lee YC, Betenbaugh MJ (2003) Engineering sialic acid synthetic ability into insect cells: identifying metabolic bottlenecks and devising strategies to overcome them. Biochemistry 42:15215–15225
69. Aumiller JJ, Hollister JR, Jarvis DL (2003) A transgenic insect cell line engineered to produce CMP-sialic acid and sialylated glycoproteins. Glycobiology 13:497–507
70. Chang GD, Chen CJ, Lin CY, Chen HC, Chen H (2003) Improvement of glycosylation in insect cells by mammalian glycosyltransferases. J Biotechnol 102:61–71
71. Okada T, Ihara H, Ito R, Nakano M, Matsumoto K, Yamaguchi Y, Taniguchi N, Ikeda Y (2010) N-Glycosylation engineering of lepidopteran insect cells by the introduction of the β1,4-N-acetylglucosaminyltransferase III gene. Glycobiology 20:1147–1159
72. Aumiller JJ, Mabashi-Asazuma H, Hillar A, Shi X, Jarvis DL (2012) A new glycoengineered insect cell line with an inducibly mammalianized protein N-glycosylation pathway. Glycobiology 22:417–428
73. Palmberger D, Wilson IBH, Berger I, Grabherr R, Rendic D (2012) SweetBac: a new approach for the production of mammalianised glycoproteins in insect cells. PLoS One 7:e34226
74. Kim NY, Baek JY, Choi HS, Chung IS, Shin S, Lee JI, Choi JY, Yang JM (2012) Short-hairpin RNA-mediated gene expression interference in *Trichoplusia ni* cells. J Microbiol Biotechnol 22:190–198
75. Mabashi-Asazuma H, Kuo CW, Khoo KH, Jarvis DL (2014) A novel baculovirus vector for the production of nonfucosylated recombinant glycoproteins in insect cells. Glycobiology 24: 325–340
76. Kati T, Kako N, Kikuta K, Miyazaki T, Kondi S, Yagi H, Kato K, Park EY (2017) N-Glycan modification of a recombinant protein via coexpression of human glycosyltransferases in silkworm pupae. Sci Rep 7:1409
77. Legardinier S, Klett D, Poirier JC, Combarnous Y, Cahoreau C (2005) Mammalian-like nonsialyl complex-type N-glycosylation of equine gonadotropins in mimic™ insect cells. Glycobiology 15:776–790
78. Lin SC, Jan JT, Dionne B, Butler M, Huang MS, Wu CY, Wong CH, Wu SC (2013) Different immunity elicited by recombinant H5N1 hemagglutinin proteins containing pauci-mannose, high-mannose, or complex type N-glycans. PLoS One 8:e66719
79. Geisler C, Mabashi-Asazuma H, Jarvis DL (2015) An overview and history of glycol-engineering in insect expression systems. In: Castilho A (ed) Glyco-engineering: methods and protocols. Methods in Molecular Biology, vol 1321, pp 131–152
80. Hancock K, Narang S, Pattabhi S, Yushak ML, Khan A, Lin S, Plemons R, Betenbaugh MJ, Tsang VCW (2008) False positive reactivity of recombinant, diagnostic, glycoproteins produced in high five™ insect cells: effect of glycosylation. J Immunol Meth 330:130–136
81. Bantleon F, Wolf S, Seismann H, Dam S, Lorentzen A, Miehe M, Jabs F, Jakob T, Plum M, Spillner E (2016) Human IgE is efficiently produced in glycosylated and biologically active form in lepidopteran cells. Mol Immunol 72:49–56
82. Dunkle LM, Izikson R, Patriarca P, Goldenthal KL, Muse D, Callahan J, Cox MMJ (2017) Efficacy of recombinant influenza vaccine in adults 50 years of age or older. New Engl J Med 376:2427–2436

# Engineering of Yeast Glycoprotein Expression

**Charlot De Wachter, Linde Van Landuyt, and Nico Callewaert**

## Contents



**Abstract** Yeasts are valuable hosts for recombinant protein production, as these unicellular eukaryotes are easy to handle, grow rapidly to a high cell density on cost-effective defined media, often offer a high space–time yield, and are able to perform posttranslational modifications. However, a key difference between yeasts and

---

Author contributed equally with all other contributors.Charlot De Wachter and Linde Van Landuyt

C. De Wachter, L. Van Landuyt, and N. Callewaert (✉)
VIB-UGent Center for Medical Biotechnology, Ghent, Belgium

Department of Biochemistry and Microbiology, Ghent University, Ghent, Belgium
e-mail: Nico.Callewaert@vib-ugent.be

mammalian cells involves the type of glycosylation structures, which hampers the use of yeasts for the production of many biopharmaceuticals. Glycosylation is not only important for the folding process of most recombinant proteins; it has a large impact on pharmacokinetics and pharmacodynamics of the therapeutic proteins as well. Yeasts' hypermannosylated glycosyl structures in some cases can evoke immune responses and lead to rapid clearance of the therapeutic protein from the blood. This chapter highlights the efforts made so far regarding the glyco-engineering of *N*- and *O*-type glycosylation, removing or reducing yeast-specific glycans. In some cases, this is combined with the introduction of humanized glycosylation pathways. After many years of patient development to overcome remaining challenges, these efforts have now culminated in effective solutions that should allow yeasts to reclaim the primary position in biopharmaceutical manufacturing that they enjoyed in the early days of biotechnology.

**Graphical Abstract**

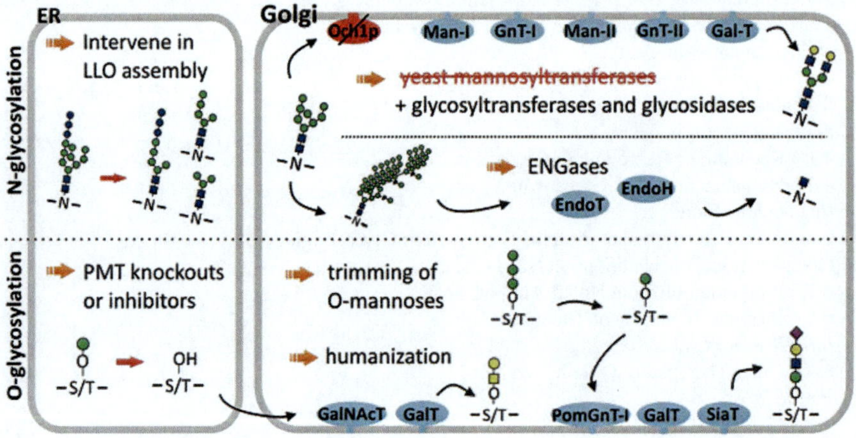


**Keywords** Fungi, *N*-glycosylation engineering, *O*-glycosylation engineering, *Pichia pastoris*, Recombinant protein expression, *Saccharomyces cerevisiae*, Yeast

## Abbreviations

| | |
|---|---|
| ADCC | Antibody-dependent cell-mediated cytotoxicity |
| ALG | Asparagine-linked glycosylation |
| AOX1/2 | Alcohol oxidase 1/2 |
| Asn | Asparagine |
| AtFXGER1 | *A. thaliana* GDP-4-keto-6-deoxy-mannose-3,5-epimerase |
| Bmtp | β-mannosyltransferase |

| | |
|---|---|
| CDC | Complement-dependent cytotoxicity |
| CHO | Chinese hamster ovary |
| CI-MPR | Cation independent-mannose-6-phosphate receptor |
| CNX/CRT | Calnexin/calreticulin |
| CRISPR/Cas9 | Clustered regularly interspaced short palindromic repeats/CRISPR-associated protein 9 |
| EDEM | ER degradation-enhanced α-1,2-mannosidase protein |
| Endo H/T | ENGase H/T |
| ENGase | Endo-β-*N*-acetylglucosaminidase |
| ER | Endoplasmatic |
| ERAD | ER-associated degradation |
| ERT | Enzyme replacement therapy |
| Fringe | Human β-1,3-*N*-acetylglucosaminyltransferase |
| GAA | Acid glucosidase α |
| Gal | Galactose |
| GalE | UDP-galactose-4-epimerase |
| GalNAc | *N*-acetylgalactosamine |
| GalT | Galactosyltransferase |
| Glc | Glucose |
| GlcNAc | *N*-acetylglucosamine |
| GLS-I/II | Glucosidase I/II |
| GM-CSF | Granulocyte-macrophage colony stimulating factor |
| GnT-I/II/III/IV | *N*-acetylglucosaminyltransferase I/II/III/IV |
| HBV | Hepatitis B virus |
| HER2 | Human epidermal growth factor receptor 2 |
| hO-FucT-1 | Human *O*-fucosyltransferase-1 |
| HPV | Human papillomavirus |
| HR | Homologous recombination |
| LLO | Lipid-linked oligosaccharide |
| mAbs | Monoclonal antibodies |
| Man | Mannose |
| Man-I/II | Mannosidase I/II |
| Mnn | (Phospho)mannosyltransferase |
| M-Pol I | Mannan-polymerase complex |
| NHEJ | Nonhomologous end joining |
| ORF | Open reading frame |
| OST | Oligosaccharyltransferase complex |
| P | Phosphate |
| PMT | Protein-*O*-mannosyltransferase |
| PomGnT-I | Protein-*O*-linked mannose β-1,2-*N*-acetylglucosaminyltransferase-I |
| POT | Protozoan single-subunit oligosaccharyltransferase |
| ppGalNAcT | Polypeptide:*N*-acetylglucosaminyltransferase |
| Pro | Proline |

| | |
|---|---|
| Rft1p | Flippase |
| rhEPO | Recombinant human erythropoietin |
| Ser | Serine |
| sgRNA | Short guide RNA |
| SiaT | Sialyltransferase |
| Thr | Threonine |

# 1 Introduction

## *1.1 Rationale for Using Yeast in Protein Expression*

To meet the individual requirements of each biopharmaceutical, different host systems have been optimized for efficient recombinant protein production. Prokaryotic organisms, especially *Escherichia coli*, are mainly used for the production of proteins that do not depend on eukaryotic posttranslational modifications for their folding, function, or stability. These organisms are well studied, and a lot of well-characterized manipulation techniques have been established. The first biopharmaceutical on the market, human insulin (Humulin®, Eli Lilly & Co., Indianapolis, IN), was produced in *E. coli*.

Another type of host system that shares the ease of handling and high production yield/cell density with prokaryotic organisms are yeasts. Moreover, unicellular eukaryotes are able to perform posttranslational modifications, including proteolytic processing of signal peptides, disulfide bond formation, subunit assembly, phosphorylation, acetylation, acylation, and glycosylation, that prokaryotes can perform only to a certain extent. Besides this, yeast and fungi differ in their ability to secrete recombinant proteins, whereas prokaryotes express recombinant proteins mainly intracellularly. Moreover, secretion facilitates the postproduction processing as most yeasts, compared to mammalian cells or filamentous fungi, secrete only very few endogenous proteins, which eases purification [1]. Yeasts are grown in cheap, defined chemical media without the need for animal-derived products.

The most important reason for the small number of biopharmaceuticals on the market produced by yeasts is their nonoptimal glycosylation pattern. Yeasts modify glycoproteins with high-mannose *N*-glycans (*Pichia pastoris*) to hypermannosyl *N*-glycan structures (*Saccharomyces cerevisiae*). Moreover, *P. pastoris* can incorporate β-1,2-mannose residues, and *S. cerevisiae* incorporates terminal α-1,3-mannoses, both of which could be immunogenic [2, 3]. Yeast-derived high-mannose *N*-glycans interact with specific receptors (i.e., C-type lectins) on the liver and lymph node endothelial cells, dendritic cells, and macrophages, which leads to fast serum clearance of the recombinant proteins [4]. To date, most therapeutic glycoproteins have been produced in mammalian cells, like Chinese hamster ovary (CHO) cells. These produce glycoproteins modified with humanlike hybrid- and complex-type *N*-glycans. Production in mammalian cells is complicated and expensive and requires animal-derived products with a risk of viral or prion contamination. The space–time yield of mammalian cell–based production processes is often lower than in yeasts and costs of production are higher. For immunoglobulin G manufacture, enormous efforts have been invested by the industry over the past two decades, which has resulted in very substantial improvements in production economics for these highly stable molecules. However, these processes are now approaching theoretical limits and are, moreover, not easily transferrable to the production of other, more labile, molecules. With increasing demand for biopharmaceuticals, new applications that require different molecular formats or massively increased scale, and increasing cost pressures as biopharmaceuticals are developed for increasingly common diseases, yeast as a manufacturing host is attracting renewed attention. To enable a next-step change in biopharmaceutical manufacturing, considerable effort has already been devoted and continues to be devoted to engineering yeast strains in such a way that they are devoid of yeast-type immunogenic glycans, sometimes combined with the introduction of humanlike glycosylation or application-customized glycosylation.

## *1.2* Saccharomyces cerevisiae *Versus* Pichia pastoris

*S. cerevisiae*, or baker's yeast, has a long history in the food industry, with its applications in bread baking and brewery. With the introduction of genetic engineering, another era started for this yeast as it started being used for recombinant protein production. A huge knowledge base was built up regarding the microbiology, genetics, molecular and cellular biology, stress response, and metabolism of this organism. The complete genome was sequenced in 1996 by an international cooperative venture involving scientists in Europe, North America, and Japan [5]. It was the first completely sequenced genome of a eukaryote. *S. cerevisiae* mainly uses glucose as a carbon source. As a facultative anaerobe, this yeast is able to switch to anaerobic consumption of glucose, however implying the concomitant production of toxic ethanol. *S. cerevisiae* expresses glycoproteins modified with hypermannosyl *N*-glycan structures consisting of α-1,2/3/6-mannoses and phosphomannoses (Fig. 3a), often comprising more than 100 mannose residues. Other biotechnologically important yeasts are *Kluyveromyces lactis*, which is able to use lactose as a sole carbon source and is used widely in the dairy industry, and *Yarrowia lipolytica*, a model organism for hydrophobic C-source catabolism due to its ability to grow on hydrophobic substrates like alkanes and fatty acids [6].

As a methylotrophic yeast, *P. pastoris* (formal nomenclature: *Komagataella phaffii*) is able to use methanol as a sole carbon and energy source, obviating the need to secrete enzymes like carbon source–procuring enzymes such as cellulases, as is the case for filamentous fungi. The presence of methanol induces expression of proteins involved in methanol metabolism, such as alcohol oxidase 1 and 2 (AOX1 and AOX2). Recombinant protein production controlled by their promoters results in an inducible expression system either with high expression levels (*AOX1* promoter) or lower expression levels (*AOX2* promoter) of the protein of interest. The genome of *P. pastoris* was reported in 2009 [7], and since then, promoter engineering efforts have also resulted in fermentable carbon source limitation-derepressed promoters, which make it possible to avoid methanol feeds where desired [8, 9]. Also, *P. pastoris* is a Crabtree-negative yeast species that, unlike, for example, *S. cerevisiae*, has a strong preference for respiratory growth in glucose- and oxygen-rich environments, avoiding the production of ethanol, as is the case for fermentative yeasts [10]. By avoiding the build-up of toxic ethanol, Crabtree-negative species can grow to very high cell densities [11]. In comparison with *S. cerevisiae*, *P. pastoris* modifies its glycoproteins with shorter glycans (oligomannoses) with a lower degree of polymerization, including α-1,2/3/6-, β-1,2-mannose residues or phosphomannoses [12, 13]. This makes *P. pastoris* a somewhat more suitable host for the production of glycoproteins compared to *S. cerevisiae*. *Hansenula polymorpha* is another methylotrophic yeast that is able not only to use methanol but also to express a pathway for nitrate assimilation. This yeast type is attractive for some applications in industry due to its thermotolerance up to 50°C [14, 15].

## 2 *N*-Glycosylation in Yeast

More than 70% of the biopharmaceuticals produced are glycoproteins, emphasizing the importance of glycosylation as a co- and posttranslational modification. Mostly two types of glycosylation modify glycoproteins: *N*-glycosylation and *O*-glycosylation (Sect. 4). Not only is *N*-glycosylation important in the folding process of most recombinant proteins, it also has a large impact on the pharmacokinetics and pharmacodynamics of the therapeutic proteins. Biopharmaceuticals modified with oligomannose-type *N*-glycans are prone to rapid clearance through Kuppfer cells (liver-resident macrophages) by binding to the mannose receptor present on the cell surface [4, 16]. Biopharmaceuticals carrying glycans fully modified with terminal sialic acid, however, show longer half-lives by reduced clearance [17]. Next to hepatic clearance, proteins with a molecular mass <30–50 kDa are rapidly cleared by the kidneys [18]. To avoid this, glycosylation of the protein can increase its hydrodynamic volume, reducing the renal clearance, as used in engineered EPO variants [19]. Next to the impact on clearance, glycosylation of the protein may offer protection against proteolytic degradation, for example, as shown for granulocyte colony stimulating factor [20–22]. Fc glycosylated IgG-type antibodies show increased resistance to proteolytic degradation by papain compared to nonglycosylated variants, with the highest degree of resistance obtained when carrying GlcNAc-terminal residues [23]. Whether these differences are relevant in terms of therapeutic use is unstudied.

*N*-glycosylation occurs on asparagine residues in an asparagine-X-serine/threonine (Asn-X-Ser/Thr) context, where X is any amino acid except for proline (Pro). The initial steps of *N*-glycosylation synthesis are common to almost all eukaryotes.

Briefly, a $Glc_3Man_9GlcNAc_2$ lipid-linked oligosaccharide (LLO) is assembled in the endoplasmic reticulum (ER) by several glycosyltransferases encoded by asparagine-linked glycosylation (*ALG*) genes (Fig. 4a). This precursor is cotranslationally transferred to a nascent polypeptide chain by the oligosaccharyltransferase complex (OST). Deglucosylation of the *N*-glycan by glucosidase I and II leads to the formation of a monoglucosylated structure that can bind calnexin/calreticulin, assisting in protein folding (Fig. 1) [24]. Subsequently, one α-1,2-mannose is removed by an ER-residing α-1,2-mannosidase. Further, α-1,2-mannosidase processing by Htm1p in the ER can expose a terminal α-1,6-mannose residue, which is a signal for the degradation of proteins that have not reached their proper fold [25]. Correctly folded proteins are transported to the Golgi apparatus, where their *N*-glycans are further modified, but in a species-specific manner. Mammalian cells trim the $Man_8GlcNAc_2$ *N*-glycan further by α-mannosidases to obtain a substrate for the generation of hybrid- and complex-type *N*-glycans by glycosyltransferases in the Golgi apparatus. Instead of reducing the $Man_8GlcNAc_2$ *N*-glycans, yeasts elongate this *N*-glycan further, starting with the introduction of an α-1,6-mannose residue. Further elongation with α-1,2/3/6-mannoses and phosphomannoses occurs to obtain hypermannosylated *N*-glycans, which can be capped in a species-specific manner (e.g., with α-1,3-mannose residues in *S. cerevisiae*). *N*-glycosylation is very heterogeneous because it is a non-template-driven process, with multiple differentially expressed glycosyltransferases sometimes competing for the same substrates and resulting from the action of all of these enzymes over a very short time frame during passage of the substrate glycoprotein through the Golgi apparatus. Moreover, interactions of the glycan with the particular protein environment to which it is attached can also influence the accessibility of its different branches to the glycosyltransferases and, hence, rates of conversion [26].

Because of the negative impact of yeast-type *N*-glycosylation on biopharmaceuticals in terms of immunogenicity and clearance, glyco-engineering of yeast strains was introduced to remove yeast-type glycosylation and to generate hybrid- and complex-type (human) *N*-glycans (Fig. 2).

## 2.1 Engineering N-Glycosylation in Yeast

Glyco-engineering of yeast cells includes the removal of yeast-specific glycosylation, sometimes followed by the construction of hybrid- or complex-type (human) glycans (Fig. 2). Next to the modification of the type of *N*-glycan structures on the protein, considerable effort has been devoted to reducing the macro- and microheterogeneity present on glycoproteins. To minimize macroheterogeneity, meaning the structural diversity due to differential occupation of *N*-glycosylation sites, different strategies have been explored to optimize the efficiency of cotranslational *N*-glycan transfer to glycoproteins. One strategy involves regulating the molecular flux in the dolichol pathway. For example, overexpression of a *S. cerevisiae* cis-prenyltransferase, a key enzyme in dolichol synthesis, in

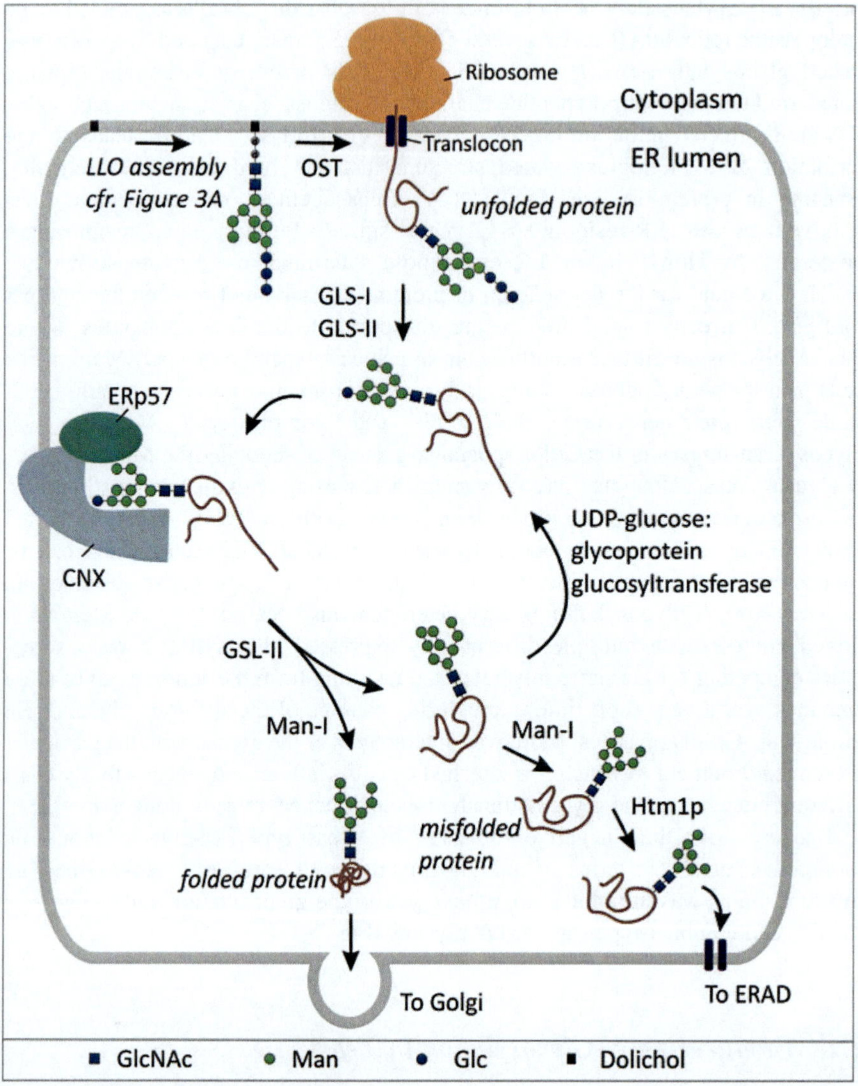


**Fig. 1** Calnexine/calreticulin cycle for protein folding. The OST transfers the LLO, consisting of $Glc_3Man_9GlcNAc_2$, to a nascent polypeptide chain. Deglucosylation of the *N*-glycan by glucosidase I (GLS-I) and glucosidase II (GLS-II) leads to the formation of a monoglucosylated structure that can bind calnexin/calreticulin (CNX/CRT), assisting in protein folding. Reglucosylation might occur by a UDP-glucose:glycoprotein glucosyltransferase in case of a misfolded protein to reinitiate the folding process. *N*-glycan trimming by an ER-residing α-1,2-mannosidase (Man-I) to $Man_8GlcNAc_2$ occurs and has most often been completed by the time a correctly folded protein is transported to the Golgi for further modification of the *N*-glycans. Proteins that remain engaged in folding for a longer time are further trimmed by Htm1p [i.e., equivalent to the ER degradation-enhancing α-1,2-mannosidase-like protein (EDEM) in mammals], which generates a glycan structure that targets proteins for ER-associated degradation (ERAD)

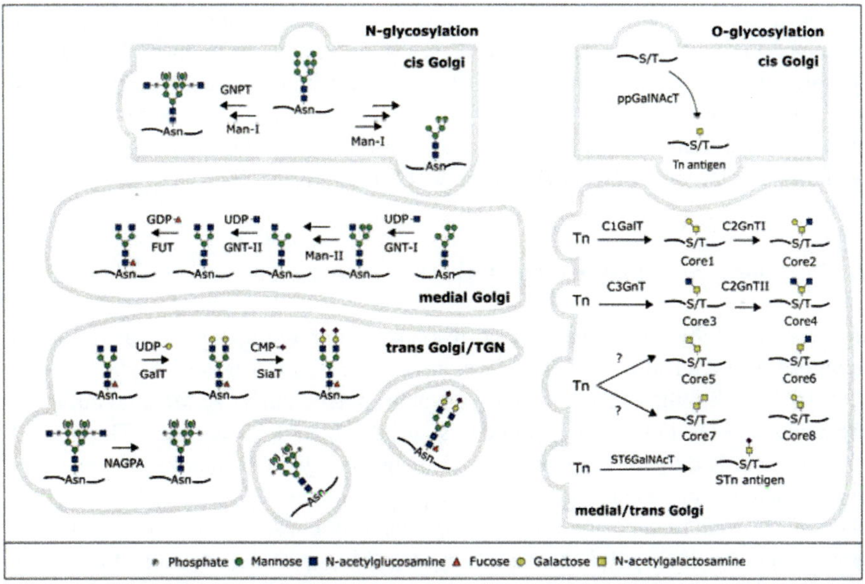


**Fig. 2** Overview of human *N*- and *O*-glycosylation in the Golgi apparatus. On the left side, the synthesis of a human glycoprotein containing a complex-type biantennary *N*-glycan is shown. In the cis Golgi, mannosidase-I (Man-I) leads to a $Man_5GlcNAc_2$ that can be modified in the medial Golgi by *N*-acetylglucosaminyltransferase I (GnT-I), mannosidase-II (Man-II), *N*-acetylglucosaminyltransferase II (GnT-II), and a fucosyltransferase (FUT). Afterward, galactosyltransferase (GalT) and sialyltransferase (SiaT) may perform their function on *N*-glycosylated proteins. Moreover, *N*-glycans of lysosomal proteins can be decorated with mannose-6-phosphate due to the action of *N*-acetyl-1-phosphotransferase (GNTP) in the cis Golgi and *N*-acetylglucosamine-1-phosphodiester-α-*N*-acetylglucosaminidase (NAGPA) in the trans Golgi. The right side of the figure shows the mucin-type *O*-glycosylation. Polypeptide-GalNAc-transferase (ppGalNAcT) initiates *O*-glycosylation in the Golgi, which is followed by the action of core 1 galactosyltransferase (C1GalT), core 2 *N*-acetylglucosaminidase I (C2GnTI), core 3 *N*-acetylglucosaminidase (C3GnT), core 2 *N*-acetylglucosaminidase II (C2GnTII), or sialyltransferase (ST6GalNAcT) to generate core 1–4 or the sialyl-Tn-antigen (STn antigen), respectively. Next to this, so far unknown enzymes synthesize core 5–8 *O*-glycans

*Trichoderma reesei* resulted in increased glycosylation levels of secreted proteins [27]. Other strategies involve overexpression of proteins involved in the OST complex (Sect. 2.1.2) [28, 29] or engineering of the acceptor tripeptide sequon for optimal transfer conditions, as Asn-X-Thr is two to three times more efficiently glycosylated than Asn-X-Ser [30]. The microheterogeneity of the carbohydrate moiety refers to different structures that can be present on a given glycosylation site of endogenous glycoproteins.

A first step in the humanization of *N*-glycosylation in yeast is the removal of the high-mannose and hypermannosyl structures. Two main strategies are followed, one based on the elimination of yeast glycosyltransferases (Sect. 2.1.1) and the other one on interference in the assembly of the LLO (Sect. 2.1.2). A subsequent step is the introduction of different glycosyltransferases and glycosidases to obtain hybrid- and complex-type *N*-glycans. A recent, third approach, which efficiently converts yeast *N*-glycosylation into a type that is often function-neutral, is the expression of an endo-β-*N*-acetylglucosaminidase, capable of removing high-mannose *N*-glycans and resulting in a largely deglycosylated product (Sect. 2.1.3).

### 2.1.1 Approach 1: Elimination of Yeast Glycosyltransferases

In the early 1990s the main enzyme responsible for the elongation of the high-mannose *N*-glycan in *S. cerevisiae* was revealed as an α-1,6-mannosyltransferase (Och1p), which initiates the α-1,6-polymannose outer chain [31]. Mannan polymerase complexes (M-Pol) I and II extend this further with α-1,6-mannoses. The chain is further elaborated by the addition of α-1,2-mannoses by α-1,2-mannosyltransferases and phosphomannoses in a process that requires both the *MNN6* and *MNN4* (which encodes for Mnn4p, a positive regulator of mannosylphosphate transferase Mnn6p) genes and terminal α-1,3-mannoses by the Mnn1p α-1,3-mannosyltransferase (Fig. 3a). *S. cerevisiae* Δ*och1* Δ*mnn1* Δ*mnn4* strains mainly modify their glycoproteins with $Man_8GlcNAc_2$ *N*-glycans, removing the immunogenic mannan *N*-glycans, but still retaining the rapidly cleared, high-mannose *N*-glycan [32]. Introduction of an α-1,2-mannosidase gene of *Aspergillus saitoi* in this triple mutant Δ*och1* Δ*mnn1* Δ*mnn4* strain resulted in the first yeast strain capable of producing some level of the human-compatible sugar chain $Man_5GlcNAc_2$ [33]. Although the enzyme was successfully retained in the ER using an HDEL tag, only 27% of the *N*-glycans of an endogenous protein (carboxypeptidase Y) were trimmed from $Man_8GlcNAc_2$ to $Man_5GlcNAc_2$. Because $Man_5GlcNAc_2$ is the substrate used to build human complex-type *N*-glycans, a homogeneous conversion to $Man_5GlcNAc_2$ is critical.

Furthermore, these manipulations have a big impact on *S. cerevisiae* yeast cells, resulting in severe growth defects and decreased protein productivity. Mutagenesis-based genetic diversity was introduced to perform screenings to obtain Δ*och1* Δ*mnn1* Δ*mnn4* yeast strains capable of more efficient production of glycoproteins [34]. In some mutants with restored functionality, the reduction of cell wall strength as a consequence of the deletion of the outer chain of the *N*-glycans was likely compensated by an increase in the glucan layer of the cell wall, as indicated by an elevated level of glucose in the cell wall [34]. Recent research has shown that disruption of the mannan glycan structures (Δ*och1* Δ*mnn9* strain) causes cell wall integrity defects, which causes cell stress and severe growth impairments [35]. At the same time, this strain shows upregulation of genes in the secretory pathway involved in protein folding (*KAR2* and *SSA1*), vesicular trafficking (*BOS1*, *ERV25*, *SNC2*, and *SSO1*), and the ERAD pathway (*DER1* and *HRD3*), increasing specific protein secretion levels. An increase in recombinant-specific protein secretion was also shown in *Δmnn10* deletion strains [36, 37]. In contrast to *S. cerevisiae*, no effect on the growth rate could be detected in the *Δoch1* strain of *Y. lipolytica* [38].

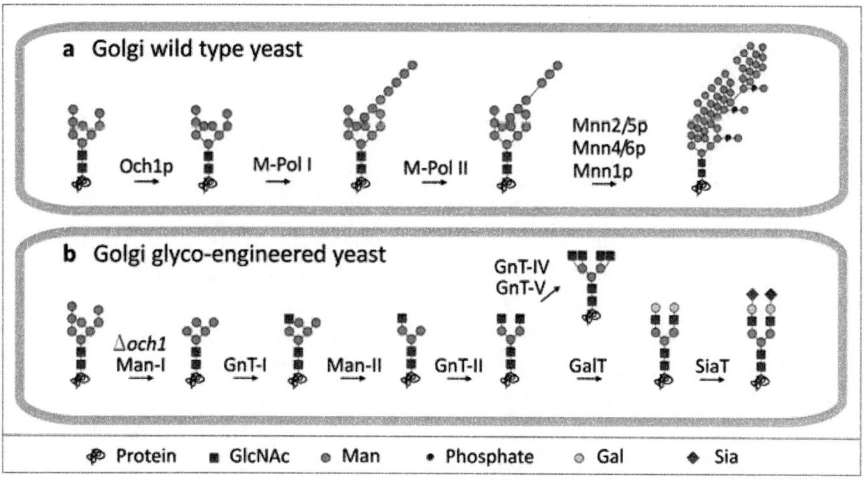


**Fig. 3** Golgi *N*-glycosylation: wild-type versus glyco-engineered yeast. (**a**) Overview of wild-type *N*-glycosylation in model ascomycetous yeast *S. cerevisiae*. The $Man_8GlcNAc_2$ *N*-glycan obtained in the ER is further elongated with yeast-specific glycosyltransferases, starting with an α-1,6-mannose by Och1p. Mannan polymerase complexes (M-Pol) I and II extend this further with α-1,6-mannoses. Further elongation with α-1,2/3-mannoses and phosphomannoses by (phospho)mannosyltransferases (Mnn) results in hypermannosylated *N*-glycans. (**b**) Elimination of yeast-specific glycosyltransferases like Och1p results in shorter high-mannose *N*-glycans. Insertion of mannosidase (Man) I, *N*-acetylglucosaminyltransferase (GnT) I, Man-II, GnT-II, GnT-IV, GnT-V, galactosyltransferase (GalT), or sialyltransferase (SiaT) results in glycoproteins modified with humanlike complex-type *N*-glycans. This has most completely been implemented in *P. pastoris*

$Man_5GlcNAc_2$ *N*-glycans have been obtained by overexpression of ER-retained *T. reesei* α-1,2-mannosidase. Additional *Δmnn9* knockout did not improve the glycosylation profile compared to single *Δoch1* deletion [38].

In *P. pastoris*, hypermannosylation occurs less frequently and to a lesser extent compared to *S. cerevisiae*. The yeast also lacks α-1,3-mannosyltransferase activity, avoiding the presence of terminal, immunogenic α-1,3-mannoses on the *N*-glycans. Disruption of the *OCH1* gene in *P. pastoris* with a knock-in strategy results in glycoproteins modified with $Man_{8\text{-}12}GlcNAc_2$ *N*-glycans [39, 40]. No severe growth defects have been observed, but it has been revealed that the knock-in event results in the inadvertent expression of an *N*-terminally truncated Och1p, which appears to be sufficient to avoid the growth defects later observed with full *OCH1* knockout in this organism. Whether this is due to rest–activity of the truncated Och1p or to, for example, a stabilizing effect of the truncated Och1p in presumed Golgi protein complexes is unclear at this time. In any case, this serendipitous finding has formed the basis for the production of a strain in which this knock-in event has been genetically stabilized, resulting in the so-called SuperMan5 *P. pastoris* strain, which is commercially available from Research Corporation Technologies, Tucson, AZ, USA [41, 42]. The full knockout of *OCH1* in *P. pastoris* results in a growth defect, but it has not been well documented whether this is further worsened or rather improved by further *N*-glycan engineering. In any case, researchers at Merck-GlycoFi, Kenilworth, NJ, USA (where full *OCH1* knockout was used as the strain basis) have filed a patent application that describes a compensatory mutation in the *ATT1* gene, which improves the *OCH1* knockout phenotype [43, 44]. Similar results were obtained recently by mutating the *CWP1* gene in an *OCH1* knockout [45]. Subsequent overexpression of an HDEL-tagged α-1,2-Man-I from *T. reesei* was successful in converting the *N*-glycans to the smaller $Man_5GlcNAc_2$ structure [40, 46]. Another successful approach involves introducing a *C. elegans* Man-I fused to ScMns1p ER-targeting signal [39]. The obtained $Man_5GlcNAc_2$ *N*-glycan structure is a starting point for the further introduction of glycosyltransferases and glycosidases yielding mammalian complex-type *N*-glycans (Fig. 3b).

The first step in obtaining complex-type *N*-glycans is the addition of a GlcNAc residue by *N*-acetylglucosaminyltransferase I (GnT-I). Overexpression of human GnT-I, targeted to the early to medial Golgi compartment using a ScKre2p signal sequence, generates almost complete conversion of a $Man_5GlcNAc_2$ *N*-glycan to $GlcNAcMan_5GlcNAc_2$ structure [40]. Using an alternative approach, Choi et al. succeeded in generating this hybrid-type *N*-glycan by introducing human GnT-I retained in the cis Golgi using a ScMnn9p targeting signal. However, coexpression of an additional UDP-GlcNAc transporter (*K. lactis*) was necessary to optimize this conversion [39]. In both studies, different yeast strains and different localization signals for both Man-I and GnT-I were used, which could have had an impact on GlcNAc transfer efficiency.

Further humanization of the *N*-glycosylation pathway implies the introduction of biantennary *N*-glycans by substituting the α-1,3-mannose and α-1,6-mannose residue of the hybrid-type for a second GlcNAc residue. To obtain this, Hamilton et al. used a combinatorial library of several mannosidase II (Man-II) and *N*-acetylglucosaminyltransferase II (GnT-II) catalytic domains fused to more than 60 fungal type II membrane localization signals [47]. Introduction of *Drosophila melanogaster* Man-II and *Rattus norvegicus* GnT-II, both coupled to a ScMnn2p medial Golgi targeting signal, resulted in a strain capable of producing $GlcNAc_2Man_3GlcNAc_2$-modified glycoproteins with the highest homogeneity and a production yield comparable to the wild-type GS115 strain. Obtaining $GlcNAcMan_3GlcNAc_2$ *N*-glycans seems to be a tricky and inefficient point in the engineering process, but it is largely resolved by modifying the generated terminal α-1,6-mannose with a GlcNAc residue. The reason for this remains unclear, but it was observed that this terminal α-1,6-mannose is a substrate for endogenous glycosyltransferases, resulting in novel structures that might interfere in cell wall biogenesis [48].

Overexpression of β-1,4-galactosyltransferase (GalT) in the Golgi compartment is necessary for the subsequent modification of GlcNAc terminal residues with a β-galactose residue. The presence of UDP-Gal, necessary as a donor substrate for Gal transfer, in the Golgi of *S. cerevisiae* was shown [49]. Based on this evidence, one can assume that this is also the case for *P. pastoris*. However, Vervecken et al. obtained only a conversion of 10% of $GlcNAcMan_5GlcNAc_2$ to $GalGlcNAcMan_5GlcNAc_2$ [40]. The expression of a fusion protein composed of the human GalT-I catalytic domain and a UDP-galactose 4-epimerase (GalE) of *Schizosaccharomyces pombe*, retained in the Golgi using the ScMnn2p signal sequence, resulted in efficient generation of biantennary $Gal_2GlcNAc_2Man_3GlcNAc_2$ *N*-glycans without the need for a UDP-Gal transporter [50]. This structure is the one found on the conserved *N*-glycosylation site (Asn297) of human IgG, except for the presence of a core α-1,6-fucose residue on native IgGs. The absence of this fucose residue is advantageous for antibody-dependent cell-mediated cytotoxicity (ADCC)-dependent antibody functionality (e.g., in cancer and viral infection treatments) (Sect. 5).

The final step of human glycosylation, on most therapeutic glycoproteins, is terminal sialylation of the *N*-glycans. This is important because desialylated glycoproteins are recognized, internalized, and degraded by the ASGP-R on hepatocytes. Because serum glycoproteins are desialylated over time and as such reflect a protein's age, it follows that the ASGP-R plays a crucial role in maintaining serum glycoprotein homeostasis. Moreover, the ASGP-R rather prefers multivalent and multimeric ligands over single galactose residues. Biantennary *N*-glycans have 100- to 1,000-fold lower affinity in comparison to triantennary *N*-glycans [51]. Since yeasts lack the capability to biosynthesize the cytidine monophosphate *N*-acetylneuraminic acid (CMP-Neu5Ac) precursor, the CMP-Neu5Ac Golgi transporter, and the sialyltransferase to transfer Neu5Ac to terminal galactose residues, the introduction of five enzymes in a $Gal_2GlcNAc_2Man_3GlcNAc_2$ engineered strain is necessary to obtain sialylation. Hamilton et al. succeeded in constructing one vector containing these five genes (*H. sapiens* UDP-*N*-acetylglucosamine-2-epimerase/*N*-acetylmannosamine kinase, *H. sapiens* *N*-acetylneuraminate-9-phosphate synthase, *H. sapiens* CMP-Neu5Ac synthase, *M. musculus* CMP-Neu5Ac transporter, and *M. musculus* α-2,6-sialyltransferase), resulting in biantennary sialylation of recombinant human erythropoietin (rhEPO) [52]. Moreover, this engineered rhEPO triggered receptor signaling in vitro equally well as a form of the protein containing tri- and tetra-antennary sialylated *N*-glycans (Darbepoetin/Aranesp®) produced in CHO cells [53]. In vivo activity was optimized by PEGylation of the engineered rhEPO, avoiding fast clearance due to smaller *N*-glycans. Merck for a while appeared to be gearing up to test this molecule in a clinical setting, but the program has now apparently been abandoned for unspecified reasons (Sect. 6.4).

Part of such glyco-engineering technology has recently become commercially available as *Pichia* GlycoSwitch® (Research Corporation Technologies, Tucson, AZ, USA), allowing for the production of glycoproteins carrying $Gal_2GlcNAc_2Man_3GlcNAc_2$ *N*-glycans [54]. The starting point is the SuperMan5 strain (Biogrammatics, Carlsbad, CA, USA), improved for genetic stability [41, 42]. Expansion of this technology to generate triantennary *N*-glycans is possible by introducing human GnT-IV in the Golgi, which transfers a β-1,4-GlcNAc to the α-1,3-mannose of the $Man_3GlcNAc_2$ core [55]. There are two caveats with the use of this "humanization" technology. First, the genetic stability of strains with large numbers of transgenes can be difficult to obtain. Recently, since the availability of the *Pichia* genome sequence [7], more flexibility has been available with respect to the sites of genomic integration, and strategies such as random rather than homologous recombination can be used to achieve higher levels of stability. Second, inactivating only

*OCH1* is insufficient to avoid synthesis of all fungal type *N*-glycans. Indeed, low levels of α-1,6-branch formation can still be detected in *OCH1* knock-in/knockout strains, and *N*-glycans can still be modified with phosphomannosyl residues. Moreover, sometimes the human-type intermediates are inadvertently recognized by yeast glycosyltransferases, resulting in the formation of novel structures. This is obviously undesirable. GlycoFi/Merck's researchers have attempted to overcome this to some extent through stacking of more and more glycosyltransferase gene knockouts in strains while still not completely solving the problem. This makes the entire concept of humanization rather unwieldy, and more efficient solutions are likely needed.

### 2.1.2 Approach 2: Interference in Lipid-Linked Oligosaccharide Assembly

Whereas the first approach to disrupting the hypermannosyl *N*-glycan chain occurs in the Golgi by elimination of glycosyltransferases such as Och1p, the second approach is based on interference in the assembly of the LLO precursor in the ER. This step is highly conserved between almost all eukaryotes and comprises the assembly of the $Glc_3Man_9GlcNAc_2$ LLO as a precursor for cotranslational transfer to the nascent protein chain. Unravelling of this LLO assembly pathway revealed the role of several glycosyltransferases encoded by asparagine-linked glycosylation (*ALG*) genes (Fig. 4a) [56].

In *S. cerevisiae*, the enzyme responsible for converting $Man_5GlcNAc_2$-PP-Dol to $Man_6GlcNAc_2$-PP-Dol at the luminal side of the ER (Dol-P-Man:$Man_5GlcNAc_2$-PP-Dol α-1,3-mannosyltransferase) is encoded by the *ALG3* gene (Fig. 4b) [57]. Δ*alg3* mutants in both *S. cerevisiae* and *P. pastoris*, however, do not lead to the accumulation of $Man_5GlcNAc_2$ only but also show *N*-glycans larger in size containing mannoses and structures recalcitrant to mannosidase digests [58, 59]. This $Man_5GlcNAc_2$ *N*-glycan may be a substrate for Och1p, resulting in the addition of α-1,6-mannoses and requiring the combination of this strategy with *OCH1* deletion.

After the deletion of *PpALG3* in the Δ*och1 P. pastoris* strain, a Man-I catalytic domain, fused to the Sec12p yeast localization signal, was inserted into the *P. pastoris* genome. This led to the trimming of the *N*-glycan structure to $Man_3GlcNAc_2$, which serves as a substrate for GnT-I, which was targeted to the Golgi after fusion with the localization signal of Mnn9p [50]. This approach obviates the use of Man-II, necessary in the human pathway to obtain the substrate for GnT-II. Introduction of rat GnT-II and fused human GalT-I and *S. pombe* UDP-galactose 4-epimerase to a single localization signal resulted in the successful generation of complex-type $Gal_2GlcNAc_2Man_3GlcNAc_2$ biantennary *N*-glycans. The engineered cells had a small reduction in growth rate, but protein production yield was comparable to that of the parental wild-type yeast [50]. To further improve on the homogeneity of the *N*-glycan structure, knockouts in genes involved in phosphomannosylation (Δ*pno1* and Δ*mnn4B*) and β-mannosylation (Δ*bmt2*) (Sect. 3) were helpful [52, 60]. Moreover, introducing the glycosyltransferases

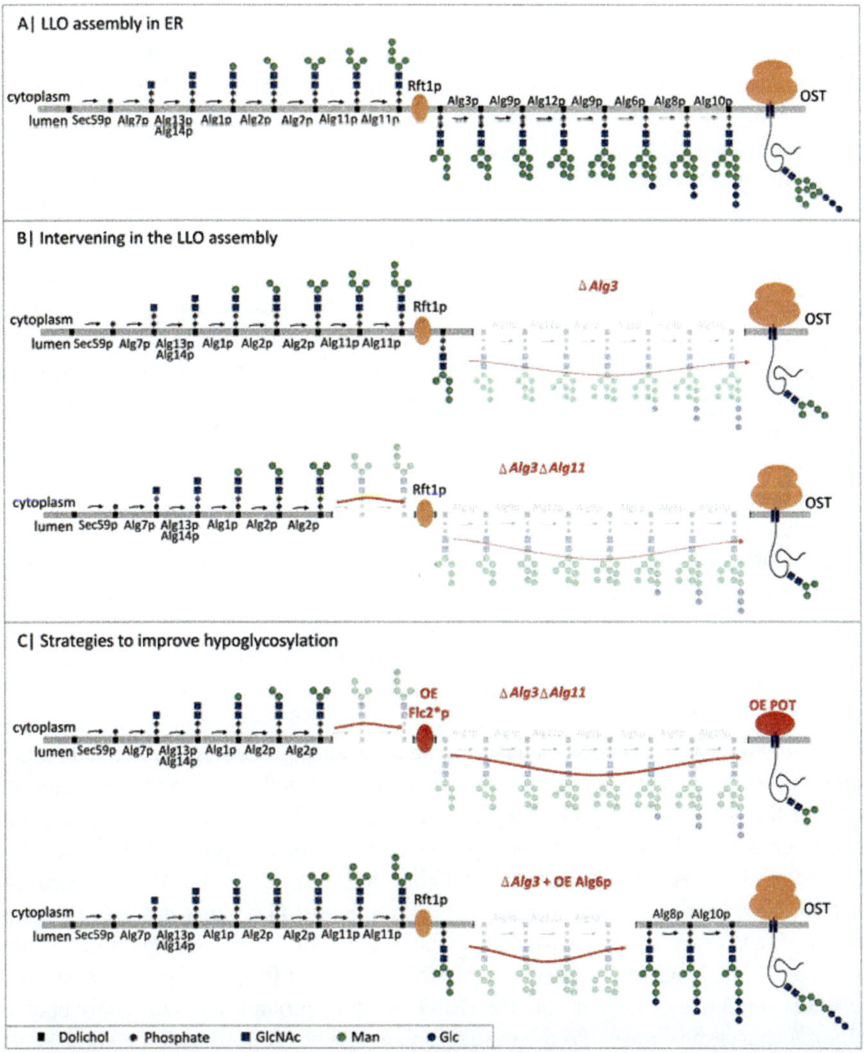


**Fig. 4** Interference in LLO assembly in ER. (**a**) Assembly of the LLO involves the cooperation of several glycosyltransferases, encoded by the *ALG* genes, a flippase (Rft1p) to catalyze flipping of the LLO to the ER lumen, and an OST, which coordinates cotranslational transfer of the oligosaccharide to the nascent protein chain. (**b**) One approach to remove high-mannose yeast-specific *N*-glycans involves the disruption of specific *ALG* genes like *ALG3* and *ALG11*, leading to a shorter $Man_5GlcNAc_2$ or $Man_3GlcNAc_2$ LLO, respectively. These shorter LLOs are transferred to nascent polypeptide chains and are substrates for further elaboration to hybrid- and complex-type *N*-glycans. (**c**) Overexpression of a C-terminally truncated flippase Flc2*p and of a protozoan single-subunit oligosaccharyltransferase (POT) are necessary to improve *N*-glycan transfer of these shorter LLOs to tackle the severe hypoglycosylation from which these strains suffer. Another approach, demonstrated in *Y. lipolytica*, involves overexpression of Alg6p, which enhances the transfer of Glc residues to $Man_5GlcNAc_2$ *N*-glycans in a Δ*alg3* strain. The presence of Glc residues facilitates the transfer of the LLO to the protein by the OST. Overexpression of GLS-II heterodimeric protein is required to remove these glucosyl residues efficiently after transfer; together with the overexpression of an α-1,2-mannosidase, glycoproteins modified with the $Man_3GlcNAc_2$ core *N*-glycan were obtained

needed to synthesize the $GlcNAc_2Man_3GlcNAc_2$ *N*-glycan-producing strain reduced the presence of structures recalcitrant to mannosidases, suggesting depletion of substrate for these yet unknown transferases when the terminal mannoses are both capped with GlcNAc [50]. This observation is similar to that made with respect to the Man-II-dependent Golgi engineering route.

Another intervention in the LLO assembly in *S. cerevisiae* was carried out by an additional deletion in the *ALG11* gene, eliminating a functional GDP-Man: $Man_3GlcNAc_2$-PP-dolichol α-1,2-mannosyltransferase. This enzyme is localized to the cytosolic side of the ER and catalyzes the transfer of α-1,2-mannose to $Man_3GlcNAc_2$-PP-Dol and $Man_4GlcNAc_2$-PP-Dol. Consequently, Δ*alg11* Δ*alg3* mutants of *S. cerevisiae* are capable of producing glycoforms modified with $Man_3GlcNAc_2$ *N*-glycans but suffer from severe hypoglycosylation and growth retardation (Fig. 4b) [61, 62]. This truncated LLO is a poor substrate for the endogenous flippase Rft1p, which catalyzes flipping of the LLO across the ER membrane into the lumen. Overexpression of a C-terminally truncated version of the endogenous ER-localized Flc2p (Flc2*p) can functionally replace Rft1p. Moreover Flc2*p shows a more relaxed substrate specificity, resulting in improved flipping of the truncated $Man_3GlcNAc_2$-PP-Dol [28, 61]. Besides improving the availability of the truncated LLO in the ER lumen, optimization of the subsequent transfer to the nascent protein chain is necessary to tackle the hypoglycosylation. In contrast to the multisubunit OST of higher eukaryotes, protozoa express single-subunit OSTs (POTs) that can catalyze the transfer of the oligosaccharide chain independently of an OST complex. Overexpression of *Leishmania major* STT3D (*Lm*STT3D) improved *N*-glycosylation in yeast strains, producing a truncated LLO, suggesting a more relaxed substrate specificity compared to the original catalytic subunit Stt3 (Fig. 4c) [28]. Overexpression of *Lm*STT3D also evoked a significant improvement of *N*-glycan site occupancy of recombinant proteins produced in *P. pastoris* [29]. Another approach to tackling hypoglycosylation, demonstrated in *Y. lipolytica*, involves the overexpression of dolichol-P-Glc:$Man_9GlcNAc_2$-PP-Dol glucosyltransferase (Alg6p), which enhances the transfer of Glc residues to the $Man_5GlcNAc_2$ LLO in a Δ*alg3* mutant strain (Fig. 4c) [63]. The presence of Glc residues facilitates transfer of the LLO to the protein by the endogenous oligosaccharyltransferase. To then remove these glucosyl residues efficiently after transfer to the protein, overexpression of the glucosidase II heterodimeric protein is required [63] and was successfully accomplished, together with α-1,2-mannosidase expression, to yield glycoproteins modified uniformly with the universal $Man_3GlcNAc_2$ core *N*-glycan.

Recently, more homogeneity of the $Man_3GlcNAc_2$ *N*-glycans in a Δ*alg3* Δ*alg11* *S. cerevisiae* strain was obtained by deletion of Mnn1p [64]. Besides the improvement of homogeneity, this also eliminated the presence of the potentially immunogenic terminal α-1,3-mannoses. Complex-type *N*-glycans have been obtained in this strain by expression of Kre2p-GnT-I and Mnn2p-GnT-II fusion proteins, and their relative abundance increased by the overexpression of the UDP-GlcNAc transporter of *K. lactis* [64]. Next to this complex-type sugar, a $Man_4GlcNAc_2$ *N*-glycan remains present and could not be removed by deletion of Mnn1p nor Mnn2p.

Production of glycoproteins modified with complex-type *N*-glycans was also shown in the industrially important, thermotolerant *H. polymorpha*. Biantennary $Gal_2GlcNAc_2Man_3GlcNAc_2$ structures have been obtained in a *Δalg3 Δalg11 Δoch1* strain overexpressing GnT-I, GnT-II, and GalT-I [62, 65, 66]. To provide sufficient donor substrates, the UDP-GlcNAc transporter of *K. lactis* and the *S. pombe* GalE are overexpressed, together with the overexpression of HpRft1p, to improve flipping of the truncated LLO to the ER lumen [62]. So far, in vivo sialylation could not be obtained, but successful in vitro sialylation is described using recombinant sialyltransferases and CMP-Neu5Ac as a donor substrate [67]. Recent research showed significant improvement of glycosylation efficiency and homogeneity in *H. polymorpha* by the overexpression of Hac1p [68]. This protein is a transcription factor mediating the unfolded protein response induced by misfolded proteins or ER stress. Previously described increases in heterologous protein expression levels (protein-dependent) by overexpression of Hac1p [69, 70] might be explained partially by enhanced protein processing involving glycosylation activity.

As a final note in this section describing ways of humanizing *N*-linked glycosylation, it is important to realize that only the Golgi-based pathway in *P. pastoris* has been put to the test of a high-cell-density fermentation so far. It remains to be seen whether this or any other of the described concepts will yield strains that are sufficiently stable and robust to deliver reliable production processes for glycoproteins modified with human-type glycans devoid of undesired yeast-produced structures. Substantial work remains ahead to achieve this longstanding goal of biopharmaceutical production science.

### 2.1.3 Approach 3: Overexpression of Endo-β-*N*-Acetylglucosaminidases

A third and recent approach is based on the removal of high-mannose yeast *N*-glycans by endo-β-*N*-acetylglucosaminidases (ENGases). These cleave the β-1,4-glycosidic bond between the two GlcNAc residues in the core of high-mannose and hybrid-type *N*-glycans, leaving behind a single GlcNAc residue. A *P. pastoris* strain expressing ENGase H (EndoH) from *Streptomyces plicatus* has been used to deglycosylate recombinant proteins produced in a second strain [71]. This was implemented either in a cofermentation process or using postfermentation methods, avoiding the potential problem that coexpression might cause competition for cellular resources, leading to a decrease in the yield of the protein of interest. Similar results were obtained in our lab using another ENGase, EndoT from *T. reesei* [72]. This fungus natively de-*N*-glycosylates its secreted glycoproteins. We recently implemented cosecretion of EndoT with target glycoproteins in *P. pastoris* (Laukens et al., manuscript in preparation) and were only successful after significant engineering to make this compatible with robust growth characteristics and lack of lysis of the engineered cells. Such deglycosylation-based engineering is suitable for proteins that need *N*-glycans for folding but in which the glycans are not needed for the intended functionality of the protein. Furthermore, this approach, at least in *P. pastoris*, efficiently removes potentially immunogenic (and partially as yet unknown) endogenous glycan modifications that the yeast adds to its *N*-glycan branches.

## 3 β-Mannose Depletion

β-mannoses were first identified in *C. albicans* [73] but were also seen in *P. pastoris* in 2000, following structural characterization of *P. pastoris* mannans [13]. It was shown that some *N*-glycans were resistant to α-1,2-mannosidase treatment. This subset accounted for 2% of the total glycan pool and was identified with nuclear magnetic resonance [74]. However, *S. cerevisiae* does not contain β-mannosylated glycoproteins, at least not to the extent that this was analyzed. The β-mannosylated structures in *C. albicans* were shown to raise antibodies in infected patients and when administered in vaccine formulations [75–78]. It should be noted that this does not necessarily mean that a low level of similar modifications on *P. pastoris*–produced proteins would be problematically immunogenic: parenterally administered proteins are highly purified and do not contain adjuvantic molecules, in contrast to the case in the context of the entire yeast cell or in vaccines. In fact, tolerance can be induced through nonadjuvanted protein administration. Nevertheless, this potential immunogenicity has inspired research and nine and four β-mannosyltransferase genes were identified in *C. albicans* and *P. pastoris*, respectively [79]. Furthermore, independent deletion mutants of the individual β-mannosyltransferases (Bmtp) were made in *C. albicans* and *P. pastoris*. This revealed that PpBmt1p adds β-mannoses on the outer chain of *N*-glycans, while PpBmt2p primarily generates β-mannoses on the core glycans (i.e., $Man_5GlcNAc_2$) but had no activity on the outer chain. Bmt3p had no detectable activity on either core or outer-chain glycans, and PpBmt4p was proposed as acting as a capping activity to β-mannose structures on both the core and the outer-chain glycans. Therefore, PpBmt2p was the only enzyme that was determined to be responsible for the initiation of β-mannosylation of the core glycan.

*PpBMT2* knockout resulted in a reduction of α-1,2-mannosidase-resistant glycoforms and yielded $Man_5GlcNAc_2$ *N*-glycans in *P. pastoris*. Although this resulted in a reduction of β-mannose residues, additional knockouts were still needed to abolish the latter modification completely. When going through double (*Ppbmt2Δbmt1Δ*, *Ppbmt2Δbmt3Δ* and *Ppbmt2Δbmt4Δ*), triple (*Ppbmt2Δbmt4Δbmt3Δ* and *Ppbmt2Δbmt4Δbmt1Δ*), and quadruple (*Ppbmt2Δbmt4Δbmt1Δbmt3Δ*) mutants, a progressively higher degree in removal of β-mannoses could be observed. Moreover, rhEPO produced in other triple mutant strains (*Ppbmt2Δbmt1Δbmt3Δ* and *Ppbmt2Δbmt3Δbmt1Δ*) and in the quadruple mutants showed no cross-reactivity with an anti-host-cell antigen antibody. This suggests that the reactive epitope from the *P. pastoris* glycans had been removed [80].

In recent work, we observed a different class of β-mannosyl-modified *N*-glycans also in a *P. pastoris* strain designed to produce $Man_5GlcNAc_2$. Here, we found that subsequent introduction of human GnT-I could outcompete the formation of these undesired structures, and we will report in the near future on these studies (Laukens et al., submitted for publication).

## 4 *O*-Glycosylation in Yeast

Similar to *N*-glycosylation, *O*-glycans are important for the pharmacokinetics and pharmacodynamics of proteins and may protect proteins against proteolytic degradation as well. Unlike *N*-glycosylation, where a specific consensus sequence is recognized for *N*-glycan attachment, prediction of the site for *O*-glycosylation is less well understood. Besides the need for a Ser or Thr residue for the attachment of an *O*-glycan, cis-peptide sequences might also be needed for efficient coupling of an *O*-glycan. Since most proteins contain numerous Ser/Thr residues, it remains unclear whether and which *O*-glycan a biopharmaceutical may contain. Moreover, the initiating glycosyltransferases for yeast-*O*-mannosyl glycans are of an entirely different enzyme family than the initiating protein-*O*-GalNAc transferases for human mucin-type *N*-glycans (Fig. 2). It should not be expected that sites of yeast *O*-glycosylation are necessarily the same as those for human-type *O*-glycans. Furthermore, yeast-type *O*-glycans are structurally different from human-type *O*-glycans, introducing concerns about the immunogenicity of *O*-glycosylated biopharmaceuticals produced by yeast. To address these problems, probably the most efficient manner is to change the sequence of the protein to avoid yeast-type *O*-glycans altogether. However, as the consensus sequence is rather unpredictable, this strategy depends on the ability to experimentally determine the sites of *O*-glycosylation, a less than trivial task. New developments in glycopeptide mass spectrometry have improved our capabilities in this area. This mutagenesis approach, of course, may again result in immunogenicity and altered protein properties, especially if multiple sites need to be adapted, but can often be manageable (it should be recalled that, as a rule of thumb, all protein therapeutics indeed generate an immune response to some extent). Humanization of *O*-glycan structures might serve as a partial solution (Sect. 4.2); however, this does not remove the uncertainty regarding the amount and position of *O*-glycan attachment. Another approach to address the *O*-glycan immunogenicity involves the reduction of the native glycosylation chains (Sect. 3), and one of the most often used biopharmaceuticals, insulin, is indeed produced today using such a yeast *O*-glycosylation suppression strategy.

In general, yeasts and fungi will exclusively perform *O*-mannosylation. In *S. cerevisiae*, transfer of a single mannose from dolichol-P-mannose to Ser or Thr residues in the ER is catalyzed by protein-*O*-mannosyltransferases (PMT) [81–85]. *S. cerevisiae* contains a highly redundant *PMT* gene family, consisting of three subgroups, *PMT1* (containing *PMT1*, *PMT5*, and *PMT7*), *PMT2* (containing *PMT2*, *PMT3*, and *PMT6*), and *PMT4* (as the sole member of this group), encoding proteins with different protein substrate specificities. After the addition of a single mannose residue, the *O*-glycan will be extended with additional α-mannose residues in the Golgi by α-1,2-mannosyltransferases, which need $Mn^{2+}$ and use GDP-Man as donor residue. ScKtr1p, ScKtr3p, and ScKre2p/Mnt1p are three known α-1,2-mannosyltransferases in this process [86]. The two former ones add the second α-1,2-mannose while the last one mainly attaches the third mannose residue to form mannobiose and mannotriose. Transfer of subsequent α-1,3-mannose residues is done by ScMnn1p, ScMnt2p, and ScMnt3p α-1,3-mannosyltransferases [87, 88]. In *P. pastoris*, α-1,2-mannoses of the *O*-glycan can be modified with terminal β-mannose residues or may be branched with mannose-6-phosphate residues [89, 90] (Fig. 5). Further modifications (e.g., β-mannosylation or galactosylation) often resemble those seen on *N*-glycans of the particular species and are indeed species-specific [91, 92].

## 4.1 *Reducing or Removing O-Glycans in Yeast*

### 4.1.1 Making Pmt Knockouts

A first approach to intervene in *O*-glycosylation in yeast is to prevent its initiation by eliminating the PMT activities. These key enzymes initiate *O*-mannosylation on

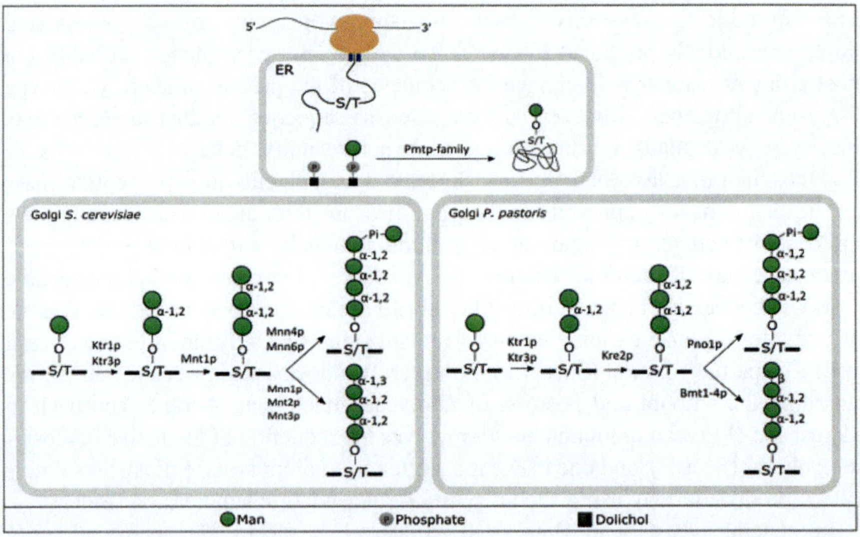


**Fig. 5** Wild-type *O*-glycosylation in *S. cerevisiae* and *P. pastoris*. In the ER, serine (S) and threonine (T) residues can be recognized by Pmt1/2p that add α-mannose residues. In the Golgi, Ktr1p/Ktr3p catalyzes the transfer of a second α-1,2-mannose, after which Mnt1p/Kre2p may add a third α-1,2-mannose. Additionally, Mnn4p/Mnn6p or Pno1p adds phosphomannose residues while competing with Mnn1p, Mnt2p, Mnt3p, or Bmt1–4 for the addition of α-1,3-mannose residues in *S. cerevisiae* or β-mannoses in *P. pastoris*. The differences in *O*-glycosylation structure may depend on the yeast strain (as shown in this figure) but also on its growth condition. Note that all enzymes are actually transmembrane proteins, not shown here as such for practical reasons

nascent secretory or membrane proteins in the ER lumen during or after protein translocation [93, 94]. The first PMT was isolated from *S. cerevisiae,* leading to cloning of the *PMT1* gene [95, 96] and the identification of six other homologous PMTs [97–99]. The seven different isoforms were eventually grouped into three subfamilies, referred to as *PMT1* (ScPmt1/5/7p), *PMT2* (ScPmt2/3/6p), and *PMT4* (ScPmt4p) [95, 98]. The *PMT1* and *PMT2* family members are highly redundant and show significant differences in three conserved sequence motifs (i.e., A, B, and C) when compared to the *PMT4* subfamily [100]. Moreover, *PMT1* and two members form heterodimers with one another and recognize different acceptor proteins compared to *PMT4*, which homodimerizes to be maximally active [98, 101, 102]. After these discoveries, individual knockouts of the *PMT1–4* genes were readily obtained in *S. cerevisiae,* but combination mutants such as *pmt1pmt2pmt4* and *pmt2pmt3pmt4,* were lethal [98].

More recently, the *PMT* genes in *P. pastoris* were identified and characterized by two independent studies [103, 104]. *P. pastoris* possesses homologs to *S. cerevisiae*'s *PMT1*, *2*, *4*, *5*, and *6* but none to *PMT3* or *PMT7*. These are grouped into subfamilies comparable to those in *S. cerevisiae*, forming similar dimers. A *PMT1* knockout resulted in 60% reduction of *O*-mannosylation on insulin produced in *P. pastoris* [103]. Moreover, an additional disruption in the *PMT2* gene resulted in serious impediment of the cell viability. Knocking out *PMT2* furthermore resulted in a reduction of *O*-mannosylation chain length in addition to the lower occupancy of the *O*-glycans. Finally, it has been shown that a combination of PMT knockouts with PMT inhibitors (Sect. 4.1.2) works synergistically and results in a stronger reduction in the degree of *O*-mannosylation (Fig. 6) [104].

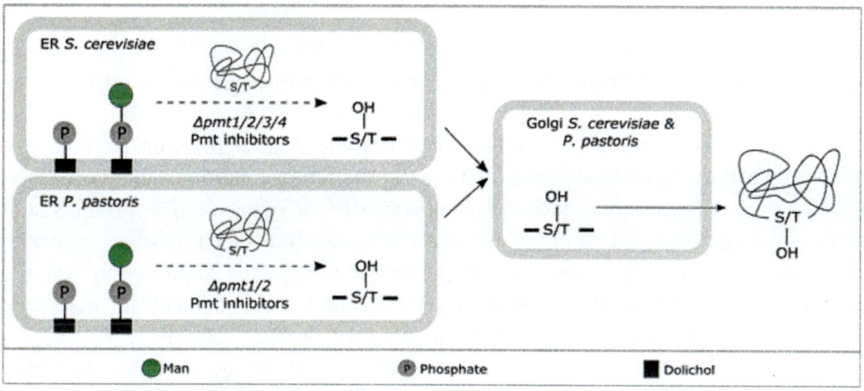


**Fig. 6** Removing or reducing *O*-glycans in yeast. In *S. cerevisiae*, knocking out the phosphomannosyltransferases 1–4 (Pmt1–4p) or supplying a PMT inhibitor (e.g., Rhodanine-3-acetic acid) in the culture medium may prevent the initial addition of α-mannose to serine (S) or threonine (T). Moreover, this avoids further modification of the glycoproteins in the Golgi and may give rise to a reduced *O*-glycan level. The same is true for the *Pmt1/2* knockout *P. pastoris* strain. It should be noted that using these knockout strains or Pmt inhibitors will not completely abolish *O*-glycans from (heterologous) produced proteins and that yeast-specific *O*-glycans may still be present

In *H. polymorpha*, five paralogous PMTs are involved in *O*-glycosylation (*Hp*Pmt1/2/4/5/6p). Knockout of *Hppmt1* resulted in a decrease in *O*-glycosylation of the chitinase protein and an increased temperature sensitivity [105, 106]. No remarkable effects were detected in single *Hppmt5Δ* and *Hppmt6Δ* mutants, while double mutations *Hppmt1pmt5Δ* and *Hppmt1pmt6Δ* affected cell wall integrity owing to reduced *O*-glycosylation of the surface glycoproteins *Hp*Wsc1p and *Hp*Mid2p, combined with sensitivity to cell wall stress inducers [106]. However, the most crucial enzymes for *O*-glycosylation of surface proteins seem to be Pmt2p and Pmt4p. No successful knockout for Pmt2p could be obtained so far, while disruption of Pmt4p in a Δ*Pmt1* background of *H. polymorpha* is synthetically lethal [106, 107].

#### 4.1.2 PMT Inhibitors

Another approach to reducing *O*-glycan occupancy and length involves the use of benzylidene thiazolidinediones such as rhodanine-3-acetic acid [108, 109]. These agents block Pmt1p activity in *C. albicans* [110] and broadly inhibit the general PMT activities in *S. cerevisiae* [111] and certain members of the PMT families in *P. pastoris* [104]. As these compounds broadly inhibit the formation of *O*-linked mannoproteins, this results in loss of cell wall integrity. Because PMT inhibitors lead to cell swelling and eventually death, concentrations of these compounds should be empirically determined and closely monitored during fermentations. Furthermore, PMT inhibitors should be used only during induction of protein expression, preventing lethality during growth to high cell densities (Fig. 6).

#### 4.1.3 Expression of Mannosidases to Limit *O*-Mannose Chain Length

Using the knockout strains or inhibitor concentrations that are more or less compatible with yeast cell growth and feasible protein production conditions, the previously mentioned strategies only reduce the occupancy of *O*-glycans on glycoproteins and generally do not completely eliminate them. In some cases (such as when only one *O*-glycan with low site occupancy is present), such partial suppression can be sufficient to push the modification below 1% of the protein molecules. Therefore, it becomes a trace contaminant like many other protein variants (e.g., due to oxidation, deamidation, mistranslation) that are almost always produced to some extent in any recombinant protein production technology.

Nevertheless, another, possibly complementary, approach to *O*-glycan reduction is enzymatic trimming of *O*-mannosyl groups attached to proteins during downstream processing. However, it should be noted that such enzymatic downstream processing of biopharmaceuticals is often costly (Fig. 7) [112].

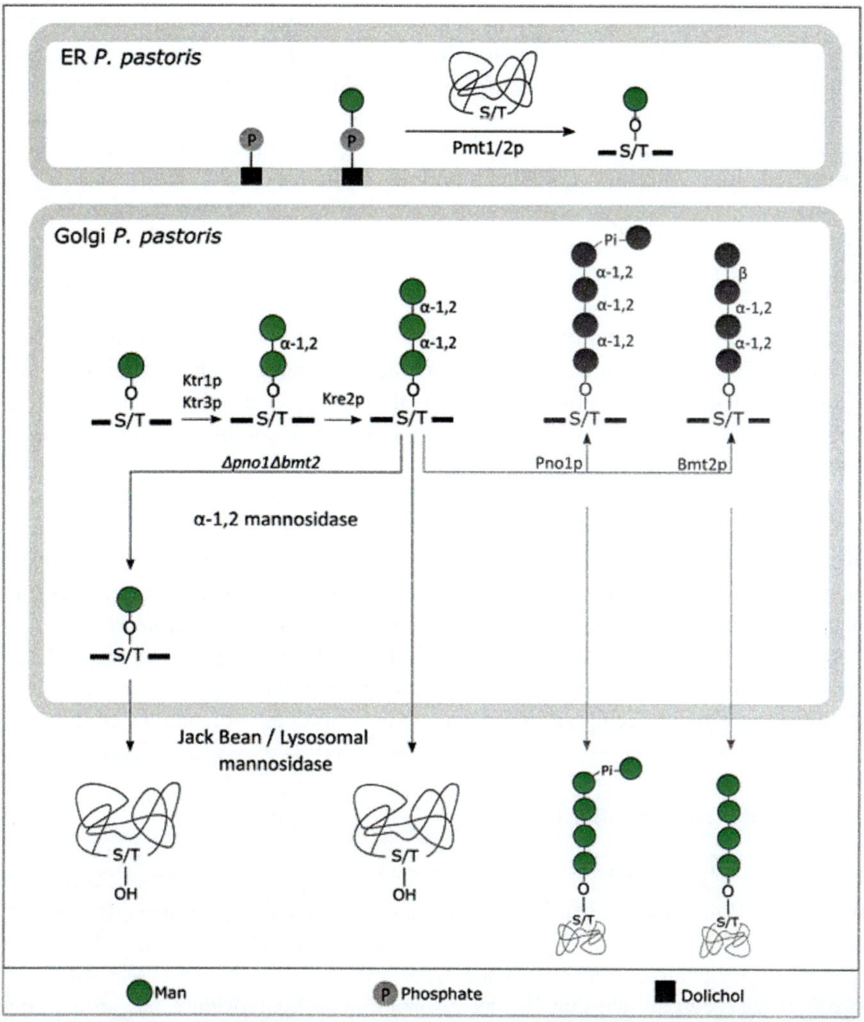


**Fig. 7** Strategies of *O*-glycosylation chain length reduction in *P. pastoris*. *Pichia* strains may have (β-mannosyltransferase 2) Bmt2p or (phosphomannosyltransferase) Pno1p disruptions and result in *O*-glycosylated proteins, sensitive to mannosidases such as α-1,2-mannosidase, that can be overexpressed in the Golgi. This results in single *O*-mannosylated glycoproteins. However, Jack Bean or lysosomal mannosidase may hydrolyze all remaining α-mannose residues in the Golgi. Moreover, these latter enzymes may be used for mannose in vitro hydrolysis as well. It should be noted that using this strategy for *O*-glycosylation reduction may still result in the production of proteins with yeast-specific *O*-glycans (i.e., Man-Pi-Man and β-Man containing structures)

### α-1,2-Mannosidase

Various glyco-engineering strategies have been developed to eliminate certain yeast-specific glycans of recombinant glycoproteins or to reengineer them to human-type

*N*-glycans (Sect. 2) [39, 40, 47, 52, 54, 113]. Additionally, also phospho- and β-mannose-depleted *P. pastoris* strains (Sect. 3) that have been shown to result in (single) α-1,2-mannose-remaining *O*-glycans could be obtained [46, 54, 114]. To reduce the amount of remaining α-1,2-mannose residues, overexpression of a recombinant α-1,2-mannosidase from *T. reesei* was introduced and eventually cosecreted along with the recombinant protein of interest (Fig. 7) [114].

### Lysosomal Mannosidases

In contrast to α-1,2-mannosidase from *T. reesei* and α-1,2/3-mannosidase from *X. manihotis*, α-1,2/3/6-mannosidase from Jack Bean was able to hydrolyze the Man-*O*-Ser/Thr linkage on top of the removal of the other mannose residues [115–118]. This can be done on wild-type *S. cerevisiae* and glyco-engineered *P. pastoris*-produced glycoproteins.

The efficacy of Jack Bean mannosidase can be limited due to steric inaccessibility of the single *O*-linked mannose, owing to the conformation of the protein or presence of nondigestible modifications (in particular β-mannose). Furthermore, Jack Bean mannosidase is currently only available as a crude plant extract, posing several issues including sourcing, reagent quality, and contaminating protease activity. Recently, the amino acid sequence of Jack Bean mannosidase was elucidated for 98% by peptide *de novo* sequencing [119], so recombinant versions of Jack Bean mannosidase are now being produced and are in the process of being tested (Fig. 7).

### Other Lysosomal Mannosidases

A screening was set up by Hopkins et al. for broad specific lysosomal hydrolases (i.e., α-1,2/3/6-mannosidase). These enzymes normally reside in the lysosome of eukaryotic cells where they degrade mannose-containing glycans. The lack of α-D-mannosidase in humans results in the lysosomal storage disease α-mannosidosis [120]. It has been shown that human lysosomal α-D-mannosidase has an activity toward free *N*-linked glycans similar to that of Jack Bean α-1,2/3/6-mannosidase [121]. The human lysosomal mannosidase could degrade remaining *O*-linked α-mannose structures after genetic elimination of β- and phosphomannoses. However, as was expected for similar sterical reasons as with Jack Bean mannosidase, human lysosomal mannosidase could not provide universal degradation of remaining mannoses (Fig. 7).

Altogether, these approaches can bring the problem of yeast *O*-glycosylation to within a manageable/acceptable range in simple cases where only one or a few *O*-glycosylation sites are present, such as is the case with insulin. This yeast-produced biopharmaceutical is produced in a *Δpmt1/2* knockout strain of *S. cerevisiae* [122], and in *P. pastoris*, work is ongoing to determine the optimal PMT knockouts for insulin production [103]. These interventions do not eliminate the *O*-glycan completely but bring it from approximately 5% to below 1% of the molecules, at which point this becomes a trace variant that is acceptable and clearly causes no demonstrated issues with safety of this chronically administered drug. This illustrates that the problem is manageable and should not be a cause for abandoning yeast as an expression host for moderately *O*-glycosylated proteins, especially if other arguments to use yeast are strong.

## 4.2 Humanization of O-Glycans in Yeast

### 4.2.1 Mucin Type

Amano et al. described the engineering of *S. cerevisiae* to obtain mucin-type glycoproteins modified with core 1 glycans (Galβ1,3GalNAc1-*O*-Ser/Thr) [123]. The inherent *O*-mannosylation pathway was inhibited by a rhodanine-3-acetic acid derivative (Sect. 4.1.2). Subsequent introduction of *Bacillus subtilis* UDP-galactose 4-epimerase enabled the synthesis of UDP-Gal and UDP-GalNAc, both of which are transported across the Golgi membrane by overexpression of a human UDP-Gal transporter. Next, human polypeptide:*N*-acetylgalactosaminyltransferase (ppGalNAcT) and *D. melanogaster* core 1 β-1,3-galactosyltransferase (DmGalT) were introduced, both fused to the Golgi-targeting *N*-terminal sequence of ScMnn9p for localization in the cis-Golgi (Fig. 8). Subsequent sialylation of this terminal Gal residue could be obtained in vitro; so far, no data on in vivo sialylation have been reported for *S. cerevisiae*. Hamilton et al., however, already described in vivo sialylation of *O*-glycans in

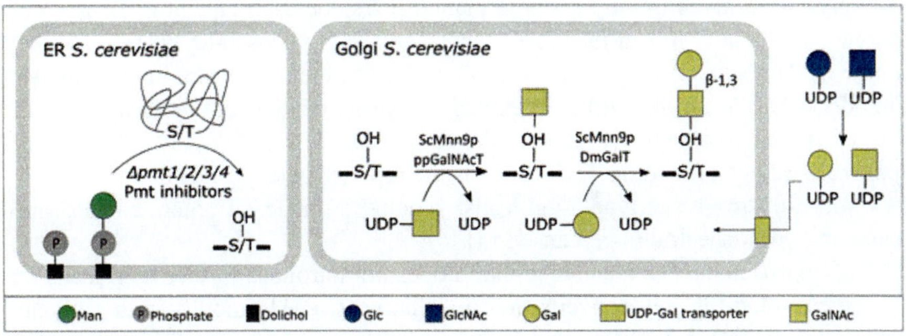


**Fig. 8** Mucin-type *O*-glycosylation engineering in *S. cerevisiae*. Strains producing reduced levels of *O*-glycans can be further engineered to produce glycoproteins with humanized mucin-type *O*-glycans. This was performed by the overexpression of *Bacillus subtilis* UDP-galactose 4-epimerase and a human UDP-Gal transporter. UDP-Gal was then used for the generation of Galβ1,3GalNAc1-*O*-Ser/Thr glycoproteins by overexpressing the human polypeptide:*N*-acetylgalactosaminyltransferase (ppGalNAcT) and *D. melanogaster* core 1 - β-1,3-galactosyltransferase (DmGalT). It should be noted that using this strategy for *O*-glycosylation humanization might still result in the production of proteins with yeast-specific *O*-glycans. Note that all enzymes are actually transmembrane proteins, not shown as such for practical reasons

*P. pastoris* using the strain engineered for in vivo sialylation of *N*-glycans (Fig. 9) (Sect. 4.2.2) [52, 114]. This suggests that in vivo sialylation might be feasible in *S. cerevisiae* as well (Fig. 8). Thus far, the reported experience with these strains remains very limited, and it is unclear whether they will be useful in biomanufacturing and for which biopharmaceutical products. On a critical note, one could justifiably argue that replacing one problematic pathway (i.e., yeast *O*-mannosylation) by two competing and low site-occupancy pathways (i.e., incomplete suppression of *O*-mannosylation pathways and incomplete mucin-type *O*-glycosylation) may worsen rather than solve the problem. In addition, the analytical challenge for such products would be more complex, as would be the possibility of purifying away the *O*-mannosylated fraction. Overall, we feel that there would have to be a very compelling pharmacological reason for producing a pharmaceutical in such strains rather than in mammalian cells. For example, production of differentially *O*-glycosylated cancer vaccine antigens could be such a reason, although mammalian cell-based alterations are available [124].

### 4.2.2 α-Dystroglycan-Type *O*-Glycans

While mucin-type *O*-glycans start with the addition of GalNAc to a Ser or Thr residue, α-dystroglycan-type *O*-glycans contain mannose as the first residue attached, catalyzed by protein *O*-mannosyltransferases related to the yeast's endogenous ones. As described earlier, yeast-endogenous extensions of yeast *O*-glycans can be removed by coexpression of Golgi-targeted α-1,2-mannosidase in a phospho- and β-mannose deficient *P. pastoris* strain (Sect. 3). This results in partial formation of *O*-glycans containing only a single mannose residue. Subsequent expression of murine protein-*O*-linked-mannose β-1,2-*N*-acetylglucosaminyltransferase I (PomGnT-I) results in the transfer of a GlcNAc residue to this single *O*-mannose. This disaccharide can be further extended by the sequential actions of β-1,4-GalT and α-2,6-SiaT to obtain sialylated *O*-linked glycans (Fig. 9) [17, 114]. So far, this approach has been tested on TNFR2:Fc as the reporter glycoprotein and resulted in 61% of α-dystroglycan-type *O*-glycans, among a range of intermediates and remaining yeast-specific *O*-glycans [114].

This approach has the advantage that it does not introduce further *O*-glycans on the target protein. It only converts the sites that yeast PMTs modify with a glycan that may be more compatible with parental use in humans (although that will need to be demonstrated, as the sites of yeast-initiated *O*-glycans are very unlikely to be modified with the rather rare α-dystroglycan-type *O*-glycan). In addition, the required genetic modification is exceedingly complex, and the approach's efficiency and robustness in scalability and across products remain unexplored.

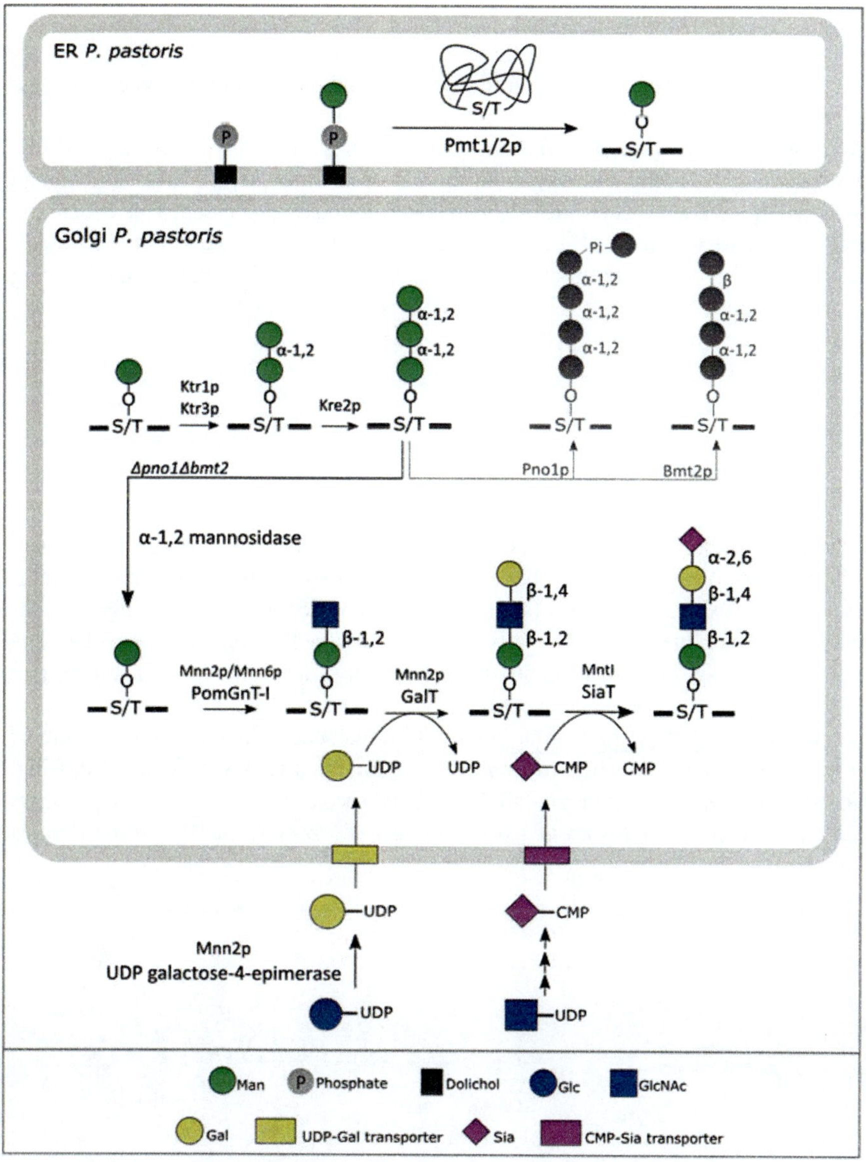


**Fig. 9** α-dystroglycan-type *O*-glycosylation engineering in *P. pastoris*. In the phospho- and β-mannose-deficient *Pichia* strain, Golgi-localized protein-*O*-linked-mannose β-1,2-*N*-acetylglucosaminyltransferase I (PomGnT-I), galactosyltransferase (GalT), and sialyltransferase (SiaT) can be overexpressed, leading to the expression of dystroglycan-type *O*-glycosylated proteins. It should be noted that using this strategy for *O*-glycosylation humanization might still result in the production of proteins with yeast-specific *O*-glycans. Note that all enzymes are transmembrane proteins, not shown here as such for practical reasons

### 4.2.3 *O*-Fucosylation

In a further specialized application, the *O*-fucosylation pathway has been successfully engineered in *S. cerevisiae*. Therefore, again in a context of drug-induced yeast PMT inhibition, *Arabidopsis thaliana* GDP-mannose-4,6-dehydratase (AtMUR1), *A. thaliana* GDP-4-keto-6-deoxy-mannose-3,5-epimerase (AtFXGER1), and human *O*-fucosyltransferase-1 (hO-FucT-1) were heterologously expressed. This eventually led to the *O*-fucosylation of the factor-VII EGF-domain, a protein that is naturally modified with this rare modification in human cells [125, 126]. Further engineering with a human β-1,3-*N*-acetylglucosaminyltransferase (Fringe) resulted in the elongation of the *O*-fucose with β-1,3-GlcNAc (Fig. 10) [127].

## 5 Filamentous Fungi

Filamentous fungi are, like yeasts, microorganisms that can grow to high cell density in fermenters. Yeasts and fungi have short development times to generate recombinant protein–expressing clones (about twice as long for filamentous fungi than for yeasts) and are inexpensive to grow when compared to mammalian cells. The filamentous fungi of protein production interest have a resorptive metabolism, that is, they digest macromolecular growth substrates extracellularly. In contrast to *S. cerevisiae* and *P. pastoris*, fungi therefore often secrete an enormous amount of proteins into their production medium, which complicates downstream processing. Commonly investigated fungi entail *Aspergillus* species (e.g., *A. awamori*, *A. niger*, and *A. oryzae*), *Trichoderma* species (e.g., *T. reesei*), and *Myceliophthora*

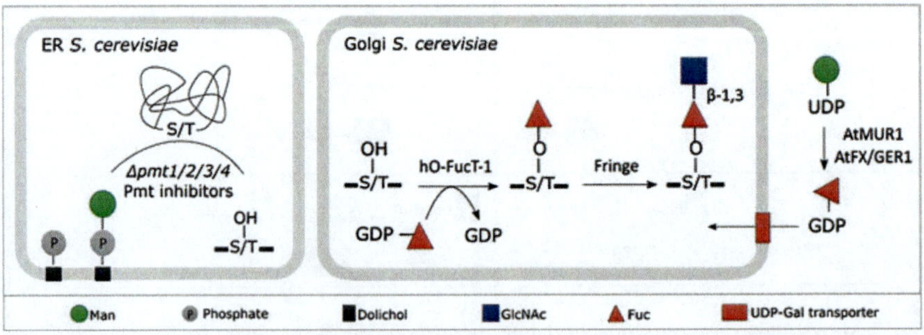


**Fig. 10** *O*-fucosylation engineering in *S. cerevisiae*. Strains expressing proteins with reduced levels of *O*-glycans can be engineered to produce GlcNAcβ-1,3Fuc-containing proteins. Therefore, *Arabidopsis thaliana* GDP-mannose-4,6-dehydratase (AtMUR1) and *A. thaliana* GDP-4-keto-6-deoxy-mannose-3,5-epimerase (AtFXGER1) are overexpressed combined with Golgi-localized human *O*-fucosyltransferase-1 (hO-FucT-1) and β-1,3-*N*-acetylglucosaminyltransferase (Fringe). It should be noted that using this strategy for *O*-glycosylation humanization may still result in the production of proteins with yeast-specific *O*-glycans. Note that all enzymes are actually transmembrane proteins, not shown as such for practical reasons

*thermophila*. These organisms are regarded as nontoxic and nonpathogenic and have received the Generally Regarded as Safe status by the US Food and Drug Administration. These filamentous fungi have long been attractive for their apparent, very effective secreted protein production machinery. However, this machinery has until recently mainly been used for the secretion of the fungus' own enzymes, which have evolved to be resistant to the fungus' secreted proteases, or to secrete proteins that naturally or after engineering were sufficiently protease-resistant. Yet, with the discovery of novel efficient genome engineering tools and the availability of genome sequences, multiple protease knockouts are now feasible, which may open up prospects of biopharmaceutical production in these organisms. Time will tell whether processes more efficient than with yeasts can be developed with the filamentous fungi. These strategies will make it possible to optimize strains further in terms of glycosylation in the future and to produce proteins containing specific types or humanized glycans, much informed by how this has been accomplished in the past for yeasts.

The *N*-glycosylation profile of fungi is similar to that of yeast, in the sense that they produce high-mannose *N*-glycans. However, contrary to yeast, these *N*-glycans are generally smaller. *Aspergilli* and *Trichoderma* species have high-mannose-type *N*-glycans that can be phosphorylated (e.g., *T. reesei*). Up to 11 mannose residues per *N*-glycan have been identified in *T. reesei*, while *A. niger* proteins may contain up to 24 mannose residues on one *N*-glycan. Unlike proteins from these species, *M. thermophila* does not appear to produce highly mannosylated glycoproteins. The most common *N*-glycan in *M. thermophila* is reportedly $Man_3GlcNAc_2$, which, to a lesser extent, may contain additional HexNAc and Hex residues (i.e., $Hex_{0\text{-}4}HexNAc_{0\text{-}6}Man_3GlcNAc_2$) [128]. However, further analysis is definitely warranted before fully concluding on this, as the methods used were quite suboptimal.

Filamentous fungi often contain two kinds of α-1,2-mannosidases, one of which is similar to the mammalian Golgi α-1,2-mannosidase that trims $Man_8GlcNAc_2$ to $Man_5GlcNAc_2$ [129]. However, it appears that this enzyme is secreted in the growth medium rather than retained in the Golgi and, thus, that the majority of $Man_8GlcNAc_2$ to $Man_5GlcNAc_2$ processing occurs post secretion. In addition, these mannosidases often generate only a small proportion of $Man_5GlcNAc_2$ in the secreted protein *N*-glycan profile. Another mannosidase, residing in the ER, catalyzes the hydrolysis of the $Man_9GlcNAc_2$ to the $Man_8GlcNAc_2$ glycan, as in virtually all eukaryotes [130]. Moreover, fungal *N*-glycans can also contain glucose, galactofuranose, and phosphomonoesters and -diesters. Less than 1% of cellobiohydrolase I, produced in *T. reesei* and modified in vitro with human GnT-I, human β-1,4-GalT, and rat α-2,6-SiaT, contained hybrid-type sugars [131]. In *A. nidulans* and *A. niger*, overexpression of an α-1,2-mannosidase from *C. elegans*, fused to the Sec12p *P. pastoris* leader sequence, produced a large amount of $Man_5GlcNAc_2$ *N*-glycans. However, considerable amounts of $Man_{7\text{-}8}GlcNAc_2$ glycans were still present as well. In the strain that additionally expresses GnT-I, fused to the *mnnJ* leader sequence, fungal *N*-glycans were almost completely lost and $GlcNAcMan_5GlcNAc_2$ *N*-glycosylated proteins were obtained. In an alternative approach, knockout of the *ALGC* and *ALG3* genes led to the production of

$Man_{3-6}GlcNAc_2$ glycosylated proteins after which in vitro digestion with α-1,2-mannosidase could result in homogeneous $Man_3GlcNAc_2$ [132, 133]. From these studies it is apparent that the strategies developed in yeasts are indeed likely to be translatable to filamentous fungi.

# 6 Examples of Recombinant Proteins Expressed in Glyco-Engineered Yeast Strains: Glycosylation Considerations

## 6.1 Production of Monoclonal Antibodies

Monoclonal antibodies (mAbs) constitute a large portion of biopharmaceuticals on the market. Their production relies heavily on posttranslational modifications, for example, disulfide bonds are necessary for correct folding and glycosylation influences the effector function and mAb stability. Complete deglycosylation of an IgG1 mAb results in reduced thermal stability for the CH2 domain and more susceptibility for proteolytic cleavage and aggregation [134]. The type of *N*-glycan present on Asn297 in the Fc region of the heavy chain has an impact on complement-dependent cytotoxicity (CDC) and ADCC [135, 136]. Currently, licensed therapeutic recombinant mAbs are mainly produced in CHO, NS0, and Sp2/0 cells and are predominantly modified with core fucosylated biantennary *N*-glycans with variable levels of galactosylation [137]. Monoclonal Abs bearing nonfucosylated *N*-glycans, however, show enhanced ADCC by increased activity of natural killer cell–mediated killing. Moreover, sialylated mAbs may suppress inflammation and reduce ADCC, resulting in a more anti-inflammatory effector function [138]. Modification of mAbs with high-mannose *N*-glycans results in fast serum clearance in humans [4, 16].

Enormous efforts have been devoted to producing cell lines that lack fucosylation activity in mammalian cells. The GlycArt® technology (Roche, Basel, Switzerland), for example, involves overexpression of GnT-III to add a bisecting GlcNAc to *N*-glycans, which is known to inhibit further fucosylation of the *N*-glycan [139]. Another approach focuses on the knockout of the α-1,6-fucosyltransferase to avoid the addition of a fucose residue (POTELLIGENT® technology) [140]. In yet another approach, inhibition of GDP-fucose synthesis is carried out by knocking out GDP-mannose 4,6-dehydratase [141]. Another promising approach to controlling the *N*-glycosylation type of mAbs involves production in glyco-engineered yeast strains. Since yeasts do not modify their glycoproteins with fucosylated and sialylated structures, an increased ADCC activity can be obtained. By use of the glyco-engineering strategies described earlier, humanized biantennary mAbs can be obtained showing optimal effector functions [60]. Zhang et al. describe the expression of an antihuman epidermal growth factor receptor 2 (HER2) mAb in a glyco-engineered *P. pastoris* strain, an analog to trastuzumab (Herceptin®, Roche, Basel, Switzerland) produced by CHO cells [142]. This *P. pastoris*-produced anti-HER2 is reportedly mainly modified with complex-type *N*-glycans carrying terminal GlcNAc or Gal residues. It is completely devoid of fucose residues and shows a remainder of $Man_5GlcNAc_2$ and hybrid-type glycoforms. A comparative study shows in vitro and in vivo properties similar to those of as the CHO-produced trastuzumab. Moreover, *P. pastoris*-produced anti-HER2 shows increased ADCC activity, probably due to the lack of core fucose residues [142].

Next to an enhanced ADCC, *P. pastoris*-produced mAbs show a more homogeneous glycosylation profile compared to the large heterogeneity of complex-type *N*-glycans in CHO cells [143]. So far, only a very small number of studies are available, and it is therefore too early to conclude whether the thus far reported yeast glycan engineering technologies will allow for robust scaling for multiple antibody products while retaining the favorable glycosylation characteristics. With its ease of handling and low production cost, *P. pastoris* may become very important for the production of "bio-better" therapeutic antibodies compared to conventional production in mammalian cells.

Finally, also a human IgG antibody against tumor necrosis factor α was expressed by the filamentous fungus *Myceliophthora thermophila* at the g/L level (Sect. 5) [144]. Glycosylation engineering of this organism has not yet been reported but could likely follow paths similar to those previously reported for other yeasts and fungi.

## 6.2 Enzyme Replacement Therapies

Lysosomal storage diseases are orphan diseases that, in some cases, can be treated with enzyme replacement therapy (ERT). These enzymes are recombinantly produced, mainly in CHO cells, but also human fibroblast carcinoma cells and plant cells such as carrot cells [145]. For the treatment of Pompe disease, the ERT, acid glucosidase α (GAA), is mainly targeted to the muscle cells via the cation-independent mannose-6-phosphate receptor (CI-MPR). To increase the mannose-6-phosphate to substantial levels, the Mnn4p ortholog of *S. cerevisiae* (i.e., PNO1) was overexpressed in *P. pastoris*, which resulted in glycoproteins carrying *N*-glycans, of which 80% contained at least one mannose-6-phosphate. These findings were confirmed for the ortholog in *O. minuta* (i.e., MNN4) [146] and *Y. lipolytica* (i.e., MPO1) [147]. Moreover, owing to the discovery of a novel *Cellulosimicrobium cellulans* phosphomannosylhydrolase enzyme, it became possible to uncap mannose residues that shielded the phosphates of the *N*-glycan and prevented efficient binding to CI-MPR [113]. An additional mannosidase from this bacterium enables removal of further terminal α-mannose residues. Besides GAA, also α-galactosidase A and *N*-acetylgalactosamine-6-sulfate sulfatase have been produced in *P. pastoris* [148, 149].

## 6.3 Vaccines

Because vaccines are administered intramuscularly and intended to be immunogenic, the presence of potentially immunogenic yeast-glycan elements is less of a concern than with intravenously injected pharmaceutical products, as long, of course, as these glycans do not dominate the immune response and do not cause safety issues. For yeast-produced vaccines, this generally is not the case, and some of the most important vaccines are produced in yeasts, being vaccines against the human hepatitis B virus (HBV) and the human papillomavirus (HPV). So far, the FDA has approved two HBV vaccines produced in *S. cerevisiae*: Engerix-B® (GlaxoSmithKline Biologicals, Rixensart, Belgium) and Recombivax HB® (Merck and Co., Inc., Kenilworth, NJ, USA). Several vaccines for HBV are available in other markets, such as AgB® (Laboratoria Pablo Cassará, Buenos Aires, Argentina), Hepavax-Gene® (Green Cross Vaccine Corp., Seoul, Korea), Biovac-B® (Worckhardt, Bombay, India) and Gene Vac-B® (Serum Institute of India, Poona, India), which are produced in *H. polymorpha*, and Shanvac®-B (Shantha Biotechnics Ltd., Hyderabad, India), which is produced in *P. pastoris*. For HPV, the FDA has approved two *S. cerevisiae*-produced vaccines: Gardasil® and Gardasil 9® (Merck and Co Inc., Kenilworth, NJ, USA).

## 6.4 Other Therapeutic Proteins

Leukine® (sargramostim, Partner Therapeutics Inc., Boston, MA, USA) is a recombinant human granulocyte-macrophage colony-stimulating factor (GM-CSF) produced in *S. cerevisiae*, carrying its wild-type glycosylation. It was approved by the FDA in 1991 for therapeutic use in the prevention of neutropenia after chemotherapy in acute myelogenous leukemia, in myeloid reconstitution after bone marrow transplantation, and in bone marrow transplantation failure or engraftment delay. These are acute, often subcutaneously and intravenously given, single-dose treatments, where the presence of yeast-type glycans can be tolerated. Regranex® (becaplermin, OMJ Pharmaceuticals Inc., San German, Puerto Rico) is a human platelet-derived growth factor recombinantly produced in *S. cerevisiae*. This gel is approved for the topical treatment of skin ulcers (from diabetes), as it promotes recruitment of macrophages, endothelial cells, and fibroblasts to increase healing. Kalbitor® (DX-88 ecallantide, Dyax, Cambridge, MA, USA) is a recombinant kallikrein inhibitor protein produced in *P. pastoris*. The protein was approved by the FDA in 2009 for the treatment of acute attacks of hereditary angioedema and is used in the prevention of blood loss during surgery. Recombinant hEPO was recently produced by glyco-engineered *P. pastoris* [53]. The strain is able to decorate proteins with humanized biantennary *N*-glycans with terminal sialic acid residues (Sect. 2.1.2). This recombinant hEPO was shown to have increased in vitro efficacy but requires PEGylation to achieve a half-life similar to that of an existing hEPO, Aranesp®

(darbepoetin, Amgen, Thousand Oaks, CA, USA), from CHO cells (which carries multiple branched sialylated *N*-glycans). In addition to all of this, insulin, a single *O*-glycosylated protein, can also be produced in yeast. It can be produced in a glyco-engineered yeast *S. cerevisiae* strain (i.e., Δpmt1/2) (Novolin, Novo Nordisk, Bagsvaerd, Denmark) [122] to reduce the *O*-glycosylation level. Furthermore, insulin is also produced in *P. pastoris* (Insugen, Biocon, Bangalore, India) [150, 151], as well as in *H. polymorpha* [152, 153]. Additionally, ThromboGenics NV (Leuven, Belgium) developed Jetrea® (Ocriplasmin) in *P. pastoris*. This protease is used for the treatment of symptomatic vitreomacular adhesion, as it degrades protein components of the vitreous body and the vitreoretinal interface (e.g., laminin, collagen, and fibronectin). Another therapeutic produced in *P. pastoris* entails recombinant albumin (Medway, Mitsubishi Tanabe Pharma Corporation, Osaka, Japan) [154, 155]. Several other products are in late stage of clinical trials like Nanobody® ALX-0061 (vobarilizumab, Ablynx, Ghent, Belgium), which is a recombinant anti-IL6 receptor single-domain antibody fragment, and Nanobody® ALX-0171 (Ablynx, Ghent, Belgium), which is a recombinant anti-RSV single-domain antibody fragment, both produced in *P. pastoris.*

# 7 Future Prospects

## *7.1 Genome Engineering in Yeast*

Classical genome engineering methods to knockout a gene rely on the replacement of the gene by a selection marker cassette based on homologous recombination (HR). In contrast to *S. cerevisiae*, HR in many fungal organisms including *P. pastoris* is a very inefficient process [156]. This makes the introduction of targeted genome modifications very challenging and laborious. Introducing targeted single- or double-strand DNA breaks induces the cellular DNA repair machinery, increasing the efficiency of HR. Storici et al. describe the introduction of DNA double-strand breaks in *S. cerevisiae* by an inducible I-SceI nuclease targeted to a previously integrated I-SceI homing endonuclease site, obtaining recombination frequencies of up to 20% [157, 158]. Näätsaari et al. describe a *ku70* mutant strain in which the normal function of Ku70p, essential for nonhomologous end joining (NHEJ), is eliminated. A defect in NHEJ mainly yields transformants in which DNA strand breaks are repaired through HR [156].

A booming technology that is highly promising for targeted genome engineering is the CRISPR/Cas9 (Clustered Regularly Interspaced Short Palindromic Repeats/CRISPR-associated protein 9) system. The Cas9 protein (from *Streptococcus pyogenes*) is an endonuclease that needs to be targeted to the nucleus of the cell and is guided by a short guide RNA (gRNA) toward a complementary, specific site in the DNA to introduce a DNA double-strand break. This DNA double-strand break is repaired by the NHEJ repair mechanism of the cell, which often generates short insertions or deletions in the open reading frame (ORF) of the gene. By cotransforming a linear DNA molecule with homology to the flanking sequences of the cut site, integrations into the genome can be obtained by homology-directed repair, making it possible to completely replace an ORF by another sequence or to insert sequences into the genome. The latter strategy might be optimized by using a Cas9 nickase variant that introduces DNA single-strand breaks, reducing NHEJ repair. Successful genome engineering in yeast using CRISPR technology was first obtained in *S. cerevisiae* and *S. pombe* [159, 160]. Weninger et al. describe targeting of the *OCH1* gene in *P. pastoris* using CRISPR/Cas9 to be less efficient than targeting of other genes involved in the methanol pathway (*AOX1*, *TMR1*, and *MMP1*) [161]. The efficiency, however, is still approximately 50 times higher than previously obtained with conventional knockout cassettes, showing the high value of the method.

This technology may become very important for facilitating knockouts in yeast strains, deleting unwanted glycosyltransferases and inserting glycosyltransferases or glycosidases of interest to modify the *N*- and *O*-glycosylation pathways.

## 8 Conclusion

Over the past decade, much promising progress has been made in the glyco-engineering of various yeast strains, resulting in either a reduction or elimination of undesirable yeast-specific glycan structures. In addition, the introduction of humanlike *N*- and *O*-glycosylation pathways in these yeasts was successfully obtained, leading to the expression of glycoproteins modified with complex-type *N*-glycans or humanlike *O*-glycans. So far, it remains impossible to engineer a strain completely devoid of yeast-specific *N*- and *O*-glycans, and further work is required. One should, however, not forget the significant impact that these glycosylation alterations cause on strain viability. Finding a balance between different genetic modifications will be of key importance to maintain the high space–time yields that make fungal hosts attractive in the first place. Furthermore, much work is ahead of us in investigating the scaling of production processes with glyco-engineered strains. Such work has started in several labs. With the increasing demands on cost effectiveness of recombinant therapeutic protein production, we feel that the time and technology are now probably more right than ever to have a major impact in biomanufacturing in the decade ahead.

**Acknowledgments** C. De Wachter holds a strategic basic research fellowship of the Flanders Institute for Innovation by Science and Technology (IWT). L. Van Landuyt holds a strategic basic research fellowship of the Flanders Fund for Scientific Research (FWO-Vlaanderen, Application 1S54817N). This work was funded in part by an ERC consolidator grant ("GlycoTarget", 616966), GOA Project 01G01412 (Ghent University), and BOF17-GOA-018, as well as by Grant G041417N of FWO-Vlaanderen.

## References


1. Mattanovich D, Branduardi P, Dato L, Gasser B, Sauer M, Porro D (2012) Recombinant protein production in yeasts. Methods Mol Biol 824:329–358
2. Dean N (1999) Asparagine-linked glycosylation in the yeast Golgi. Biochim Biophys Acta 1426:309–322
3. Han Y, Kanbe T, Cherniak R, Cutler JE (1997) Biochemical characterization of Candida albicans epitopes that can elicit protective and nonprotective antibodies. Infect Immun 65:4100–4107
4. Goetze AM, Liu YD, Zhang Z, Shah B, Lee E, Bondarenko PV, Flynn GC (2011) High-mannose glycans on the Fc region of therapeutic IgG antibodies increase serum clearance in humans. Glycobiology 21:949–959
5. Goffeau A, Barrell BG, Bussey H, Davis RW, Dujon B, Feldmann H, Galibert F, Hoheisel JD, Jacq C, Johnston M, Louis EJ, Mewes HW, Murakami Y, Philippsen P, Tettelin H, Oliver SG (1996) Life with 6000 genes. Science 274:546, 563–546, 567
6. Fickers P, Benetti P-H, Waché Y, Marty A, Mauersberger S, Smit MS, Nicaud J-M (2005) Hydrophobic substrate utilisation by the yeast Yarrowia lipolytica, and its potential applications. FEMS Yeast Res 5:527–543
7. De Schutter K, Lin Y-C, Tiels P, Van Hecke A, Glinka S, Weber-Lehmann J, Rouzé P, Van de Peer Y, Callewaert N (2009) Genome sequence of the recombinant protein production host Pichia pastoris. Nat Biotechnol 27:561–566
8. Mellitzer A, Ruth C, Gustafsson C, Welch M, Birner-Grünberger R, Weis R, Purkarthofer T, Glieder A (2014) Synergistic modular promoter and gene optimization to push cellulase secretion by Pichia pastoris beyond existing benchmarks. J Biotechnol 191:187–195
9. Ruth C, Zuellig T, Mellitzer A, Weis R, Looser V, Kovar K, Glieder A (2010) Variable production windows for porcine trypsinogen employing synthetic inducible promoter variants in Pichia pastoris. Syst Synth Biol 4:181–191
10. Crabtree HG (1929) Observations on the carbohydrate metabolism of tumours. Biochem J 23:536–545
11. Cereghino GPL, Cereghino JL, Ilgen C, Cregg JM (2002) Production of recombinant proteins in fermenter cultures of the yeast Pichia pastoris. Curr Opin Biotechnol 13:329–332
12. Hirose M, Kameyama S, Ohi H (2002) Characterization of N-linked oligosaccharides attached to recombinant human antithrombin expressed in the yeast Pichia pastoris. Yeast 19:1191–1202
13. Vinogradov E, Petersen BO, Duus JO (2000) Isolation and characterization of non-labeled and 13C-labeled mannans from Pichia pastoris yeast. Carbohydr Res 325:216–221
14. Madeira-Lopes A, Cabeça-Silva C (1984) The dependence on temperature of thermal death, growth and yield of Candida tropicalis. Z Allg Mikrobiol 24:133–135
15. Ishchuk OP, Voronovsky AY, Abbas CA, Sibirny AA (2009) Construction of Hansenula polymorpha strains with improved thermotolerance. Biotechnol Bioeng 104:911–919
16. Alessandri L, Ouellette D, Acquah A, Rieser M, Leblond D, Saltarelli M, Radziejewski C, Fujimori T, Correia I (2012) Increased serum clearance of oligomannose species present on a human IgG1 molecule. MAbs 4:509–520
17. Liu L, Gomathinayagam S, Hamuro L, Prueksaritanont T, Wang W, Stadheim TA, Hamilton SR (2013) The impact of glycosylation on the pharmacokinetics of a TNFR2:Fc fusion protein expressed in glycoengineered Pichia Pastoris. Pharm Res 30:803–812
18. Kompella U, Lee V (1991) Pharmacokinetics of peptide and protein drugs. In: Lee VH (ed) Peptide and protein drug delivery. Marcel Dekker, New York, pp 391–484
19. Sinclair AM, Elliott S (2005) Glycoengineering: the effect of glycosylation on the properties of therapeutic proteins. J Pharm Sci 94:1626–1635
20. Carter CRD, Whitmore KM, Thorpe R (2004) The significance of carbohydrates on G-CSF: differential sensitivity of G-CSFs to human neutrophil elastase degradation. J Leukoc Biol 75:515–522

21. Nissen C (1994) Glycosylation of recombinant human granulocyte colony stimulating factor: implications for stability and potency. Eur J Cancer 30A(Suppl 3):S12–S14
22. Solá RJ, Griebenow K (2009) Effects of glycosylation on the stability of protein pharmaceuticals. J Pharm Sci 98:1223–1245
23. Raju TS, Scallon B (2007) Fc glycans terminated with N-acetylglucosamine residues increase antibody resistance to papain. Biotechnol Prog 23:964–971
24. Hammond C, Helenius A (1994) Folding of VSV G protein: sequential interaction with BiP and calnexin. Science 266:456–458
25. Xie W, Kanehara K, Sayeed A, Ng DTW (2009) Intrinsic conformational determinants signal protein misfolding to the Hrd1/Htm1 endoplasmic reticulum-associated degradation system. Mol Biol Cell 20:3317–3329
26. Hang I, Lin C, Grant OC, Fleurkens S, Villiger TK, Soos M, Morbidelli M, Woods RJ, Gauss R, Aebi M (2015) Analysis of site-specific N-glycan remodeling in the endoplasmic reticulum and the Golgi. Glycobiology 25:1335–1349
27. Perlińska-Lenart U, Bańkowska R, Palamarczyk G, Kruszewska JS (2006) Overexpression of the Saccharomyces cerevisiae RER2 gene in Trichoderma reesei affects dolichol dependent enzymes and protein glycosylation. Fungal Genet Biol 43:422–429
28. Parsaie Nasab F, Aebi M, Bernhard G, Frey AD (2013) A combined system for engineering glycosylation efficiency and glycan structure in Saccharomyces cerevisiae. Appl Environ Microbiol 79:997–1007
29. Choi B-K, Warburton S, Lin H, Patel R, Boldogh I, Meehl M, Meehl M, d'Anjou M, Pon L, Stadheim TA, Sethuraman N (2012) Improvement of N-glycan site occupancy of therapeutic glycoproteins produced in Pichia pastoris. Appl Microbiol Biotechnol 95:671–682
30. Kasturi L, Chen H, Shakin-Eshleman SH (1997) Regulation of N-linked core glycosylation: use of a site-directed mutagenesis approach to identify Asn-Xaa-Ser/Thr sequons that are poor oligosaccharide acceptors. Biochem J 323(Pt 2):415–419
31. Nakayama K, Nagasu T, Shimma Y, Kuromitsu J, Jigami Y (1992) OCH1 encodes a novel membrane bound mannosyltransferase: outer chain elongation of asparagine-linked oligosaccharides. EMBO J 11:2511–2519
32. Nakanishi-Shindo Y, Nakayama K, Tanaka A, Toda Y, Jigami Y (1993) Structure of the N-linked oligosaccharides that show the complete loss of alpha-1,6-polymannose outer chain from och1, och1 mnn1, and och1 mnn1 alg3 mutants of Saccharomyces cerevisiae. J Biol Chem 268:26338–26345
33. Chiba Y, Suzuki M, Yoshida S, Yoshida A, Ikenaga H, Takeuchi M, Jigami Y, Ichishima E (1998) Production of human compatible high mannose-type (Man5GlcNAc2) sugar chains in Saccharomyces cerevisiae. J Biol Chem 273:26298–26304
34. Abe H, Takaoka Y, Chiba Y, Sato N, Ohgiya S, Itadani A, Hirashima M, Shimoda C, Jigami Y, Nakayama K (2009) Development of valuable yeast strains using a novel mutagenesis technique for the effective production of therapeutic glycoproteins. Glycobiology 19:428–436
35. Tang H, Wang S, Wang J, Song M, Xu M, Zhang M, Shen Y, Hou J, Bao X (2016) N-hypermannose glycosylation disruption enhances recombinant protein production by regulating secretory pathway and cell wall integrity in Saccharomyces cerevisiae. Sci Rep 6:25654
36. Bartkevičiūtė D, Sasnauskas K (2004) Disruption of the MNN10 gene enhances protein secretion in Kluyveromyces lactis and Saccharomyces cerevisiae. FEMS Yeast Res 4:833–840
37. Wang T-Y, Huang C-J, Chen H-L, Ho P-C, Ke H-M, Cho H-Y, Ruan S-K, Hung K-Y, Wang I-L, Cai Y-W, Sung H-M, Li W-H, Shih M-C (2013) Systematic screening of glycosylation- and trafficking-associated gene knockouts in Saccharomyces cerevisiaeidentifies mutants with improved heterologous exocellulase activity and host secretion. BMC Biotechnol 13:71
38. De Pourcq K, Vervecken W, Dewerte I, Valevska A, Van Hecke A, Callewaert N (2012) Engineering the yeast Yarrowia lipolytica for the production of therapeutic proteins homogeneously glycosylated with Man8GlcNAc2 and Man5GlcNAc2. Microb Cell Fact 11:53

39. Choi B-K, Bobrowicz P, Davidson RC, Hamilton SR, Kung DH, Li H, Miele RG, Nett JH, Wildt S, Gerngross TU (2003) Use of combinatorial genetic libraries to humanize N-linked glycosylation in the yeast Pichia pastoris. Proc Natl Acad Sci U S A 100:5022–5027
40. Vervecken W, Kaigorodov V, Callewaert N, Geysens S, De Vusser K, Contreras R (2004) In vivo synthesis of mammalian-like, hybrid-type N-glycans in Pichia pastoris. Appl Environ Microbiol 70:2639–2646
41. Gehlsen K, Chappell T (2014) Pichia pastoris strains for producing predominantly homogeneous glycan structure. WO/2014/066479
42. Krainer FW, Gmeiner C, Neutsch L, Windwarder M, Pletzenauer R, Herwig C, Altmann F, Glieder A, Spadiut O (2013) Knockout of an endogenous mannosyltransferase increases the homogeneity of glycoproteins produced in Pichia pastoris. Sci Rep 3:3279
43. Jiang B, Argyros R, Bukowski J, Nelson S, Sharkey N, Kim S, Copeland V, Davidson RC, Chen R, Zhuang J, Sethuraman N, Stadheim TA (2015) Inactivation of a GAL4-like transcription factor improves cell fitness and product yield in glycoengineered Pichia pastoris strains. Appl Environ Microbiol 81:260–271
44. Jiang B, Argyros R, Nelson S, Davidson R, Chen R, Zhuang J (2013) Engineered lower eukaryotic host strains for recombinant protein expression. WO/2013/062940
45. Weinhandl K, Ballach M, Winkler M, Ahmad M, Glieder A, Birner-Gruenberger R, Fotheringham I, Escalettes F, Camattari A (2016) Pichia pastoris mutants as host strains for efficient secretion of recombinant branched chain aminotransferase (BCAT). J Biotechnol 235:84–91
46. Callewaert N, Laroy W, Cadirgi H, Geysens S, Saelens X, Min Jou W, Contreras R (2001) Use of HDEL-tagged Trichoderma reesei mannosyl oligosaccharide 1,2-alpha-D-mannosidase for N-glycan engineering in Pichia pastoris. FEBS Lett 503:173–178
47. Hamilton SR, Bobrowicz P, Bobrowicz B, Davidson RC, Li H, Mitchell T, Nett JH, Rausch S, Stadheim TA, Wischnewski H, Wildt S, Gerngross TU (2003) Production of complex human glycoproteins in yeast. Science 301:1244–1246
48. Laukens B, De Wachter C, Callewaert N (2015) Engineering the Pichia pastoris N-glycosylation pathway using the GlycoSwitch technology. Methods Mol Biol 1321:103–122
49. Roy SK, Yoko-o T, Ikenaga H, Jigami Y (1998) Functional evidence for UDP-galactose transporter in Saccharomyces cerevisiae through the in vivo galactosylation and in vitro transport assay. J Biol Chem 273:2583–2590
50. Bobrowicz P, Davidson RC, Li H, Potgieter TI, Nett JH, Hamilton SR, Stadheim TA, Miele RG, Bobrowicz B, Mitchell T, Rausch S, Renfer E, Wildt S (2004) Engineering of an artificial glycosylation pathway blocked in core oligosaccharide assembly in the yeast Pichia pastoris: production of complex humanized glycoproteins with terminal galactose. Glycobiology 14:757–766
51. Lodish HF (1991) Recognition of complex oligosaccharides by the multi-subunit asialoglycoprotein receptor. Trends Biochem Sci 16:374–377
52. Hamilton SR, Davidson RC, Sethuraman N, Nett JH, Jiang Y, Rios S, Bobrowicz P, Stadheim TA, Li H, Choi B-K, Hopkins D, Wischnewski H, Roser J, Mitchell T, Strawbridge RR, Hoopes J, Wildt S, Gerngross TU (2006) Humanization of yeast to produce complex terminally sialylated glycoproteins. Science 313:1441–1443
53. Nett JH, Gomathinayagam S, Hamilton SR, Gong B, Davidson RC, Du M, Hopkins D, Mitchell T, Mallem MR, Nylen A, Shaikh SS, Sharkey N, Barnard GC, Copeland V, Liu L, Evers R, Li Y, Gray PM, Lingham RB, Visco D, Forrest G, DeMartino J, Linden T, Potgieter TI, Wildt S, Stadheim TA, d'Anjou M, Li H, Sethuraman N (2012) Optimization of erythropoietin production with controlled glycosylation-PEGylated erythropoietin produced in glycoengineered Pichia pastoris. J Biotechnol 157:198–206
54. Jacobs PP, Geysens S, Vervecken W, Contreras R, Callewaert N (2009) Engineering complex-type N-glycosylation in Pichia pastoris using GlycoSwitch technology. Nat Protoc 4:58–70

55. Bobrowicz P, Hamilton S, Gerngross T, Wildt S, Choi B-K, Nett J, Davidson R (2004) Production of modified glycoproteins having multiple antennary structures. WO/2004/074461
56. Breitling J, Aebi M (2013) N-linked protein glycosylation in the endoplasmic reticulum. Cold Spring Harb Perspect Biol 5:a013359
57. Sharma CB, Knauer R, Lehle L (2001) Biosynthesis of lipid-linked oligosaccharides in yeast: the ALG3 gene encodes the Dol-P-Man:Man5GlcNAc2-PP-Dol mannosyltransferase. Biol Chem 382:321–328
58. Aebi M, Gassenhuber J, Domdey H, te Heesen S (1996) Cloning and characterization of the ALG3 gene of Saccharomyces cerevisiae. Glycobiology 6:439–444
59. Davidson RC, Nett JH, Renfer E, Li H, Stadheim TA, Miller BJ, Miele RG, Hamilton SR, Choi B-K, Mitchell TI, Wildt S (2004) Functional analysis of the ALG3 gene encoding the Dol-P-Man: Man5GlcNAc2-PP-Dol mannosyltransferase enzyme of P. pastoris. Glycobiology 14:399–407
60. Li H, Sethuraman N, Stadheim TA, Zha D, Prinz B, Ballew N, Bobrowicz P, Choi B-K, Cook WJ, Cukan M, Houston-Cummings NR, Davidson R, Gong B, Hamilton SR, Hoopes JP, Jiang Y, Kim N, Mansfield R, Nett JH, Rios S, Strawbridge R, Wildt S, Gerngross TU (2006) Optimization of humanized IgGs in glycoengineered Pichia pastoris. Nat Biotechnol 24:210–215
61. Helenius J, Ng DTW, Marolda CL, Walter P, Valvano MA, Aebi M (2002) Translocation of lipid-linked oligosaccharides across the ER membrane requires Rft1 protein. Nature 415:447–450
62. Song H, Qian W, Wang H, Qiu B (2010) Identification and functional characterization of the HpALG11 and the HpRFT1 genes involved in N-linked glycosylation in the methylotrophic yeast Hansenula polymorpha. Glycobiology 20:1665–1674
63. De Pourcq K, Tiels P, Van Hecke A, Geysens S, Vervecken W, Callewaert N (2012) Engineering Yarrowia lipolytica to produce glycoproteins homogeneously modified with the universal Man3GlcNAc2 N-glycan core. PLoS One 7:e39976
64. Piirainen MA, Boer H, de Ruijter JC, Frey AD (2016) A dual approach for improving homogeneity of a human-type N-glycan structure in Saccharomyces cerevisiae. Glycoconj J 33:189–199
65. Qian W, Song H, Liu Y, Zhang C, Niu Z, Wang H, Qiu B (2009) Improved gene disruption method and Cre-loxP mutant system for multiple gene disruptions in Hansenula polymorpha. J Microbiol Methods 79:253–259
66. Wang H, Song H-L, Wang Q, Qiu B-S (2013) Expression of glycoproteins bearing complex human-like glycans with galactose terminal in Hansenula polymorpha. World J Microbiol Biotechnol 29:447–458
67. Cheng J, Huang S, Yu H, Li Y, Lau K, Chen X (2010) Trans-sialidase activity of Photobacterium damsela alpha2,6-sialyltransferase and its application in the synthesis of sialosides. Glycobiology 20:260–268
68. Moon H-Y, Cheon SA, Kim H, Agaphonov MO, Kwon O, Oh D-B, Kim J-Y, Kang HA (2015) Hansenula polymorpha Hac1p is critical to protein N-glycosylation activity modulation, as revealed by functional and transcriptomic analyses. Appl Environ Microbiol 81:6982–6993
69. Guerfal M, Ryckaert S, Jacobs PP, Ameloot P, Van Craenenbroeck K, Derycke R, Callewaert N (2010) The HAC1 gene from Pichia pastoris: characterization and effect of its overexpression on the production of secreted, surface displayed and membrane proteins. Microb Cell Fact 9:49
70. Valkonen M, Penttilä M, Saloheimo M (2003) Effects of inactivation and constitutive expression of the unfolded- protein response pathway on protein production in the yeast Saccharomyces cerevisiae. Appl Environ Microbiol 69:2065–2072
71. Wang F, Wang X, Yu X, Fu L, Liu Y, Ma L, Zhai C (2015) High-level expression of endo-β-N-acetylglucosaminidase H from Streptomyces plicatus in Pichia pastoris and its application for the deglycosylation of glycoproteins. PLoS One 10:e0120458
72. Stals I, Samyn B, Sergeant K, White T, Hoorelbeke K, Coorevits A, Devreese B, Claeyssens M, Piens K (2010) Identification of a gene coding for a deglycosylating enzyme in Hypocrea jecorina. FEMS Microbiol Lett 303:9–17

73. Shibata N, Ichikawa T, Tojo M, Takahashi M, Ito N, Okubo Y, Suzuki S (1985) Immunochemical study on the mannans of Candida albicans NIH A-207, NIH B-792, and J-1012 strains prepared by fractional precipitation with cetyltrimethylammonium bromide. Arch Biochem Biophys 243:338–348
74. Gomathinayagam S, Mitchell T, Zartler ER, Heiss C, Azadi P, Zha D, Houston-Cummings NR, Jiang Y, Li F, Giaccone E, Porambo RJ, Anderson CL, Sethuraman N, Li H, Stadheim TA (2011) Structural elucidation of an alpha 1,2 mannosidase resistant oligosaccharide produced in Pichia pastoris. Glycobiology 21:1606–1615
75. Ueno K, Okawara A, Yamagoe S, Naka T, Umeyama T, Utena-Abe Y, Tarumoto N, Niimi M, Ohno H, Doe M, Fujiwara N, Kinjo Y, Miyazaki Y (2013) The mannan of Candida albicans lacking β-1,2-linked oligomannosides increases the production of inflammatory cytokines by dendritic cells. Med Mycol 51:385–395
76. Miyakawa Y, Kuribayashi T, Kagaya K, Suzuki M, Nakase T, Fukazawa Y (1992) Role of specific determinants in mannan of Candida albicans serotype A in adherence to human buccal epithelial cells. Infect Immun 60:2493–2499
77. Masuoka J (2004) Surface glycans of Candida albicans and other pathogenic fungi: physiological roles, clinical uses, and experimental challenges. Clin Microbiol Rev 17:281–310
78. Singleton DR, Masuoka J, Hazen KC (2005) Surface hydrophobicity changes of two Candida albicans serotype B mnn4delta mutants. Eukaryot Cell 4:639–648
79. Mille C, Bobrowicz P, Trinel P-A, Li H, Maes E, Guerardel Y, Fradin C, Martínez-Esparza M, Davidson RC, Janbon G, Poulain D, Wildt S (2008) Identification of a new family of genes involved in β-1,2-mannosylation of glycans in Pichia pastoris and Candida albicans. J Biol Chem 283:9724–9736
80. Hopkins D, Gomathinayagam S, Rittenhour AM, Du M, Hoyt E, Karaveg K, Mitchell T, Nett JH, Sharkey NJ, Stadheim TA, Li H, Hamilton SR (2011) Elimination of β-mannose glycan structures in Pichia pastoris. Glycobiology 21:1616–1626
81. Behrens NH, Leloir LF (1970) Dolichol monophosphate glucose: an intermediate in glucose transfer in liver. Proc Natl Acad Sci U S A 66:153–159
82. Orlean P, Albright C, Robbins PW (1988) Cloning and sequencing of the yeast gene for dolichol phosphate mannose synthase, an essential protein. J Biol Chem 263:17499–17507
83. Sharma CB, Babczinski P, Lehle L, Tanner W (1974) The role of dolicholmonophosphate in glycoprotein biosynthesis in Saccharomyces cerevisiae. Eur J Biochem 46:35–41
84. Tanner W (1969) A lipid intermediate in mannan biosynthesis in yeast. Biochem Biophys Res Commun 35:144–150
85. Tanner W, Jung P, Behrens NH (1971) Dolicholmonophosphates: mannosyl acceptors in a particulate in vitro system of S. cerevisiae. FEBS Lett 16:245–248
86. Lussier M, Sdicu AM, Bussey H (1999) The KTR and MNN1 mannosyltransferase families of Saccharomyces cerevisiae. Biochim Biophys Acta 1426:323–334
87. Graham TR, Seeger M, Payne GS, MacKay VL, Emr SD (1994) Clathrin-dependent localization of alpha 1,3 mannosyltransferase to the Golgi complex of Saccharomyces cerevisiae. J Cell Biol 127:667–678
88. Romero PA, Lussier M, Veronneau S, Sdicu A-M, Herscovics A, Bussey H (1999) Mnt2p and Mnt3p of Saccharomyces cerevisiae are members of the Mnn1p family of α-1,3-mannosyltransferases responsible for adding the terminal mannose residues of O-linked oligosaccharides. Glycobiology 9:1045–1051
89. Jigami Y, Odani T (1999) Mannosylphosphate transfer to yeast mannan. Biochim Biophys Acta 1426:335–345
90. Trimble RB, Lubowski C, Hauer CR, Stack R, McNaughton L, Gemmill TR, Kumar SA (2004) Characterization of N- and O-linked glycosylation of recombinant human bile salt-stimulated lipase secreted by Pichia pastoris. Glycobiology 14:265–274
91. Goto M (2007) Protein O-glycosylation in fungi: diverse structures and multiple functions. Biosci Biotechnol Biochem 71:1415–1427

92. Lommel M, Strahl S (2009) Protein O-mannosylation: conserved from bacteria to humans. Glycobiology 19:816–828
93. Bause E, Lehle L (1979) Enzymatic N-glycosylation and O-glycosylation of synthetic peptide acceptors by dolichol-linked sugar derivatives in yeast. Eur J Biochem 101:531–540
94. Larriba G, Elorza MV, Villanueva JR, Sentandreu R (1976) Participation of dolichol phosphomannose in the glycosylation of yeast wall manno-proteins at the polysomal level. FEBS Lett 71:316–320
95. Strahl-Bolsinger S, Immervoll T, Deutzmann R, Tanner W (1993) PMT1, the gene for a key enzyme of protein O-glycosylation in Saccharomyces cerevisiae. Proc Natl Acad Sci U S A 90:8164–8168
96. Strahl-Bolsinger S, Tanner W (1991) Protein O-glycosylation in Saccharomyces cerevisiae. Purification and characterization of the dolichyl-phosphate-D-mannose-protein O-D-mannosyltransferase. Eur J Biochem 196:185–190
97. Gentzsch M, Strahl-Bolsinger S, Tanner W (1995) A new Dol-P-Man:protein O-D-mannosyltransferase activity from Saccharomyces cerevisiae. Glycobiology 5:77–82
98. Gentzsch M, Tanner W (1996) The PMT gene family: protein O-glycosylation in Saccharomyces cerevisiae is vital. EMBO J 15:5752–5759
99. Lussier M, Gentzsch M, Sdicu AM, Bussey H, Tanner W (1995) Protein O-glycosylation in yeast. The PMT2 gene specifies a second protein O-mannosyltransferase that functions in addition to the PMT1-encoded activity. J Biol Chem 270:2770–2775
100. Girrbach V, Zeller T, Priesmeier M, Strahl-Bolsinger S (2000) Structure-function analysis of the dolichyl phosphate-mannose: protein O-mannosyltransferase ScPmt1p. J Biol Chem 275:19288–19296
101. Gentzsch M, Tanner W (1997) Protein-O-glycosylation in yeast: protein-specific mannosyltransferases. Glycobiology 7:481–486
102. Girrbach V, Strahl S (2003) Members of the evolutionarily conserved PMT family of protein O-mannosyltransferases form distinct protein complexes among themselves. J Biol Chem 278:12554–12562
103. Govindappa N, Hanumanthappa M, Venkatarangaiah K, Kanojia K, Venkatesan K, Chatterjee A, Kusumanchi M, Dave N, Hazra P, Tiwari S, Sastry K (2013) PMT1 gene plays a major role in O-mannosylation of insulin precursor in Pichia pastoris. Protein Expr Purif 88:164–171
104. Nett JH, Cook WJ, Chen M-T, Davidson RC, Bobrowicz P, Kett W, Brevnova E, Potgieter TI, Mellon MT, Prinz B, Choi B-K, Zha D, Burnina I, Bukowski JT, Du M, Wildt S, Hamilton SR (2013) Characterization of the Pichia pastoris protein-O-mannosyltransferase gene family. PLoS One 8:e68325
105. Agaphonov MO, Romanova NV, Trushkina PM, Smirnov VN, Ter-Avanesyan MD (2002) Aggregation and retention of human urokinase type plasminogen activator in the yeast endoplasmic reticulum. BMC Mol Biol 3:15
106. Kim H, Moon HY, Lee D-J, Cheon SA, Yoo SJ, Park J-N, Agaphonov MO, Oh D-B, Kwon O, Kang HA (2013) Functional and molecular characterization of novel Hansenula polymorpha genes, HpPMT5 and HpPMT6, encoding protein O-mannosyltransferases. Fungal Genet Biol 58–59:10–24
107. Kim H, Thak EJ, Lee D-J, Agaphonov MO, Kang HA (2015) Hansenula polymorpha Pmt4p plays critical roles in O-mannosylation of surface membrane proteins and participates in heteromeric complex formation. PLoS One 10:e0129914
108. Bhatti RS, Shah S, Suresh KP, Sandhu JS (2013) Recent pharmacological developments on rhodanines and 2,4-thiazolidinediones. Int J Med Chem 2013:793260
109. Orchard MG, Neuss JC, Galley CMS, Carr A, Porter DW, Smith P, Scopes DIC, Haydon D, Vousden K, Stubberfield CR, Young K, Page M (2004) Rhodanine-3-acetic acid derivatives as inhibitors of fungal protein mannosyl transferase 1 (PMT1). Bioorg Med Chem Lett 14:3975–3978

110. Cantero PD, Lengsfeld C, Prill SK-H, Subanović M, Román E, Pla J, Ernst JF (2007) Transcriptional and physiological adaptation to defective protein-O-mannosylation in Candida albicans. Mol Microbiol 64:1115–1128
111. Arroyo J, Hutzler J, Bermejo C, Ragni E, García-Cantalejo J, Botías P, Piberger H, Schott A, Sanz AB, Strahl S (2011) Functional and genomic analyses of blocked protein O-mannosylation in baker's yeast. Mol Microbiol 79:1529–1546
112. Pastores GM (2010) Recombinant glucocerebrosidase (imiglucerase) as a therapy for Gaucher disease. BioDrugs 24:41–47
113. Tiels P, Baranova E, Piens K, De Visscher C, Pynaert G, Nerinckx W, Stout J, Fudalej F, Hulpiau P, Tännler S, Geysens S, Van Hecke A, Valevska A, Vervecken W, Remaut H, Callewaert N (2012) A bacterial glycosidase enables mannose-6-phosphate modification and improved cellular uptake of yeast-produced recombinant human lysosomal enzymes. Nat Biotechnol 30:1225–1231
114. Hamilton SR, Cook WJ, Gomathinayagam S, Burnina I, Bukowski J, Hopkins D, Schwartz S, Du M, Sharkey NJ, Bobrowicz P, Wildt S, Li H, Stadheim TA, Nett JH (2013) Production of sialylated O-linked glycans in Pichia pastoris. Glycobiology 23:1192–1203
115. Bergwerff AA, Stark W, Fendrich G, Knecht R, Blommers MJJ, Maerki W, Kragten EA, van Oostrum J (1998) Identification of Manα1-3Manα1-2Man and Man-linked phosphate on O-mannosylated recombinant leech-derived tryptase inhibitor produced by Saccharomyces cerevisiae and determination of the solution conformation of the mannosylated polypeptide. Eur J Biochem 253:560–575
116. Bretthauer RK (2007) Characterization of O-linked saccharides on glycoproteins. Methods Mol Biol 389:107–118
117. Gomathinayagam S, Hamilton SR (2014) In vitro enzymatic treatment to remove O-linked mannose from intact glycoproteins. Appl Microbiol Biotechnol 98:2545–2554
118. Ibatullin FM, Golubev AM, Firsov LM, Neustroev KN (1993) A model for cleavage of O-glycosidic bonds in glycoproteins. Glycoconj J 10:214–218
119. Kumar BSG, Pohlentz G, Schulte M, Mormann M, Kumar NS (2014) Jack bean α-mannosidase: amino acid sequencing and N-glycosylation analysis of a valuable glycomics tool. Glycobiology 24:252–261
120. Malm D, Nilssen Ø (2008) Alpha-mannosidosis. Orphanet J Rare Dis 3:21
121. Aronson NN, Kuranda MJ (1989) Lysosomal degradation of Asn-linked glycoproteins. FASEB J 3:2615–2622
122. Hubalek F, Pettersson A, Kjeldsen T, Andersen A (2014) Method for making mature insulin polypeptides. WO/2014/195452
123. Amano K, Chiba Y, Kasahara Y, Kato Y, Kaneko MK, Kuno A, Ito H, Kobayashi K, Hirabayashi J, Jigami Y, Narimatsu H (2008) Engineering of mucin-type human glycoproteins in yeast cells. PNAS 105:3232–3237
124. Steentoft C, Vakhrushev SY, Vester-Christensen MB, Schjoldager KT-BG, Kong Y, Bennett EP, Mandel U, Wandall H, Levery SB, Clausen H (2011) Mining the O-glycoproteome using zinc-finger nuclease-glycoengineered SimpleCell lines. Nat Methods 8:977–982
125. Chigira Y, Oka T, Okajima T, Jigami Y (2008) Engineering of a mammalian O-glycosylation pathway in the yeast Saccharomyces cerevisiae: production of O-fucosylated epidermal growth factor domains. Glycobiology 18:303–314
126. Kao YH, Lee GF, Wang Y, Starovasnik MA, Kelley RF, Spellman MW, Lerner L (1999) The effect of O-fucosylation on the first EGF-like domain from human blood coagulation factor VII. Biochemistry 38:7097–7110
127. Nakayama K, Maeda Y, Jigami Y (2003) Interaction of GDP-4-keto-6-deoxymannose-3,5-epimerase-4-reductase with GDP-mannose-4,6-dehydratase stabilizes the enzyme activity for formation of GDP-fucose from GDP-mannose. Glycobiology 13:673–680
128. Gusakov AV, Antonov AI, Ustinov BB (2008) N-glycosylation in Chrysosporium lucknowense enzymes. Carbohydr Res 343:48–55

129. Ichishima E, Taya N, Ikeguchi M, Chiba Y, Nakamura M, Kawabata C, Inoue T, Takahashi K, Minetoki T, Ozeki K, Kumagai C, Gomi K, Yoshida T, Nakajima T (1999) Molecular and enzymic properties of recombinant 1, 2-alpha-mannosidase from Aspergillus saitoi overexpressed in Aspergillus oryzae cells. Biochem J 339(Pt 3):589–597
130. Yoshida T, Kato Y, Asada Y, Nakajima T (2000) Filamentous fungus Aspergillus oryzae has two types of alpha-1,2-mannosidases, one of which is a microsomal enzyme that removes a single mannose residue from Man9GlcNAc2. Glycoconj J 17:745–748
131. Maras M, Saelens X, Laroy W, Piens K, Claeyssens M, Fiers W, Contreras R (1997) In vitro conversion of the carbohydrate moiety of fungal glycoproteins to mammalian-type oligosaccharides--evidence for N-acetylglucosaminyltransferase-I-accepting glycans from Trichoderma reesei. Eur J Biochem 249:701–707
132. Kainz E, Gallmetzer A, Hatzl C, Nett JH, Li H, Schinko T, Pachlinger R, Berger H, Reyes-Dominguez Y, Bernreiter A, Gerngross T, Wildt S, Strauss J (2008) N-glycan modification in Aspergillus species. Appl Environ Microbiol 74:1076–1086
133. Ward OP (2012) Production of recombinant proteins by filamentous fungi. Biotechnol Adv 30:1119–1139
134. Zheng K, Bantog C, Bayer R (2011) The impact of glycosylation on monoclonal antibody conformation and stability. MAbs 3:568–576
135. Shields RL, Lai J, Keck R, O'Connell LY, Hong K, Meng YG, Weikert SHA, Presta LG (2002) Lack of fucose on human IgG1 N-linked oligosaccharide improves binding to human Fcgamma RIII and antibody-dependent cellular toxicity. J Biol Chem 277:26733–26740
136. Shinkawa T, Nakamura K, Yamane N, Shoji-Hosaka E, Kanda Y, Sakurada M, Uchida K, Anazawa H, Satoh M, Yamasaki M, Hanai N, Shitara K (2003) The absence of fucose but not the presence of galactose or bisecting N-acetylglucosamine of human IgG1 complex-type oligosaccharides shows the critical role of enhancing antibody-dependent cellular cytotoxicity. J Biol Chem 278:3466–3473
137. Jefferis R (2009) Recombinant antibody therapeutics: the impact of glycosylation on mechanisms of action. Trends Pharmacol Sci 30:356–362
138. Beck A, Reichert JM (2012) Marketing approval of mogamulizumab: a triumph for glycoengineering. MAbs 4:419–425
139. Umaña P, Jean-Mairet J, Moudry R, Amstutz H, Bailey JE (1999) Engineered glycoforms of an antineuroblastoma IgG1 with optimized antibody-dependent cellular cytotoxic activity. Nat Biotechnol 17:176–180
140. Yamane-Ohnuki N, Kinoshita S, Inoue-Urakubo M, Kusunoki M, Iida S, Nakano R, Wakitani M, Niwa R, Sakurada M, Uchida K, Shitara K, Satoh M (2004) Establishment of FUT8 knockout Chinese hamster ovary cells: an ideal host cell line for producing completely defucosylated antibodies with enhanced antibody-dependent cellular cytotoxicity. Biotechnol Bioeng 87:614–622
141. Kanda Y, Imai-Nishiya H, Kuni-Kamochi R, Mori K, Inoue M, Kitajima-Miyama K, Okazaki A, Iida S, Shitara K, Satoh M (2007) Establishment of a GDP-mannose 4,6-dehydratase (GMD) knockout host cell line: a new strategy for generating completely non-fucosylated recombinant therapeutics. J Biotechnol 130:300–310
142. Zhang N, Liu L, Dumitru CD, Cummings NRH, Cukan M, Jiang Y, Li Y, Li F, Mitchell T, Mallem MR, Ou Y, Patel RN, Vo K, Wang H, Burnina I, Choi B-K, Huber H, Stadheim TA, Zha D (2011) Glycoengineered Pichia produced anti-HER2 is comparable to trastuzumab in preclinical study. MAbs 3:289–298
143. Potgieter TI, Cukan M, Drummond JE, Houston-Cummings NR, Jiang Y, Li F, Lynaugh H, Mallem M, McKelvey TW, Mitchell T, Nylen A, Rittenhour A, Stadheim TA, Zha D, d'Anjou M (2009) Production of monoclonal antibodies by glycoengineered Pichia pastoris. J Biotechnol 139:318–325
144. Visser H, Joosten V, Punt PJ, Gusakov AV, Olson PT, Joosten R, Bartels J, Visser J, Sinitsyn AP, Emalfarb MA, Verdoes JC, Wery J (2011) Development of a mature fungal technology

and production platform for industrial enzymes based on a Myceliophthora thermophila isolate, previously known as Chrysosporium lucknowense C1. Ind Biotechnol 7:214–223
145. Grabowski GA, Golembo M, Shaaltiel Y (2014) Taliglucerase alfa: an enzyme replacement therapy using plant cell expression technology. Mol Genet Metab 112:1–8
146. Akeboshi H, Kasahara Y, Tsuji D, Itoh K, Sakuraba H, Chiba Y, Jigami Y (2009) Production of human beta-hexosaminidase A with highly phosphorylated N-glycans by the overexpression of the Ogataea minuta MNN4 gene. Glycobiology 19:1002–1009
147. Park J-N, Song Y, Cheon SA, Kwon O, Oh D-B, Jigami Y, Kim J-Y, Kang HA (2011) Essential role of YlMPO1, a novel Yarrowia lipolytica homologue of Saccharomyces cerevisiae MNN4, in mannosylphosphorylation of N- and O-linked glycans. Appl Environ Microbiol 77:1187–1195
148. Chiba Y, Sakuraba H, Kotani M, Kase R, Kobayashi K, Takeuchi M, Ogasawara S, Maruyama Y, Nakajima T, Takaoka Y, Jigami Y (2002) Production in yeast of α-galactosidase A, a lysosomal enzyme applicable to enzyme replacement therapy for Fabry disease. Glycobiology 12:821–828
149. Rodríguez-López A, Alméciga-Díaz CJ, Sánchez J, Moreno J, Beltran L, Díaz D, Pardo A, Ramírez AM, Espejo-Mojica AJ, Pimentel L, Barrera LA (2016) Recombinant human N-acetylgalactosamine-6-sulfate sulfatase (GALNS) produced in the methylotrophic yeast Pichia pastoris. Sci Rep 6:29329
150. Gurramkonda C, Polez S, Skoko N, Adnan A, Gäbel T, Chugh D, Swaminathan S, Khanna N, Tisminetzky S, Rinas U (2010) Application of simple fed-batch technique to high-level secretory production of insulin precursor using Pichia pastoris with subsequent purification and conversion to human insulin. Microb Cell Fact 9:31
151. Mansur M, Cabello C, Hernández L, País J, Varas L, Valdés J, Terrero Y, Hidalgo A, Plana L, Besada V, García L, Lamazares E, Castellanos L, Martínez E (2005) Multiple gene copy number enhances insulin precursor secretion in the yeast Pichia pastoris. Biotechnol Lett 27:339–345
152. Baeshen MN, Bouback TAF, Alzubaidi MA, Bora RS, Alotaibi MAT, Alabbas OTO, Alshahrani SM, Aljohani AAM, Munshi RAA, Al-Hejin A, Ahmed MMM, Redwan EM, Ramadan HAI, Saini KS, Baeshen NA (2016) Expression and purification of C-peptide containing insulin using Pichia pastoris expression system. BioMed Res Int 2016:e3423685
153. Porro D, Sauer M, Branduardi P, Mattanovich D (2005) Recombinant protein production in yeasts. Mol Biotechnol 31:245–259
154. Kobayashi K, Kuwae S, Ohya T, Ohda T, Ohyama M, Ohi H, Tomomitsu K, Ohmura T (2000) High-level expression of recombinant human serum albumin from the methylotrophic yeast Pichia pastoris with minimal protease production and activation. J Biosci Bioeng 89:55–61
155. Mallem M, Warburton S, Li F, Shandil I, Nylen A, Kim S, Jiang Y, Meehl M, d'Anjou M, Stadheim TA, Choi B-K (2014) Maximizing recombinant human serum albumin production in a Mut(s) Pichia pastoris strain. Biotechnol Prog 30:1488–1496
156. Näätsaari L, Mistlberger B, Ruth C, Hajek T, Hartner FS, Glieder A (2012) Deletion of the Pichia pastoris KU70 homologue facilitates platform strain generation for gene expression and synthetic biology. PLoS One 7:e39720
157. Storici F, Snipe JR, Chan GK, Gordenin DA, Resnick MA (2006) Conservative repair of a chromosomal double-strand break by single-strand DNA through two steps of annealing. Mol Cell Biol 26:7645–7657
158. Storici F, Durham CL, Gordenin DA, Resnick MA (2003) Chromosomal site-specific double-strand breaks are efficiently targeted for repair by oligonucleotides in yeast. Proc Natl Acad Sci U S A 100:14994–14999
159. DiCarlo JE, Norville JE, Mali P, Rios X, Aach J, Church GM (2013) Genome engineering in Saccharomyces cerevisiae using CRISPR-Cas systems. Nucleic Acids Res 41:4336–4343
160. Jacobs JZ, Ciccaglione KM, Tournier V, Zaratiegui M (2014) Implementation of the CRISPR-Cas9 system in fission yeast. Nat Commun 5:5344
161. Weninger A, Hatzl A-M, Schmid C, Vogl T, Glieder A (2016) Combinatorial optimization of CRISPR/Cas9 expression enables precision genome engineering in the methylotrophic yeast Pichia pastoris. J Biotechnol 235:139–149

# Glyco-Engineering of Plant-Based Expression Systems

**Rainer Fischer, Tanja Holland, Markus Sack, Stefan Schillberg, Eva Stoger, Richard M. Twyman, and Johannes F. Buyel**

## Contents



---

R. Fischer (✉)
Indiana Biosciences Research Institute, Indianapolis, IN, USA

Institute for Molecular Biotechnology, RWTH Aachen University, Aachen, Germany
e-mail: rfischer@indianabiosciences.org

T. Holland
Institute for Molecular Biotechnology, RWTH Aachen University, Aachen, Germany

Eppendorf AG, Bioprocess Center, Juelich, Germany

M. Sack
Institute for Molecular Biotechnology, RWTH Aachen University, Aachen, Germany

Pro-SPR GmbH, Alsdorf, Germany

S. Schillberg
Fraunhofer Institute for Molecular Biology and Applied Ecology IME, Aachen, Germany

E. Stoger
Department of Applied Genetics and Cell Biology, University of Natural Resources and Life Sciences, Vienna, Austria

R. M. Twyman
TRM Ltd, York, UK

J. F. Buyel
Institute for Molecular Biotechnology, RWTH Aachen University, Aachen, Germany

Fraunhofer Institute for Molecular Biology and Applied Ecology IME, Aachen, Germany

**Abstract** Most secreted proteins in eukaryotes are glycosylated, and after a number of common biosynthesis steps the glycan structures mature in a species-dependent manner. Therefore, human therapeutic proteins produced in plants often carry plant-like rather than human-like glycans, which can affect protein stability, biological function, and immunogenicity. The glyco-engineering of plant-based expression systems began as a strategy to eliminate plant-like glycans and produce human proteins with authentic or at least compatible glycan structures. The precise replication of human glycans is challenging, owing to the absence of a pathway in plants for the synthesis of sialylated proteins and the necessary precursors, but this can now be achieved by the coordinated expression of multiple human enzymes. Although the research community has focused on the removal of plant glycans and their replacement with human counterparts, the presence of plant glycans on proteins can also provide benefits, such as boosting the immunogenicity of some vaccines, facilitating the interaction between therapeutic proteins and their receptors, and increasing the efficacy of antibody effector functions.

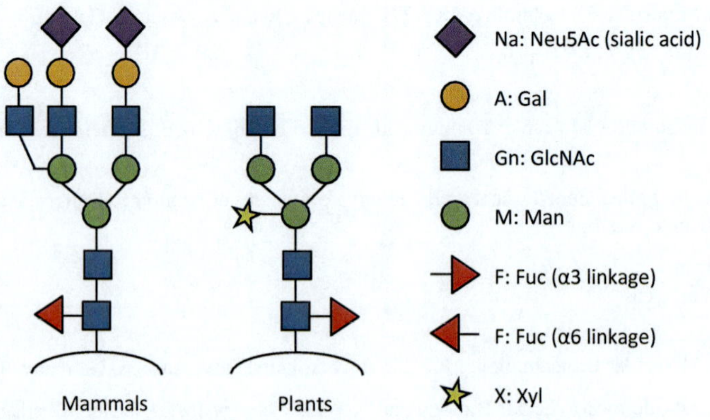


**Graphical Abstract** Typical structures of native mammalian and plant glycans with symbols indicating sugar residues identified by their short form and single-letter codes. Both glycans contain fucose, albeit with different linkages

**Keywords** Glycan, Glycoprotein, Glycotransferase, *N*-linked, *O*-linked, Pharmaceutical protein, Recombinant protein, Transgenic plant, Transient expression

# 1 Introduction

Most secreted proteins in eukaryotes are modified by glycosylation, a term that refers to the co-translational or post-translational covalent addition of oligosaccharide chains (glycans) to the polypeptide backbone [1]. The two main categories of glycosylation involve *N*-linked glycans added to the amide side chain of an asparagine residue in the consensus sequence NxS/T (where x is any amino acid except for proline) and *O*-linked glycans added to the hydroxyl groups of serine, threonine, hydroxylysine, and/or hydroxyproline residues [2, 3]. Common features of both glycosylation pathways include the sequential nature of glycan synthesis and maturation in the endoplasmic reticulum (ER)/Golgi apparatus, and the fact that all eukaryotes share common core glycans that are matured in a species-dependent and sometimes also a tissue-dependent manner, so that the resulting mature glycan structures differ when the same polypeptide is expressed in different hosts.

The host-specific differences in glycan structures are particularly important when human therapeutic proteins are expressed in heterologous cells, because the presence/absence and precise composition of the glycans can affect both the structure and function of the protein [4]. In structural terms, glycans increase the size of the protein, may alter its charge, and may have more specific effects such as influencing the way in which the polypeptide backbone folds. These factors can, in turn, affect the stability of the protein (e.g., by protecting it from proteases) and its ability to interact with other molecules, both of which contribute to its biological activity and its pharmacokinetic/pharmacodynamic properties. Finally, non-human glycans on human therapeutic proteins can trigger an immune response [4]. Like other heterologous expression platforms, plants do not naturally produce human-like glycans, so it is important to determine the extent to which plant glycans influence the structure and activity of human therapeutic proteins, and to develop strategies to alter or remove them if necessary. However, the presence of non-human glycans can be an advantage [5]. Indeed, some human therapeutic proteins with atypical glycans are found to be more efficacious than their counterparts produced in mammalian cells [6, 7].

# 2 Endogenous Glycosylation Pathways in Plants

## 2.1 N-Linked Glycosylation in Plants

All *N*-glycans in most eukaryotes are derived from the initial structure $Glc_3Man_9GlcNAc_2$, which is modified as the glycoprotein moves through the secretory pathway by the removal and addition of sugars, giving rise to a diverse array of oligosaccharides. Even so, almost all *N*-glycans share the common core structure

$Man_3GlcNAc_2$, formally defined as an *N*,*N*′-diacetyl chitobiose unit with a β-mannose residue attached to the chitobiose and two α-mannose residues linked to hydroxyl groups at positions 3 and 6 on the β-mannose residue [2]. When the protein moves through the different compartments of the Golgi body, the sugar residues added to the glycans are species-dependent. Core β(1,2)xylose and core α(1,3)fucose residues are added in plants, whereas core α(1,6)fucose and terminal sialic acid residues are added in animals. Some plant glycoproteins are also augmented with the so-called Lewis[a] epitope, which contains β(1,3) galactose and α(1,4)fucose, whereas mammalian glycoproteins often contain β(1,4)galactose combined with *N*-acetylneuraminic acid (GlcNAc) or *N*-glycolylneuraminic acid. The specific biological relevance of these different glycan structures in plants is largely unknown, although they may regulate protein turnover and interactions as they do in mammals [8].

The processing of *N*-glycans in plants leads to the formation of five major types of structure (Tables 1 and 2). These are the high-mannose type (also known as the oligo-mannose type), the short- and long-chain complex types, the hybrid type, and

**Table 1** The glycan structures attached to a model protein (phytase) in different plant species and tissues

| PHYTASE | | | | Localization | Glycosylation | References |
|---|---|---|---|---|---|---|
| Dicots | Tobacco | Seeds | Embryo | Apoplast / PSV | GnGnXF, MMXF | Arcalis et al. [9] |
| | | | Endosperm | PSV / apoplast | | |
| | | Leaves | | Apoplast | GnGnXF | |
| | Medicago | Seeds | | Apoplast | GnGnXF | Abranches et al. [10] |
| | | Leaves | | Apoplast / lytic vacuole | Lewisa, MMXF, MMX | |
| Monocots | Rice | Seeds | | PSV / prolamin bodies | GnGnXF, MMX | Drakakaki et al. [11] |
| | | Leaves | | Apoplast | GnGnXF, Lewisa | |
| | Wheat | Seeds | | PSV | GnGnXF | Arcalis et al. [12] |
| | Maize | Seeds | | PSV / zein bodies | GlcNAc, MMXF | Arcalis et al. [13] |

Reproduced from Arcalis et al. [9] with permission from Springer-Nature

PSV = protein storage vacuole, □ = GlcNAc (Gn), ○ = mannose (M), ▾ = fucose (F), ⋆ = xylose (X), ● = galactose

**Table 2** The glycan structures attached to a model protein (monoclonal antibody 2G12) in different plant species and tissues

| 2G12 | | | | SECRETED | | | RETAINED | | |
|---|---|---|---|---|---|---|---|---|---|
| | | | | Localization | Glycosylation | References | Localization | Glycosylation | References |
| Dicots | Tobacco | Seeds | Embryo | Apoplast/PSV | GnGnXF | Arcalis et al. [9] | PSV | GlcNAc | Floss et al. [17] Arcalis et al. [9] |
| | | | Endosperm | PSV/ apoplast | GnMXF | | PSV | OMT | |
| | | Leaves | | Apoplast | GnGnXF | | ER | OMT | |
| | Arabidopsis | Seeds | Embryo | Apoplast | GnGnXF | Loos et al. [14] Arcalis et al. [9] | PSV | OMT | Loos et al. [14] |
| | | | Endosperm | Apoplast | | | n.d. | | |
| | | Leaves | | n.d. | GnGnXF | Schähs et al. [15] | n.d. | n.d | n.d. |
| Monocots | Maize | Seeds | | PSV/ zein bodies (unpublished data) | GlcNAc MMXF | Ramessar et al. [16] | Zein bodies | GlcNAc OMT | Rademacher et al. [18] |

Reproduced from Arcalis et al. [9] with permission from Springer-Nature

PSV = protein storage vacuole. □ = GlcNAc, ○ = mannose, ▾ = fucose, ⋆ = xylose, ● = galactose, ER endoplasmic reticulum, n.d. not done, OMT = oligo-mannose type

the paucimannosidic type [19]. *N*-glycan biosynthesis in plants begins in the ER when the precursor oligosaccharide $Glc_3Man_9GlcNAc_2$ is transferred from a dolichol lipid carrier to the target asparagine residue as the nascent protein is translocated into the ER lumen. The three terminal glucose residues are trimmed from this precursor oligosaccharide in the ER by glycoside hydrolases (glucosidases I and II) and a single mannose residue is removed by ER mannosidase to generate the core structure $Man_8GlcNAc_2$. These steps are common to all eukaryotes and in subsequent processing steps result in the generation of high-mannose glycans, with five to nine mannose residues attached to the *N,N′*-diacetyl chitobiose unit [20, 21]. Beyond the *cis*-Golgi, the nascent glycoprotein encounters *N*-acetylglucosaminyltransferase I (GlcNAc transferase I), which is also highly conserved in plants and animals. But following the addition of GlcNAc to the α(1,3) arm, species-dependent modifications begin to occur (Fig. 1). Short-chain complex type *N*-glycans are formed when high-mannose type *N*-glycans are modified in the Golgi body, and are characterized by the presence of an α(1,3)-fucose residue attached to the proximal GlcNAc and/or a β(1,2)-xylose residue linked to the β-mannose residue of the core. These α(1,3)-fucose and β(1,2)-xylose residues are predominantly found on plant glycoproteins, but the former are also found in insects and the latter in some molluscs [23]. Between one and four α(1,2)mannose residues are removed by α-mannosidase I, converting $Man_{8\text{-}9}GlcNAc_2$ to $Man_5GlcNAc_2$ and then GlcNAc is added to the α(1,3)mannose branch of $Man_5GlcNAc_2$ by GlcNAc transferase I [24, 25]. Two further mannose residues are then trimmed by α-mannosidase II, and GlcNAc transferase II transfers the second GlcNAc to the α(1,6)mannose branch. These steps were characterized by the analysis of glycosylation mutants in *Arabidopsis thaliana* [26–28]. Long-chain complex type plant

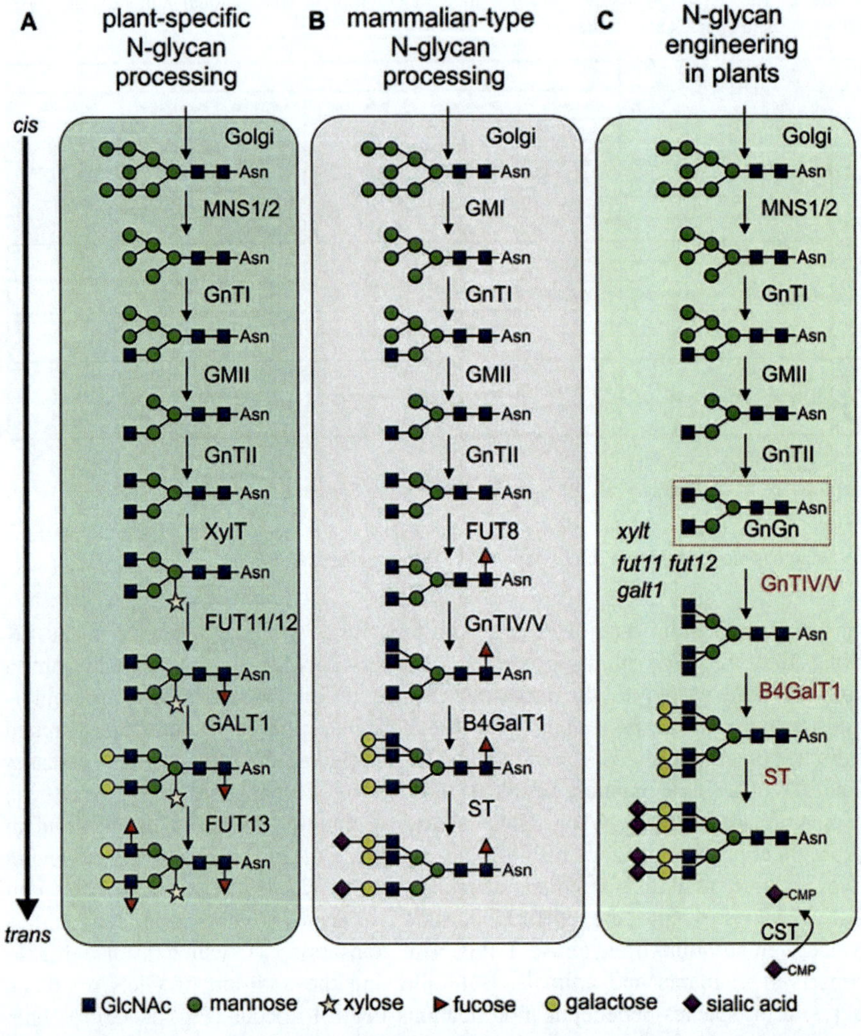


**Fig. 1** The synthesis of *N*-linked glycans. (**A**) Schematic overview of complex *N*-linked glycan processing in plants. Golgi-α-mannosidase I (MNS1/2), *N*-acetylglucosaminyltransferase I (GnTI), Golgi-α-mannosidase II (GMII), *N*-acetylglucosaminyltransferase II (GnTII), β(1,2) xylosyltransferase (XylT), core α(1,3)fucosyltransferase (FUT11/12), β(1,3)galactosyltransferase (GALT1), α(1,4)fucosyltransferase (FUT13). (**B**) Simplified schematic overview of *N*-linked glycan processing in mammalian cells. Golgi α-mannosidase I (GMI), core α(1,6)fucosyltransferase (FUT8), *N*-acetylglucosaminyltransferase IV (GnTIV) and V (GnTV), β(1,4)galactosyltransferase (B4GalT1), α(2,6)sialyltransferase (ST). (**C**) Optimized *N*-linked glycan engineering approach: the generation of *xylt*, *fut11*, *fut12*, and *galt1* knockouts results in the formation of the GnGn structure, which serves as an acceptor for GnTIV, GnTV, B4GalT1, and ST, resulting in fully processed complex *N*-linked glycans. Sialylation in plants requires the co-expression of the Golgi cytidine-5′-monophospho (CMP)-sialic acid transporter (CST) and proteins for CMP-sialic acid biosynthesis. Reproduced from Schoberer and Strasser [22] with permission from Elsevier

*N*-glycans, which contain bi-antennary groups and additional side chains of α(1,4)-fucose and β(1,3)-galactose linked to the GlcNAc units, possess terminal antennae containing the Galβ(1–3(Fucα(1–4))GlcNAc Lewis[a] structure. Long-chain complex *N*-glycans bearing the Lewis[a] antigen are generated by the addition of galactose and fucose to terminal GlcNAc residues by β(1,3)galactosyltransferase and α(1,4) fucosyltransferase, respectively. Hybrid type *N*-glycans are formed when only the α(1,3)-mannose branch of the intermediate $Man_5GlcNAc_2$ is processed, resulting in an oligosaccharide with α(1,3)fucose and/or β(1,2)xylose linked to the $GlcNAcMan_5GlcNAc_2$ core [29]. However, in plants expressing a recombinant galactosyltransferase, the same modification can occur if the enzyme gains access to the $GlcNAcMan_5GlcNAc_2$ structure too early and the terminal galactose prevents further processing by mannosidase II and GlcNAc transferase II [30]. The core structure may then be augmented by α(1,3)fucosylation and β(1,2)xylosylation as long as at least one terminal GlcNAc is present [31]. The α(1,3)fucosylation and β(1,2)xylosylation reactions occur independently, as shown by the analysis of *N*-glycans in plants where one or the other enzyme is mutated [32]. Paucimannosidic type *N*-glycans are modified oligosaccharides containing only α(1,3)fucose linked to the proximal GlcNAc and/or a β(1,2)xylose attached to the β-mannose residue of the intact $Man_3GlcNAc_2$ core or the truncated core structure $Man_2GlcNAc_2$. These are the typical *N*-linked glycan structures formed when glycoproteins are targeted to the plant vacuole [33] and then processed by the vacuolar β-*N*-acetylhexosaminidase HEXO1, or when they are secreted and trimmed by the plasma membrane β-*N*-acetylhexosaminidases HEXO2 and HEXO3 [34, 35].

*N*-glycan structures in *A. thaliana* have been investigated in detail through the analysis of individual endogenous proteins produced in wild-type plants and various glycosylation mutants, as well as the proteomic analysis of whole plants and specific tissues. In one of the most comprehensive studies thus far, total protein extracts from whole plants were digested with trypsin and the glycopeptides were enriched by converting the carbohydrates into aldehydes, which were then covalently coupled to hydrazide-derivatized beads. The peptides were subsequently released by treatment with peptide *N*-glycosidase (PNGase), analyzed by two-dimensional nano-liquid chromatography-mass spectrometry (nanoLC-MS), and the glycan acceptor sites were identified by the deamidation footprint left by PNGase [36]. This study helped to characterize the *A. thaliana* *N*-glycoproteome and provided information about glycosylation site occupancy on numerous secreted proteins carrying typical plant glycans, comprising a mixture of the five principal glycan structures described above.

## 2.2 *O*-Linked Glycosylation in Plants

*O*-linked glycosylation involves the addition of oligosaccharides to the hydroxyl oxygen on serine, threonine, hydroxylysine, and/or hydroxyproline side chains [4, 37]. The presence of *O*-linked glycans influences a range of structural and functional properties important for therapeutic proteins, including folding, solubility,

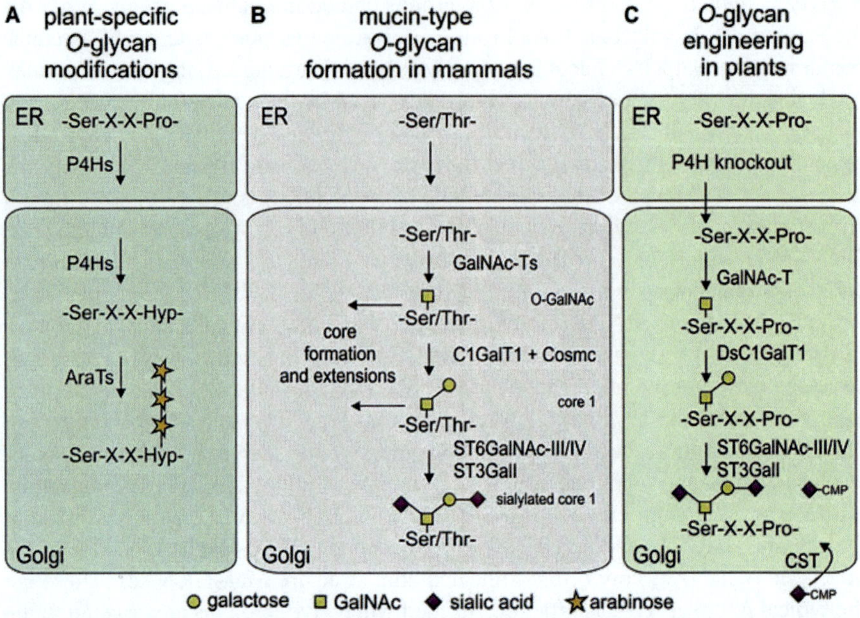


**Fig. 2** The synthesis of *O*-linked glycans. (**A**) Schematic overview of plant-type *O*-linked glycosylation. Proline residues next to *O*-linked glycosylation sites are converted to hydroxyproline (Hyp) by prolyl-4-hydroxylases (P4H). Hyp residues are further elongated (e.g., by arabinosyltransferases – AraTs). (**B**) Mucin-type *O*-linked glycan biosynthesis pathway in mammals. Polypeptide GalNAc-transferases (GalNAc-Ts), β(1,3)galactosyltransferases 1 (C1GalT1), Cosmc (chaperone), sialyltransferases (ST6GalNAcIII/IV, ST3GalI). (**C**) Mucin-type *O*-linked glycan engineering in plants. Strategies involve the knockout of P4H to prevent Hyp formation, and the expression of mammalian GalNAc-T, *Drosophila melanogaster* C1GalT1, and ST. ER Endoplasmic reticulum. Reproduced from Schoberer and Strasser [22] with permission from Elsevier

stability, biological function, and immunogenicity. There are significant differences in *O*-glycosylation between plants and animals, including the preferred acceptor site and the structure and composition of the mature glycans (Fig. 2). In animals, most *O*-linked glycoproteins are mucin-type glycoproteins in which GalNAc is added to serine and threonine residues in the Golgi body and then elaborated predominantly with GlcNAc, fucose, and galactose to create a heterogeneous population of complex *O*-glycans [38–40]. Mucin-type glycoproteins do not appear to be widely synthesized by plants, although they have been detected in rice seeds [41, 42]. *O*-linked glycosylation is common in plants and is used to regulate growth, wound healing, and plant-microbe interactions [43, 44]. However, the most abundant *O*-linked glycans are found on the hydroxyproline residues of hydroxyproline-rich glycoproteins, and this modification is unique to plants [45]. Although *O*-glycan synthesis in plants sometimes begins in the ER, it usually begins in the Golgi body with the addition of galactose or arabinose, followed by further elaboration [46, 47]. Contiguous sequences of hydroxyproline result in the addition of short unbranched arabino-oligosaccharides, as seen in the case of extensins [48]. Clustered non-contiguous hydroxyproline sequences may also be glycosylated, generally by the addition of branched arabinogalactan polysaccharides [48, 49].

# 3 *N*-Linked Glycosylation of Recombinant Proteins in Plants Without Glyco-Engineering

## *3.1 Species-Dependent and Tissue-Dependent Effects*

Many glycoproteins in animals are naturally produced as a mixture of glycoforms. The distribution of glycoforms in terms of site occupancy and glycan structure often varies between cell types, tissues, and individuals, and changes over time [50]. In the context of heterologous expression systems there are also differences in glycan structure in different taxonomic groups; for example, insect cells form shorter and less complex *N*-glycan structures than mammalian cells, typically oligomannose or paucimannose forms with core fucose structures but no terminal sialic acid residues, and the core structures include both human-like α(1,6)-linked fucose and α(1,3)-linked fucose [51]. Even cell lines from different mammals show minor variations in glycan structures, e.g., Chinese hamster ovary cells lack bisecting GlcNAc residues and add α(2,3)-linked sialic acid rather than the human-like α(2,6) linkage, and murine SP2/0 cells introduce a Galα(1,3)Gal structure that is not found in human cells [52, 53]. Therefore, it is not surprising that different taxonomic groups of plants also produce diverse glycans [45, 54]. Even so, although some green algae have thus far been shown to produce only high-mannose type glycans, other groups of plants, including liverworts, hornworts, mosses, ferns, gymnosperms, and angiosperms, appear to be able to form the five groups of *N*-glycans discussed earlier, including the Lewis[a] epitope, although the abundance of this trisaccharide varies in different tissues and at different developmental stages [4]. The Lewis[a] epitope is abundant in the seeds of dicotyledonous plants but is not found in the seeds of monocotyledonous plants, although it is produced in the vegetative organs of both [55, 56].

Although the overall glycosylation capacity of plants appears highly conserved at the species level, there are considerable differences at the level of individual tissues. The comprehensive analysis of tissue-specific effects on recombinant proteins is rarely undertaken, but antibodies are exceptional because many different antibodies have been expressed in diverse plant systems [57, 58] and some specific antibodies, such as the HIV-neutralizing monoclonal IgG 2G12, have been systematically tested in different host species and tissues [9]. Furthermore, several groups have also deliberately used model glycoproteins such as the enzyme phytase to investigate how different plants, and different tissues in the same plant, influence the glycan structures [9]. Overall the glycan structures on recombinant phytase (Table 1) and 2G12 (Table 2) are the same mix of high-mannose and complex type *N*-glycans, albeit with some variation in the quantities of different glycoforms. Although the single GlcNAc structure probably reflects the high endoglycanase activity in certain specialized tissues [18], tissue-specific differences in glycan profiles mostly represent differences in the underlying pattern of subcellular protein trafficking, in turn reflecting the spatial separation of the enzymes responsible for different glycosylation steps. Protein targeting is therefore one way in which the glycan structures of recombinant proteins produced in plants can be controlled, as discussed below.

# 4 The Impact of Protein Targeting on *N*-Linked Glycosylation

## *4.1 Secreted Proteins and the Role of the ER and Golgi Apparatus*

Whereas mammalian glycoproteins are usually synthesized as a mixture of diverse glycoforms, some plant systems tend to generate more homogeneous glycan profiles in which there is one dominant form and a mixture of less abundant minor species. The precise glycan profile appears to depend on the route through the secretory pathway (including the final destination) and the intrinsic properties of the protein. The latter is important because it may explain the apparently irreconcilable differences in glycan structures between different proteins that are targeted in the same manner.

Secreted proteins in plants are co-translationally imported into the ER, and unless directed otherwise they are eventually secreted by default to the apoplast, which is the space between the plasma membrane and the cell wall (other potential destinations include retention in the ER, deposition in storage organelles, or transport to the vacuole, as discussed in more detail below). Some proteins secreted to the apoplast are trapped under the cell wall owing to their size or physicochemical properties, whereas others can diffuse through the cell wall and can reach the environment. In the context of recombinant proteins, fully secreted proteins can be collected from the medium surrounding the plant (the culture medium for plant cell suspension cultures and aquatic plants, or the hydroponic medium or plant exudates such as nectar, mucilage, and leaf guttation fluid for terrestrial plants), whereas proteins trapped in the apoplast require assistance, such as enzymatic digestion of the cell wall or mechanical grinding of the plant tissues. The latter is also required for intracellular proteins. Notably, the apoplast surrounding each cell is not isolated but forms a supracellular compartment which allows the limited diffusion of proteins.

The different intrinsic properties of recombinant proteins can help to explain differences in their ultimate destination and their glycan profiles. For example, many proteins secreted to the apoplast (including most antibodies) move through the ER and Golgi body and are exposed to the full panel of glycosylation enzymes, resulting in a glycan profile dominated by complex type *N*-glycans [59, 60]. Recombinant proteins produced in plant cell suspension cultures and secreted to the culture medium have an even more homogeneous glycan profile, with $GlcNAc_2Man_3XylFucGlcNAc_2$ as the major form (~86% of all glycans) and only a small proportion of $GlcNAcMan_3XylFucGlcNAc_2$, because the secreted and intracellular proteins become separated [61]. In contrast, the secretion of recombinant follicle-stimulating hormone produced a glycan profile dominated by paucimannosidic structures (suggesting the activity of vacuolar or plasma membrane hexosaminidases) carrying core α1,3-fucose and/or β1,2-xylose residues that indicate transfer through the Golgi body [62]. The Golgi body is an important determinant of the species-dependent variations in glycan structures because it is the site for both complex *N*-linked glycan synthesis and mucin-type *O*-linked glycan synthesis, and the expression of human glycosyltransferases in plants has shown that the resulting glycan structures are highly dependent on the sub-Golgi targeting of such enzymes, as discussed in more detail below.

## 4.2 Proteins Retained in the ER

Secreted proteins can be retrieved from the Golgi to the ER by appending a short C-terminal peptide tag such as HDEL or KDEL, which prevents such proteins from traversing Golgi compartments containing the enzymes responsible for the synthesis of complex glycans [6, 63]. The resulting glycan profiles are therefore dominated by high-mannose glycans ($Man_{7–9}$) devoid of any plant-like complex glycan structures, but devoid of any human-like complex glycan structures too. The use of ER-retrieval tags for therapeutic proteins is a double-edged sword. The advantages are that the glycan profile is more homogeneous, and although few human proteins occur naturally with high-mannose glycans, the structures themselves are at least human-compatible (in that the same structures are naturally synthesized in human cells). Several comparative studies have also revealed that certain proteins, including most antibodies, accumulate to higher levels when they are retained in the ER rather than secreted to the apoplast, a finding that may reflect the combination of a favorable environment for folding/assembly and the absence of proteases [18, 64, 65]. On the other hand, both the C-terminal tag and the glycans themselves have the potential to form foreign epitopes in the context of heterologous proteins, and glycoproteins containing high-mannose glycans are removed from the bloodstream more rapidly than those with sialylated mammalian-type complex glycans because macrophages carry mannose receptors [66, 67].

In cereal seeds, ER-targeting has also been achieved using mRNA targeting signals (mainly those located within the 5′ and 3′ untranslated regions of the γ-zein mRNA) rather than a protein retrieval tag, to avoid changing the therapeutic protein structure [68]. The resulting mannosidic *N*-linked glycan profile provided a favorable starting point for further enzymatic processing to create the terminal residues required for efficient receptor-mediated uptake into human lysosomes [68].

## 4.3 Formation of ER-Derived Compartments

Proteins carrying C-terminal ER-retention tags will accumulate in the ER without affecting the ultrastructure of the plant cell if they are expressed at low to moderate levels. However, once the amount of protein reaches a certain threshold, the plant cell may respond by generating novel ER-derived compartments that appear to function as protective mechanisms by preventing the distortion of ER traffic [69–71]. ER-derived compartments are formed naturally in the seeds of monocotyledonous plants for the deposition of native storage proteins, which appear to induce the formation of such compartments owing to their abundance and propensity for aggregation [72, 73]. For some storage proteins, the sequences responsible for the induction of controlled polymerization are well defined, and one of the most widely known assembly sequences is located near the *N*-terminus of the 27-kDa maize storage protein γ-zein. This includes seven cysteine residues that promote inter-chain interactions and a highly repetitive amphipathic proline-rich sequence [74–76]. Proteins carrying this sequence form aggregates in the ER, which causes organelles known as protein bodies to bud off. If γ-zein is expressed in vegetative tissues it can induce the formation of ectopic protein bodies, indicating that protein body formation is an intrinsic capability of the secretory pathway, which is induced by the nature of the protein and is not dependent on the tissue [77, 78]. The aggregation-promoting sequence has been commercialized as a Zera® tag (Zip Solutions, Barcelona, Spain) and has been shown to induce the formation of storage compartments in the vegetative tissues of plants and also in fungi, insects, and mammalian cells [79]. The ability of certain KDEL-tagged proteins to induce ER-derived compartments suggests that the trigger is the accumulation of protein aggregates, which is facilitated by the γ-zein/Zera tag and may happen spontaneously for certain proteins carrying the KDEL tag, owing to the unique properties of the protein. The situation in seeds is complicated by the availability of multiple storage organelles whose prevalence varies in a species-dependent manner and for which several partially overlapping trafficking pathways may be available [70]. For example, the incorporation of ER-derived storage organelles into protein storage vacuoles is frequently observed in cereal seeds and has implications for the deposition of recombinant proteins [69]. The incomplete retention of KDEL-tagged recombinant proteins is also frequently observed in seeds, where it leads to unpredictable accumulation sites and glycan structures [12, 14, 80, 81]. Proteins expressed in cereal seed storage organelles often carry single GlcNAc residues, whereas proteins with single GlcNAc residues tend to be minor glycoforms in other tissues [13, 16, 18, 82].

## 4.4 Proteins Targeted to the Vacuole

Most plant cells contain lytic vacuoles whose function is to digest and recycle damaged or overabundant cellular macromolecules. Storage tissues may also contain storage vacuoles, which exist for the same purpose as the protein bodies described above, i.e., to stockpile proteins and other macromolecules and isolate them from the rest of the cell [83]. The protein bodies and protein storage vacuoles in seeds do not contain the same proteins, but it is not entirely clear how proteins are sorted into each compartment. Recombinant proteins containing KDEL tags are expected to accumulate in the ER, which promotes their incorporation into protein bodies, but when expressed in seeds devoid of separate ER-derived protein bodies such proteins have sometimes accumulated in the protein storage vacuoles instead [12, 13]. Other proteins expressed in seeds have also accumulated in the vacuole when their anticipated destination was the apoplast, indicating a prevailing tendency towards accumulation in protein storage vacuoles in some seed tissues [84–86]. Alternatively, specific vacuolar targeting sequences have been identified which can overrule any intrinsic properties [87–89]. Recombinant proteins targeted to vacuolar compartments are anticipated to carry paucimannosidic *N*-glycans like those found on native vacuolar proteins, but some nevertheless have been shown to carry high-mannose glycans, suggesting the route to the vacuole in some cases circumvents the Golgi body [70]. The first plant-derived therapeutic protein approved for parenteral use in humans (taliglucerase alfa, marketed as Elelyso (Prolalix BioTherapeutics, Karmiel, Israel), a recombinant form of human glucocerebrosidase indicated for Gaucher's disease) is produced in carrot cell suspension cultures and targeted to the vacuole. This ensures that the protein presents terminal mannose residues that are recognized by circulating macrophages, this recognition being necessary for the enzyme's therapeutic activity [90]. In contrast, the same enzyme produced in mammalian cells (imiglucerase, marketed as Cerezyme (Sanofi Genzyme, Cambridge, Massachusetts. USA)) carries terminal sialic acid residues. The sialic acid, galactose, and GlcNAc residues must therefore be cleaved off in vitro, adding to production costs. Elelyso is therefore one example in which the production of human therapeutic proteins with non-native glycans is an improvement rather than an impediment [7].

### *4.5 Proteins Targeted to Other Compartments*

Other plant cell compartments are largely devoid of *N*-glycans because the necessary glycosyltransferases are not present in the destination compartment or in any of the compartments *en route*. Proteins targeted to plastids or mitochondria are not *N*-glycosylated, and indeed the absence of glycans is one of the drawbacks of plastid transformation as a strategy for the production of recombinant pharmaceutical proteins, unless of course an aglycosylated protein is required [91]. Nuclear and cytoplasmic proteins do not contain *N*-glycans, but may be modified by nucleocytoplasmic *O*-GlcNAcylation, e.g., as shown for tobacco histones [92].

## 5 *O*-Linked Glycosylation of Recombinant Proteins in Plants Without Glyco-Engineering

*O*-linked glycosylation is necessary for the activity of many human proteins, including glycophorin A [93] and interleukin-5 [94]. Although *O*-linked glycosylation is carried out by plants, the modification typically involves hydroxyproline residues, whereas mammals produce predominantly mucin-type glycans added to serine and threonine residues. The very different structure and site occupancy/selectivity of plant *O*-linked glycosylation is therefore a major issue, one which has received comparatively little attention compared with that paid to *N*-linked glycans, although one of the advantages of the lack of mucin-type *O*-linked glycans in plants is that the relatively heterogeneous nature of *O*-linked glycosylation in mammals can be replaced with homogeneous *O*-linked glycans if plants are modified with specific enzymes from the pathway [22]. Few studies have specifically addressed the *O*-linked glycosylation of recombinant proteins produced in plants, but the addition of *O*-linked glycans to hydroxyproline residues has been reported on the proline-rich hinge region of a human IgA antibody expressed in maize seeds [95]. Interestingly, the same region is also *O*-glycosylated when the native protein is produced in humans, although the glycans in the native host are added to serine residues [96].

# 6 Glyco-Engineering in Plants

## *6.1 Glyco-Engineering Strategies*

The targeting of recombinant proteins to particular subcellular compartments can influence the glycan structures that are formed, but only within the repertoire of the natural capabilities of the plant cell at the resolution afforded by its compartmentalization. Therefore, it is not possible to precisely control the structure of complex glycans to the resolution of single enzyme functionalities, because several enzymes may be present in the same compartment. Similarly, it is not possible to change the glycan structures beyond those produced in the host cell. Two major engineering strategies have therefore been developed to create tailored glycans in plants, the first involving the specific removal or inhibition of particular enzymes and the second involving the introduction of additional enzymes to allow the synthesis of non-native glycans in planta. The implementation mechanism varies according to the most practicable and efficient process in each host species. Until recently, targeted mutagenesis in many plant species was laborious, and the removal of glycosyltransferase genes by gene targeting (gene knockout) was only possible in species amenable to homologous recombination (e.g., moss) or in those with readily available libraries of mutants, e.g., *A. thaliana* or *Lotus japonica* [97]. More recently, the advent of genome editing, using tools such as zinc finger nucleases, transcription activator-like effector nucleases (TALENs), and the clustered regularly interspaced short palindromic repeats (CRISPR/Cas9) system has made it much more straightforward to generate targeted mutations in plants. Until such methods became available, the easiest way to achieve the functional knockdown of glycosyltransferase genes was to use RNA interference (RNAi) or similar methods of post-transcriptional gene suppression. For the second strategy, the two main approaches are the generation of transgenic lines endowed with the ability to express non-native glycosyltransferases, or the use of transient expression to confer, temporarily, upon the host plant the ability to synthesize these enzymes along with the recombinant protein that is modified. In both cases, the appropriate intracellular localization of the enzyme is necessary to achieve the desired glycan profile [98].

## 6.2 Elimination of Plant N-Linked Glycans

The presence of core β(1,2)xylose, core α(1,3)fucose, and Lewis[a] residues on recombinant proteins produced in plants is considered undesirable because they are potentially immunogenic. Antibodies against these residues have been detected in sera from humans and other mammals [99] and can elicit IgG production when injected into humans [55]. Furthermore both β(1,2)xylose and α(1,3)fucose are IgE-binding determinants of plant allergens [100]. As discussed in the Introduction, the presence of non-human glycan epitopes is a well-known phenomenon even in the gold standard production platforms for human therapeutic proteins (rodent cell lines). Immune responses to such glycans are widely recognized [101] and have been reported even with approved drugs such as cetuximab [102]. Immune responses and allergic responses have not been reported for the parenteral administration of Elelyso, which retains its β(1,2)xylose and α(1,3)fucose residues [5, 103–106]. The topical application of plant-derived antibodies in human patients with IgE against plant *N*-linked glycans has not led to adverse effects [107–110].

Despite the absence of adverse effects, the elimination of plant glycans is still considered beneficial to pre-empt regulatory concerns. The most straightforward way to achieve the synthesis of plant glycoproteins without core β(1,2)xylose and α(1,3)fucose residues is to mutate the genes encoding the corresponding enzymes; namely, β(1,2)xylosyltransferase and α(1,3)fucosyltransferase. This was first achieved in *A. thaliana* without any noticeable effect on the phenotype [14, 15, 32, 81], and subsequently in the aquatic production hosts *Lemna minor*, a duckweed [111], and *Physcomitrella patens*, a moss [112]. In other plants that do not benefit from genome-wide mutant libraries or efficient homologous recombination pathways, including alfalfa [113], rice [114], and *Nicotiana benthamiana* [60], functionally equivalent production hosts were produced by RNAi. As anticipated, the glycoproteins produced in these hosts either completely lacked or contained only residual amounts of plant glycans, and in many cases featured biantennary structures with terminal GlcNAc residues, but without α(1,3)fucose and β(1,2)xylose as the dominant glycoforms. In *N. benthamiana*, two β(1,2)xylosyltransferases and two α(1,3)fucosyltransferases were knocked out using TALENs, without affecting growth or fertility. Endogenous proteins expressed in these ΔXF plants carried *N*-linked glycans that lacked β(1,2)xylose and had a significant reduction in core α(1,3)fucose levels (40% compared with wild-type plants). Similar *N*-linked glycans were carried by a recombinant rituximab antibody transiently expressed in the mutant plants. The remaining α(1,3)fucosyltransferase activity in the mutant line probably reflected the presence of redundant copies of the gene in the *N. benthamiana* genome [115]. As an alternative strategy, the endogenous *N. benthamiana* GlcNAc transferase I was downregulated by RNAi. Human glucocerebrosidase produced in this background contained 70–80% high-mannose *N*-glycans lacking β(1,2)xylose and α(1,3)fucose epitopes [116].

Antibodies produced in ΔXF hosts retain their antigen-binding activity and complement-dependent cytotoxicity, but show much more potent antibody-dependent cellular cytotoxicity than the same antibody produced using unmodified plants, again showing that plants can produce 'glyco-optimized' products, sometimes referred to as 'biobetters' to contrast with 'biosimilars' [82, 111, 117, 118]. The ability of plants to produce the Lewis$^a$ epitope has also been eliminated by mutating the genes encoding α(1,4)fucosyltransferase and β(1,3)galactosyltransferase, allowing the production of a recombinant human erythropoietin lacking this structure [119]. Furthermore, the proportion of complex *N*-linked glycans with terminal GlcNAc residues on a plant-derived human α1-antitrypsin was increased by using RNAi to suppress HEXO3 expression [120]. Overall, the full integration of the different steps needed to generate a plant with no plant-type fucose/xylose and no degradation of terminal GlcNAc has yet to be achieved [121]. An alternative and apparently effective strategy when large glycans are not required for the function of a protein is to knock out *N*-acetylglucosaminyltransferase I (GnTI) and overexpress an endoglycosidase [122].

## 6.3 Introduction of Human N-Linked Glycans

### 6.3.1 Core α(1,6)Fucose

The core α(1,6)fucose residue found in human glycoproteins influences the interaction between IgG and Fc receptors [123]. Antibodies devoid of α(1,6)fucose show increased antibody-dependent cellular cytotoxicity [124] and plants that lack the ability to produce core β(1,2)xylose and α(1,3)fucose residues are therefore particularly suitable as production hosts for cancer-targeting antibodies. Such plants are also suitable for the expression of heterologous enzymes that generate *N*-glycans carrying core α(1,6)fucose. For example, the expression of human α(1,6)fucosyltransferase in *N. benthamiana* ΔXF plants allowed the production of Ebola virus-specific antibodies with human-like core fucosylation [59]. The HIV-neutralizing antibody 2G12 was produced in a panel of glyco-engineered plants and displayed a spectrum of glycoforms. The ability to test different glycoforms individually revealed that the core fucose residue had no impact on antigen binding but influenced effector functions (particularly Fc binding), which, in turn, appeared to affect the antibody's neutralization potency [117], as was also recently shown for the same antibody produced in unmodified rice endosperm [82].

### 6.3.2 Multi-Antennary Complex Glycans and Bisecting GlcNAc Residues

Human glycoproteins often contain multi-antennary *N*-linked glycans and bisecting GlcNAc residues, and their abundance often correlates with increasing in vivo activity [125, 126]. However, the enzymes responsible for these modifications are not found in plants, which therefore can add only bi-antennary *N*-linked glycans to human recombinant proteins, unless they are engineered for optimized glycoprotein production [127]. Bisected *N*-linked glycans have been produced in plants by expressing mammalian β(1,4)GlcNAc transferase III, but it was necessary to target the enzyme to the *trans*-Golgi compartment to ensure that fully processed structures were synthesized [59, 128–130]. Tri-antennary glycans have been produced by expressing either β(1,4)GlcNAc transferase IV or β(1,6)GlcNAc transferase V, and by expressing both enzymes simultaneously it has been possible to produce tetra-antennary complexes [131]. The production of fully processed complex multi-antennary glycans required the enzymes to be targeted to the medial-Golgi compartment [59, 132].

### 6.3.3 Glycans Containing β(1,4)Galactose

The β(1,4)galactosylation of human proteins is a common modification that may play a role in the modulation of IgG activity [123] and is necessary as a precursor step for the addition of terminal sialic acid residues (see below). Plants are not equipped with the β(1,4)galactosyltransferase responsible for this modification and therefore cannot synthesize either β(1,4)galactosylated proteins or sialylated proteins. The production of β(1,4)galactosylated proteins in plants expressing human β(1,4)galactosyltransferase was initially only partly successful because the enzyme acted on the $GlcNAcMan_5GlcNAc_2$ glycans in the medial-Golgi compartment and the resulting β(1,4)galactosylated intermediates were not substrates for β(1,2) GlcNAc transferase II [30, 133–136]. However, as discussed above for multi-antennary glycans, targeting the enzyme to the *trans*-Golgi compartment allowed the formation of completely processed glycans, including an antibody with a dominant complex bigalactosylated glycoform [137].

### 6.3.4 Terminal Sialylation and the Lewis$^X$ Epitope

As stated above, the β(1,4)galactosylation of human proteins is a necessary step before terminal sialylation, which is an important functional requirement for many human therapeutic proteins, mostly owing to the presence of asialoglycoprotein receptors on liver cells. It is widely believed that plants do not synthesize sialylated oligosaccharides and lack the necessary donor and acceptor substrates [138, 139]. Nevertheless, cytidine-5′-monophospho (CMP)-sialic acid transporters and sialyltransferases have been detected in some plants [140, 141]. Protein sialylation in plants is challenging even when β(1,4)galactosylated structures are available, because plants lack the metabolic capacity to produce and transport the precursor CMP-*N*-acetylneuraminic acid as well as the sialyltransferase needed to transfer sialic acid from the precursor onto a terminal galactose residue. As discussed above, protein targeting was found to be essential to achieve coordinated enzyme activity [142–144]. The full sialylation of recombinant proteins required the expression and specific targeting of six mammalian proteins to carry out sialic acid synthesis, activation, translocation, and terminal transfer: murine uridine diphosphate (UDP)-*N*-acetylglucosamine-2-epimerase/*N*-acetylmannosamine kinase, human *N*-acetylneuraminic acid phosphate synthase, human CMP-*N*-acetylneuraminic acid synthase, murine CMP-sialic acid transporter, human β(1,4)galactosyltransferase, and the rat α(2,6)sialyltransferase [145]. When these six components were co-expressed with an antibody, more than 80% of the assembled antibody molecules were sialylated [145]. More recently, a combination of stably transformed plants and transient expression modules has been used to control the linkages between sialic acids and proteins, as well as the degree of polymerization, resulting in the directed formation of α(2,6) and α(2,3) linkages, and the synthesis of polysialic acid structures containing more than 40 units [146].

The β(1,4)galactosylation of human proteins is also a necessary step for the synthesis of another terminal structure known as the Lewis$^X$ epitope, which induces strong immune responses when added to a foreign antigen used as a vaccine and could therefore be useful for the generation of recombinant subunit vaccines [147]. These structures have been generated in tobacco by co-expressing the recombinant proteins with β(1,4)galactosyltransferase and α(1,3)fucosyltransferase IXa [148].

## 6.4 Engineering of O-Linked Glycans

Whereas *N*-glycans in animals and plants have similar core structures but different additional residues, the predominant *O*-glycans in animals and plants are completely distinct structures and it is much more challenging to generate modified plants that can synthesize human *O*-linked glycans. Instead, research has also focused on the exploitation of plant *O*-linked glycans to improve the performance of recombinant human proteins. For example, *O*-linked glycosylation may protect proteins from degradation in vivo in much the same way as polyethylene glycol (PEG)ylation, thus increasing the half-life of proteins without the need for in vitro chemical modification after purification [48, 149]. This has been investigated by expressing proteins with a glycotag comprising tandem repeats of a serine-hydroxyproline dipeptide [150]; the yields of several proteins, including green fluorescent protein, human interferon α2b, and human growth hormone, increased by more than 1000-fold when endowed with the tag [49, 151]. Hydroxylation was restricted to the tag, and each hydroxyproline residue in the tag was glycosylated with a variable number of arabinogalactans, but PEGylation also generates a heterogeneous population of molecules so this is not seen as a disadvantage. The longer glycotags increased the in vivo half-life of interferon α2b from 0.75 to 9.8 h and that of growth hormone from 0.41 to 2.5 h without significant loss of biological activity [48, 151]. However, studies including repeat administrations in realistic therapeutic settings remain to be carried out to confirm the utility of such approaches. The impact of internal plant-type *O*-linked glycans on protein stability has not been reported [95].

Other researchers have attempted to produce mammalian mucin-type *O*-glycans in plants by expressing the human enzyme *N*-acetylgalactosaminyltransferase II (GalNAc transferase II), which catalyzes the first step in the pathway [152]. However, transgenic *N. benthamiana* plants expressing this enzyme and infected with a MagnICON vector expressing a model substrate (the *Escherichia coli* heat-labile toxin B subunit fused to the human mucin 1 glycoprotein) produced only a minute amount of the correctly modified protein, indicating that the lack of glycosylation was due to the limited precursor supply. Accordingly, the proportion of mucin-type glycans was improved by increasing the availability of UDP-GalNAc and ensuring its efficient transfer to the Golgi body, which was achieved by co-expressing UDP-GlcNAc 4-epimerase and a UDP-GlcNAc/UDP-GalNAc transporter along with GalNAc transferase II [152].

Other efforts have been directed toward the elimination of potentially immunogenic carbohydrate epitopes containing arabinosides or arabinogalactans and toward the understanding of interactions between engineered *N*-linked and *O*-linked glycosylation pathways to avoid competition for common metabolites [153]. Tailored mucin-like structures have been produced [154], but owing to the absence of a sialic acid biosynthesis pathway in plants, it has been much more difficult to reproduce this frequent terminal modification of human mucin-type *O*-linked glycans. A fusion protein comprising erythropoietin joined to an antibody fragment (EPO-Fc) was successfully decorated with sialylated *O*-linked glycans in *N. benthamiana* by simultaneously expressing eight genes: the EPO-Fc protein, several components of the sialic acid biosynthesis pathway described above [145], two mammalian sialyltransferases, human GalNAc transferase II, and *Drosophila melanogaster* C1GALT1 [132].

A number of pharmaceutical proteins, including IgA-class antibodies, contain both *N*-linked and *O*-linked glycans and have been produced by the co-expression of enzymes required for the initiation and elongation of human *O*-linked glycans in a ΔXF background, resulting in proteins carrying disialylated mucin-type core 1 *O*-glycans [155]. This was further expanded to obtain galactose-deficient and α(2,6)sialylated *O*-glycans, mimicking the main glycans present on IgA1 molecules from patients with IgA nephropathy [156].

## 6.5 Exploiting Plant Glycosylation for Selective Product Purification

The purification of recombinant proteins from bulk plant extracts can be a challenging task [157] especially if no product-specific ligands are available for affinity chromatography, such as Protein A in the case of monoclonal antibodies [158]. Accordingly, clarification and purification can account for up to 80% of the total process costs in plant molecular farming [61, 159, 160]. However, it is possible to exploit the glycosylation of recombinant proteins to facilitate purification. In the past, immobilized glycans have been used to isolate lectins from the green juice of plants [161, 162]. The common feature of different classes of lectins is that they selectively bind to certain glycan structures [163]. For example, mistletoe viscumin preferentially binds Neu5Acα(2–6)Galββ(1–4)GlcNAc and this carbohydrate can therefore be used for lectin purification [164]. More importantly, the setup can be reversed; that is, immobilized lectins can be used to selectively enrich and purify glycoproteins carrying specific carbohydrate structures. This strategy has been successfully used for more than 20 years to purify glycoprotein-derived carbohydrates and glycoproteins [165–167]. In future, this approach could also be adapted for large-scale preparative chromatography, but the lectins used for such procedures should be simple, non-toxic, and structurally rigid to improve process performance (e.g., the number of production cycles possible using the same batch of resin) and ensure the process complies with the safety requirements of good manufacturing practice [168]. For example, viscumin consists of a toxic A-chain and a glycoprotein-binding B-chain [169]. The latter contains a rigid fold that is rich in disulfide bonds and β-sheets [170]. A non-toxic viscumin mutant may therefore be useful for the purification of glycoproteins terminating with Neu5Acα(2–6)Galβ(1–4)GlcNAc from bulk plant extracts. Genetic engineering may make it possible to alter the carbohydrate selectivity of this protein, as shown for Protein L in the case of antibodies [171]. Alternatively, proteins with modified glycan profiles could be purified by boronate affinity chromatography [172]. The selectivity of this resin reflects the differential binding of boronate covalently attached to a base resin to *cis*-diol groups found, e.g., in monosaccharides and oligosaccharides. The benefits of this method over lectin-based purification include a more stable ligand (which does not require recombinant protein expression) and a lower risk of immunogenic process-related impurities. However, boronate affinity chromatography is less selective than lectin-based resins and additional purification steps may be required to remove interfering small compounds with *cis*-diol groups, particularly free sugars. Also, this method will not help to separate the target therapeutic from endogenous plant glycoproteins, which typically are much more abundant.

## 7 Conclusions

Many recombinant proteins with a glycan structure similar to that of native human protein, or at least compatible with humans, can now be produced in plant cells. However, the intensive research that has made glyco-engineering possible in plants has also yielded some unexpected benefits of plant glycans. The immunogenicity of some of these structures can increase the visibility of plant-derived vaccines to the mammalian immune system. Plant glycans can also target antigen-presenting cells, particularly via lectins or mannose-fucose receptors on the surfaces of dendritic cells. Therapeutic proteins with plant-derived glycans are likely to be undesirable where cost-effective alternative expression platforms are available, but in certain cases such glycans achieve a functionality that can provide the basis for 'biobetter' therapeutic products, which not only benefit from the greater economy, scalability, and safety of plant-based production platforms, but also have intrinsic therapeutic benefits conferred specifically by the plant glycan structures. Where plant glycans are undesirable, several strategies are available, based on protein targeting or genetic engineering, to remove such glycans or to replace them with human-compatible structures.

## References


1. Strasser R (2016) Plant protein glycosylation. Glycobiology 26:926–939
2. Kornfeld R, Kornfeld S (1985) Assembly of asparagine-linked oligosaccharides. Annu Rev Biochem 54:631–664
3. Tarp MA, Clausen H (2008) Mucin-type *O*-glycosylation and its potential use in drug and vaccine development. Biochim Biophys Acta 1780:546–563
4. Gomord V, Fitchette AC, Menu-Bouaouiche L, Saint-Jore-Dupas C, Plasson C, Michaud D, Faye L (2010) Plant-specific glycosylation patterns in the context of therapeutic protein production. Plant Biotechnol J 8:564–587
5. Shaaltiel Y, Tekoah Y (2016) Plant specific N-glycans do not have proven adverse effects in humans. Nat Biotechnol 34:706–708
6. Bosch D, Castilho A, Loos A, Schots A, Steinkellner H (2013) N-glycosylation of plant-produced recombinant proteins. Curr Pharm Des 19:5503–5512
7. Mor TS (2015) Molecular pharming's foot in the FDA's door: Protalix's trailblazing story. Biotechnol Lett 37:2147–2150
8. Xu C, Ng DT (2015) Glycosylation-directed quality control of protein folding. Nat Rev Mol Cell Biol 16:742–752
9. Arcalis E, Stadlmann J, Rademacher T, Marcel S, Sack M, Altmann F, Stoger E (2013) Plant species and organ influence the structure and subcellular localization of recombinant glycoproteins. Plant Mol Biol 83:105–117
10. Abranches R, Marcel S, Arcalis E, Altmann F, Fevereiro P, Stoger E (2005) Plants as bioreactors: a comparative study suggests that *Medicago truncatula* is a promising production system. J Biotechnol 120:121–134
11. Drakakaki G, Marcel S, Arcalis E, Altmann F, Gonzalez-Melendi P, Fischer R, Christou P, Stoger E (2006) The intracellular fate of a recombinant protein is tissue dependent. Plant Physiol 141:578–586
12. Arcalis E, Marcel S, Altmann F, Kolarich D, Drakakaki G, Fischer R, Christou P, Stoger E (2004) Unexpected deposition patterns of recombinant proteins in post-endoplasmic reticulum compartments of wheat endosperm. Plant Physiol 136:3457–3466
13. Arcalis E, Stadlmann J, Marcel S, Drakakaki G, Winter V, Rodriguez J, Fischer R, Altmann F, Stoger E (2010) The changing fate of a secretory glycoprotein in developing maize endosperm. Plant Physiol 153:693–702
14. Loos A, Van Droogenbroeck B, Hillmer S, Grass J, Kunert R, Cao J, Robinson DG, Depicker A, Steinkellner H (2011) Production of monoclonal antibodies with a controlled N-glycosylation pattern in seeds of *Arabidopsis thaliana*. Plant Biotechnol J 9:179–192
15. Schahs M, Strasser R, Stadlmann J, Kunert R, Rademacher T, Steinkellner H (2007) Production of a monoclonal antibody in plants with a humanized N-glycosylation pattern. Plant Biotechnol J 5:657–663
16. Ramessar K, Rademacher T, Sack M, Stadlmann J, Platis D, Stiegler G, Labrou N, Altmann F, Ma J, Stöger E, Capell T, Christou P (2008) Cost-effective production of a vaginal protein microbicide to prevent HIV transmission. Proc Natl Acad Sci U S A 105:3727–3732

17. Floss DM, Sack M, Arcalis E, Stadlmann J, Quendler H, Rademacher T, Stoger E, Scheller J, Fischer R, Conrad U (2009) Influence of elastin-like peptide fusions on the quantity and quality of a tobacco-derived human immunodeficiency virus-neutralizing antibody. Plant Biotechnol J 7:899–913
18. Rademacher T, Sack M, Arcalis E, Stadlmann J, Balzer S, Altmann F, Quendler H, Stiegler G, Kunert R, Fischer R, Stoger E (2008) Recombinant antibody 2G12 produced in maize endosperm efficiently neutralizes HIV-1 and contains predominantly single-GlcNAc N-glycans. Plant Biotechnol J 6:189–201
19. Lerouge P, Cabanes-Macheteau M, Rayon C, Fischette-Lainé AC, Gomord V, Faye L (1998) N-glycoprotein biosynthesis in plants: recent developments and future trends. Plant Mol Biol 38:31–48
20. Kaushal GP, Pastuszak I, Hatanaka K, Elbein AD (1990) Purification to homogeneity and properties of glucosidase II from mung bean seedlings and suspension-cultured soybean cells. J Biol Chem 265:16271–16279
21. Szumilo T, Kaushal GP, Elbein AD (1986) Purification and properties of glucosidase I from mung bean seedlings. Arch Biochem Biophys 247:261–271
22. Schoberer J, Strasser R (2017) Plant glyco-biotechnology. Semin Cell Dev Biol 80:133–141
23. Fitchette-Lainé AC, Gomord V, Chekkafi A, Faye L (1994) Distribution of xylosylation and fucosylation in the plant Golgi apparatus. Plant J 5:673–682
24. Johnson KD, Chrispeels MJ (1987) Substrate specificities of N-acetylglucosaminyl-, fucosyl-, and xylosyltransferases that modify glycoproteins in the Golgi apparatus of bean cotyledon. Plant Physiol 84:1301–1308
25. Sturm A, Van Kuik JA, Vliegenthart JF, Chrispeels MJ (1987) Structure, position, and biosynthesis of the high mannose and the complex oligosaccharide side chains of the bean storage protein phaseolin. J Biol Chem 262:13392–13403
26. Strasser R, Steinkellner H, Borén M, Altmann F, Mach L, Glössl J, Mucha J (1999) Molecular cloning of cDNA encoding N-acetylglucosaminyltransferase II from *Arabidopsis thaliana*. Glycoconjugate J 16:787–791
27. Strasser R, Stadlmann J, Svoboda B, Altmann F, Glössl J, Mach L (2005) Molecular basis of N-acetylglucosaminyltransferase I deficiency in *Arabidopsis thaliana* plants lacking complex N-glycans. Biochem J 387:385–391
28. Strasser R, Schoberer J, Jin C, Glössl J, Mach L, Steinkellner H (2006) Molecular cloning and characterization of *Arabidopsis thaliana* Golgi alpha-mannosidase II, a key enzyme in the formation of complex N-glycans in plants. Plant J 45:789–803
29. Oxley D, Munro SL, Craik DJ, Bacic A (1996) Structure of N-glycans on the S3- and S6-allele stylar self-incompatibility ribonucleases of *Nicotiana alata*. Glycobiology 6:611–618
30. Bakker H, Rouwendal GJ, Karnoup AS, Florack DE, Stoopen GM, Helsper JP, van Ree R, van Die I, Bosch D (2006) An antibody produced in tobacco expressing a hybrid beta-1,4-galactosyltransferase is essentially devoid of plant carbohydrate epitopes. Proc Natl Acad Sci U S A 103:7577–7582
31. Zeng Y, Bannon G, Thomas VH, Rice K, Drake R, Elbein A (1997) Purification and specificity of β-1,2-xylosyltransferase, an enzyme that contributes to the allergenicity of some plant proteins. J Biol Chem 272:31340–31347
32. Strasser R, Altmann F, Mach L, Glössl J, Steinkellner H (2004) Generation of *Arabidopsis thaliana* plants with complex N-glycans lacking β 1,2-linked xylose and core α-1,3-linked fucose. FEBS Lett 561:132–136
33. Vitale A, Chrispeels MJ (1984) Transient N-acetylglucosamine in the biosynthesis of phytohemagglutinin: attachment in the Golgi apparatus and removal in protein bodies. Cell Biol 99:133–140
34. Liebminger E, Veit C, Pabst M, Batoux M, Zipfel C, Altmann F, Mach L, Strasser R (2011) Beta-N-acetylhexosaminidases HEXO1 and HEXO3 are responsible for the formation of paucimannosidic N-glycans in *Arabidopsis thaliana*. J Biol Chem 286:10793–10802

35. Strasser R, Bondili JS, Schoberer J, Svoboda B, Liebminger E, Glössl J, Altmann F, Steinkellner H, Mach L (2007) Enzymatic properties and subcellular localization of Arabidopsis beta-N-acetylhexosaminidases. Plant Physiol 145:5–16
36. Song W, Mentink RA, Henquet MG, Cordewener JH, van Dijk AD, Bosch D, America AH, van der Krol AR (2013) N-glycan occupancy of Arabidopsis N-glycoproteins. J Proteomics 93:343–355
37. Li H, d'Anjou M (2009) Pharmacological significance of glycosylation in therapeutic proteins. Curr Opin Biotechnol 20:678–684
38. Gill DJ, Chia J, Senewiratne J, Bard F (2010) Regulation of O-glycosylation through Golgi-to-ER relocation of initiation enzymes. J Cell Biol 189:843–858
39. Gill DJ, Clausen H, Bard F (2011) Location, location, location: new insights into O-GalNAc protein glycosylation. Trends Cell Biol 21:149–158
40. Tian E, Ten Hagen KG (2009) Recent insights into the biological roles of mucin-type O-glycosylation. Glycoconjugate J 26:325–334
41. Kilcoyne M, Shah M, Gerlach JQ, Bhavanandan V, Nagaraj V, Smith AD, Fujiyama K, Sommer U, Costello CE, Olszewski N, Joshi L (2009) O-glycosylation of protein subpopulations in alcohol-extracted rice proteins. J Plant Physiol 166:219–232
42. Kishimoto T, Watanabe M, Mitsui T, Hori H (1999) Glutelin basic subunits have a mammalian mucin-type O-linked disaccharide side chain. Arch Biochem Biophys 370:271–277
43. Seifert GJ, Roberts K (2007) The biology of arabinogalactan proteins. Annu Rev Plant Biol 58:137–161
44. Stulemeijer IJ, Joosten MH (2008) Post-translational modification of host proteins in pathogen-triggered defence signalling in plants. Mol Plant Pathol 9:545–560
45. Wilson IBH (2002) Glycosylation of proteins in plants and invertebrates. Curr Opin Struct Biol 12:569–577
46. Shpak E, Leykam JF, Kieliszewski MJ (1999) Synthetic genes for glycoprotein design and the elucidation of hydroxyproline-O-glycosylation codes. Proc Natl Acad Sci U S A 96: 14736–14741
47. Shpak E, Barbar E, Leykam JF, Kieliszewski MJ (2001) Contiguous hydroxyproline residues direct hydroxyproline arabinosylation in *Nicotiana tabacum*. J Biol Chem 276:11272–11278
48. Xu J, Tan L, Goodrum KJ, Kieliszewski MJ (2007) High-yields and extended serum half-life of human interferon alpha2b expressed in tobacco cells as arabinogalactan-protein fusions. Biotechnol Bioeng 97:997–1008
49. Xu J, Tan L, Lamport DT, Showalter AM, Kieliszewski MJ (2008) The O-hyp glycosylation code in tobacco and Arabidopsis and a proposed role of Hyp-glycans in secretion. Phytochemistry 69:1631–1640
50. Twyman RM (2013) Principles of proteomics2nd edn. Garland Science, Abingdon, pp 181–190
51. Harrison RL, Jarvis DL (2006) Protein N-glycosylation in the baculovirus–insect cell expression system and engineering of insect cells to produce mammalianized recombinant glycoproteins. Adv Virus Res 68:159–191
52. Raju TS (2003) Glycosylation variations with expression systems and their impact on biological activity of therapeutic immunoglobulins. BioProc Int 1(4):44–52
53. Yoo EM, Yu LJ, Wims LA, Goldberg D, Morrison SL (2010) Differences in N-glycan structures found on recombinant IgA1 and IgA2 produced in murine myeloma and CHO cell lines. MAbs 2:320–334
54. Wilson IBH, Zeleny R, Kolarich D, Staudacher E, Stroop CJ, Kamerling JP, Altmann F (2001) Analysis of Asn-linked glycans from vegetable foodstuffs: widespread occurrence of Lewis a, core alpha1,3-linked fucose and xylose substitutions. Glycobiology 11:261–274
55. Bardor M, Faveeuw C, Fitchette AC, Gilbert D, Galas L, Trottein F, Faye L, Lerouge P (2003) Immunoreactivity in mammals of two typical plant glyco-epitopes, core alpha(1,3)-fucose and core xylose. Glycobiology 13:427–434

56. Léonard R, Kolarich D, Paschinger K, Altmann F, Wilson IB (2004) A genetic and structural analysis of the N-glycosylation capabilities of rice and other monocotyledons. Plant Mol Biol 55:631–644
57. Elbers IJ, Stoopen GM, Bakker H, Stevens LH, Bardor M, Molthoff JW, Jordi WJ, Bosch D, Lommen A (2001) Influence of growth conditions and developmental stage on N-glycan heterogeneity of transgenic immunoglobulin G and endogenous proteins in tobacco leaves. Plant Physiol 126:1314–1322
58. Twyman RM, Schillberg S, Fischer R (2012) The production of vaccines and therapeutic antibodies in plants. In: Ma S, Wang A (eds) Molecular farming in plants: recent advances and future prospects. Springer, NY, pp 145–159
59. Castilho A, Bohorova N, Grass J, Bohorov O, Zeitlin L, Whaley K, Altmann F, Steinkellner H (2011) Rapid high yield production of different glycoforms of Ebola virus monoclonal antibody. PLoS One 6:e26040
60. Strasser R, Stadlmann J, Schähs M, Stiegler G, Quendler H, Mach L, Glössl J, Weterings K, Pabst M, Steinkellner H (2008) Generation of glyco-engineered Nicotiana benthamiana for the production of monoclonal antibodies with a homogeneous human-like N-glycan structure. Plant Biotechnol J 6:392–402
61. Raven N, Rasche S, Kuehn C, Anderlei T, Klöckner W, Schuster F, Henquet M, Bosch D, Büchs J, Fischer R, Schillberg S (2015) Scaled-up manufacturing of recombinant antibodies produced by plant cells in a 200 L orbitally shaken disposable bioreactor. Biotechnol Bioeng 112:308–321
62. Dirnberger D, Steinkellner H, Abdennebi L, Remy JJ, van de Wiel D (2001) Secretion of biologically active glycoforms of bovine follicle stimulating hormone in plants. Eur J Biochem 268:4570–4579
63. Webster DE, Thomas MC (2012) Post-translational modification of plant-made foreign proteins: glycosylation and beyond. Biotechnol Adv 30:410–418
64. Fiedler U, Phillips J, Artsaenko O, Conrad U (1997) Optimization of scFv antibody production in transgenic plants. Immunotechnology 3:205–216
65. Schouten A, Roosien J, van Engelen FA, de Jong GA, Borst-Vrenssen AW, Zilverentant JF, Bosch D, Stiekema WJ, Gommers FJ, Schots A, Bakker J (1996) The C-terminal KDEL sequence increases the expression level of a single-chain antibody designed to be targeted to both the cytosol and the secretory pathway in transgenic tobacco. Plant Mol Biol 30:781–793
66. Ko K, Tekoah Y, Rudd PM, Harvey DJ, Dwek RA, Spitsin S, Hanlon CA, Rupprecht C, Dietzschold B, Golovkin M, Koprowski H (2003) Function and glycosylation of plant-derived antiviral monoclonal antibody. Proc Natl Acad Sci U S A 100:8013–8018
67. Lee SJ, Evers S, Roeder D, Parlow AF, Risteli J, Risteli L, Lee YC, Feizi T, Langen H, Nussenzweig MC (2002) Mannose receptor-mediated regulation of serum glycoprotein homeostasis. Science 295:1898–1901
68. He X, Haselhorst T, von Itzstein M, Kolarich D, Packer NH, Gloster TM, Vocadlo DJ, Clarke LA, Qian Y, Kermode AR (2012) Production of α-L-iduronidase in maize for the potential treatment of a human lysosomal storage disease. Nat Commun 3:1062
69. Arcalis E, Ibl V, Peters J, Melnik S, Stoger E (2014) The dynamic behavior of storage organelles in developing cereal seeds and its impact on the production of recombinant proteins. Front Plant Sci 5:439
70. Hofbauer A, Stoger E (2013) Subcellular accumulation and modification of pharmaceutical proteins in different plant tissues. Curr Pharm Des 19:5495–5502
71. Ibl V, Stoger E (2012) The formation, function and fate of protein storage compartments in seeds. Protoplasma 249:379–392
72. Lending CR, Larkins BA (1989) Changes in the zein composition of protein bodies during maize endosperm development. Plant Cell 1:1011–1023
73. Shewry PR, Halford NG (2002) Cereal seed storage proteins: structures, properties and role in grain utilization. J Exp Bot 53:947–958

74. Geli MI, Torrent M, Ludevid D (1994) Two structural domains mediate two sequential events in γ-zein targeting: protein endoplasmic reticulum retention and protein body formation. Plant Cell 6:1911–1922
75. Llop-Tous I, Madurga S, Giralt E, Marzabal P, Torrent M, Ludevid MD (2010) Relevant elements of a maize gamma-zein domain involved in protein body biogenesis. J Biol Chem 285:35633–35644
76. Pompa A, Vitale A (2006) Retention of a bean phaseolin/maize gamma-zein fusion in the endoplasmic reticulum depends on disulfide bond formation. Plant Cell 18:2608–2621
77. Bagga S, Adams H, Kemp JD, Sengupta-Gopalan C (1995) Accumulation of 15-kilodalton zein in novel protein bodies in transgenic tobacco. Plant Physiol 107:13–23
78. Coleman CE, Herman EM, Takasaki K, Larkins BA (1996) The maize gamma-zein sequesters alpha-zein and stabilizes its accumulation in protein bodies of transgenic tobacco endosperm. Plant Cell 8:2335–2345
79. Torrent M, Llompart B, Lasserre-Ramassamy S, Llop-Tous I, Bastida M, Marzabal P, Westerholm-Parvinen A, Saloheimo M, Heifetz PB, Ludevid MD (2009) Eukaryotic protein production in designed storage organelles. BMC Biol 7:5
80. He X, Haselhorst T, von Itzstein M, Kolarich D, Packer NH, Kermode AR (2012) Influence of an ER-retention signal on the N-glycosylation of recombinant human alpha-L-iduronidase generated in seeds of Arabidopsis. Plant Mol Biol 79:157–169
81. Loos A, Van Droogenbroeck B, Hillmer S, Grass J, Pabst M, Castilho A, Kunert R, Liang M, Arcalis E, Robinson DG, Depicker A, Steinkellner H (2011) Expression of antibody fragments with a controlled N-glycosylation pattern and induction of endoplasmic reticulum-derived vesicles in seeds of Arabidopsis. Plant Physiol 155:2036–2048
82. Vamvaka E, Twyman RM, Murad AM, Melnik S, Teh AYH, Arcalis E, Altmann F, Stoger E, Rech E, Ma JKC, Christou P, Capell T (2016) Rice endosperm produces an underglycosylated and potent form of the HIV-neutralizing monoclonal antibody 2G12. Plant Biotechnol J 14: 97–108
83. Paris N, Stanley CM, Jones RL, Rogers JC (1996) Plant cells contain two functionally distinct vacuolar compartments. Cell 85:563–572
84. Nicholson L, Gonzalez-Melendi P, van Dolleweerd C, Tuck H, Perrin Y, Ma JK, Fischer R, Christou P, Stoger E (2005) A recombinant multimeric immunoglobulin expressed in rice shows assembly-dependent subcellular localization in endosperm cells. Plant Biotechnol J 3: 115–127
85. Reggi S, Marchetti S, Patti T, De Amicis F, Cariati R, Bembi B, Fogher C (2005) Recombinant human acid β-glucosidase stored in tobacco seed is stable, active and taken up by human fibroblasts. Plant Mol Biol 57:101–113
86. Wright KE, Prior F, Sardana R, Altosaar I, Dudani AK, Ganz PR, Tackaberry ES (2001) Sorting of glycoprotein B from human cytomegalovirus to protein storage vesicles in seeds of transgenic tobacco. Transgenic Res 10:177–181
87. Frigerio L, de Virgilio M, Prada A, Faoro F, Vitale A (1998) Sorting of phaseolin to the vacuole is saturable and requires a short C-terminal peptide. Plant Cell 10:1031–1042
88. Neuhaus JM, Sticher L, Meins Jr F, Boller T (1991) A short C-terminal sequence is necessary and sufficient for the targeting of chitinases to the plant vacuole. Proc Natl Acad Sci U S A 88:10362–10366
89. Sonnewald U, Brauer M, von Schaewen A, Stitt M, Willmitzer L (1991) Transgenic tobacco plants expressing yeast-derived invertase in either the cytosol, vacuole or apoplast: a powerful tool for studying sucrose metabolism and sink/source interactions. Plant J 1:95–106
90. Grabowski GA, Golembo M, Shaaltiel Y (2014) Taliglucerase alfa: an enzyme replacement therapy using plant cell expression technology. Mol Genet Metab 112:1–8
91. Bock R (2015) Engineering plastid genomes: methods, tools, and applications in basic research and biotechnology. Annu Rev Plant Biol 66:211–241

92. Delporte A, De Zaeytijd J, De Storme N, Azmi A, Geelen D, Smagghe G, Guisez Y, Van Damme EJ (2014) Cell cycle-dependent O-GlcNAc modification of tobacco histones and their interaction with the tobacco lectin. Plant Physiol Biochem 83:151–158
93. Remaley AT, Ugorski M, Wu N, Litzky L, Burger SR, Moore JS, Fukuda M, Spitalnik SL (1991) Expression of human glycophorin A in wild type and glycosylation-deficient Chinese hamster ovary cells. Role of N- and O-linked glycosylation in cell surface expression. J Biol Chem 266:24176–24183
94. Kodama S, Tsujimoto M, Tsuruoka N, Sugo T, Endo T, Kobata A (1993) Role of sugar chains in the in vitro activity of recombinant human interleukin 5. Eur J Biochem 211:903–908
95. Karnoup AS, Turkelson V, Anderson WH (2005) O-linked glycosylation in maize-expressed human IgA1. Glycobiology 15:965–981
96. Novak J, Tomana M, Kilian M, Coward L, Kulhavy R, Barnes S, Mestecky J (2000) Heterogeneity of O-glycosylation in the hinge region of human IgA1. Mol Immunol 37: 1047–1056
97. Pedersen CT, Loke I, Lorentzen A, Wolf S, Kamble M, Kristensen SK, Munch D, Radutoiu S, Spillner E, Roepstorff P, Thaysen-Andersen M, Stougaard J, Dam S (2017) *N*-glycan maturation mutants in *Lotus japonicus* for basic and applied glycoprotein research. Plant J 91: 394–407
98. Strasser R, Altmann F, Steinkellner H (2014) Controlled glycosylation of plant-produced recombinant proteins. Curr Opin Biotechnol 30:95–100
99. Gomord V, Chamberlain P, Jefferis R, Faye L (2005) Biopharmaceutical production in plants: problems, solutions and opportunities. Trends Biotechnol 23:559–565
100. van Ree R (2002) Carbohydrate epitopes and their relevance for the diagnosis and treatment of allergic diseases. Int Arch Allerg Immunol 129:189–197
101. Hamadeh RM, Jarvis GA, Galili U, Mandrell RE, Zhou P, Griffiss JM (1992) Human natural anti-gal IgG regulates alternative complement pathway activation on bacterial surfaces. J Clin Invest 89:1223–1235
102. Chung CH, Mirakhur B, Chan E, Le QT, Berlin J, Morse M, Murphy BA, Satinover SM, Hosen J, Mauro D, Slebos RJ, Zhou Q, Gold D, Hatley T, Hicklin DJ, Platts-Mills TA (2008) Cetuximab-induced anaphylaxis and IgE specific for galactose-alpha-1,3-galactose. N Engl J Med 358:1109–1117
103. Pastores GM, Shankar SP, Petakov M, Giraldo P, Rosenbaum H, Amato DJ, Szer J, Chertkoff R, Brill-Almon E, Zimran A (2016) Enzyme replacement therapy with taliglucerase alfa: 36-month safety and efficacy results in adult patients with Gaucher disease previously treated with imiglucerase. Am J Hematol 91:661–665
104. Rup B, Alon S, Amit-Cohen BC, Brill Almon E, Chertkoff R, Tekoah Y, Rudd PM (2017) Immunogenicity of glycans on biotherapeutic drugs produced in plant expression systems-the taliglucerase alfa story. PLoS One 12:e0186211
105. Zimran A, Durán G, Giraldo P, Rosenbaum H, Giona F, Petakov M, Terreros Muñoz E, Solorio-Meza SE, Cooper PA, Varughese S, Alon S, Chertkoff R (2016) Long-term efficacy and safety results of taliglucerase alfa through 5 years in adult treatment-naïve patients with Gaucher disease. Blood Cells Mol Dis, published online ahead of print. https://doi.org/10.1016/j.bcmd.2016.07.002
106. Zimran A, Durán G, Mehta A, Giraldo P, Rosenbaum H, Giona F, Amato DJ, Petakov M, Muñoz ET, Solorio-Meza SE, Cooper PA, Varughese S, Chertkoff R, Brill-Almon E (2016) Long-term efficacy and safety results of taliglucerase alfa up to 36 months in adult treatment-naïve patients with Gaucher disease. Am J Hematol 91:656–660
107. Ma JK, Hikmat BY, Wycoff K, Vine ND, Chargelegue D, Yu L, Hein MB, Lehner T (1998) Characterization of a recombinant plant monoclonal secretory antibody and preventive immunotherapy in humans. Nat Med 4:601–606
108. Ma JKC, Drossard J, Lewis D, Altmann F, Boyle J, Christou P, Cole T, Dale P, van Dolleweerd CJ, Isitt V, Katinger D, Lobedan M, Mertens H, Paul MJ, Rademacher T, Sack M, Hundleby PAC, Stiegler G, Stoger E, Twyman RM, Vcelar B, Fischer R (2015)

Regulatory approval and a first-in-human phase I clinical trial of a monoclonal antibody produced in transgenic tobacco plants. Plant Biotechnol J 13:1106–1120

109. Mari A, Ooievaar-de Heer P, Scala E, Giani M, Pirrotta L, Zuidmeer L, Bethell D, van Ree R (2008) Evaluation by double-blind placebo-controlled oral challenge of the clinical relevance of IgE antibodies against plant glycans. Allergy 63:891–896
110. Zeitlin L, Olmsted SS, Moench TR, Co MS, Martinell BJ, Paradkar VM, Russell DR, Queen C, Cone RA, Whaley KJ (1998) A humanized monoclonal antibody produced in transgenic plants for immunoprotection of the vagina against genital herpes. Nat Biotechnol 16:1361–1364
111. Cox KM, Sterling JD, Regan JT, Gasdaska JR, Frantz KK, Peele CG, Black A, Passmore D, Moldovan-Loomis C, Srinivasan M, Cuison S, Cardarelli PM, Dickey LF (2006) Glycan optimization of a human monoclonal antibody in the aquatic plant *Lemna minor*. Nat Biotechnol 24:1591–1597
112. Koprivova A, Stemmer C, Altmann F, Hoffmann A, Kopriva S, Gorr G, Reski R, Decker EL (2004) Targeted knockouts of Physcomitrella lacking plant-specific immunogenic N-glycans. Plant Biotechnol J 2:517–523
113. Sourrouille C, Marquet-Blouin E, D'Aoust MA, Kiefer-Meyer MC, Seveno M, Pagny-Salehabadi S, Bardor M, Durambur G, Lerouge P, Vezina L, Gomord V (2008) Down-regulated expression of plant-specific glycoepitopes in alfalfa. Plant Biotechnol J 6:702–721
114. Shin YJ, Chong YJ, Yang MS, Kwon TH (2011) Production of recombinant human granulocyte macrophage-colony stimulating factor in rice cell suspension culture with a human-like N-glycan structure. Plant Biotechnol J 9:1109–1119
115. Li J, Stoddard TJ, Demorest ZL, Lavoie PO, Luo S, Clasen BM, Cedrone F, Ray EE, Coffman AP, Daulhac A, Yabandith A, Retterath AJ, Mathis L, Voytas DF, D'Aoust MA, Zhang F (2016) Multiplexed, targeted gene editing in *Nicotiana benthamiana* for glyco-engineering and monoclonal antibody production. Plant Biotechnol J 14:533–542
116. Limkul J, Iizuka S, Sato Y, Misaki R, Ohashi T, Ohashi T, Fujiyama K (2016) The production of human glucocerebrosidase in glyco-engineered *Nicotiana benthamiana* plants. Plant Biotechnol J 14:1682–1694
117. Forthal DN, Gach JS, Landucci G, Jez J, Strasser R, Kunert R, Steinkellner H (2010) Fc-glycosylation influences Fcγ receptor binding and cell-mediated anti-HIV activity of monoclonal antibody 2G12. J Immunol 185:6876–6882
118. Schuster M, Jost W, Mudde GC, Wiederkum S, Schwager C, Janzek E, Altmann F, Stadlmann J, Stemmer C, Gorr G (2007) *In vivo* glyco-engineered antibody with improved lytic potential produced by an innovative non-mammalian expression system. Biotechnol J 2:700–708
119. Parsons J, Altmann F, Arrenberg CK, Koprivova A, Beike AK, Stemmer C, Gorr G, Reski R, Decker EL (2012) Moss-based production of asialo-erythropoietin devoid of Lewis A and other plant-typical carbohydrate determinants. Plant Biotechnol J 10:851–861
120. Shin YJ, Castilho A, Dicker M, Sádio F, Vavra U, Grünwald-Gruber C, Kwon TH, Altmann F, Steinkellner H, Strasser R (2017) Reduced paucimannosidic N-glycan formation by suppression of a specific β-hexosaminidase from *Nicotiana benthamiana*. Plant Biotechnol J 15: 197–206
121. Jansing J, Sack M, Augustine S, Fischer R, Bortesi L (2018) CRISPR/Cas9-mediated knockout of six glycosyltransferase genes in Nicotiana benthamiana for the production of recombinant proteins lacking β-1,2-xylose and core α-1,3-fucose. Plant Biotechnol J (in press)
122. Piron R, Santens F, De Paepe A, Depicker A, Callewaert N (2015) Using GlycoDelete to produce proteins lacking plant-specific *N*-glycan modification in seeds. Nat Biotechnol 33: 1135–1137
123. Jefferis R (2009) Glycosylation as a strategy to improve antibody-based therapeutics. Nat Rev Drug Discov 8:226–234
124. Shinkawa T, Nakamura K, Yamane N, Shoji-Hosaka E, Kanda Y, Sakurada M, Uchida K, Anazawa H, Satoh M, Yamasaki M, Hanai N, Shitara K (2003) The absence of fucose but not

the presence of galactose or bisecting N-acetylglucosamine of human IgG1 complex-type oligosaccharides shows the critical role of enhancing antibody-dependent cellular cytotoxicity. J Biol Chem 278:3466–3473
125. Takeuchi M, Inoue N, Strickland TW, Kubota M, Wada M, Shimizu R, Hoshi S, Kozutsumi H, Takasaki S, Kobata A (1989) Relationship between sugar chain structure and biological activity of recombinant human erythropoietin produced in Chinese hamster ovary cells. Proc Natl Acad Sci U S A 86:7819–7822
126. Yuen CT, Storring PL, Tiplady RJ, Izquierdo M, Wait R, Gee CK, Gerson P, Lloy P, Cremata JA (2003) Relationships between the N-glycan structures and biological activities of resins produced using different culture conditions and purification procedures. J Haematol 12:511–526
127. Castilho A, Gattinger P, Grass J, Jez J, Pabst M, Altmann F, Gorfer M, Strasser R, Steinkellner H (2011) N-glycosylation engineering of plants for the biosynthesis of glycoproteins with bisected and branched complex N-glycans. Glycobiology 21:813–823
128. Frey AD, Karg SR, Kallio PT (2009) Expression of rat beta(1,4)-N-acetylglucosaminyltransferase III in *Nicotiana tabacum* remodels the plant-specific N-glycosylation. Plant Biotechnol J 7:33–48
129. Karg SR, Frey AD, Kallio PT (2010) Reduction of N-linked xylose and fucose by expression of rat beta1,4-N-acetylglucosaminyltransferase III in tobacco BY-2 cells depends on Golgi enzyme localization domain and genetic elements used for expression. J Biotechnol 146:54–65
130. Rouwendal GJ, Wuhrer M, Florack DE, Koeleman CA, Deelder AM, Bakker H, Stoopen GM, van Die I, Helsper JP, Hokke CH, Bosch D (2007) Efficient introduction of a bisecting GlcNAc residue in tobacco N-glycans by expression of the gene encoding human N-acetylglucosaminyltransferase III. Glycobiology 17:334–344
131. Nagels B, Van Damme EJ, Pabst M, Callewaert N, Weterings K (2011) Production of complex multiantennary N-glycans in Nicotiana benthamiana plants. Plant Physiol 155:1103–1112
132. Castilho A, Neumann L, Gattinger P, Strasser R, Vorauer-Uhl K, Sterovsky T, Altmann F, Steinkellner H (2013) Generation of biologically active multi-sialylated recombinant human EPO-Fc in plants. PLOS One 8:e54836
133. Bakker H, Bardor M, Molthoff JW, Gomord V, Elbers I, Stevens LH, Jordi W, Lommen A, Faye L, Lerouge P, Bosch D (2001) Galactose-extended glycans of antibodies produced by transgenic plants. Proc Natl Acad Sci U S A 98:2899–2904
134. Fujiyama K, Furukawa A, Katsura A, Misaki R, Omasa T, Seki T (2007) Production of mouse monoclonal antibody with galactose-extended sugar chain by suspension cultured tobacco BY2 cells expressing human beta(1,4)-galactosyltransferase. Biochem Biophys Res Commun 358:85–91
135. Misaki R, Kimura Y, Palacpac NQ, Yoshida S, Fujiyama K, Seki T (2003) Plant cultured cells expressing human β1,4-galactosyltransferase secrete glycoproteins with galactose-extended N-linked glycans. Glycobiology 13:199–205
136. Palacpac NQ, Yoshida S, Sakai H, Kimura Y, Fujiyama K, Yoshida T, Seki T (1999) Stable expression of human beta1,4-galactosyltransferase in plant cells modifies N-linked glycosylation patterns. Proc Natl Acad Sci U S A 96:4692–4697
137. Strasser R, Castilho A, Stadlmann J, Kunert R, Quendler H, Gattinger P, Jez J, Rademacher T, Altmann F, Mach L, Steinkellner H (2009) Improved virus neutralization by plant-produced anti-HIV antibodies with a homogeneous beta1,4-galactosylated N-glycan profile. J Biol Chem 284:20479–20485
138. Seveno M, Bardor M, Paccalet T, Gomord V, Lerouge P, Faye L (2004) Glycoprotein sialylation in plants? Nat Biotechnol 22:1351–1352 Author reply 1352–1353
139. Zeleny R, Kolarich D, Strasser R, Altmann F (2006) Sialic acid concentrations in plants are in the range of inadvertent contamination. Planta 224:222–227

140. Bakker H, Routier F, Ashikov A, Neumann D, Bosch D, Gerardy-Schahn R (2008) A CMP-sialic acid transporter cloned from *Arabidopsis thaliana*. Carbohydr Res 343: 2148–2152
141. Shah MM, Fujiyama K, Flynn CR, Joshi L (2003) Sialylated endogenous glycoconjugates in plant cells. Nat Biotechnol 21:1470–1471
142. Castilho A, Pabst M, Leonard R, Veit C, Altmann F, Mach L, Glössl J, Strasser R, Steinkellner H (2008) Construction of a functional CMP-sialic acid biosynthesis pathway in Arabidopsis. Plant Physiol 147:331–339
143. Misaki R, Fujiyama K, Seki T (2006) Expression of human CMP-*N*-acetylneuraminic acid synthetase and CMP-sialic acid transporter in tobacco suspension-cultured cell. Biochem Biophys Res Commun 339:1184–1189
144. Wee EG, Sherrier DJ, Prime TA, Dupree P (1998) Targeting of active sialyltransferase to the plant Golgi apparatus. Plant Cell 10:1759–1768
145. Castilho A, Strasser R, Stadlmann J, Grass J, Jez J, Gattinger P, Kunert R, Quendler H, Pabst M, Leonard R, Altmann F, Steinkellner H (2010) In planta protein sialylation through overexpression of the respective mammalian pathway. J Biol Chem 285:15923–15930
146. Kallolimath S, Castilho A, Strasser R, Grünwald-Gruber C, Altmann F, Strubl S, Galuska CE, Zlatina K, Galuska SP, Werner S, Thiesler H, Werneburg S, Hildebrandt H, Gerardy-Schahn R, Steinkellner H (2016) Engineering of complex protein sialylation in plants. Proc Natl Acad Sci U S A 113:9498–9503
147. Wang J, Zhang Y, Wei J, Zhang X, Zhang B, Zhu Z, Zou W, Wang Y, Mou Z, Ni B, Wu Y (2007) Lewis X oligosaccharides targeting to DC-SIGN enhanced antigen-specific immune response. Immunology 121:174–182
148. Rouwendal GJ, Florack DE, Hesselink T, Cordewener JH, Helsper JP, Bosch D (2009) Synthesis of Lewis X epitopes on plant N-glycans. Carbohydr Res 344:1487–1493
149. Werle M, Bernkop-Schnurch A (2006) Strategies to improve plasma half life time of peptide and protein drugs. Amino Acids 30:351–367
150. Xu J, Ge X, Dolan MC (2011) Towards high-yield production of pharmaceutical proteins with plant cell suspension cultures. Biotechnol Adv 29:278–299
151. Xu J, Okada S, Tan L, Goodrum KJ, Kopchick JJ, Kieliszewski MJ (2010) Human growth hormone expressed in tobacco cells as an arabinogalactan-protein fusion glycoprotein has a prolonged serum life. Transgenic Res 19:849–867
152. Daskalova SM, Radder JE, Cichacz ZA, Olsen SH, Tsaprailis G, Mason H, Lopez LC (2010) Engineering of *N. benthamiana* plants for production of N-acetylgalactosamine-glycosylated proteins—towards development of a plant-based platform for production of protein therapeutics with mucin type O-glycosylation. BMC Biotechnol 10:62
153. Strasser R (2012) Challenges in O-glycan engineering of plants. Front Plant Sci 3:218
154. Strasser R (2013) Engineering of human-type O-glycosylation in Nicotiana benthamiana plants. Bioengineered 4:191–196
155. Dicker M, Tschofen M, Maresch D, König J, Juarez P, Orzaez D, Altmann F, Steinkellner H, Strasser R (2016) Transient glyco-engineering to produce recombinant IgA1 with defined N- and O-glycans in plants. Front Plant Sci 7:18
156. Dicker M, Maresch D, Strasser R (2016) Glyco-engineering for the production of recombinant IgA1 with distinct mucin-type O-glycans in plants. Bioengineered 22:1–6
157. Buyel JF, Fischer R (2014) Generic chromatography-based purification strategies accelerate the development of downstream processes for biopharmaceutical proteins produced in plants. Biotechnol J 9:566–577
158. CMC Biotech Working Group (2009) A-Mab: a case study in bioprocess development. CASSS, Emeryville, pp 1–278
159. Buyel JF, Twyman RM, Fischer R (2015) Extraction and downstream processing of plant-derived recombinant proteins. Biotechnol Adv 33:902–913
160. Wilken LR, Nikolov ZL (2012) Recovery and purification of plant-made recombinant proteins. Biotechnol Adv 30:419–433

161. Antonyuk V, Grama S, Plichta Z, Magorivska I, Horak D, Stoika R (2015) Use of specific polysaccharide-immobilized monodisperse poly(glycidyl methacrylate) core-silica shell microspheres for affinity purification of lectins. Biomed Chromatogr 29:783–787
162. Zhu BCR, Laine RA (1989) Purification of acetyllactosamine-specific tomato lectin by erythroglycan-Sepharose affinity chromatography. Prep Biochem 19:341–350
163. Souza MA, Carvalho FC, Ruas LP, Ricci-Azevedo R, Roque-Barreira MC (2013) The immunomodulatory effect of plant lectins: a review with emphasis on ArtinM properties. Glycoconj J 7:641–657
164. Muthing J, Meisen I, Bulau P, Langer M, Witthohn K, Lentzen H, Neumann U, Peter-Katalinic J (2004) Mistletoe lectin I is a sialic acid-specific lectin with strict preference to gangliosides and glycoproteins with terminal Neu5Ac alpha 2-6Gal beta 1-4GlcNAc residues. Biochemistry 11:2996–3007
165. Endo T (1996) Fractionation of glycoprotein-derived oligosaccharides by affinity chromatography using immobilized lectin columns. J Chromatogr A 720:251–261
166. Fanayan S, Hincapie M, Hancock WS (2012) Using lectins to harvest the plasma/serum glycoproteome. Electrophoresis 33:1746–1754
167. Hortin GL (1990) Isolation of glycopeptides containing O-linked oligosaccharides by lectin affinity-chromatography on jacalin agarose. Anal Biochem 191:262–267
168. Fischer R, Schillberg S, Hellwig S, Twyman RM, Drossard J (2012) GMP issues for recombinant plant-derived pharmaceutical proteins. Biotechnol Adv 2:434–439
169. Kourmanova AG, Soudarkina OJ, Olsnes S, Kozlov JV (2004) Cloning and characterization of the genes encoding toxic lectins in mistletoe (*Viscum album* L). Eur J Biochem 12:2350–2360
170. Niwa H, Tonevitsky AG, Agapov II, Saward S, Pfuller U, Palmer RA (2003) Crystal structure at 3 Å of mistletoe lectin I, a dimeric type-II ribosome-inactivating protein, complexed with galactose. Eur J Biochem 13:2739–2749
171. Boes A, Spiegel H, Delbruck H, Fischer R, Schillberg S, Sack M (2011) Affinity purification of a framework 1 engineered mouse/human chimeric IgA2 antibody from tobacco. Biotechnol Bioeng 12:2804–2814
172. Li YC, Pfuller U, Larsson EL, Jungvid H, Galaev IY, Mattiasson B (2001) Separation of mistletoe lectins based on the degree of glycosylation using boronate affinity chromatography. J Chromatogr A 925:115–121

# Bacterial Glycoengineering as a Biosynthetic Route to Customized Glycomolecules

**Laura E. Yates, Dominic C. Mills, and Matthew P. DeLisa**

## Contents



**Abstract** Bacteria have garnered increased interest in recent years as a platform for the biosynthesis of a variety of glycomolecules such as soluble oligosaccharides, surface-exposed carbohydrates, and glycoproteins. The ability to engineer commonly used laboratory species such as *Escherichia coli* to efficiently synthesize non-native sugar structures by recombinant expression of enzymes from various

---

L. E. Yates, D. C. Mills, and M. P. DeLisa (✉)
School of Chemical and Biomolecular Engineering, Cornell University, Ithaca, NY, USA
e-mail: md255@cornell.edu

carbohydrate biosynthesis pathways has allowed for the facile generation of important products such as conjugate vaccines, glycosylated outer membrane vesicles, and a variety of other research reagents for studying and understanding the role of glycans in living systems. This chapter highlights some of the key discoveries and technologies for equipping bacteria with the requisite biosynthetic machinery to generate such products. As the bacterial glyco-toolbox continues to grow, these technologies are expected to expand the range of glycomolecules produced recombinantly in bacterial systems, thereby opening up this platform to an even larger number of applications.

## Graphical Abstract

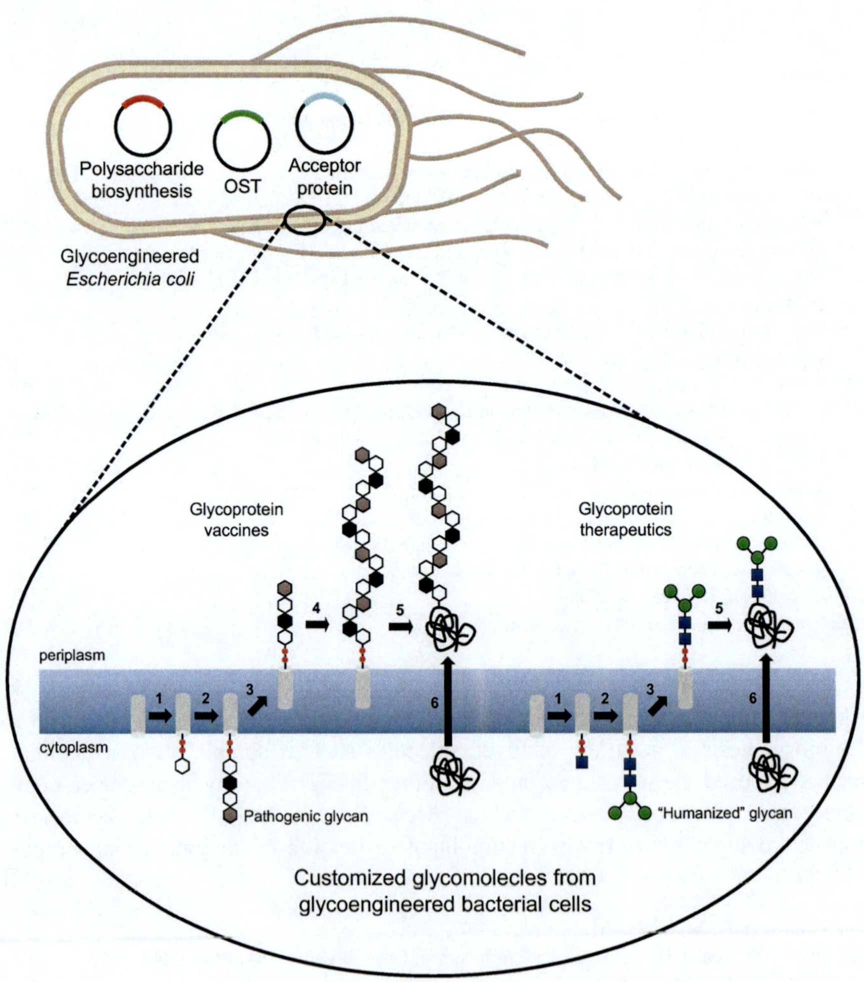

**Keywords** Bacterial oligosaccharyltransferase, Bacterial polysaccharides, Bacterial protein glycosylation, Carbohydrate biosynthesis pathways, Conjugate vaccines, Glycoengineering, Glycosyltransferase

## Abbreviations

| | |
|---|---|
| ABC-transporter | ATP-binding cassette transporter |
| CPS | Capsular polysaccharide |
| diNAcBac | Bacillosamine |
| ECA | Enterobacterial common antigen |
| EPA | Exotoxin A from *Pseudomonas aeruginosa* |
| EPO | Erythropoietin |
| Gal | Galactose |
| GalNAc | *N*-Acetylgalactosamine |
| $Gb_3$ | Globotriaosylceramide |
| $Gb_4$ | Globotetraosylceramide |
| Glc | Glucose |
| GlcNAc | *N*-Acetylglucosamine |
| GM | Monosialotetrahexosylganglioside |
| HA | Hyaluronic acid |
| hGH | Human growth hormone |
| Hib | *Haemophilus influenza* type b |
| IgG | Immunoglobulin G |
| LacNAc | *N*-Acetyllactosamine |
| $Le^X$ | Lewis X antigen |
| $Le^y$ | Lewis Y antigen |
| LOS | Lipo-oligosaccharide |
| LPS | Lipopolysaccharide |
| Man | Mannose |
| MBP | Maltose binding protein |
| NCAM | Neural cell adhesion molecule |
| NeuNAc | *N*-Acetylneuraminic acid |
| OMV | Outer membrane vesicle |
| PEG | Polyethylene glycol |
| PolySia | Polysialic acid |
| S-layer | Surface layer |
| STEC | Shigatoxin producing *Escherichia coli* |
| STX | Shiga toxin |
| T-antigen | Thomsen–Friedenreich antigen |
| Und-PP | Undecaprenyl pyrophosphate |

# 1 Bacteria as a Platform for Polysaccharide and Glycoconjugate Production

In recent years there has been growing interest in developing bacterial species as hosts for glycoengineering applications involving the biosynthesis of structurally diverse polysaccharides, which can be produced as free glycans or as conjugates to lipids or proteins. The most obvious advantage of this approach is the much simpler and cheaper culturing conditions required for maintenance of bacterial cells when compared to a eukaryotic cell culture. However, bacteria are highly proficient producers of carbohydrates, with more than 140 unique monosaccharide base types identified in bacterial species, in contrast to the 14 base types produced by mammalian species [1]. Many of these bacterial monosaccharides are then assembled into an even more diverse array of polysaccharides, often as part of surface structures such as capsular polysaccharide (CPS) and the *O*-antigen component of lipopolysaccharide (LPS), which are often important virulence factors in pathogenic species. In *Escherichia coli* alone, 187 unique *O*-antigen structures and 80 CPS structures have been identified to date [2–4]. Other bacterial polysaccharides have important structural functions (e.g., peptidoglycan), or play a role in adaptation to environmental conditions by mechanisms such as osmoregulation (e.g., enterobacterial common antigen, ECA) [5].

The pathways responsible for production of mono- and polysaccharides are frequently well defined in bacteria, especially in commonly used host species such as *E. coli* [6]. Furthermore, with the exception of the ubiquitous structural polysaccharide peptidoglycan, bacterial polysaccharides are typically not essential for viability, meaning biosynthesis pathways are amenable to genetic manipulation and deletion. For example, metabolic engineering studies have identified routes to enhance the availability of relevant nucleotide-activated sugars, leading to improved polysaccharide yields [7]. As a result of these and other related efforts, bacteria have been shown to represent a tractable, well-defined platform for engineering the biosynthesis of polysaccharides.

Although the ability of bacteria to produce polysaccharides and glycolipids is established, it was long believed that bacteria were incapable of modifying proteins with carbohydrate moieties, a process known as glycosylation. However, this paradigm was overturned in the 1970s with the identification of glycosylated surface layer (S-layer) proteins in *Halobacterium salinarum*, *Clostridium thermosaccharolyticum*, and *Clostridium thermohydrosulfuricum* [8, 9]. Although examples of bacterial protein glycosylation remain relatively uncommon, in the past 15 years a diverse array of systems has been discovered and characterized, including examples of sequential and en bloc transfer of both *N*-linked and *O*-linked glycans [10–13].

From an engineering perspective, perhaps the most significant advance came in 2002 with the functional transfer of a complete protein *N*-glycosylation system from the gastrointestinal pathogen *Campylobacter jejuni* into a laboratory strain of *E. coli*, which is naturally incapable of protein glycosylation [14]. The versatility of this system was further enhanced by a series of experiments demonstrating the modularity of the bacterial glycosylation machinery, which was found to tolerate a number of different glycan structures and protein substrates [15–17]. Importantly, the new-found ability to generate glycoproteins in a genetically tractable host organism such as *E. coli* provided a unique opportunity both to understand and to exploit the glycosylation process in ways that were not previously possible with eukaryotic systems. This is because, even though the pathways involved in the production of protein-linked polysaccharides in eukaryotic cells are well understood, the essential nature of many of these mechanisms limits the potential for manipulation.

## 2 Polysaccharide Production in Bacteria

Enzymatic synthesis of polysaccharides uses nucleotide-activated sugars as glycosyl donors to supply the necessary energy for the reaction. In bacteria, these nucleotide sugars are typically only present in the cytoplasm where they are synthesized. Consequently, all initial polysaccharide biosynthesis in bacteria also takes place within the cytoplasm. The majority of polysaccharides are synthesized by one of three pathways: the Wzy-dependent pathway, the ATP-binding cassette (ABC) transporter-dependent pathway, and the synthase-dependent pathway (Fig. 1), although shorter oligosaccharides may be formed by the direct action of glycosyltransferases on a substrate such as lipid A in the case of the LPS core or lipooligosaccharides (LOSs) [18].

The Wzy-dependent pathway involves the sequential action of glycosyltransferases on a lipid anchor, undecaprenyl diphosphate (Und-PP), on the inner leaflet of the cytoplasmic membrane, followed by translocation of a completed subunit across the membrane by the flippase Wzx. The subunits then undergo polymerization by the polymerase Wzy. The number of repeat units is modulated somewhat by Wzz, the chain-length regulator, although the resulting polymers are not strictly uniform in length. Completed polysaccharides are then removed from Und-PP and transferred to a target location, which differs depending on the species in question and the type of polysaccharide produced [19]. Common examples of polysaccharides produced by this mechanism include the majority of *O*-antigen polysaccharides and a significant proportion of capsules, as well as specific examples such as ECA, a surface polysaccharide common to most *Enterobacteriacae*, but limited to this family [5].

In contrast, the ABC transporter-dependent pathway involves the assembly of the entire polysaccharide on a lipid anchor at the inner face of the cytoplasmic membrane, before the chain is capped to indicate completion, and the entire structure is transported across the membrane by the ABC-transporter complex [20]. As with Wzy-dependent systems, however, the polysaccharide is then removed from the lipid anchor and transferred to a permanent point of attachment. Polysaccharides assembled by this method typically form *O*-antigen polysaccharides or capsules.

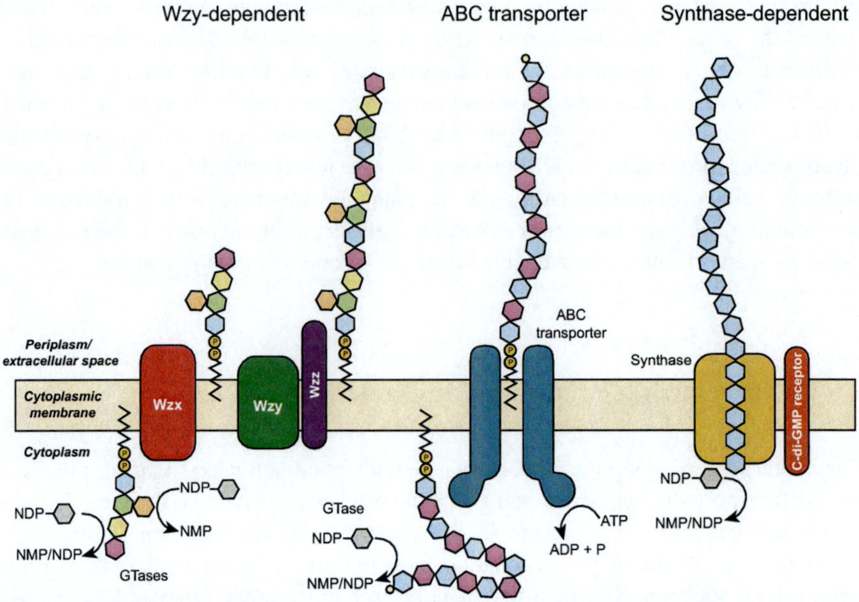


**Fig. 1** Biosynthesis of bacterial polysaccharides. The majority of bacterial polysaccharides are assembled by one of three mechanisms, the Wzy-dependent, the ABC transporter-dependent, or the synthase-dependent pathway. The key protein components for each mechanism are indicated, and are located in the inner membrane of Gram-negative organisms or the membrane of Gram-positive organisms. Polysaccharides are synthesized from nucleotide diphosphate (NDP) sugars. For the Wzy-dependent pathway, multiple glycosyltransferases (GTases) in the cytoplasm synthesize oligosaccharides on Und-PP. Oligosaccharides typically contain diverse monosaccharides and may be branched; consequently this assembly mechanism is responsible for the production of most high-complexity sugars. The completed oligosaccharide repeat unit is transported across the relevant membrane by the translocase or flippase enzyme Wzx. Multiple repeat units are then linked together by the polymerase enzyme Wzy to form a repeating heteropolymer. The final length of the polymer may be controlled by the chain length regulator Wzz. In the ABC transporter-dependent pathway, a homopolymer or simple heteropolymer is assembled on Und-PP on the cytoplasmic face of the membrane, often by just a single GTase. The completed polysaccharide is capped with a moiety such as a phosphate group, and transported through the membrane by the ATP-binding cassette (ABC) transporter. For synthase-dependent biosynthesis, the polysaccharide is simultaneously polymerized and transported across the membrane. In the absence of a membrane anchor, a receptor protein for a signaling molecule such as bis-(3′-5′)-cyclic dimeric guanosine monophosphate (c-di-GMP) may play a role in initiation of polysaccharide assembly. In Gram-negative organisms, polysaccharides are frequently transported across the outer membrane by an additional export system to enable surface display

Synthase-dependent polysaccharide assembly is unique in that it can occur in the presence or absence of a lipid anchor. A transmembrane glycosyltransferase simultaneously catalyzes formation of the polymer and translocation across the membrane [21]. Polysaccharides produced by this mechanism may be attached to the exterior of the cell, but more frequently they are released into the extracellular environment to form non-covalently associated exopolysaccharides such as hyaluronic acid (HA), alginate, or cellulose.

## 3 Bioengineering of Secreted Oligosaccharides in Bacteria

Small, soluble oligosaccharides play many important roles in biological systems, and as such have a multitude of potential uses in research, medicine, and industry. However, because of the extremely high heterogeneity of such structures, together with low yield and complex purification when isolating from natural sources, engineered production has been the focus of much research. Chemical synthesis is complex and costly, and the resulting oligosaccharides are subject to the same issues regarding heterogeneity, limiting their usefulness without significant downstream purification. Chemo- and in vitro-enzymatic methods have also been widely explored, and have shown great improvements with respect to yield and structural homogeneity, but isolation of the required enzymes is a demanding process, and the necessary nucleotide-activated sugars are extremely expensive to supply for such large-scale synthesis. Consequently, production beyond the milligram scale, especially for larger tri- and tetrasaccharides, remains unfeasible by this method.

The development of a metabolically-engineered *E. coli* strain that could produce human milk oligosaccharides in a fermentation process represented a significant advance within the field [22]. The engineered strain utilizes glycerol as an affordable carbon source, relying on native metabolic pathways within the bacterium to produce a continuous supply of the required nucleotide sugars. The approach also relies on the presence of a soluble acceptor sugar in the cytoplasm as an assembly platform. In this case, lactose, which can be imported from the growth medium, was used. However, methods for the in situ synthesis of acceptor sugars have also been developed [23]. Such engineered strains have been shown to produce quantities of up to 34 g/L of secreted oligosaccharide, and the scalable nature of production means the manufacture of kilogram quantities of sugar are entirely feasible [24]. This approach has since been used for the production of more than 25 different oligosaccharides ranging from disaccharides to pentasaccharides, including structures known to have immunomodulatory effects or to be associated with cancer in humans [7]. It should be pointed out that transporters for milk oligosaccharides larger than about four or five residues are unknown. As a result, these molecules accumulate in the bacterial cytoplasm, resulting in feedback inhibition and, more importantly, requiring purification from bacterial lysates. Overcoming this bottleneck is necessary for the development of a food- or pharmaceutical-grade production process. It is also worth mentioning that in 2015 the U.S. FDA granted approval for 2-fucosyllactose, one of the most abundant human milk oligosaccharides, produced by bacterial fermentation, to be used as an ingredient in infant and toddler formula (see https://www.accessdata.fda.gov/scripts/fdcc/?set=GRASNotices&id=571). This was followed by European approval in 2017.

## 4 Bioengineering of Exopolysaccharides in Bacteria

Many exopolysaccharides produced by bacteria have significant commercial value [25], the most widely studied of which are listed in Table 1. Some of these polymers occur naturally in bacteria, and others have been engineered via heterologous gene expression, particularly in cases where the original source or isolation method was undesirable. One example is HA, an extremely hydrophilic polymer of alternating β-D-glucuronic acid and β-D-*N*-acetyl-glucosamine residues that is a desirable material in medicine and cosmetics because of its high water retention capacity and lack of toxicity. Initially, this polysaccharide was purified from rooster combs, although the majority of production is now achieved via microbial fermentation [26]. Native bacterial production of HA was first achieved from *Streptococcus zooepidemicus* [27] but, because of co-production of the streptolysin exotoxin, recombinant production remained a priority. Indeed, recombinant HA was eventually achieved using the host organism *Bacillus subtilis* [28] and subsequently *E. coli* [29]. Such approaches achieve yields of ~10 g/L, which is thought to be near the production limit because of the effect of the exopolysaccharide on the viscosity of the growth medium [29]. Key advances have come instead in the area of polymer length regulation, allowing for better control of physiochemical properties, and achieved largely through metabolic engineering and tighter control of the availability of the precursor nucleotide sugars [30].

In other cases, such as the commercially valuable xanthan, metabolic engineering has enabled yields of up to 50 g/L, also thought likely to be the highest level feasible for bioreactor processing [31]. Further increases require additional engineering strategies to alter the molecular structure of the polysaccharide and reduce the resulting viscosity via modifications such as limiting polymer length or altering the degree of acylation or pyruvylation of a compound [25]. Bacterial production also offers unprecedented levels of purity when compared to extraction methods from other sources – for example, cellulose free from the common plant contaminants lignin and hemicellulose [32]. Furthermore, with the growing understanding of the pathways behind bacterial synthesis of such exopolysaccharides and recent advances in bioinformatics and systems biology, it may soon be possible to engineer bacteria to produce entirely novel polysaccharides with useful chemical properties. Indeed, a metabolic engineering approach was recently used to synthesize a variant form of cellulose containing a proportion of *N*-acetylglucosamine (GlcNAc) monomers in addition to the usual glucose. This modification resulted in the production of a biopolymer that is far more readily biodegradable than the standard form [33].

**Table 1** Extensively studied bacterial exopolysaccharides: composition, sources and uses

| EPS | Components | Organism | Main applications[a] |
|---|---|---|---|
| Cellulose | Glucose | *Gluconacetobacter xylinus* | Foods (indigestible fiber)<br>Wound healing<br>Engineered blood vessels<br>Audio speaker diaphragms |
| Xanthan | Glucose<br>Mannose<br>Glucuronic acid<br>Acetate<br>Pyruvate | *Xanthomonas campestris* | Foods<br>Petroleum industry<br>Pharmaceuticals<br>Cosmetics and personal care products<br>Agriculture |
| Alginate | Guluronic acid<br>Mannuronic acid<br>Acetate | *Pseudomonas aeruginosa, Azotobacter vinelandii* | Surgical dressings<br>Wound management<br>Controlled drug release |
| Gellan | Glucose<br>Rhamnose<br>Glucuronic acid<br>Acetate<br>Glycerate | *Sphingomonas paucimobilis* | Foods<br>Pet food<br>Pharmaceuticals<br>Agar substitute |
| Dextran | Glucose | *Leuconostoc mesenteroides* | Foods<br>Blood volume expander<br>Chromatographic media |
| Curdlan | Glucose | *Agrobacterium tumefaciens, Alcaligenes faecalis* | Foods<br>Pharmaceuticals<br>Heavy metal removal<br>Concrete additive |
| Hyaluronic acid | Glucuronic acid<br>GlcNAc | *S. zooepidemicus, B. subtilis* | Medicine<br>Solid culture media |
| Succinoglycan | Glucose<br>Galactose<br>Acetate<br>Pyruvate<br>Succinate | *Sinorhizobium meliloti* | Food<br>Oil recovery |
| Levan | Fructose | *B. subtilis, Zymomonas mobilis* | Food (prebiotic)<br>Medicines<br>Cosmetics |

[a]Summarized from [25]

# 5 Bioengineering of Intracellular and Cell-Associated Polysaccharides in Bacteria

The most widely manipulated cellular polysaccharide biosynthesis system is probably the LPS pathway (Fig. 2), in part because of the significance of this polysaccharide in pathogenesis, but also because of the conserved mechanistic nature of the pathway combined with the highly variable glycan structures produced.

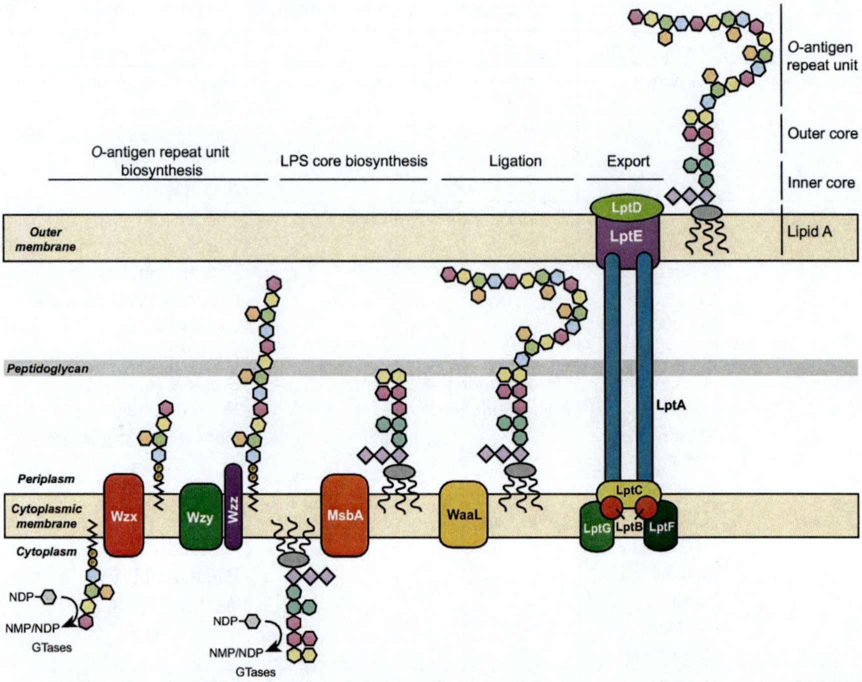


**Fig. 2** LPS biosynthesis. Multiple glycosyltransferases (GTases) transfer NDP-sugars to the nascent oligosaccharide to form an *O*-antigen repeat unit on Und-PP in the cytoplasm of the cell. The completed oligosaccharide is transported across the inner membrane by the flippase, Wzx, and multiple repeat units are joined together by the polymerase enzyme Wzy to form the completed *O*-antigen portion of the LPS. The structure shown is representative and does not indicate a specific serotype. Simultaneously, the LPS core, comprising lipid A, and the inner and outer core sugars are assembled in the cytoplasm. The inner core consists of two or three Kdo monosaccharides (*diamond shapes*) which are added during synthesis of the lipid A molecule, and three heptose sugars (*heptagon shapes*) which are added by the sequential action of three GTases. The outer core shown is an R1 structure, consisting of three glucose and two galactose residues (*hexagons*), and is assembled by the sequential action of a further five GTases. The completed LPS core is transported across the inner membrane by the ABC transporter MsbA. The *O*-antigen repeat unit is removed from the Und-PP membrane anchor and attached to the first galactose on the R1 outer core by the ligase enzyme WaaL. The entire LPS structure is then extracted from the inner membrane and transported across the periplasm and through the outer membrane to the extracellular face by the Lpt protein complex, where lipid A becomes a component of the outer face of the outer membrane with the polysaccharide displayed on the surface of the cell

The tendency for genes responsible for production of a bacterial polysaccharide to be organized as a single, continuous operon, especially in the case of *O*-antigens and CPS, has greatly facilitated the transfer of polysaccharide coding loci from their native species into a heterologous host, typically *E. coli*. Early methods generally centered on the generation of a cosmid library from fragmented genomic DNA, followed by screening of individual cosmids at the genomic or phenotypic level to locate clones conferring production of the polysaccharide of interest. This approach has been used to produce a variety of *O*-antigens from Gram-negative organisms including *Pseudomonas aeruginosa*, *Salmonella typhimurium*, and *Yersinia enterocolitica* in an *E. coli* strain background [34]. A similar approach has also been used for the production of CPS from the Gram-positive organism *Streptococcus pneumoniae* in the Gram-positive host *Lactococcus lactis* [35]. Cloning of sequenced, annotated polysaccharide biosynthetic loci has enabled production in *E. coli* of polysaccharides from diverse Gram-negative species such as *Bukholderia pseudomallei* [36] and *Francisella tularensis* [37]. A further advance was the recent demonstration that various CPS structures from the Gram-positive bacterium *S. pneumoniae* could be produced in a Gram-negative host, namely *E. coli*, using the en bloc transfer of the entire CPS coding locus [38, 39]. The recombinant CPS structures are produced essentially as an *O*-antigen in *E. coli*, and some features of processing appear to be borrowed from the host, including the action of the *O*-antigen ligase WaaL in attaching the polymerized polysaccharide to the outer core on lipid A, and subsequent transport to the outer surface of the cell. These findings demonstrated an unexpected cross-compatibility between systems from two disparate sources, and highlighted the mechanistic similarity of CPS biosynthesis in Gram-positive bacteria and *O*-antigen biosynthesis in Gram-negative bacteria.

The wide availability of whole-genome sequences and a thorough understanding of the mechanisms behind bacterial polysaccharide biosynthesis have recently led to a more informed approach to the production of heterologous polysaccharides. A recent study produced two different *Staphylococcus aureus* CPS structures by expressing combinations of *P. aeruginosa* and *S. aureus* glycosyltransferases in *E. coli*, with sugar precursors provided by a combination of *P. aeruginosa* enzymes along with native enzymes in the *E. coli* host. The resulting glycans were confirmed by MALDI-TOF/TOF tandem mass spectrometry analysis as having the same structure as the native CPS, and were recognized by capsular serotype-specific typing antiserum [40]. Hence, bacterial glycosyltransferase enzymes may be regarded as modular entities defined only by function, opening up a new approach to polysaccharide bioengineering in host species such as *E. coli*. This insight also facilitates the engineering of bacterial glycans in cases where information regarding the biosynthesis of a target polysaccharide (and/or its intermediates) is incomplete or incompatible with further processing as a result of assembly on a lipid other than Und-PP. For example, the Vi polysaccharide of *Salmonella enterica* serovar Typhi is currently licensed as a purified polysaccharide vaccine for typhoid fever, but represents an interesting candidate for further development as a glycoconjugate. Unfortunately, recombinant production of this polysaccharide is challenging because the lipid on which it is assembled in the native host is not currently known. To circumvent this issue, Wetter et al. modified the *E. coli* O121 *O*-antigen, a structure well-known to build on Und-PP, to resemble the Vi polysaccharide. Following transfer of the resulting Vi-like polysaccharide to a carrier protein, a glycoconjugate was produced that elicited antibodies immunoreactive with *E. coli* O121 LPS [41].

## 6 Bioengineering of Eukaryotic Polysaccharides on the LPS Core in Bacteria

The ability to expand the bacterial polysaccharide production system to engineer structures beyond prokaryotic polysaccharides is crucial if this approach is to become broadly applicable and useful. Several human-like glycans have been assembled on a truncated LPS outer core structure. Typical mutations involve the disruption of the second glycosyltransferase enzyme of the outer core, resulting in an intact lipid A molecule, coupled to a complete inner core structure, but with only a single glucose residue from the outer core added to the second heptose residue of the inner core (see Fig. 2). This exposed glucose then becomes the attachment site for recombinant glycans, and the Lpt export system translocates the resulting LOS structure to the surface of the cell, ensuring the recombinant glycan is exposed [42].

The human glycosphingolipid globotriaosylceramide ($Gb_3$) is the receptor for Shiga-toxin (Stx), a potent $AB_5$ toxin produced by pathogenic species such as *Shigella dysenteriae* and *E. coli* O157. This receptor is composed of a trisaccharide, Gal(α1–4)Gal(β1–4)Glc, and is present on many eukaryotic cell types, but is found at highest concentrations in renal tissue and in microvascular endothelial cells [43]. An analogous structure to the $Gb_3$ receptor is produced by *Neisseria spp.* as a component of LOS and is representative of a common strategy used by mucosal pathogens whereby surface display of host glycan epitopes aids immune evasion [44]. Expression of the glycosyltransferases LtxC from *Neisseria meningitidis*, and LtxE from *Neisseria gonorrhoeae* in *E. coli* resulted in the production of a novel LPS-associated $Gb_3$ polysaccharide structure. When administered to mice, the engineered *E. coli* were found to protect against challenge with a Shiga-toxin producing *E. coli* (STEC) strain, suggesting an effective molecular mimic of the toxin binding site had been recreated that sequestered the secreted toxin [43]. An analogous approach has been used to engineer *E. coli* cells that express molecular mimics for other receptors implicated in bacterial toxin binding – globotetraosylceramide ($Gb_4$) and the gangliosides $GM_1$ and $GM_2$ [45, 46]. These engineered bacterial strains have also proven efficacious in animal models for the treatment of toxin-associated bacterial infections such as cholera and STEC.

A similar approach was used to produce the ganglioside $GM_3$ epitope, NeuNAcα (2,3)Galβ(1,4), as an attachment to the exposed glucose residue of truncated lipid A [47]. This feat was accomplished by expressing the *Neisseria* enzymes SiaB, a CMP-sialic acid synthetase, together with the galactosyltransferase LgtE and the sialic acid transferase Lst, which together generated a $GM_3$-like structure that was displayed on the surface of the cell. This strain may be useful for investigating the effects of sialic acid-containing bacterial LOS structures and their role in development of post-infection autoimmune diseases such as Guillain-Barre syndrome. Other human-like glycans with a role in bacterial attachment have also been expressed in *E. coli*, including fucosylated oligosaccharides: the blood group H, Lewis X ($Le^X$) and Lewis Y ($Le^Y$) antigens [48], and poly-*N*-acetyllactosamine [49]. Fucose is a common component of human glycans, and is thought to play a role in the binding of various pathogenic bacteria including *P. aeruginosa* and *C. jejuni*, and it is envisioned that these strains may prove useful for studying specific bacterial interactions with human receptors, as well as revealing the design of competitive inhibitors for novel probiotic-based therapies.

A further example of a eukaryotic glycan that may also be produced as a bacterial mimic is polysialic acid (PolySia), a linear homopolymer of α-2,8-linked sialic acid residues. In humans this glycan is most notably found as an elaboration of the *N*-linked glycan on neural cell adhesion molecule (NCAM), but is also expressed by *E. coli* K1 and *N. meningitidis* group B as the K1 capsule and CPS A, respectively [50]. Because of its occurrence on these pathogens as well as its enhanced expression on some malignant tumors [51, 52], PolySia represents an intriguing target for vaccine or therapeutic antibody development. By expressing a combination of glycosyltransferases from *N. gonorrhoeae*, *C. jejuni*, and *E. coli*, Valentine and co-workers were able to produce PolySia directly on the LPS core of an *E. coli* strain not normally capable of synthesizing this structure. Interestingly, where the aforementioned $GM_3$ production study supplied NeuAc via the growth medium and relied on a single synthetase enzyme to convert the sugar into the nucleotide activated form CMP-NeuAc [47], the authors reconstituted the entire biosynthesis pathway capable of converting the readily available housekeeping sugar UDP-GlcNAc into CMP-NeuAc [53], highlighting the flexibility and versatility of bacteria as hosts for glycoengineering.

## 7 Bioengineering of Eukaryotic Polysaccharides on the Lipid Anchor Und-PP in Bacteria

Because direct conjugation to the LPS core is not always possible or desirable, alternative sites for polysaccharide assembly have also been explored, such as the common lipid anchor Und-PP. In *E. coli* K-12, the ECA and *O*-antigen biosynthesis pathways involve installation of a GlcNAc residue on Und-PP by an initiating glycosyltransferase called WecA. By introducing glycosyltransferases from the *Haemophilus influenzae* LOS biosynthesis pathway that were capable of modifying this Und-PP-linked GlcNAc in the recombinant system, a tetrasaccharide resembling the $Le^X$ antigen (minus the fucose residue) was assembled on Und-PP [54]. The use of this lipid as a carrier enabled subsequent conjugation of the glycan to a protein using an oligosaccharyltransferase-mediated mechanism that is described in greater detail below. To complete the $Le^X$ structure, the purified glycoconjugate was subjected to in vitro enzymatic elaboration to add the fucose residue [54]. The use of engineered bacteria to produce $Le^X$ containing glycoproteins is significant because these proteins are known to function as immunomodulatory molecules [55–57], and have been shown to ameliorate symptoms associated with autoimmune disorders in animal models [58].

Another human-like glycan produced in a similar manner is the Thomsen–Friedenreich antigen (T antigen), a Galβ1-3GalNAc disaccharide. Valentine et al. [53] used UndPP-linked GlcNAc as a primer for producing the T antigen disaccharide. This was accomplished by addition of two heterologous glycosyltransferases and a nucleotide sugar epimerase to ensure availability of the required substrate UDP-GalNAc. Because T antigen is overexpressed on a number of malignancies, including breast, colon, prostate, and stomach cancers [59], recombinant biosynthesis could yield highly immunogenic glycoconjugates that elicit antibodies against this important glycan epitope.

A final example of engineering human-like glycans in a bacterial host involved the bottom-up creation of a eukaryotic *N*-glycan biosynthesis pathway. Specifically, the conserved core of all human *N*-glycans, the oligosaccharide $Man_3GlcNAc_2$, was successfully produced on Und-PP by co-expression of four eukaryotic glycosyltransferases, including the yeast uridine diphosphate-GlcNAc transferases Alg13 and Alg14 and the mannosyltransferases Alg1 and Alg2 [60]. By including a bacterial oligosaccharyltransferase PglB from *C. jejuni*, glycans were successfully transferred to eukaryotic target proteins as discussed below. The $Man_3GlcNAc_2$ structure has been shown to be the minimal structure required for efficacy of a glycoprotein therapeutic [61], and is the predominant glycoform conjugated to proteins expressed in a baculovirus host system. Furthermore, as the conserved core of human *N*-glycans, this structure has enormous potential as a precursor for further modification, either in vivo or in vitro.

## 8 Glycoprotein Expression in Bacterial Hosts: Current Applications and Future Opportunities

The above findings demonstrate the remarkable versatility of bacterial systems for the biosynthesis of a vast array of carbohydrate structures. However, to exploit the full potential of carbohydrates, it is often necessary to conjugate these structures to additional biomolecules such as proteins. Two different mechanisms are responsible for making the majority of proteins that become covalently modified with sugar molecules (i.e., glycoproteins). These mechanisms are defined based on the amino acid residue onto which the glycan is installed. In *N*-linked glycosylation, the glycan is attached to the nitrogen atom of an asparagine residue, whereas in *O*-linked glycosylation the sugar moiety is attached to the oxygen atom of either a serine or a threonine side chain. Although both types of glycosylation were long believed to occur exclusively in eukaryotes, multiple bacterial machineries for the generation of both types of modifications have been discovered over the last 15 years. These bacterial glycosylation systems, or hybrids thereof, have opened the door to using bacteria for the production of two important classes of glycoproteins – (1) glycoconjugate vaccines, whereby immunogenic carbohydrates from pathogens including bacteria and viruses are linked to proteins and (2) therapeutic proteins that are glycosylated in their natural form and require modification for full function, for example, monoclonal antibodies.

Glycoconjugates are amongst the most successful vaccines generated to date, eliciting a robust T-cell-dependent immune response and conferring protection across all age groups [62]. For three important bacterial pathogens in particular, *H. influenza* type B (Hib), *S. pneumoniae*, and *N. meningitidis*, glycoconjugates have proven to be highly effective in countries where they have been introduced [63, 64]. The standard production method for these conjugates involves the separate generation and purification of the protein and the carbohydrate moiety, chemical activation thereof, and conjugation as well as subsequent purification of the resulting glycoprotein [65]. Even though it is an established and accepted method, there are several drawbacks to this approach. First, it requires culturing large volumes of a pathogenic species of interest for the generation of the native carbohydrate, followed by harvesting and purification of the carbohydrate. Depending on the biosafety level of the species of interest, as well as the ease of culturing, this step can present a major hurdle regarding the expansion of the technique to novel pathogenic species. Second, the activation and chemical conjugation steps required to couple the glycan to the carrier protein can be technically challenging and inefficient, resulting in low yields, as well as a heterogeneous population of glycoproteins with different numbers of target glycans attached at different locations throughout the protein. Therefore, alternative methods for generating glycoconjugates that overcome some of these limitations are desired.

In addition to glycoconjugate vaccines, many proteins of therapeutic interest are also glycoproteins. In fact, 70% of therapeutic proteins approved by regulatory agencies or currently in clinical and preclinical trials are decorated with glycans in their native form [66]. Historically, this has limited the use of *E. coli* to proteins and peptides not natively glycosylated, such as insulin and homologues thereof, or to those that are natively glycosylated but are functional without the addition of the glycan moiety, such as human growth hormone (hGH) and interferon α [67]. It should be pointed out that these proteins often require additional post-translational modifications such as the addition of polyethylene glycol (PEG) to increase serum half-life [68]. Although some notable breakthroughs have been made [60], the routine use of *E. coli* as a production platform for therapeutic glycoproteins and glycopeptides requires further engineering of glycosylation pathways in this host.

# 9 *N*-Linked Glycoprotein Expression in Bacteria

The discovery of an *N*-glycosylation machinery in the human intestinal bacterial pathogen *C. jejuni* [12] and the subsequent functional transfer of the complete machinery into the more tractable species *E. coli* [14] demonstrated for the first time that bacteria could be an alternative source of recombinant *N*-glycoproteins. Subsequent studies showed that a single enzyme, an oligosaccharyltransferase named *Cj*PglB (PglB from *C. jejuni*), was responsible for transferring the glycan to the acceptor protein. Interestingly, this enzyme was shown to share sequence homology with the STT3 catalytic subunit of the eukaryotic oligosaccharyltransferase enzyme complex [14]. A functional study of the genes within the glycosylation locus demonstrated that the substrate glycan was assembled on the lipid carrier Und-PP [69], in a fashion similar to the *O*-antigen biosynthesis pathway present in many Gram-negative species of bacteria [70]. It was further demonstrated that the *Cj*PglB enzyme possesses remarkably relaxed glycan substrate

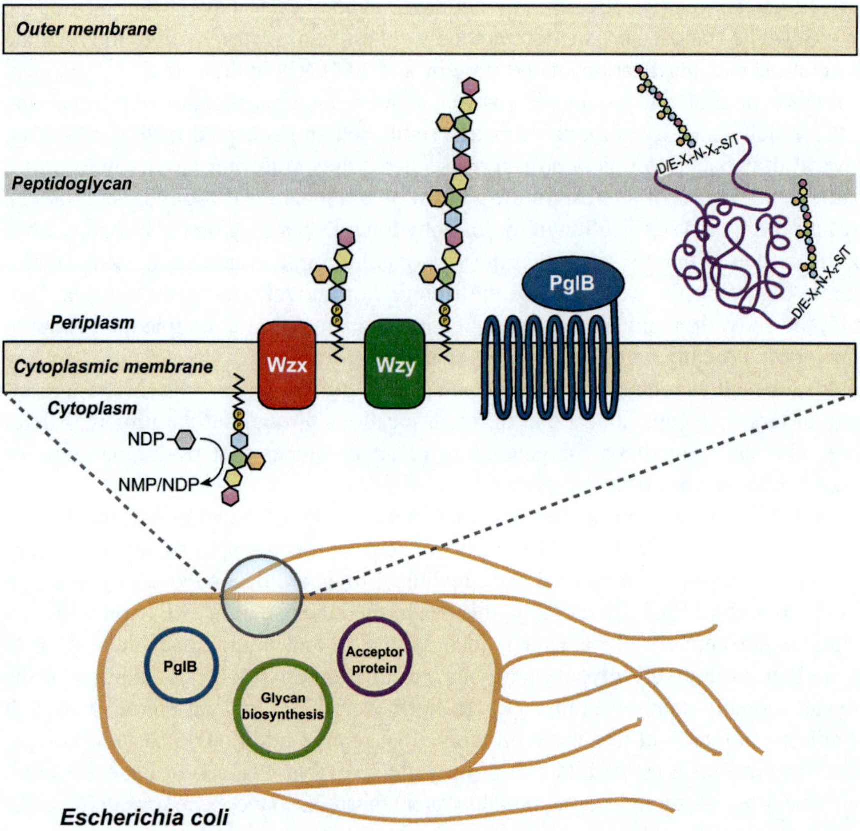


**Fig. 3** Recombinant protein glycosylation in *E. coli* using the bacterial oligosaccharyltransferase. Co-expression of three components is required for recombinant glycosylation in *E. coli*: (1) the glycan biosynthetic locus for the production of the carbohydrate of interest on the lipid carrier undecaprenol pyrophosphate; (2) the oligosaccharyltransferase (e.g., *Cj*PglB); and (3) the acceptor protein of interest that has been engineered with a signal peptide for export into the periplasm and an acceptor sequon (e.g., D/E-$X_1$-N-$X_2$-S/T, where X can be any amino acid except proline) for glycosylation by the oligosaccharyltransferase. Sequons can be engineered into an exposed, flexible loop or at either the *N*- or the *C*-terminus of the protein. The glycoprotein can then be purified from the bacterial cells using standard methods

specificity. That is, in addition to its native substrate oligosaccharide – a heptasaccharide glycan with the structure diNAcBacGalNAc$_5$Glc [71] – the enzyme was also able to recognize much larger polysaccharides such as structurally different bacterial *O*-antigens and transfer these to proteins [15]. Around the same time, a five amino acid glycosylation sequon for *Cj*PglB was discovered [16], which could be engineered either into flexible secondary structures within a protein [16] or at either the *N*- or the *C*-terminus [72]. Altogether, these studies provided the requisite ingredients for making customized recombinant bacterial glycoproteins, where potentially any protein of interest could be modified with any glycan moiety at a desired position by co-expression of *Cj*PglB, the glycan of interest assembled on Und-PP, and the desired acceptor protein modified to contain one or more glycosylation sequon(s) (Fig. 3).

## 10 Customized *N*-Glycoproteins Produced Recombinantly in *E. coli*

To date, the predominant class of glycoproteins produced using the above components consists of conjugates in which bacterial surface glycan structures are site-specifically linked to immunogenic carrier proteins. In the majority of published cases, the glycans are *O*-antigen polysaccharides built on Und-PP (see above for in-depth discussion of the different methods used for the recombinant production of these structures) and installed on the carrier protein by *Cj*PglB. Table 2 summarizes the glycoconjugate vaccine candidates generated and tested to date. Although multiple studies have demonstrated the generation of specific, and potentially protective antibody responses against *E. coli*-derived glycoconjugate vaccine candidates, it is particularly noteworthy that two have been successfully tested in Phase I trials. The first is a conjugate vaccine candidate against *S. dysenteriae* type 1 composed of the *O*-antigen glycan coupled to the exotoxin A of *P. aeruginosa* (EPA). Testing of this vaccine candidate in healthy adults at two different doses with or without co-adminstration of adjuvant revealed it to be well-tolerated and capable of eliciting statistically significant antigen-specific humoral immune responses [74]. A second conjugate vaccine candidate comprised of the *Shigella flexneri* 2a *O*-antigen conjugated to EPA was also tested in healthy adults, with similar results regarding tolerance and immunogenicity [77]. Hence, recombinant production of glycogonjugates in *E. coli* appears to be a promising alternative to the traditional methods used for biomanufacturing conjugate vaccines.

Glycoconjugate proteins produced recombinantly in *E. coli* have found uses in other applications as well. For instance, bacterial glycoconjugates have been successfully used as diagnostic tools for human and bovine brucellosis [81, 82] as well as for the Shiga-toxin-producing *E. coli* serotypes O157, O145, and O121 [83]. Additionally, Shang and co-workers generated a glycoconjugate comprised of the maltose binding protein (MBP) and the *E. coli* O86:B7 *O*-antigen, which bears structural similarity to the blood group B antigen epitope. This glycoconjugate functioned as a ‘molecular sponge’ to lower the levels of blood group B antibodies in plasma without negatively affecting the clotting function of the plasma [84].

Although there is a great deal of promise for glycoconjugates where the sugar moiety is derived from an immunogenic bacterial glycan, these types of glycans are not useful in applications where the goal is to install native, eukaryotic glycans onto therapeutic proteins. Several attempts have been made to leverage the bacterial protein glycosylation machinery for the generation of glycoproteins carrying mammalian glycans. Perhaps the most notable example is Valderrama-Rincon et al. [60]

**Table 2** List of bacterial glycoconjugate vaccine candidates produced using a bacterial glycoengineering approach

| Target species | Carrier protein | Carbohydrate | Animal model | Safety in humans | Immunogenicity in humans | References |
|---|---|---|---|---|---|---|
| *S. dysenteriae* type 1 | *C. jejuni* model glycoprotein AcrA, Exotoxin A from *P. aeruginosa* (EPA) | *O*-antigen | | X | X | [73–75] |
| *S. flexneri* 2a | EPA | *O*-antigen | | X | X | [76, 77] |
| *F. tularensis* | EPA | *O*-antigen | Mice | | | [37] |
| *B. pseudomallei* | AcrA | *O*-antigen | Mice | | | [36] |
| *Brucella abortus* | AcrA | *O*-antigen of *Y. enterocolitica* O9[a] | Mice | | | [78] |
| *E. coli* serotypes O1, O2, O6 and O25 | EPA | *O*-antigens of the four strains | Mice, Rabbits, Rats | | | [79] |
| *E. coli* O157:H7 | Maltose binding protein (MBP) | *O*-antigen[b] | Mice | | | [80] |
| *S. typhi* | EPA | Vi capsule (*O*-antigen of *E. coli* O121)[c] | Mice | | | [41] |
| *S. aureus* serotypes 5 and 8 | EPA and *S. aureus* α toxin | Capsular polysaccharide | Mice | | | [40] |

[a]Glycoconjugate generated by recombinant expression of the glycosylation machinery in *Y. enterocolitica* serotype O9, taking advantage of the structural identity of the *O*-antigen of this species with the target species
[b]Glycoconjugate generated by recombinant expression of the glycosylation machinery directly in the target *E. coli* serotype
[c]Glycoconjugate generated by recombinant expression of a related *O*-antigen structure modified to resemble the target glycan

who demonstrated the complete recombinant assembly and transfer to protein of the eukaryotic *N*-linked core glycan $GlcNAc_2Man_3$ (see above for description of the approach used for biosynthesis of the glycan). Transfer of the $GlcNAc_2Man_3$ glycan to asparagine residues in several different target proteins including the Fc domain of human immunoglobulin G (IgG) was achieved with *Cj*PglB which, as mentioned above, has fairly relaxed specificity toward the glycan substrate. One can imagine an extension of this glycan, either in vivo or potentially in vitro, to generate additional structures found in mammalian *N*-glycans. It should be pointed out that, despite successful eukaryotic protein glycosylation, the yield of glycosylated proteins in this seminal report was reported to be ~50 μg/L, which amounted to only a small fraction (<1%) of each expressed protein under the conditions tested [60]. It was proposed that increasing these levels would likely require, among other things, strategies for relieving enzymatic and metabolic bottlenecks and/or optimizing the glycosylation enzymes. Indeed, a flow cytometric approach was recently used to optimize pathway enzyme expression in a manner that resulted in enhanced production of lipid-linked $Man_3GlcNAc_2$ [85]. In turn, the yield of glycosylated acceptor proteins produced by these optimized strains appeared to be very efficient, with nearly 100% of the acceptor protein undergoing conversion to the glycosylated form. Moreover, yields of ~14 mg/L were achieved in the best cases, representing an improvement of two orders of magnitude compared to our earlier report, and rivaling the yield (25 mg/L) reported for *E. coli*-based production of carrier proteins glycosylated with bacterial polysaccharides [73].

In an alternative approach, post-processing of a purified pre-form of the glycoconjugate outside of the bacterial cell can be performed to generate the final product. For instance, the same $GlcNAc_2Man_3$ glycan structure was installed on a protein by a combination of recombinant in vivo glycosylation of the protein with the *Campylobacter lari* heptasaccharide glycan, $GalNAc_5GlcNAc$, followed by in vitro enzymatic trimming of the glycan down to a single GlcNAc residue, and finally transglycosylation of the trimmed glycan with a preassembled $Man_3GlcNAc$ sugar to obtain the final structure [86]. However, for large-scale production, the cost burden of cGMP-compliant precursors for the transglycosylation reaction likely limits the applicability of this approach.

A similar combined method of in vitro and in vivo glycosylation and modification was used to install the blood group antigen $Le^x$ on a protein [54]. The recombinantly expressed tetrasaccharide $GalNAc_2Gal_2$ was produced on Und-PP in *E. coli* and this glycan was subsequently transferred in vivo to an acceptor protein using *Cj*PglB. Following purification, in vitro fucosylation was performed to yield the final $Le^x$ glycan on the protein. Although these combined in vivo and in vitro methods of glycoprotein biosynthesis are potentially less applicable to large scale production of glycoproteins, they nevertheless expand the range of glycan modifications on proteins, which may be beneficial for the generation of glycoproteins carrying sugars that are potentially too challenging for the expression and transfer in vivo alone.

## 11 Expanding Glycosylation Through Identification of Alternative Oligosaccharyltransferases

Although *Cj*PglB remains one of the best-characterized bacterial oligosaccharyltransferases, there are two main limitations that restrict its use for the coupling of designer glycans to acceptor proteins. First, compared to the canonical eukaryotic glycosylation sequon, N-X-S/T (where X can be any amino acid except proline), used by eukaryotic oligosaccharyltransferases, *Cj*PglB requires an extended sequon (D/E-$X_1$-N-$X_2$-S/T) for the attachment of glycans to proteins [16]. One consequence of this requirement is that, at a minimum, these five amino acids need to be engineered into the protein of interest either by addition of the residues as a terminal or internal tag or by changing of a native stretch of amino acids to render it a substrate for glycosylation. If these modifications are added to either of the termini, it can be speculated that this does not have a major impact on the overall structure and function of the protein. However, it may be desirable to engineer the site of glycan attachment into the protein, in which case these modifications may interfere with protein folding and/or function. Another consequence is that native *N*-glycoproteins of mammalian origin need to have their shorter sequons extended to include a D or E in the −2 position to be glycosylated by *Cj*PglB. It should be noted that the need to extend sequons in this manner may change the properties and immunogenicity of the modified protein.

To address this limitation, several groups have used bioinformatics to identify orthologues of *Cj*PglB, which were then functionally characterized in glyco-competent *E. coli* cells [87–91]. From these studies, oligosaccharyltransferases were identified from two species of *Desulfovibrio* that did not require the negatively charged amino acid at position −2 and were therefore able to glycosylate the shorter eukaryotic N-X-S/T sequon [87, 90]. Of these, only the PglB orthologue of *Desulfovibrio gigas* was able to modify the native QYNST sequon in the Fc domain of human IgG [90], suggesting that additional factors govern acceptor-site specificity and must be satisfied to allow for the installation of glycans onto shorter eukaryotic sequons. Additionally, the orthologue from *Desulfovibrio desulfuricans* showed markedly lower efficiency in transferring the *E. coli* O7 *O*-antigen polysaccharide [87], suggesting that this enzyme may not be as flexible as *Cj*PglB regarding the glycan structure. As no other polysaccharides were tested as substrates for the *D. desulfuricans* PglB, it is unclear whether the low efficiency of transfer of the O7 *O*-antigen is specific to this substrate or an inherent property of the enzyme. The ability of the orthologues from *Desulfovibrio vulgaris* and *D. gigas* to transfer mono- and polysaccharides was not tested, so it remains unclear whether these enzymes may be useful in the generation of custom glycoconjugates.

In parallel with the functional characterization of *Cj*PglB orthologues, a directed evolution approach has been applied to *Cj*PglB with the goal of relaxing the acceptor-sequon specificity. Using the crystal structure of the closely related PglB enzyme of *C. lari* [92] as a guide, combined with a high-throughput genetic screen using a secreted acceptor protein, a library of *Cj*PglB mutants was screened for the ability of the enzyme to glycosylate non-canonical acceptor protein sites [93]. This screen identified three *Cj*PglB variants that no longer required the negatively charged residue at the −2 position. The three mutants glycosylated a eukaryotic protein at its native N-X-S/T sequon, suggesting that these enzymes may be useful for authentically glycosylating eukaryotic proteins and peptides. Although the glycan specificity was not specifically tested, the fact that the mutants were derived from *Cj*PglB suggests that the relaxed glycan specificity of the parent enzyme remains.

A second limitation of *Cj*PglB is the requirement of the native enzyme for an acetamido group at the monosaccharide that constitutes the reducing end of the oligo- or polysaccharide [17]. Many glycans of interest do not terminate in a glycan that conforms to this requirement, such as most capsular glycans of *S. pneumoniae* serotypes that terminate in either galactose or glucose residues [94]. Although a natural variant among the orthologues of *Cj*PglB enzymes from other species may lack this requirement, evidence for this has yet to be reported. In fact, two studies analyzing the protein *N*-glycan diversity within the *Campylobacter* genus and in one species of *Helicobacter* identified exclusively sugars containing an acetamido group at the reducing end [95, 96], suggesting that this is a shared feature among many of the bacterial species that possess protein *N*-glycosylation machineries. The same appears to be true for the sugar attached to an identified glycoprotein in *D. gigas*, which was *N*-glycosylated with a disaccharide of GlcNAc and *N*-acetylallosamine [97]. To address this issue, one study used structure-guided mutagenesis to engineer a *Cj*PglB variant that was able to transfer two *O*-antigens from *S. typhimurium* that both contain non-acetylated sugars (galactose residues) at the reducing end [98]. This work demonstrates that the glycan specificity of *Cj*PglB can be engineered to a certain extent, and suggests that in the future it should be possible to transfer virtually any glycan to any protein using modified versions of *Cj*PglB.

## 12 Alternative Routes for Bacterial Protein *N*-Linked Glycosylation

A novel family of bacterial enzymes has recently emerged that may be of potential use in bacterial glycoengineering. In contrast to the enzymes described in the previous section, these enzymes: (1) are active in the bacterial cytoplasm, not the periplasm; (2) use nucleotide-activated glycans instead of lipid-linked glycans as a substrate; and (3) recognize the shorter, bacterial N-X-S/T glycosylation sequon [99]. The first member of the family was discovered in *H. influenzae* and was shown to be involved in the glycosylation of the high molecular weight adhesin protein HMW1 [11]. The glycans attached to the adhesin protein were identified predominantly as hexose sugars, and glycosylation of the adhesin protein was demonstrated to be important for correct secretion of the adhesin as well as adhesion of the bacteria to airway epithelial cells [100]. Further members of the family have been identified in several other species of bacteria [99], and in vitro experiments confirmed activity of the orthologues from *Y. enterocolitica* and *Actinobacillus pleuropneumoniae* [101]. The preferred substrate for the *A. pleuropneumoniae* enzyme (termed ApNGT) was demonstrated to be UDP-Glc [101], and a downstream gene was shown to encode a glycosyltranferase enzyme that was able to extend the Glc moiety installed by ApNGT with further Glc residues. Additionally, when expressed in *E. coli*, ApNGT was shown to glycosylate recombinantly co-expressed auto-transporter proteins from the same species (the enzyme's native substrate), as well as co-expressed human erythropoietin (EPO) and several native *E. coli* proteins [102]. A polypeptide modified with a glucose moiety by ApNGT was also successfully elaborated through in vitro transglycosylation mediated by endoglycosidase enzymes [103]. This suggests that ApNGT and other enzymes from this family may be useful tools for installation of a priming glucose residue on proteins of interest, followed by either in vitro or in vivo elaboration of the glycan. It can also be imagined that directed evolution of the enzyme from this family may allow for the modulation of the carbohydrate specificity in a similar way to *Cj*PglB.

## 13 Customized *O*-Glycoproteins Produced Recombinantly in *E. coli*

In addition to the bacterial *N*-glycosylation mechanisms discussed above, pathways that lead to the modification of serine or threonine residues (*O*-linked glycosylation) have also been identified in several bacterial species. These mechanisms are more commonly found in bacteria than their *N*-glycosylation counterparts [104], and are currently being pursued for recombinant protein glycosylation. The following section highlights similarities and differences between the *N*- and *O*-linked pathways.

Over the last decade, *O*-glycosylation machineries that share mechanistic similarities with the *N*-glycosylation pathways described above have been identified and characterized in several bacterial species [104]. It was initially observed that the type IV pilus subunit protein PilA in *P. aeruginosa* strain 1244 was modified with a glycan in a manner dependent on the product of the gene adjacent to *pilA* named PilO/TfpO [10]. A similar machinery was identified in *N. meningitidis*, whereby deletion of a gene termed *pglL* led to the loss of a carbohydrate moiety from the pilus subunit protein PilE [105]. Interestingly, both the *P. aeruginosa* PilO/TfpO and the *N. meningitidis* PglL proteins showed homology to *O*-antigen ligase proteins that are involved in transfer of the *O*-antigen subunit from the lipid carrier Und-PP onto the lipid A moiety during LPS biogenesis [106]. This suggested that these enzymes may use Und-PP-linked glycans as substrate. Analysis of the glycan structure present on *P. aeruginosa* PilA showed the presence of a single *O*-antigen repeat unit, further strengthening the hypothesis that Und-PP-linked glycans may be the substrate for this enzyme family [107]. When PilO/TfpO and PilA from *P. aeruginosa* (or PglL and PilE from *N. meningitidis*) were recombinantly co-expressed in *E. coli* along with a Und-PP-linked oligo- or polysaccharide, transfer of the glycan to the pilin protein was observed [108]. These results not only demonstrated recombinant activity of this new family of bacterial *O*-oligosaccharyltransferase enzymes, but also confirmed the substrate identity as Und-PP-linked glycans. Further analysis of the glycan specificity of PglL demonstrated a remarkable promiscuity with regards to the glycan. Diverse glycan structures were shown to be transferred to PilE by PglL in vivo including structures containing a Gal residue at the reducing end such as the *S. typhimurium* LT2 *O*-antigen and the disaccharide-pentapeptide peptidoglycan building block, none of which are substrates for the *C. jejuni* oligosaccharyltransferase *Cj*PglB [109]. Additionally, in vitro glycosylation experiments revealed that the enzyme displayed flexibility toward the lipid carrier [109, 110]. Altogether, these characteristics suggest that this enzyme is a very promising tool for the generation of designer glycoproteins with *O*-linked sugars.

To date, however, the biotechnological use of this enzyme family has been hampered by one major bottleneck. Unlike in the case of *Cj*PglB, there is a lack of a consensus sequon for glycosylation that would allow for the 'tagging' of any protein as a substrate for *O*-glycosylation. Analysis of the *O*-glycome of several organisms that possess PglL-like *O*-glycosylation systems identified multiple glycosylated proteins, and although these helped to determine that the amino acid residues around the glycan attachment site were rich in serine, proline, and alanine, they did not reveal the presence of any consensus sequence [111–114]. Toward a more universal glycosylation strategy, Qutyan and coworkers showed that a C-terminal fusion of *E. coli* alkaline phosphatase with the final 15 amino acids from the C-terminus of PilA was glycosylated by PilO/TfpO when expressed in *P. aeruginosa*; however, the observed glycosylation was not very efficient [115]. Additionally, although it has been shown that PilO/TfpO has relatively relaxed specificity and was able to transfer multiple different serotype *O*-glycans of *P. aeruginosa* [116], the enzyme was only able to transfer a single *O*-antigen subunit both in the native organism and recombinantly in *E. coli* [108, 116]. Hence, alternative PilO/TfpO *O*-oligosaccharyltransferases need to be identified or engineered for transferring longer polysaccharides, which are often desirable for glycoengineering purposes. This issue appears to have been solved recently by Pan and co-workers [117] who reported the development and optimization of an *O*-linked 'glycosylation tag' consisting of an eight amino acid motif flanked by two approximately ten amino acid sequences containing mainly hydrophilic residues. This tag was successfully fused to both the *N*- and *C*-termini of three potential vaccine carrier proteins – the cholera toxin B subunit, exotoxin A from *P. aeruginosa*, and the detoxified variant of diphtheria toxin CRM197 – and glycosylated with two different sugars including the *S. typhimurium* LT2 *O*-antigen, which, as discussed above, is not a substrate for *Cj*PglB. Recombinant *O*-glycoproteins produced with this method were tested in a series of animal experiments and elicited a glycan-specific antibody response [117]. The ability to tag proteins for PglL-dependent *O*-glycosylation opens up this enzyme family for biotechnological applications, in particular in cases where the glycan of interest may not be an optimal substrate for *N*-glycosylation by *Cj*PglB.

## 14 Alternative Routes for Bacterial Protein *O*-Linked Glycosylation

Many bacterial species possess *O*-glycosylated flagellar proteins, with the glycosylation patterns ranging from a single glycan at a single site to multiple glycans attached to different sites on the protein [118]. These glycans are installed in a processive manner, with individual glycosyltransferases adding the glycans sequentially to the protein. This mechanism is similar to the installation of *O*-linked glycans in eukaryotic mucin-like glycosylation [119]. It could therefore be hypothesized that enzymes from these machineries could potentially be used/engineered to install mucin-like glycans on human proteins. The successful recombinant installation of the first monosaccharide of the core of human mucin-like glycan, a GalNAc residue, has been demonstrated in the cytoplasm of *E. coli* using a recombinantly expressed human GalNAc transferase enzyme [120]. However, no further elaboration of this priming glycan with other sugars has been demonstrated.

## 15 Alternative Therapeutic Bacterial Conjugates

Although some unconjugated polysaccharides are currently licensed as vaccines, they often elicit a T-cell independent immune response stimulated by the extensive cross-linking of receptors on the surface of B cells. As such, they are poorly immunogenic in children less than 2 years of age and elderly patients, greatly limiting their usefulness [121]. Although protein conjugation is the most widely studied approach to counter this problem, the field of bacterial glycobiology is opening up alternative approaches to boost the immunogenicity of carbohydrate epitopes.

One such approach is based on bacterial outer membrane vesicles (OMVs), which are small (20–200 nm) liposomes released from the outer membrane of nearly all Gram-negative bacterial species. These vesicles are non-replicating versions of their bacterial 'parent', and contain many of the same components as the bacterial outer membrane, including membrane proteins, CPS, and LOS and LPS, as well as some of the luminal components of the bacterial periplasm [122]. OMVs have garnered interest as vaccine candidates because vesicles from several bacterial pathogens have been shown to possess potent immunogenic capacities [123–125]. Intriguingly, OMVs also appear to possess intrinsic adjuvant properties, potentially removing the need to include adjuvants in the formulation [126, 127]. OMVs derived directly from pathogenic *N. meningitidis* have been successfully incorporated into a commercial vaccine formulation, the recently licensed Bexsero [128, 129]. Native OMVs have been further engineered to carry additional immunogenic proteins, which are recombinantly displayed on the surface of the OMV through genetic fusion to outer membrane proteins or in the OMV lumen through periplasmic expression [126, 130]. Importantly, robust immune responses against these recombinant immunogens have been demonstrated [126, 130]. Three recent reports highlight a novel bacterial glycoengineering approach to OMV-based vaccines whereby immunogenic glycans are recombinantly displayed on the exterior of OMVs. The approach takes advantage of the following: (1) the fact that standard laboratory strains of *E. coli* have lost the ability to produce a native *O*-antigen glycan because of the insertion of an IS element in the second glycosyltransferase gene *wbbL* [131] whereas the rest of the mechanism including the flippase and ligase genes remain intact; (2) the ability to express recombinantly non-native polysaccharides in *E. coli*; (3) the fact that the *O*-antigen ligase WaaL has relative relaxed glycan specificity and efficiently transfers engineered glycans from Und-PP to the lipid A-core in cells that lack the native *O*-antigen [132]; and importantly (4) the recombinant *O*-antigen is efficiently transported to the cell surface and packaged into released OMVs. Using this approach, *E. coli*-derived glycosylated OMVs (glycOMVs) have been decorated with the *O*-antigens of eight Gram-negative bacterial species, including *F. tularensis* [133], PolySia [53], the CPS of *S. pneumoniae* serotype 14, and the *N*-linked heptasaccharide of *C. jejuni* [39]. Following immunization, the glycOMVs carrying the *F. tularensis O*-antigen were shown to elicit significant serum titers of class-switched, glycan-specific IgG antibodies in mice, and prolonged survival upon challenge with the highly virulent *F. tularensis* subsp. *tularensis* (type A) strain Shu S4 [133]. Likewise, glycOMVs decorated with PolySia also elicited glycan-specific IgG antibodies in mouse immunization studies, and the serum antibodies had potent bactericidal activity, killing *N. meningitidis* serogroup B bacteria that possess a PolySia capsular glycan [53]. GlycOMVs carrying the *S. pneumoniae* serotype 14 CPS also elicited glycan-specific antibodies in mice, and the serum antibodies were shown to possess potent bactericidal properties when tested in an opsonophagocytic assay. In fact, the bacterial killing of the serum from mice vaccinated with the glycOMVs carrying the capsular glycan was as efficient as the serum from mice that had been vaccinated with the commercial glycoconjugate vaccine Prevnar13® [39]. Finally, glycOMVs displaying the *C. jejuni N*-linked glycan were shown to significantly lower levels of *C. jejuni* colonization in chickens [39]. The expansion of the technology to cover further species or serotypes is thought to be relatively straightforward, simply requiring the recombinant expression of a pathogen-specific glycan structure on the surface of *E. coli* cells.

A related approach to glycOMV vaccines is the development of whole-cell vaccines displaying recombinant glycan epitopes. This strategy also leverages the fact that recombinant polysaccharides assembled on Und-PP are often efficiently transferred to lipid A and displayed as recombinant chimeric LPS on the surface of Gram-negative bacteria. This approach has been evaluated using several different species of Gram-negative bacteria as hosts (*S. enterica* serovar Typhi, *S. enterica* serovar Typhimurium, and *E. coli*) carrying biosynthesis gene clusters for immunogenic carbohydrates of *S. dysenteriae* serotype O1 [134], shiga-toxin producing *E. coli* serotype O111 [135], and *C. jejuni* [136]. In contrast to glycOMV vaccine candidates, these whole-cell vaccine candidates replicate. Although it is desirable to control their ability to replicate, a balance needs to be found between controlling the replication of the bacteria and ensuring they persist long enough in the vaccinated organisms to generate a desired immune response. Genetic inactivation of the *aroA* gene encoding a 5-enolpyruvylshikimate-3-phosphate synthetase, involved in the shikimate pathway that directly connects glycolysis to the synthesis of aromatic amino acids [137], is a commonly used strategy to attenuate live bacterial vaccine candidates. This is particularly useful in species of *Salmonella* as these mutants are able to grow in rich media in vitro but become self-limiting in vivo, where aromatic amino acids are not freely available [138]. However, recent data suggest that deletion of *aroA*, at least in *S. enterica* serovar Typhimurium, can lead to additional effects in cellular physiology that may have an influence on the behavior of the recombinant bacteria within the host [139]. Nonetheless, attenuated, glycan epitope-expressing bacteria offer an additional opportunity for glycoengineering of vaccine candidates, in particular in areas where minimal cost of production may be a priority, such as in poultry and other livestock vaccines.

## 16 Concluding Remarks and Outlook

In summary, bacterial expression systems have been successfully used for the production of a variety of carbohydrate structures ranging from small secreted oligosaccharides to repeating polymers of high molecular weight, and spanning structures found in all kingdoms of life. Furthermore, the characterization of both *N*- and *O*-linked protein glycosylation systems in a variety of bacterial species has greatly enhanced the potential of bacterial systems for the generation of therapeutically relevant glycoconjugates. These bacterial conjugation systems have been employed to generate well-defined therapeutic compounds, including the first conjugate vaccines produced entirely in bacteria as well as novel immunogenic entities such as glycosylated outer membrane vesicles. Two of these bacterially-derived glycoconjugates have recently undergone successful Phase I clinical trials, and new candidates are also emerging.

Owing to their versatility and ease of manipulation, bacteria are ideal hosts for the production of a diverse array of structurally defined polysaccharides and glycoconjugates that are of interest as medical and industrial products. Furthermore, the low costs associated with the culturing of bacterial strains, especially *E. coli*, opens up this technology to a far wider range of laboratories than existing chemical/chemoenzymatic synthesis methods or mammalian cell culture approaches. The findings from a recent report commissioned by the National Academy of Sciences states that "glycans play roles in almost every biological process and are involved in every major disease" and further asserts that "the development of transformative methods for the facile synthesis of carbohydrates and glycoconjugates should be a high priority" [140]. Bacterial glycoengineering represents an emerging field with the potential to play a major role in meeting these goals.

## References

1. Herget S, Toukach PV, Ranzinger R, Hull WE, Knirel YA, von der Lieth CW (2008) Statistical analysis of the Bacterial Carbohydrate Structure Data Base (BCSDB): characteristics and diversity of bacterial carbohydrates in comparison with mammalian glycans. BMC Struct Biol 8:35
2. Senchenkova SN, Guo X, Naumenko OI, Shashkov AS, Perepelov AV, Liu B, Knirel YA (2016) Structure and genetics of the O-antigens of Escherichia coli O182-O187. Carbohydr Res 435:58–67
3. Stenutz R, Weintraub A, Widmalm G (2006) The structures of Escherichia coli O-polysaccharide antigens. FEMS Microbiol Rev 30(3):382–403
4. Whitfield C (2006) Biosynthesis and assembly of capsular polysaccharides in Escherichia coli. Annu Rev Biochem 75:39–68
5. Kuhn HM, Meier-Dieter U, Mayer H (1988) ECA, the enterobacterial common antigen. FEMS Microbiol Rev 4(3):195–222
6. Aoki-Kinoshita KF, Kanehisa M (2015) Glycomic analysis using KEGG GLYCAN. Methods Mol Biol 1273:97–107
7. Ruffing A, Chen RR (2006) Metabolic engineering of microbes for oligosaccharide and polysaccharide synthesis. Microb Cell Factories 5:25
8. Sleytr UB (1975) Heterologous reattachment of regular arrays of glycoproteins on bacterial surfaces. Nature 257(5525):400–402
9. Sleytr UB, Thorne KJ (1976) Chemical characterization of the regularly arranged surface layers of Clostridium thermosaccharolyticum and Clostridium thermohydrosulfuricum. J Bacteriol 126(1):377–383
10. Castric P (1995) pilO, a gene required for glycosylation of Pseudomonas aeruginosa 1244 pilin. Microbiology 141(5):1247–1254
11. Grass S, Buscher AZ, Swords WE, Apicella MA, Barenkamp SJ, Ozchlewski N, St Geme 3rd JW (2003) The Haemophilus influenzae HMW1 adhesin is glycosylated in a process that requires HMW1C and phosphoglucomutase, an enzyme involved in lipooligosaccharide biosynthesis. Mol Microbiol 48(3):737–751
12. Szymanski CM, Yao R, Ewing CP, Trust TJ, Guerry P (1999) Evidence for a system of general protein glycosylation in Campylobacter jejuni. Mol Microbiol 32(5):1022–1030
13. Thibault P, Logan SM, Kelly JF, Brisson JR, Ewing CP, Trust TJ, Guerry P (2001) Identification of the carbohydrate moieties and glycosylation motifs in Campylobacter jejuni flagellin. J Biol Chem 276(37):34862–34870
14. Wacker M, Linton D, Hitchen PG, Nita-Lazar M, Haslam SM, North SJ, Panico M, Morris HR, Dell A, Wren BW, Aebi M (2002) N-Linked glycosylation in Campylobacter jejuni and its functional transfer into E. coli. Science 298(5599):1790–1793
15. Feldman MF, Wacker M, Hernandez M, Hitchen PG, Marolda CL, Kowarik M, Morris HR, Dell A, Valvano MA, Aebi M (2005) Engineering N-linked protein glycosylation with diverse O antigen lipopolysaccharide structures in Escherichia coli. Proc Natl Acad Sci U S A 102 (8):3016–3021
16. Kowarik M, Young NM, Numao S, Schulz BL, Hug I, Callewaert N, Mills DC, Watson DC, Hernandez M, Kelly JF, Wacker M, Aebi M (2006) Definition of the bacterial N-glycosylation site consensus sequence. EMBO J 25(9):1957–1966
17. Wacker M, Feldman MF, Callewaert N, Kowarik M, Clarke BR, Pohl NL, Hernandez M, Vines ED, Valvano MA, Whitfield C, Aebi M (2006) Substrate specificity of bacterial oligosaccharyltransferase suggests a common transfer mechanism for the bacterial and eukaryotic systems. Proc Natl Acad Sci U S A 103(18):7088–7093
18. Kalynych S, Morona R, Cygler M (2014) Progress in understanding the assembly process of bacterial O-antigen. FEMS Microbiol Rev 38(5):1048–1065
19. Raetz CR, Whitfield C (2002) Lipopolysaccharide endotoxins. Annu Rev Biochem 71:635–700

20. Cuthbertson L, Kos V, Whitfield C (2010) ABC transporters involved in export of cell surface glycoconjugates. Microbiol Mol Biol Rev 74(3):341–362
21. Whitney JC, Howell PL (2013) Synthase-dependent exopolysaccharide secretion in Gram-negative bacteria. Trends Microbiol 21(2):63–72
22. Priem B, Gilbert M, Wakarchuk WW, Heyraud A, Samain E (2002) A new fermentation process allows large-scale production of human milk oligosaccharides by metabolically engineered bacteria. Glycobiology 12(4):235–240
23. Samain E, Drouillard S, Heyraud A, Driguez H, Geremia RA (1997) Gram-scale synthesis of recombinant chitooligosaccharides in Escherichia coli. Carbohydr Res 302(1–2):35–42
24. Drouillard S, Mine T, Kajiwara H, Yamamoto T, Samain E (2010) Efficient synthesis of 6′-sialyllactose, 6,6′-disialyllactose, and 6′-KDO-lactose by metabolically engineered E. coli expressing a multifunctional sialyltransferase from the Photobacterium sp. JT-ISH-224. Carbohydr Res 345(10):1394–1399
25. Schmid J, Sieber V, Rehm B (2015) Bacterial exopolysaccharides: biosynthesis pathways and engineering strategies. Front Microbiol 6:496
26. Liu L, Liu Y, Li J, Du G, Chen J (2011) Microbial production of hyaluronic acid: current state, challenges, and perspectives. Microb Cell Factories 10:99
27. Thonard JC, Migliore SA, Blustein R (1964) Isolation of hyaluronic acid from broth cultures of streptococci. J Biol Chem 239:726–728
28. Widner B, Behr R, Von Dollen S, Tang M, Heu T, Sloma A, Sternberg D, Deangelis PL, Weigel PH, Brown S (2005) Hyaluronic acid production in Bacillus subtilis. Appl Environ Microbiol 71(7):3747–3752
29. Yu H, Stephanopoulos G (2008) Metabolic engineering of Escherichia coli for biosynthesis of hyaluronic acid. Metab Eng 10(1):24–32
30. Jia Y, Zhu J, Chen X, Tang D, Su D, Yao W, Gao X (2013) Metabolic engineering of Bacillus subtilis for the efficient biosynthesis of uniform hyaluronic acid with controlled molecular weights. Bioresour Technol 132:427–431
31. Seviour RJ, McNeil B, Fazenda ML, Harvey LM (2011) Operating bioreactors for microbial exopolysaccharide production. Crit Rev Biotechnol 31(2):170–185
32. Klemm D, Heublein B, Fink HP, Bohn A (2005) Cellulose: fascinating biopolymer and sustainable raw material. Angew Chem Int Ed Engl 44(22):3358–3393
33. Yadav V, Paniliatis BJ, Shi H, Lee K, Cebe P, Kaplan DL (2010) Novel in vivo-degradable cellulose-chitin copolymer from metabolically engineered Gluconacetobacter xylinus. Appl Environ Microbiol 76(18):6257–6265
34. Goldberg JB, Hatano K, Meluleni GS, Pier GB (1992) Cloning and surface expression of Pseudomonas aeruginosa O antigen in Escherichia coli. Proc Natl Acad Sci U S A 89 (22):10716–10720
35. Gilbert C, Robinson K, Le Page RW, Wells JM (2000) Heterologous expression of an immunogenic pneumococcal type 3 capsular polysaccharide in Lactococcus lactis. Infect Immun 68(6):3251–3260
36. Garcia-Quintanilla F, Iwashkiw JA, Price NL, Stratilo C, Feldman MF (2014) Production of a recombinant vaccine candidate against Burkholderia pseudomallei exploiting the bacterial N-glycosylation machinery. Front Microbiol 5:381
37. Cuccui J, Thomas RM, Moule MG, D'Elia RV, Laws TR, Mills DC, Williamson D, Atkins TP, Prior JL, Wren BW (2013) Exploitation of bacterial N-linked glycosylation to develop a novel recombinant glycoconjugate vaccine against Francisella tularensis. Open Biol 3 (5):130002
38. Kay EJ, Yates LE, Terra VS, Cuccui J, Wren BW (2016) Recombinant expression of Streptococcus pneumoniae capsular polysaccharides in Escherichia coli. Open Biol 6 (4):150243
39. Price NL, Goyette-Desjardins G, Nothaft H, Valguarnera E, Szymanski CM, Segura M, Feldman MF (2016) Glycoengineered outer membrane vesicles: a novel platform for bacterial vaccines. Sci Rep 6:24931

40. Wacker M, Wang L, Kowarik M, Dowd M, Lipowsky G, Faridmoayer A, Shields K, Park S, Alaimo C, Kelley KA, Braun M, Quebatte J, Gambillara V, Carranza P, Steffen M, Lee JC (2014) Prevention of Staphylococcus aureus infections by glycoprotein vaccines synthesized in Escherichia coli. J Infect Dis 209(10):1551–1561
41. Wetter M, Kowarik M, Steffen M, Carranza P, Corradin G, Wacker M (2013) Engineering, conjugation, and immunogenicity assessment of Escherichia coli O121 O antigen for its potential use as a typhoid vaccine component. Glycoconj J 30(5):511–522
42. Merritt JH, Ollis AA, Fisher AC, DeLisa MP (2013) Glycans-by-design: engineering bacteria for the biosynthesis of complex glycans and glycoconjugates. Biotechnol Bioeng 110 (6):1550–1564
43. Paton AW, Morona R, Paton JC (2000) A new biological agent for treatment of Shiga toxigenic Escherichia coli infections and dysentery in humans. Nat Med 6(3):265–270
44. Cress BF, Englaender JA, He W, Kasper D, Linhardt RJ, Koffas MA (2014) Masquerading microbial pathogens: capsular polysaccharides mimic host-tissue molecules. FEMS Microbiol Rev 38(4):660–697
45. Focareta A, Paton JC, Morona R, Cook J, Paton AW (2006) A recombinant probiotic for treatment and prevention of cholera. Gastroenterology 130(6):1688–1695
46. Hostetter SJ, Helgerson AF, Paton JC, Paton AW, Cornick NA (2014) Therapeutic use of a receptor mimic probiotic reduces intestinal Shiga toxin levels in a piglet model of hemolytic uremic syndrome. BMC Res Notes 7:331
47. Ilg K, Yavuz E, Maffioli C, Priem B, Aebi M (2010) Glycomimicry: display of the GM3 sugar epitope on Escherichia coli and Salmonella enterica sv Typhimurium. Glycobiology 20 (10):1289–1297
48. Yavuz E, Maffioli C, Ilg K, Aebi M, Priem B (2011) Glycomimicry: display of fucosylation on the lipo-oligosaccharide of recombinant Escherichia coli K12. Glycoconj J 28(1):39–47
49. Mally M, Fontana C, Leibundgut-Landmann S, Laacisse L, Fan YY, Widmalm G, Aebi M (2013) Glycoengineering of host mimicking type-2 LacNAc polymers and Lewis X antigens on bacterial cell surfaces. Mol Microbiol 87(1):112–131
50. Moe GR, Bhandari TS, Flitter BA (2009) Vaccines containing de-N-acetyl sialic acid elicit antibodies protective against neisseria meningitidis groups B and C. J Immunol 182 (10):6610–6617
51. Komminoth P, Roth J, Lackie PM, Bitter-Suermann D, Heitz PU (1991) Polysialic acid of the neural cell adhesion molecule distinguishes small cell lung carcinoma from carcinoids. Am J Pathol 139(2):297–304
52. Livingston BD, Jacobs JL, Glick MC, Troy FA (1988) Extended polysialic acid chains (n greater than 55) in glycoproteins from human neuroblastoma cells. J Biol Chem 263 (19):9443–9448
53. Valentine JL, Chen L, Perregaux EC, Weyant KB, Rosenthal JA, Heiss C, Azadi P, Fisher AC, Putnam D, Moe GR, Merritt JH, DeLisa MP (2016) Immunization with outer membrane vesicles displaying designer glycotopes yields class-switched, glycan-specific antibodies. Cell Chem Biol 23(6):655–665
54. Hug I, Zheng B, Reiz B, Whittal RM, Fentabil MA, Klassen JS, Feldman MF (2011) Exploiting bacterial glycosylation machineries for the synthesis of a Lewis antigen-containing glycoprotein. J Biol Chem 286(43):37887–37894
55. Atochina O, Daly-Engel T, Piskorska D, McGuire E, Harn DA (2001) A schistosome-expressed immunomodulatory glycoconjugate expands peritoneal Gr1(+) macrophages that suppress naive CD4(+) T cell proliferation via an IFN-gamma and nitric oxide-dependent mechanism. J Immunol 167(8):4293–4302
56. Srivastava L, Tundup S, Choi BS, Norberg T, Harn D (2014) Immunomodulatory glycan lacto-N-fucopentaose III requires clathrin-mediated endocytosis to induce alternative activation of antigen-presenting cells. Infect Immun 82(5):1891–1903
57. van Die I, van Vliet SJ, Nyame AK, Cummings RD, Bank CM, Appelmelk B, Geijtenbeek TB, van Kooyk Y (2003) The dendritic cell-specific C-type lectin DC-SIGN is a receptor for

Schistosoma mansoni egg antigens and recognizes the glycan antigen Lewis x. Glycobiology 13(6):471–478
58. Atochina O, Harn D (2006) Prevention of psoriasis-like lesions development in fsn/fsn mice by helminth glycans. Exp Dermatol 15(6):461–468
59. Heimburg-Molinaro J, Lum M, Vijay G, Jain M, Almogren A, Rittenhouse-Olson K (2011) Cancer vaccines and carbohydrate epitopes. Vaccine 29(48):8802–8826
60. Valderrama-Rincon JD, Fisher AC, Merritt JH, Fan YY, Reading CA, Chhiba K, Heiss C, Azadi P, Aebi M, DeLisa MP (2012) An engineered eukaryotic protein glycosylation pathway in Escherichia coli. Nat Chem Biol 8(5):434–436
61. Van Patten SM, Hughes H, Huff MR, Piepenhagen PA, Waire J, Qiu H, Ganesa C, Reczek D, Ward PV, Kutzko JP, Edmunds T (2007) Effect of mannose chain length on targeting of glucocerebrosidase for enzyme replacement therapy of Gaucher disease. Glycobiology 17 (5):467–478
62. Vella M, Pace D (2015) Glycoconjugate vaccines: an update. Expert Opin Biol Ther 15 (4):529–546
63. Grijalva CG, Nuorti JP, Arbogast PG, Martin SW, Edwards KM, Griffin MR (2007) Decline in pneumonia admissions after routine childhood immunisation with pneumococcal conjugate vaccine in the USA: a time-series analysis. Lancet 369(9568):1179–1186
64. Ladhani SN (2012) Two decades of experience with the Haemophilus influenzae serotype b conjugate vaccine in the United Kingdom. Clin Ther 34(2):385–399
65. Lees A, Puvanesarajah V, Frasch CE (2008) Conjugation chemistry. In: Siber GR, Klugman KP, Makela PH (eds) Pneumococcal vaccines: the impact of conjugate vaccines. ASM Press, Washington DC
66. Sethuraman N, Stadheim TA (2006) Challenges in therapeutic glycoprotein production. Curr Opin Biotechnol 17(4):341–346
67. Ferrer-Miralles N, Domingo-Espin J, Corchero JL, Vazquez E, Villaverde A (2009) Microbial factories for recombinant pharmaceuticals. Microb Cell Factories 8:17
68. Bailon P, Won CY (2009) PEG-modified biopharmaceuticals. Expert Opin Drug Deliv 6 (1):1–16
69. Linton D, Dorrell N, Hitchen PG, Amber S, Karlyshev AV, Morris HR, Dell A, Valvano MA, Aebi M, Wren BW (2005) Functional analysis of the Campylobacter jejuni N-linked protein glycosylation pathway. Mol Microbiol 55(6):1695–1703
70. Hug I, Feldman MF (2011) Analogies and homologies in lipopolysaccharide and glycoprotein biosynthesis in bacteria. Glycobiology 21(2):138–151
71. Young NM, Brisson JR, Kelly J, Watson DC, Tessier L, Lanthier PH, Jarrell HC, Cadotte N, St Michael F, Aberg E, Szymanski CM (2002) Structure of the N-linked glycan present on multiple glycoproteins in the Gram-negative bacterium, Campylobacter jejuni. J Biol Chem 277(45):42530–42539
72. Fisher AC, Haitjema CH, Guarino C, Celik E, Endicott CE, Reading CA, Merritt JH, Ptak AC, Zhang S, DeLisa MP (2011) Production of secretory and extracellular N-linked glycoproteins in Escherichia coli. Appl Environ Microbiol 77(3):871–881
73. Ihssen J, Kowarik M, Dilettoso S, Tanner C, Wacker M, Thony-Meyer L (2010) Production of glycoprotein vaccines in Escherichia coli. Microb Cell Factories 9:61
74. Hatz CF, Bally B, Rohrer S, Steffen R, Kramme S, Siegrist CA, Wacker M, Alaimo C, Fonck VG (2015) Safety and immunogenicity of a candidate bioconjugate vaccine against Shigella dysenteriae type 1 administered to healthy adults: a single blind, partially randomized Phase I study. Vaccine 33(36):4594–4601
75. Ravenscroft N, Haeuptle MA, Kowarik M, Fernandez FS, Carranza P, Brunner A, Steffen M, Wetter M, Keller S, Ruch C, Wacker M (2016) Purification and characterization of a Shigella conjugate vaccine, produced by glycoengineering Escherichia coli. Glycobiology 26(1):51–62
76. Kampf MM, Braun M, Sirena D, Ihssen J, Thony-Meyer L, Ren Q (2015) In vivo production of a novel glycoconjugate vaccine against Shigella flexneri 2a in recombinant Escherichia coli: identification of stimulating factors for in vivo glycosylation. Microb Cell Factories 14:12

77. Riddle MS, Kaminski RW, Di Paolo C, Porter CK, Gutierrez RL, Clarkson KA, Weerts HE, Duplessis C, Castellano A, Alaimo C, Paolino K, Gormley R, Gambillara Fonck V (2016) Safety and immunogenicity of a candidate bioconjugate vaccine against Shigella flexneri 2a administered to healthy adults: a single blind, randomized phase I study. Clin Vaccine Immunol 23(12):908–917
78. Iwashkiw JA, Fentabil MA, Faridmoayer A, Mills DC, Peppler M, Czibener C, Ciocchini AE, Comerci DJ, Ugalde JE, Feldman MF (2012) Exploiting the Campylobacter jejuni protein glycosylation system for glycoengineering vaccines and diagnostic tools directed against brucellosis. Microb Cell Factories 11:13
79. van den Dobbelsteen GP, Fae KC, Serroyen J, van den Nieuwenhof IM, Braun M, Haeuptle MA, Sirena D, Schneider J, Alaimo C, Lipowsky G, Gambillara-Fonck V, Wacker M, Poolman JT (2016) Immunogenicity and safety of a tetravalent E. coli O-antigen bioconjugate vaccine in animal models. Vaccine 34(35):4152–4160
80. Ma Z, Zhang H, Shang W, Zhu F, Han W, Zhao X, Han D, Wang PG, Chen M (2014) Glycoconjugate vaccine containing Escherichia coli O157:H7 O-antigen linked with maltose-binding protein elicits humoral and cellular responses. PLoS One 9(8):e105215
81. Ciocchini AE, Rey Serantes DA, Melli LJ, Iwashkiw JA, Deodato B, Wallach J, Feldman MF, Ugalde JE, Comerci DJ (2013) Development and validation of a novel diagnostic test for human brucellosis using a glyco-engineered antigen coupled to magnetic beads. PLoS Negl Trop Dis 7(2):e2048
82. Ciocchini AE, Serantes DA, Melli LJ, Guidolin LS, Iwashkiw JA, Elena S, Franco C, Nicola AM, Feldman MF, Comerci DJ, Ugalde JE (2014) A bacterial engineered glycoprotein as a novel antigen for diagnosis of bovine brucellosis. Vet Microbiol 172(3–4):455–465
83. Melli LJ, Ciocchini AE, Caillava AJ, Vozza N, Chinen I, Rivas M, Feldman MF, Ugalde JE, Comerci DJ (2015) Serogroup-specific bacterial engineered glycoproteins as novel antigenic targets for diagnosis of Shiga toxin-producing-escherichia coli-associated hemolytic-uremic syndrome. J Clin Microbiol 53(2):528–538
84. Shang W, Zhai Y, Ma Z, Yang G, Ding Y, Han D, Li J, Zhang H, Liu J, Wang PG, Liu XW, Chen M (2016) Production of human blood group B antigen epitope conjugated protein in Escherichia coli and utilization of the adsorption blood group B antibody. Microb Cell Factories 15(1):138
85. Glasscock CJ, Jaroentomeechai T, Yates LE, Wilson JD, Merritt JH, Lucks JB, DeLisa MP (2018) A flow cytometric approach to engineering *Escherichia coli* for improved eukaryotic protein glycosylation. Metab Eng 47:488–495 (in review)
86. Schwarz F, Huang W, Li C, Schulz BL, Lizak C, Palumbo A, Numao S, Neri D, Aebi M, Wang LX (2010) A combined method for producing homogeneous glycoproteins with eukaryotic N-glycosylation. Nat Chem Biol 6(4):264–266
87. Ielmini MV, Feldman MF (2011) Desulfovibrio desulfuricans PglB homolog possesses oligosaccharyltransferase activity with relaxed glycan specificity and distinct protein acceptor sequence requirements. Glycobiology 21(6):734–742
88. Jervis AJ, Langdon R, Hitchen P, Lawson AJ, Wood A, Fothergill JL, Morris HR, Dell A, Wren B, Linton D (2010) Characterization of N-linked protein glycosylation in Helicobacter pullorum. J Bacteriol 192(19):5228–5236
89. Mills DC, Jervis AJ, Abouelhadid S, Yates LE, Cuccui J, Linton D, Wren BW (2016) Functional analysis of N-linking oligosaccharyl transferase enzymes encoded by deep-sea vent proteobacteria. Glycobiology 26(4):398–409
90. Ollis AA, Chai Y, Natarajan A, Perregaux E, Jaroentomeechai T, Guarino C, Smith J, Zhang S, DeLisa MP (2015) Substitute sweeteners: diverse bacterial oligosaccharyltransferases with unique N-glycosylation site preferences. Sci Rep 5:15237
91. Schwarz F, Lizak C, Fan YY, Fleurkens S, Kowarik M, Aebi M (2011) Relaxed acceptor site specificity of bacterial oligosaccharyltransferase in vivo. Glycobiology 21(1):45–54
92. Lizak C, Gerber S, Numao S, Aebi M, Locher KP (2011) X-ray structure of a bacterial oligosaccharyltransferase. Nature 474(7351):350–355

93. Ollis AA, Zhang S, Fisher AC, DeLisa MP (2014) Engineered oligosaccharyltransferases with greatly relaxed acceptor-site specificity. Nat Chem Biol 10(10):816–822
94. Bentley SD, Aanensen DM, Mavroidi A, Saunders D, Rabbinowitsch E, Collins M, Donohoe K, Harris D, Murphy L, Quail MA, Samuel G, Skovsted IC, Kaltoft MS, Barrell B, Reeves PR, Parkhill J, Spratt BG (2006) Genetic analysis of the capsular biosynthetic locus from all 90 pneumococcal serotypes. PLoS Genet 2(3):e31
95. Jervis AJ, Butler JA, Lawson AJ, Langdon R, Wren BW, Linton D (2012) Characterization of the structurally diverse N-linked glycans of Campylobacter species. J Bacteriol 194 (9):2355–2362
96. Nothaft H, Scott NE, Vinogradov E, Liu X, Hu R, Beadle B, Fodor C, Miller WG, Li J, Cordwell SJ, Szymanski CM (2012) Diversity in the protein N-glycosylation pathways within the Campylobacter genus. Mol Cell Proteomics 11(11):1203–1219
97. Santos-Silva T, Dias JM, Dolla A, Durand MC, Goncalves LL, Lampreia J, Moura I, Romao MJ (2007) Crystal structure of the 16 heme cytochrome from Desulfovibrio gigas: a glycosylated protein in a sulphate-reducing bacterium. J Mol Biol 370(4):659–673
98. Ihssen J, Haas J, Kowarik M, Wiesli L, Wacker M, Schwede T, Thony-Meyer L (2015) Increased efficiency of Campylobacter jejuni N-oligosaccharyltransferase PglB by structure-guided engineering. Open Biol 5(4):140227
99. McCann JR, St Geme 3rd JW (2014) The HMW1C-like glycosyltransferases--an enzyme family with a sweet tooth for simple sugars. PLoS Pathog 10(4):e1003977
100. Grass S, Lichti CF, Townsend RR, Gross J, St Geme 3rd JW (2010) The Haemophilus influenzae HMW1C protein is a glycosyltransferase that transfers hexose residues to asparagine sites in the HMW1 adhesin. PLoS Pathog 6(5):e1000919
101. Schwarz F, Fan YY, Schubert M, Aebi M (2011) Cytoplasmic N-glycosyltransferase of Actinobacillus pleuropneumoniae is an inverting enzyme and recognizes the NX(S/T) consensus sequence. J Biol Chem 286(40):35267–35274
102. Naegeli A, Neupert C, Fan YY, Lin CW, Poljak K, Papini AM, Schwarz F, Aebi M (2014) Molecular analysis of an alternative N-glycosylation machinery by functional transfer from Actinobacillus pleuropneumoniae to Escherichia coli. J Biol Chem 289(4):2170–2179
103. Lomino JV, Naegeli A, Orwenyo J, Amin MN, Aebi M, Wang LX (2013) A two-step enzymatic glycosylation of polypeptides with complex N-glycans. Bioorg Med Chem 21 (8):2262–2270
104. Iwashkiw JA, Vozza NF, Kinsella RL, Feldman MF (2013) Pour some sugar on it: the expanding world of bacterial protein O-linked glycosylation. Mol Microbiol 89(1):14–28
105. Power PM, Seib KL, Jennings MP (2006) Pilin glycosylation in Neisseria meningitidis occurs by a similar pathway to wzy-dependent O-antigen biosynthesis in Escherichia coli. Biochem Biophys Res Commun 347(4):904–908
106. Whitfield C, Amor PA, Koplin R (1997) Modulation of the surface architecture of Gram-negative bacteria by the action of surface polymer:lipid A-core ligase and by determinants of polymer chain length. Mol Microbiol 23(4):629–638
107. Castric P, Cassels FJ, Carlson RW (2001) Structural characterization of the Pseudomonas aeruginosa 1244 pilin glycan. J Biol Chem 276(28):26479–26485
108. Faridmoayer A, Fentabil MA, Mills DC, Klassen JS, Feldman MF (2007) Functional characterization of bacterial oligosaccharyltransferases involved in O-linked protein glycosylation. J Bacteriol 189(22):8088–8098
109. Faridmoayer A, Fentabil MA, Haurat MF, Yi W, Woodward R, Wang PG, Feldman MF (2008) Extreme substrate promiscuity of the Neisseria oligosaccharyl transferase involved in protein O-glycosylation. J Biol Chem 283(50):34596–34604
110. Musumeci MA, Hug I, Scott NE, Ielmini MV, Foster LJ, Wang PG, Feldman MF (2013) In vitro activity of Neisseria meningitidis PglL O-oligosaccharyltransferase with diverse synthetic lipid donors and a UDP-activated sugar. J Biol Chem 288(15):10578–10587

111. Elhenawy W, Scott NE, Tondo ML, Orellano EG, Foster LJ, Feldman MF (2016) Protein O-linked glycosylation in the plant pathogen Ralstonia solanacearum. Glycobiology 26 (3):301–311
112. Iwashkiw JA, Seper A, Weber BS, Scott NE, Vinogradov E, Stratilo C, Reiz B, Cordwell SJ, Whittal R, Schild S, Feldman MF (2012) Identification of a general O-linked protein glycosylation system in Acinetobacter baumannii and its role in virulence and biofilm formation. PLoS Pathog 8(6):e1002758
113. Lithgow KV, Scott NE, Iwashkiw JA, Thomson EL, Foster LJ, Feldman MF, Dennis JJ (2014) A general protein O-glycosylation system within the Burkholderia cepacia complex is involved in motility and virulence. Mol Microbiol 92(1):116–137
114. Vik A, Aas FE, Anonsen JH, Bilsborough S, Schneider A, Egge-Jacobsen W, Koomey M (2009) Broad spectrum O-linked protein glycosylation in the human pathogen Neisseria gonorrhoeae. Proc Natl Acad Sci U S A 106(11):4447–4452
115. Qutyan M, Henkel M, Horzempa J, Quinn M, Castric P (2010) Glycosylation of pilin and nonpilin protein constructs by Pseudomonas aeruginosa 1244. J Bacteriol 192(22):5972–5981
116. DiGiandomenico A, Matewish MJ, Bisaillon A, Stehle JR, Lam JS, Castric P (2002) Glycosylation of Pseudomonas aeruginosa 1244 pilin: glycan substrate specificity. Mol Microbiol 46 (2):519–530
117. Pan C, Sun P, Liu B, Liang H, Peng Z, Dong Y, Wang D, Liu X, Wang B, Zeng M, Wu J, Zhu L, Wang H (2016) Biosynthesis of conjugate vaccines using an O-linked glycosylation system. MBio 7(2):e00443–e00416
118. Nothaft H, Szymanski CM (2010) Protein glycosylation in bacteria: sweeter than ever. Nat Rev Microbiol 8(11):765–778
119. Kudelka MR, Ju T, Heimburg-Molinaro J, Cummings RD (2015) Simple sugars to complex disease--mucin-type O-glycans in cancer. Adv Cancer Res 126:53–135
120. Henderson GE, Isett KD, Gerngross TU (2011) Site-specific modification of recombinant proteins: a novel platform for modifying glycoproteins expressed in E. coli. Bioconjug Chem 22(5):903–912
121. De Gregorio E, Rappuoli R (2014) From empiricism to rational design: a personal perspective of the evolution of vaccine development. Nat Rev Immunol 14(7):505–514
122. Kulp A, Kuehn MJ (2010) Biological functions and biogenesis of secreted bacterial outer membrane vesicles. Annu Rev Microbiol 64:163–184
123. Alaniz RC, Deatherage BL, Lara JC, Cookson BT (2007) Membrane vesicles are immunogenic facsimiles of Salmonella typhimurium that potently activate dendritic cells, prime B and T cell responses, and stimulate protective immunity in vivo. J Immunol 179(11):7692–7701
124. Ellis TN, Leiman SA, Kuehn MJ (2010) Naturally produced outer membrane vesicles from Pseudomonas aeruginosa elicit a potent innate immune response via combined sensing of both lipopolysaccharide and protein components. Infect Immun 78(9):3822–3831
125. Schild S, Nelson EJ, Camilli A (2008) Immunization with Vibrio cholerae outer membrane vesicles induces protective immunity in mice. Infect Immun 76(10):4554–4563
126. Chen DJ, Osterrieder N, Metzger SM, Buckles E, Doody AM, DeLisa MP, Putnam D (2010) Delivery of foreign antigens by engineered outer membrane vesicle vaccines. Proc Natl Acad Sci U S A 107(7):3099–3104
127. Sanders H, Feavers IM (2011) Adjuvant properties of meningococcal outer membrane vesicles and the use of adjuvants in Neisseria meningitidis protein vaccines. Expert Rev Vaccines 10 (3):323–334
128. Gorringe AR, Pajon R (2012) Bexsero: a multicomponent vaccine for prevention of meningococcal disease. Hum Vaccin Immunother 8(2):174–183
129. Holst J, Martin D, Arnold R, Huergo CC, Oster P, O'Hallahan J, Rosenqvist E (2009) Properties and clinical performance of vaccines containing outer membrane vesicles from Neisseria meningitidis. Vaccine 27(Suppl 2):B3–B12
130. Muralinath M, Kuehn MJ, Roland KL, Curtiss 3rd R (2011) Immunization with Salmonella enterica serovar Typhimurium-derived outer membrane vesicles delivering the pneumococcal

protein PspA confers protection against challenge with Streptococcus pneumoniae. Infect Immun 79(2):887–894
131. Liu D, Reeves PR (1994) Escherichia coli K12 regains its O antigen. Microbiology 140 (1):49–57
132. Han W, Wu B, Li L, Zhao G, Woodward R, Pettit N, Cai L, Thon V, Wang PG (2012) Defining function of lipopolysaccharide O-antigen ligase WaaL using chemoenzymatically synthesized substrates. J Biol Chem 287(8):5357–5365
133. Chen L, Valentine JL, Huang CJ, Endicott CE, Moeller TD, Rasmussen JA, Fletcher JR, Boll JM, Rosenthal JA, Dobruchowska J, Wang Z, Heiss C, Azadi P, Putnam D, Trent MS, Jones BD, DeLisa MP (2016) Outer membrane vesicles displaying engineered glycotopes elicit protective antibodies. Proc Natl Acad Sci U S A 113(26):E3609–E3618
134. Xu DQ, Cisar JO, Osorio M, Wai TT, Kopecko DJ (2007) Core-linked LPS expression of Shigella dysenteriae serotype 1 O-antigen in live Salmonella Typhi vaccine vector Ty21a: preclinical evidence of immunogenicity and protection. Vaccine 25(33):6167–6175
135. Wang L, Curd H, Reeves PR (1999) Immunization of mice with live oral vaccine based on a Salmonella enterica (sv Typhimurium) aroA strain expressing the Escherichia coli O111 O antigen. Microb Pathog 27(1):55–59
136. Nothaft H, Davis B, Lock YY, Perez-Munoz ME, Vinogradov E, Walter J, Coros C, Szymanski CM (2016) Engineering the Campylobacter jejuni N-glycan to create an effective chicken vaccine. Sci Rep 6:26511
137. Bentley R (1990) The shikimate pathway--a metabolic tree with many branches. Crit Rev Biochem Mol Biol 25(5):307–384
138. Ruby T, McLaughlin L, Gopinath S, Monack D (2012) Salmonella's long-term relationship with its host. FEMS Microbiol Rev 36(3):600–615
139. Felgner S, Frahm M, Kocijancic D, Rohde M, Eckweiler D, Bielecka A, Bueno E, Cava F, Abraham WR, Curtiss 3rd R, Haussler S, Erhardt M, Weiss S (2016) aroA-Deficient Salmonella enterica Serovar Typhimurium is more than a metabolically attenuated mutant. MBio 7 (5):e01220–e01216
140. National-Research-Council (2012) Transforming glycoscience: a roadmap for the future. The National Academies Press, Washington, DC

# Advances in the Chemical Synthesis of Carbohydrates and Glycoconjugates

**Ankita Malik, Peter H. Seeberger, and Daniel Varón Silva**

## Contents



**Abstract** Carbohydrates are functional and structural biomolecules with structures ranging from monosaccharides to polysaccharides. They are naturally found as pure glycans or attached to lipids and proteins forming glycoconjugates. The biosynthesis of carbohydrates is not genetically controlled. The regulation takes place by the expression of enzymes that transfer and hydrolyze the glycan units, leading to glycocojugates having complex mixtures of glycan structures. Chemical synthesis emerged as the best strategy to obtain defined glycan and glycoconjugates and

The original version of this chapter was revised: Abbreviations included as new additional correction in this chapter.

A. Malik, P. H. Seeberger, and D. Varón Silva (✉)
Max Planck Institute of Colloids and Interfaces, Biomolecular Systems, Potsdam, Germany

Department of Chemistry and Biochemistry, Freie Universität Berlin, Berlin, Germany
e-mail: daniel.varon@mpikg.mpg.de

overcome the challenging purification processes. Here, we review the recent advances in the synthesis of oligosaccharides using manual and automated methods. The chapter covers the methods for the preparation of building blocks and control of stereoselectivity and regioselectivity during glycosylations. Finally, it also presents the strategies to obtain natural and non-natural glycoconjugates with lipids and proteins.

**Graphical Abstract**

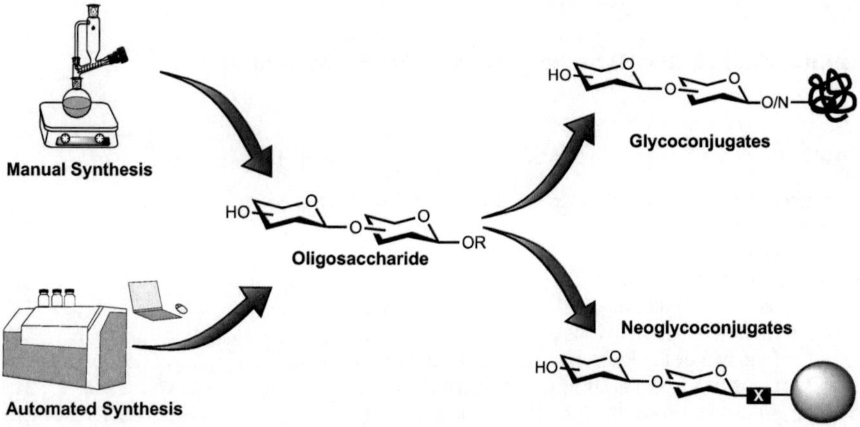


**Keywords** Carbohydrates, Chemical synthesis, Glycoconjugates, Glycoproteins, Glycan modifications

## Abbreviations

| | |
|---|---|
| Ac | Acetyl |
| $Ac_2O$ | Acetic anhydride |
| AcCl | Acetylchloride |
| AcOH | Acetic acid |
| AgOTf | Silver triflate |
| AllylBr | Allyl bromide |
| Asn | Asparagine |
| $BF_3.Et_2O$ | Boron trifluoride diethyl etherate |
| Bn | Benzyl |
| BnBr | Benzylbromide |
| Boc | *tert*-Butoxycarbonyl |
| Bu | Butyl |
| Bz | Benzoyl |
| $CCl_3CN$ | Trichloroacetonitrile |
| Cer | Ceramide |
| Cl-Ac | Chloro-acetyl |

| | |
|---|---|
| CSA | Camphor sulfonic acid |
| DAG | Diacylglycerol |
| DBU | 1,8-diazabicycloundec-7-ene |
| DCC | *N,N′*-Dicyclohexylcarbodiimide |
| DCM | Dichloromethane |
| DDQ | 2,3-Dichloro-5,6-dicyano-1,4-benzoquinone |
| DIPEA | *N,N*-Diisopropylethyl amine |
| DMAP | 4-Dimethylaminopyridine |
| DMF | Dimethylformamide |
| DMSO | Dimethyl sulfoxide |
| DMT | Dimethoxytrityl |
| $Et_2O$ | Diethyl ether |
| $Et_3N$ | Triethylamine |
| EtNP | Phosphoethanolamine |
| Et | Ethyl |
| Fmoc | 9-Fluorenylmethoxycarbonyl |
| GAG | Glycosaminoglycan |
| Gal | Galactose |
| Glc | Glucose |
| GlcNAc | *N*-acetylglucosamine |
| GPI-APs | Glycosylphosphatidylinositols anchored proteins |
| GPIs | Glycosylphosphatidylinositols |
| HF-Pyridine | Hydrogen fluoride in pyridine complex |
| IAD | Intramolecular aglycon delivery |
| Ino | Inositol |
| Lev | Levulinoyl |
| LG | Leaving group |
| Man | Mannose |
| MeCN/ACN | Acetonitrile |
| Me | Methyl |
| MeOH | Methanol |
| MP | 4-Methylphenyl |
| NaOAc | Sodium Acetate |
| NaOMe | Sodium Methoxide |
| Nap | 2-Naphthylmethyl |
| NIS | *N*-iodosuccinimide |
| NMR | Nuclear magnetic resonance |
| PBS | Phosphate buffer saline |
| PCR | Polymerase chain reaction |
| Pd/C | Palladium on activated charcoal |
| $Ph_3P$ | Triphenylphosphine |
| $PhCH(OMe)_2$ | Benzaldehyde dimethyl acetal |
| PhMe | Toluene |
| PhS | Thiophenyl |

| | |
|---|---|
| Pico | Picoloyl |
| PMB | 4-Methoxybenzyl |
| PivCl | Pivaloyl chloride |
| TBAF | Tetrabutylammonium fluoride |
| TBDPS | *tert*-Butyldiphenylsilyl |
| TBS | *tert*-Butyldimethylsilyl |
| TBSOTf | *tert*-Butyldimethylsilyltrifluoromethanesulfonate |
| TCA | Trichloroacetamide |
| $Tf_2O$ | Trifluoromethansulfonic acid anhydride, triflic anhydride |
| TFA | Trifluoroacetic acid |
| Thr | Threonine |
| TIPS | Triisopropylsilyl |
| TMS | Trimethylsilyl |
| TMSOTf | Trimethylsilyltrifluoromethanesulfonate |
| Tol | Tolyl |
| Troc | Trichloromethyl-oxy-carbonyl |
| Trt | Triphenylmethyl, trityl |
| Zn | Zinc |

Carbohydrates are natural molecules having an important role in the biological processes of cells, as an energy source and structural materials. Carbohydrates are also covalently attached to proteins, peptides, and lipids forming glycoconjugates having different functions and properties [1]. The structure and composition of carbohydrates on glycoconjugates are not genetically controlled and be ranging from monosaccharides to polysaccharides. Glycoconjugates generally contain a heterogeneous mixture of glycans that result from differential expression and activity of the enzymes participating in their biosynthesis and degradation. This heterogeneity hinders the evaluation of defined carbohydrate structures in biological processes.

There are no amplification procedures for carbohydrates comparable to the expression of proteins in cell lines or to amplification of DNA using PCR techniques. However, these molecules have been accessible in homogeneous form, high complexity, and good quantity using chemical [2], chemo-enzymatic [3, 4], or enzymatic synthesis [5]. Carbohydrates obtained using these methods include oligo- and polysaccharide fragments, regioisomers, natural products, and glycomimetics. These molecules have found application in deciphering the biological role of carbohydrates, in material science and biomedicine [6], development of high-affinity ligands [7], mapping of immunogenic carbohydrate epitopes in polysaccharides [8], and to introduce labeling in living organisms [9]. Further, synthetic glycans are important building blocks for preparing glycoconjugates and the synthesis of glycopolymers and glycosylated nanoparticles.

In contrast to proteins and nucleic acids that are linear oligomers of amino acids and nucleotides connected via an amide or phosphodiester bond without stereochemical requirements; oligosaccharides and polysaccharides form both linear and

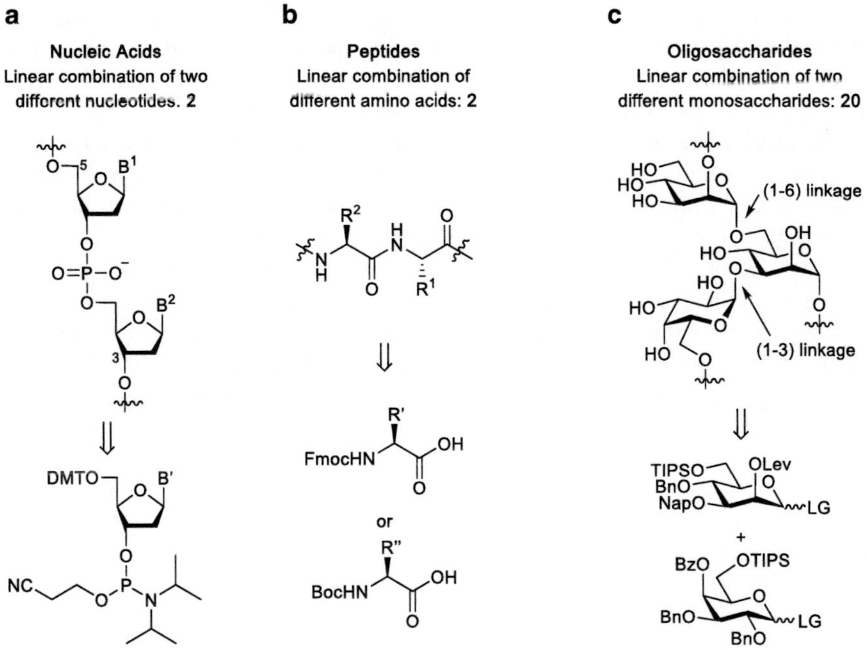


**Fig. 1** Linkages and combinational complexity found in the three major biopolymers. (**a**) Nucleic acids, (**b**) Proteins, (**c**) Oligosaccharides and Polysaccharides

branch structures with linkages having a defined stereochemistry (Fig. 1). Despite the progress during the last decades and a plethora of reported methods for the assembly of glycans and their linkage to other biomolecules, there are no standardized protocols for the synthesis of glycans and the production of natural and non-natural glycoconjugates.

In this chapter, we review the recent advances in the synthesis of glycans focusing on chemical methods in both, in solution and on a solid phase, and illustrate the main methods to obtain glycoconjugates. We discuss the challenges and the efforts towards the development of simple and efficient methods for automated glycan assembly and the application of synthetic glycans to obtain glycoconjugates with proteins and lipids.

## 1 Chemical Synthesis of Oligosaccharides

The control of regioselectivity and stereoselectivity during glycosylation reactions are two fundamental challenges in the chemical synthesis of oligosaccharides. A large set of protecting groups, leaving groups, reaction conditions, and glycosylation promoters have been established to overcome these difficulties and get access to glycans having any desired structure and modification. The control of regioselectivity is performed mainly by using protecting groups to mask the undesired hydroxy groups in the glycosyl acceptor and donor. But, regioselectivity during glycosylations is also accessible based on the reactivity of the hydroxy groups in the acceptor. In contrast, the formation of the correct stereochemistry in the new glycosidic bond is controlled by using protecting groups that participate in the steric and electronic stabilization of intermediates or by modifying the reaction conditions such as the temperature, solvents, and presence of additives.

The efficiency of the synthesis of an oligosaccharide depends on the target structure, the methodology selected for the assembly, and the diversity of monosaccharides present in the structure. Independently of the strategy, the assembly of oligosaccharides containing high monosaccharide and linkage variability is a time demanding and costly process that requires the synthesis of multiple building blocks. Thus, the design and development of new strategies to prepare building blocks on a large scale became a key process that is continuing evolving during the last years. New advances include the optimized introduction of protecting groups that affect the reactivity, the reaction conditions, and the stereochemical outcome of glycosylations.

## 2 Synthesis of Building Blocks

The design of a building block starts with the selection of protecting and leaving groups. Depending on the role played in the synthesis, these groups can be either temporary or permanent groups [10]. Permanent protecting groups mask functionalities that do not require any modification during the assembly. Temporary protecting groups block the positions to modify in the following step or a late step of the synthesis. To install protecting groups in a regioselective manner, chemists rely on the reactivity of the different functionalities and alcohols being in equatorial and axial positions. Using current methods is possible to distinguish between hemiacetals, primary and secondary alcohols, amines, carboxylic acids (Fig. 2). Recent strategies rely on the difference in the reactivity of active intermediates to exchange these protecting groups in a regioselective manner using one-pot or multistep processes.

Phosphorylation, acylation, sulfation, and alkylation are typical modifications of carbohydrates. The synthesis of oligosaccharides having these groups and branched structures linked to the glycan core requires multiple temporary protecting groups being orthogonal to each other. Some temporary protecting groups are frequent during the synthesis of building blocks, but they are rarely present in prolonged synthesis and advanced oligosaccharide intermediates. The most common orthogonal protecting groups for hydroxyl groups include benzyl ether derivatives, silyl ethers, esters, orthoesters, and acetals. (Fig. 3 and Table 1).

Some protecting groups have been introduced in carbohydrate synthesis during the last years to expand the classical set of protecting groups [11]. These groups

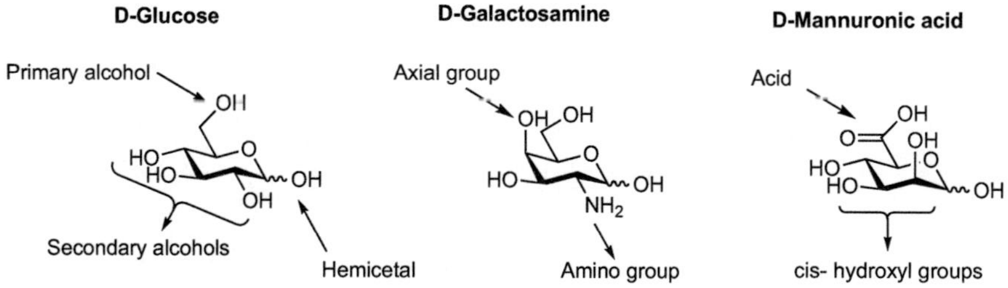


**Fig. 2** Functional groups on monosaccharides. Different functionalities on monosaccharides used for the selective installation of protecting group

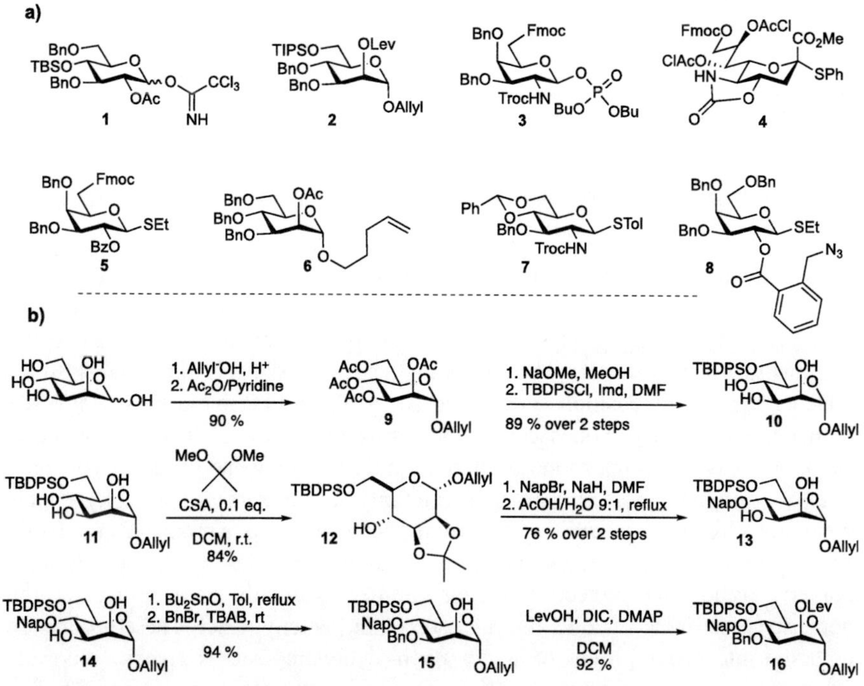


**Fig. 3** (**a**) Example of building blocks (BBs) used for the synthesis of glycans. (**b**) Synthesis of mannose building blocks for the assembly of GPIs [15]

include participating groups such as picoloyl (Pico) esters, 2-*O*-alkoxysilyl protecting groups for intramolecular aglycon delivery (IAD) [12], and groups having chiral auxiliaries to increase the stereoselectivity of glycosylations [13, 14].

The removal of permanent protecting groups is generally the final step of the synthesis. Therefore, the selection of this group depends on the composition of the target molecule and the functionalities required in the final product. Acetyl and benzoyl esters and benzyl ethers are still the most frequently used permanent protecting groups. However, the need to introduce functionalities into glycans that

**Table 1** Commonly used orthogonal protecting groups

| Type | Protecting group | Removal Conditions[a,b] |
|---|---|---|
| Ether | Allyl | Pd-catalyzed reactions |
| | PMB (p-methoxybenzyl)<br>Nap (2-(naphthyl)methyl) | DDQ or acids (TFA) |
| | Trityl (triphenyl-methyl) | Acid (TFA) |
| Ester | Lev (levulinoyl) | Base ($NH_2$-$NH_2$) |
| | Ac (acetyl)<br>Bz (benzoyl) | Base ($CH_3ONa$) |
| Carbonates | Fmoc (9-fluorenylmethyloxycarbonyl) | Base (pyridine) |
| Silyl ethers | TIPS (*tri*-isopropylsilyl)<br>TBS (*tert*-butyl-dimethylsilyl)<br>TBDPS (*tert*-butyl-diphenyl-silyl) | Fluoride/ acids (TBAF/TFA)<br>Lewis acids ($Sc(OTf)_3$) |
| Acetals | Isopropylidene | Acids ($CH_3CO_2H$) |
| | Benzylidene[c] | Acids (TfOH) |

[a]In parenthesis, common reagent for the removal
[b]There is a multiple methods for the installation and removal of these groups reported in the literature [10]
[c]This acetal can be open selectively leaving one hydroxyl group protected

are not compatible with hydrogenolysis and basic conditions for the removal of these groups has stimulated the use of other groups in glycan synthesis. Prominent examples of protecting groups introduced for permanent protection include the *para*-methoxybenzyl (PMB) [16], 2-naphthylmethyl ether [17], and acetals [18].

Depending on the position in the oligosaccharide and the synthetic strategy, the anomeric center of the monosaccharide building blocks has a linker, a protecting group, or a leaving group. Monosaccharides at the reducing end generally contain a linker or orthogonal protecting group. Commonly used linkers are short alkyl chains containing a hydroxyl group on one extreme and an alkyne or a protected amino or a thiol function on the other. Internal monosaccharide building blocks usually have anomeric orthogonal protecting groups. Some commonly used groups at the anomeric center include azide, fluoride, acetyl, or pentenyl ether. These groups are converted into active groups or active glycosyl donors such as glycosyl fluoride, tricloroacetimidate, or a phosphate (Fig. 4). These orthogonal groups are installed mostly for attachment of oligosaccharide building blocks in convergent strategies. However, they are also groups useful for the synthesis of active glycans for reactions with amino acids and lipids required for the preparation of glycoconjugates.

More recently, Hong and colleagues introduced the one-pot multistep synthesis of protected monosaccharide units for the rapid synthesis of protected building blocks. This method involves the initial conversion of the anomeric hemiacetal into a thioglycoside or methyl glycoside and the following silylation of the remaining hydroxyl groups. The silylated hydroxyl groups have differentiated reactivity or are easily removed under the acid conditions used to install other protecting groups, including benzylidene acetals and regioselective etherification and esterification (Fig. 5) [19].

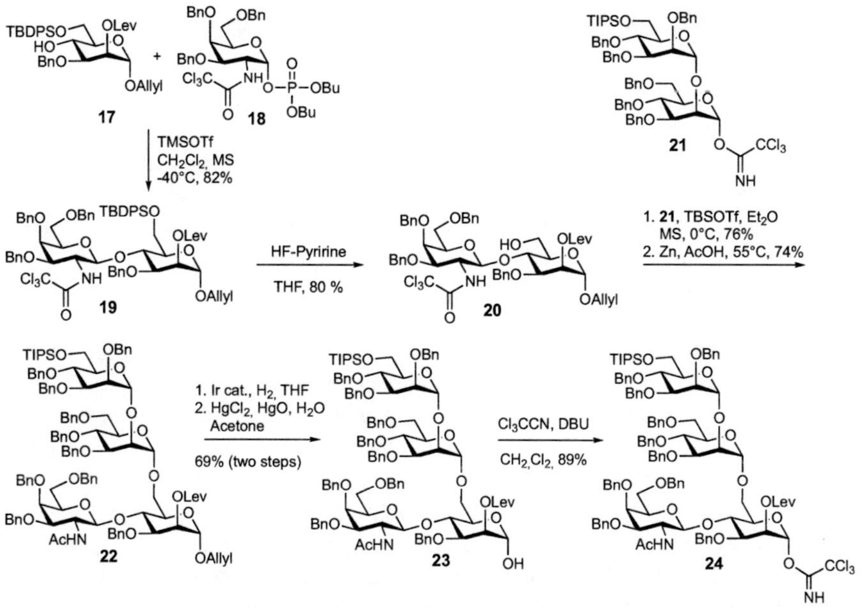


**Fig. 4** Application of allyl ether as temporary protecting group. Synthesis of the oligomannoses part **24** of the glycosylphosphatidylinositol anchoring *Toxoplasma gondii* proteins [15]

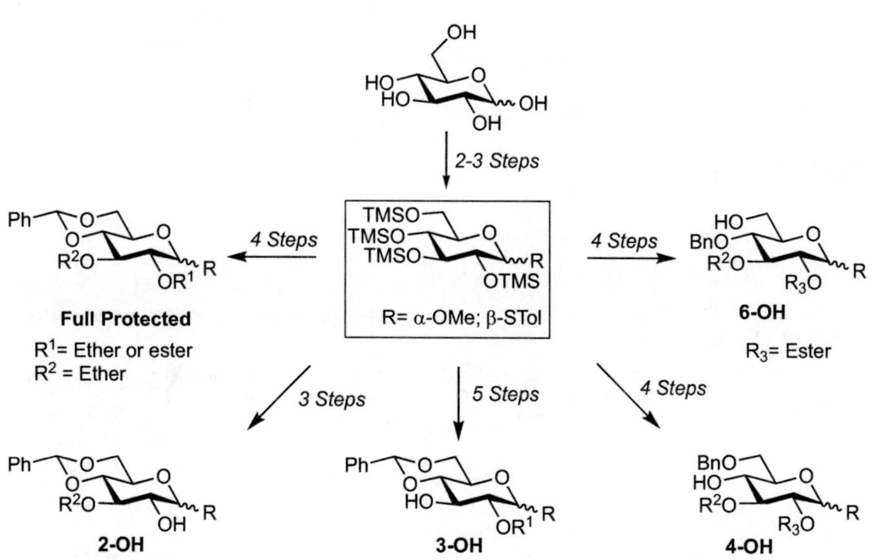


**Fig. 5** One-pot synthesis of protected monosaccharide. Installation of orthogonal protecting groups starting from a per-*O*-silylated compounds [19]

## 2.1 Glycosylation Reactions and Stereoselectivity

A big challenge in the synthesis of oligosaccharides is still the control of stereoselectivity during the glycosylation and the formation of products having only a 1,2-*trans*- or 1,2-*cis*-linkage. As already mentioned, the production of a single stereoisomer depends on different factors. The reaction temperature, solvent, glycosylation promotor, reactivity of the leaving group, and the protecting group in both the glycosyl acceptor and particularly the presence or absence of participation in the donor are the most important. Often these parameters need to be established and optimized to favor the formation of one specific product.

In glycosylation reactions, the promotor activates the leaving group at the anomeric position of the glycosyl donor in the first step of the process. These promoters are generally silver salts ($AgCO_3$, $AgClO_4$ and $Ag_2O$), salts or esters of trifluoromethanesulfonic acid (AgOTf, $Cu(OTf)_2$) [20], Lewis acids (NIS/TfOH, $Tf_2O$, $BF_3$-$Et_2O$, $SnCl_4$), TfOH and its derivatives (TMSOTf, TBSOTf and MeOTf), or the recently introduced use of other metal salts such as AgCl, $AuCl_3$ and $CuCl_2$ [21]. Upon activation, the leaving group departs, and the oxocarbenium ion is formed (Fig. 6). This ion can be stabilized by adjacent protecting groups and react with the hydroxyl group of the acceptor approaching from either the top or the bottom face to form a 1,2-*cis*- or 1,2-*trans*-configured glycosylation product.

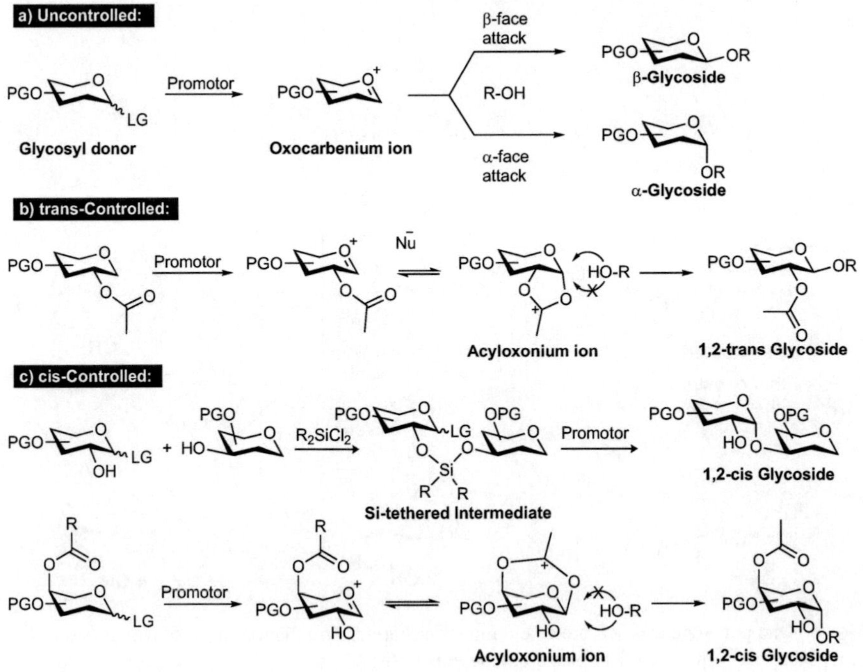


**Fig. 6** Strategies for controlling the outcome in glycosylation reactions

The reaction temperature can affect the stereoselectivity of glycosylations and lead to products under thermodynamic or kinetic control. Without the participation of protecting groups, high temperature favors the formation of the more stable anomer by thermodynamic control. In contrast, low temperature and short reaction time lead to the kinetically controlled product, which is mainly the less steric hindrance anomer (Fig. 6).

The selection of the protecting group can affect selectivity during glycosylations. Therefore, it should well consider the use of groups that can undergo interaction with the oxocarbenium anion and neighboring group participation. The carbonyl group of ester, amide, or carbonate function at the C-2 position of the glycosyl donor interacts with the oxocarbenium ion forming an active cyclic intermediate. The formed cyclic acyloxonium ion hinders the attack of the nucleophile from the side that the protecting group is localized (Fig. 6b). As a consequence, the attack from the nucleophile (glycosyl acceptor) can take place only from one face forming the 1,2-*trans*-product [22].

The formation of 1,2-*cis*-linkages is more challenging and requires additional considerations during the glycosylation reaction, i.e. β-mannopyranosides and α-glucopyranosides [23]. Some strategies to increase the selectivity to 1,2-*cis* products are the intramolecular aglycon delivery (IAD) [12] and the use of fluorides [24] and bromides as leaving groups to favor the progress of the glycosylation via an $S_N2$-type mechanism. Other strategies include the use of remote group participation from the protecting group at C-3, C-4, or C-6 position of the glycosyl donor [25]. These strategies have been developed over the past years and are efficiently applied to ensure the synthesis of complex saccharides having 1,2-*cis* linkages (Fig. 6c) [25].

Changes in other reaction conditions such as solvent and presence of additives can favor the formation of an α- or a β-product. Ethers can interact with the oxocarbenium anion and hinder the attach of the nucleophile from the equatorial face favoring the production of the thermodynamically stable α-linkage. This effect is called inversion of the anomeric effect [26]. By contrary, acetonitrile induces the formation of an α-nitrilium-nitrile-conjugate with strongly activated donors that favors the formation of equatorial β-product [27]. Theoretical studies using quantum-mechanical calculation and molecular dynamic simulations suggest other mechanisms involving oxocarbenium-counterion and the conformation of intermediates [28]. However, further studies are still required to confirm these models and to establish a clear relationship between solvent and stereoselectivity.

## 3 Assembly of Oligosaccharides

Depending on the structure, the synthesis of an oligosaccharide is possible using a linear or a convergent strategy. Linear strategies are convenient for oligosaccharides having repeating units such as fragments of polysaccharides [29]. They are also ideal for automated protocols on solid phase [30–32] or using tags [33]. Convergent strategies are applied mostly in the solution phase and to obtain branched structures or glycans having a complex monosaccharide composition. There are multiple examples of oligosaccharides obtained by both strategies. They include *N*-glycans from proteins [34], glycosylphosphatidylinositols [15], and repeating units of bacterial polysaccharides (Fig. 7) [35].

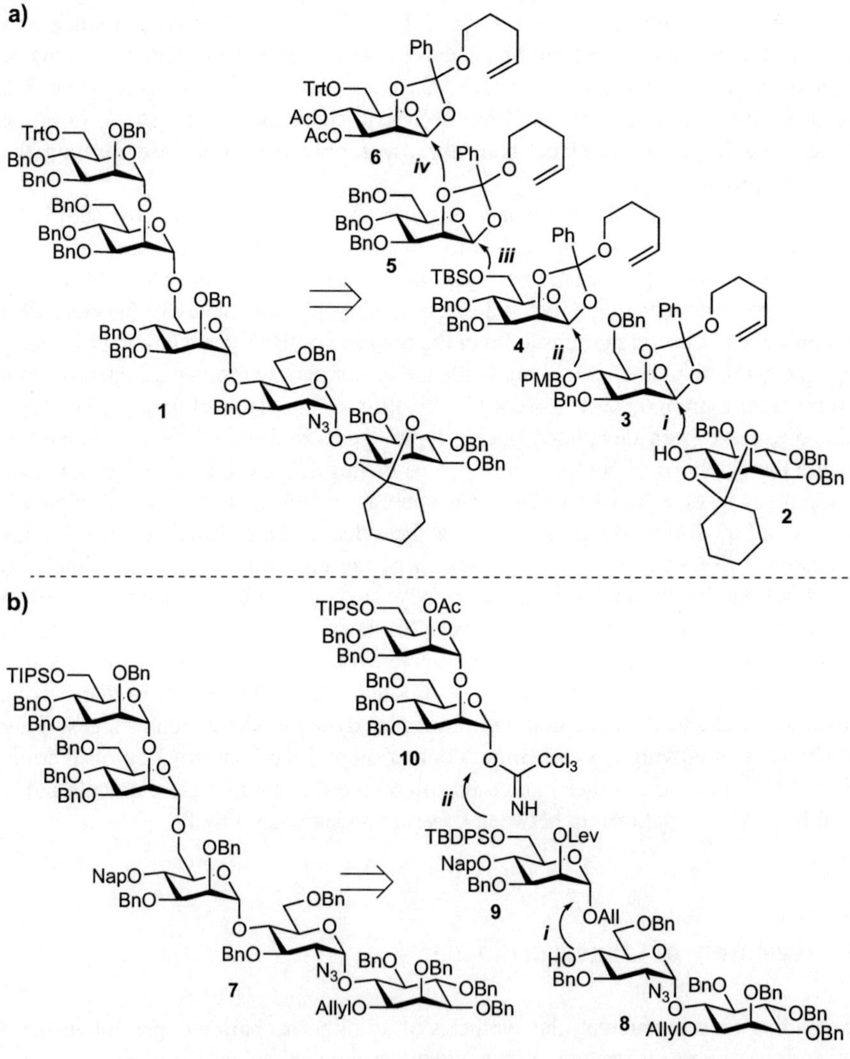


**Fig. 7** Examples of a linear and a convergent synthesis of the core from glycosylphosphatidylinositol anchors. (**a**) linear assembly; [37] (**b**) convergent assembly [15]

# 4 Solution-Phase Synthesis of Glycans

Linear or convergent methods to assemble oligosaccharides involve sequential deprotection and glycosylation steps. Successive elongation cycles with monosaccharides deliver the oligosaccharide by a linear strategy. Each elongation cycle includes glycosylation and removal of the temporary protecting group until completion of the desired structure. In a convergent method, large and complex fragments are synthesized from monosaccharides and then assembled to an oligosaccharide (Fig. 6b). The two strategies have advantages and disadvantages, and their applicability depends on the target structure and availability of monosaccharides. The major drawback of a solution-phase synthesis, especially in linear strategies, is the need for purification after each reaction step and the concomitant loss of material. In the synthesis of large oligosaccharides, the high number and difficulty of the purification processes imply the loss of precious advanced intermediates. Therefore, synthesis using solid-phase methods and diverse one-pot multistep assembly of oligosaccharides emerged to reduce the number of purification steps and the handling of intermediates [36].

## *4.1 Modification of Glycans*

Similar to other biomolecules, relevant glycans can contain site-specific modifications that increase the complexity of the oligosaccharides and contribute to the biological activity of these molecules. These modifications include the presence of functionalities such as amides, amines, acids, ketones or the derivatization of the hydroxyl groups such as methylation, esterification (acetylation, acylation), sulfation, and phosphorylations [38]. These modifications are naturally introduced by specialized enzymes that recognize specific functional groups and specific sites on oligosaccharides and polysaccharides. These modifications are also called as post-glycosylational modifications (PGMs) and are present on all types of glycans (Fig. 8) [39].

Most of the PGMs are labile to the glycosylation conditions and the removal of some protecting groups. Therefore, these modifications are generally added at the late-stage of the synthesis or require of appropriated positioning of orthogonal protecting groups. A particular challenge in obtaining modified oligosaccharides has been the synthesis of glycosaminoglycans (GAGs). They are complex glycopolymers containing sulfation at different positions as a major PGMs. The synthesis of GAGs has been investigated by different strategies and methodologies. In addition to the general difficulties related to the assembly of the carbohydrate core and a defined sulfation pattern, the synthesis of GAGs requires the insertion of carboxylates and amines [40]. The carboxylates are introduced by using protected uronic acid building blocks or they can be generated after completion of the glycan assembly by oxidation of selectively deprotected primary hydroxyl groups. In

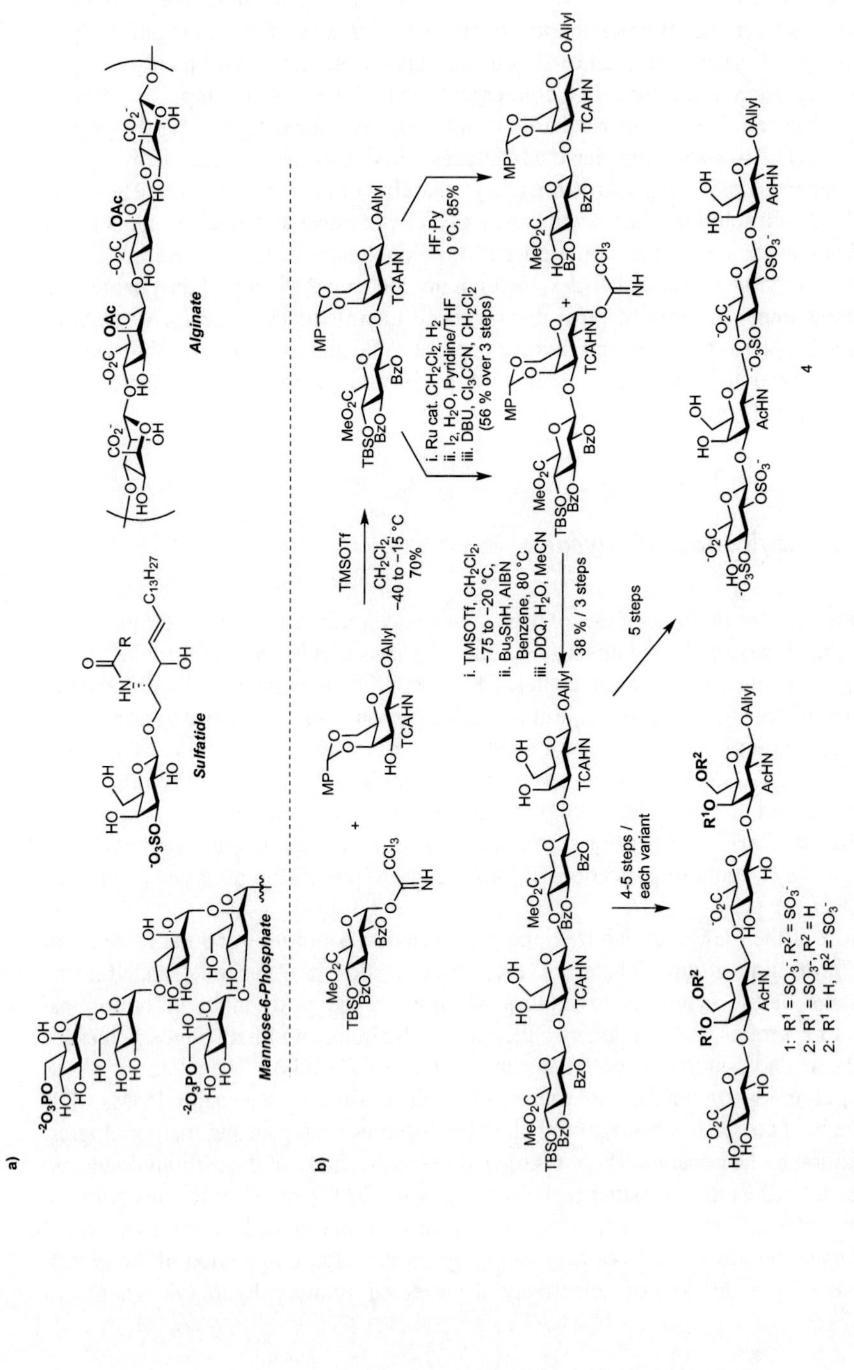


**Fig. 8** **(a)** Different post-glycosylational modifications, **(b)** Example of a modular, convergent approach used to obtain chondroitin sulfate structures having distinct sulfation motifs [40]

contrast, the amino group is mostly introduced as part of the building blocks and is kept protected during the glycan assembly. Upon assembly of the oligosaccharide, the orthogonal protecting groups are removed and if required, the free positions are sulfated (Fig. 8) [40].

## 4.2 Removal of the Protecting Groups, Final Deprotection

The removal of all protecting groups is generally the final step in the synthesis of oligosaccharides, it is also called global deprotection. This process takes place after complete assembly of the oligosaccharide and installation of the desired glycan modifications and is followed by purification of the desired products. The global deprotection is generally a multistep process involving treatment with oxidants, acids, bases, catalyzed reactions or a combination of them. Typical deprotection conditions are hydrogenolysis or Birch reduction to remove benzyl ethers [41], saponification with sodium salts to remove benzoyl and acetyl esters, or treatment with an acid such as trifluoracetic acid and diluted HCl to hydrolyze substituted benzyl ethers, acetals, and other acid labile groups [16, 17]. Recently, oxidative conditions have also been established to remove ether groups such as benzyl and 2-naphthylmethyl ether from glycans without affecting reduction-labile functional groups [42, 43].

## 4.3 Solid-Phase Synthesis of Oligosaccharides

Despite multiple advances in the synthesis of oligosaccharides using convergent strategies in the solution phase, the assembly of large glycans is still a time-consuming process that requires multiple deprotection and glycosylation steps and low yields. Therefore, solid-phase synthesis (SPS) of glycans emerged as an alternative to facilitate the manipulations and the assembly of large oligosaccharides. Similar to the synthesis of peptides and nucleic acids, the SPS of glycans is operationally easy and rapid to perform. The main advantages are the uncomplicated separation of the growing oligosaccharide from the reagents and solubilized side products after each reaction has been completed, and the easy and reproducible control of the conditions. This methodology has also been efficiently employed for the automation of the oligosaccharide synthesis process [30, 31].

## 4.4 Automated Assembly of Oligosaccharides

Automated glycan assembly (AGA) follows the process described in Fig. 9. The process starts with the glycosylation of the solid support functionalized with a linker

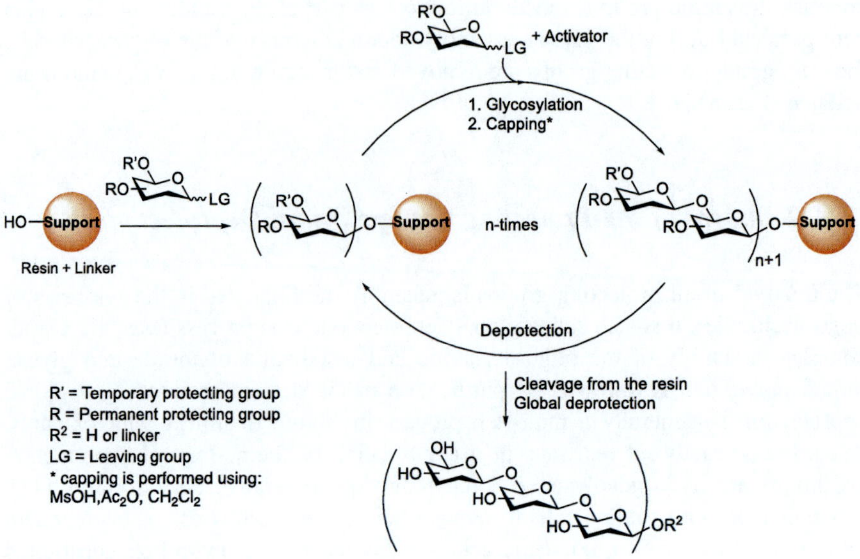


**Fig. 9** Synthesis of oligosaccharides using automated glycan assembly

using a fully protected monosaccharide. Following, unreacted groups on the support are blocked by capping and the temporary protecting group is removed from the first monosaccharide. This reaction releases a hydroxyl group that acts as an acceptor in the next step. Glycosylation with the next monosaccharide and capping of the unreacted groups starts the elongation process. Removal of the orthogonal protecting group, glycosylation, and capping are repeated until the desired structure is completed. Finally, the glycan is released from the solid support and all protecting groups are removed to obtain the desired product.

The conditions for the glycosylation and deprotection reactions carried out in solid phase are generally similar to the conditions used in the solution phase. However, some additional requirements are necessary to enhance the yields of each process near to completion since non-purification steps are possible within a synthesis. Glycosylation reactions in AGA are executed using a large excess of reagents (3–10 M equivalents) and are performed using optimized protecting groups, leaving groups and solvents [31, 32, 44, 45].

Automated synthesis of oligosaccharides is commonly carried out on insoluble supports, such as polystyrene resins or in resins containing PEG chains. These resins offer a high chemical stability against the reagents and the conditions used during the glycan assembly. They also have good swelling properties in organic solvents to facilitate the interaction of the reagents with the growing oligosaccharide. In addition to insoluble polymers, oligosaccharides have also be synthesized using other supports such as gold nanoparticles [46], or on a high surface area porous gold [47].

In addition to the support, the development of suitable linkers for attaching the growing oligosaccharides to the solid support has been an important factor for establishing the AGA. These linkers are chemically inert to the manipulation during the synthesis and to the conditions used during glycosylation and deprotection steps. Furthermore, the linkers should release the oligosaccharide by a chemoselective reaction after completion of the assembly. The connection between the linker and the oligosaccharide is generally easy to break without affecting the connections whiting the oligosaccharide. Typical linkers are bifunctional units connecting at one side the oligosaccharide's reducing end via an ether linkage and to the solid support via an ester, amide or a carbamate linkage at the other side. These linkers can contain additional active functionalities that are used to release the synthesized glycans from the support (i.e. presence of double bonds or photoactive groups). New traceless photolabile linkers deliver oligosaccharides with a free hemiacetal at the reducing end, opening the possibility to attach synthetic AGA products to any other functionality without spacers [48].

In the pioneering reports on automated glycan assembly, Seeberger and coworkers utilized linkers having a double bond to remove the glycan from the solid support using metathesis [49]. By using this linker, the glycan is released as a 4-pentenyl glycoside, a group that can be easily activated for hydrolysis and to introduce other functionalities at the reducing end of oligosaccharides. Other linkers release the glycans attached to a spacer having an amino or carboxylate group that is used for linking other modifications to glycans, for the attachment of glycans to surfaces or for chemoselective reactions with cross-linkers to obtain glycoconjugates (Fig. 10). Photolabile cleavable linkers, base labile linkers, and linkers that can be activated to release active glycans were established for glycosylation to amino acids [50, 51].

There are no differences in the requirements of building blocks for AGA and the solution phase. However, the lack of purification after each step demands that in an automated process all the reactions are completed with very high yields and with excellent stereoselectivity. To this aim, glycosylation involving neighboring group participation and the use of phosphate and thioglycoside show the best results. Building blocks containing benzyl ethers and the combination of carbonate and esters as temporary, orthogonal protecting groups are the most common combination in this strategy (Fig. 11).

Capping reactions avoid the formation of structures that have similar physicochemical properties to the product by preventing the elongation of deletion structures

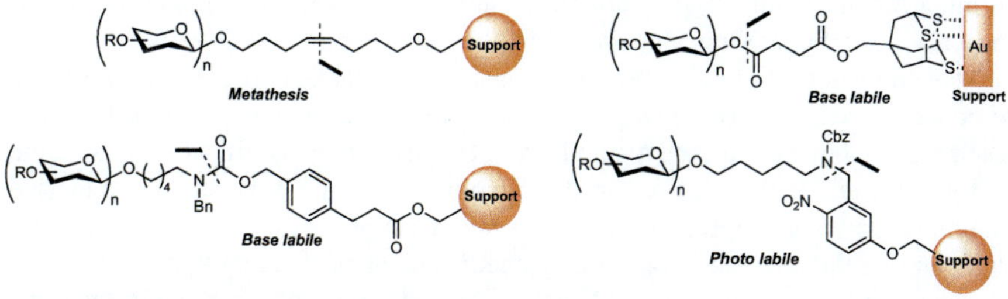


**Fig. 10** Different linkers used for solid-phase glycan synthesis

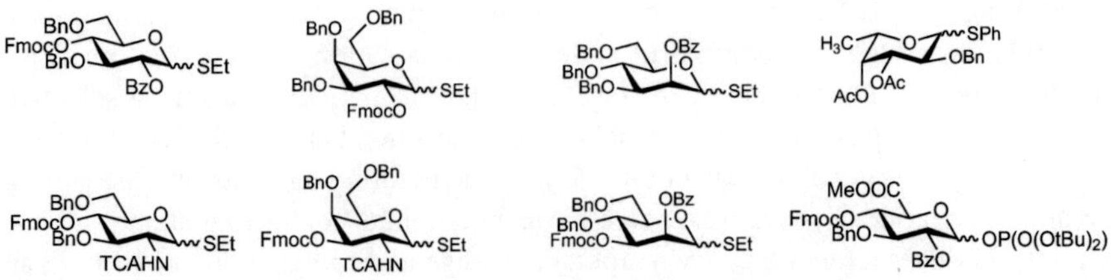


**Fig. 11** Example of building blocks used for automated glycan assembly [31]

due to incomplete reactions. After each glycosylation reaction, the unreacted hydroxyl groups are blocked with a protecting group that remains attached throughout the process. Two advantages are obtained from this step, uncomplete structures remain at the capped size and the reagents will be used only for the extension of the desired molecule reducing the consumption of the active reagents in the formation of undesired products. Due to its easy introduction and stability during the removal of the orthogonal groups (Lev and Fmoc), acetylation under acidic conditions is the favorite method in AGA for this process (Fig. 9) [52].

# 5 Synthesis of Glycoconjugates

Carbohydrates forming part of natural glycoconjugates such as glycoproteins, glycopeptides, and glycolipids are important for the activity of these molecules. Therefore, the development of methods to attach oligosaccharides to proteins and lipids is necessary to elucidate the role of carbohydrates in the biological activity of glycoconjugates. This has been a long process, however, there are some methods nowadays for the synthesis of natural glycoconjugates such as glycolipids and glycoproteins [53].

Natural glycoproteins exist generally a mixture of molecules having the same peptide sequence displaying diverse oligosaccharide structures at the glycosylation positions. These so-called glycoforms can have different physical and biochemical properties [54, 55]. Carbohydrates are covalently linked to the protein mainly via nitrogen or oxygen atoms at the side chain of asparagine, serine or threonine residues forming *N*- and *O*-linked glycoproteins. *S*-linked, *P*-linked, or *C*-linked glycoproteins also exist but are less abundant [56].

A plethora of strategies has been investigated and established during the last years to obtain naturally glycosylated proteins using molecular biology, chemical synthesis or the combination of both methods [57]. However, there is still a lack of suitable and generalized methods to obtain these molecules. Some protocols to synthesize natural *O*- and *N*-glycoproteins are fully synthetic strategies that combine carbohydrate and peptide synthesis or semi-synthetic strategies that required of synthetic peptides and glycopeptides and expressed proteins that are connected by chemoselective ligation reactions. These strategies have been limited to the synthesis of small glycoproteins, generally below 20 kDa [58–61]. Therefore, there is still a need for new methods to obtain glycoproteins of high molecular weight and multiple glycosylation sites.

Chemical glycosylation methods and chemoselective reactions are not restricted to the synthesis of natural glycoproteins, they can also be used for generating neoglycoproteins, which are glycoproteins containing unnatural linkages between protein and oligosaccharides. Neoglycoproteins are often selected as the best alternative to investigate the function of carbohydrates because their synthesis is easier to perform using a chemoselective reaction between an activated glycan and the lateral chains of natural and no-natural amino acids present on the protein.

## 5.1 Synthesis of Neoglycoconjugates

Big efforts and developments have been performed during the last decades to establish strategies for the incorporation of glycans into proteins. Depending on the functional groups in glycans and amino acids involved in the process, diverse types of linkages can be formed between the carbohydrate and the protein in neoglycoproteins (Fig. 12).

Carbohydrate-protein linkages have been mostly synthesized by using the inherent reactivity of the amino group of lysine and the thiol of the cysteine side chains to achieve chemoselective reactions (Fig. 13) [61, 62]. The resulting linkages are usually very different from the linkage in natural glycoproteins and include a spacer, but they are generally stable to biological conditions allowing the biological evaluation of neoglycoconjugates. This strategy has been the favorite strategy applied to

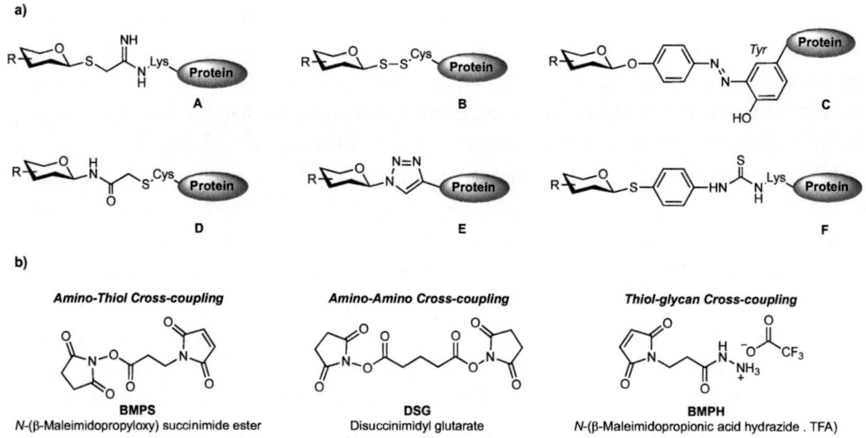


**Fig. 12** Example of linkages used for the synthesis of neoglycoconjugates. (**a**) Some typical linkages obtained in neoglycoconjugates, (**b**) examples of linkers using for conjugation of glycans to proteins

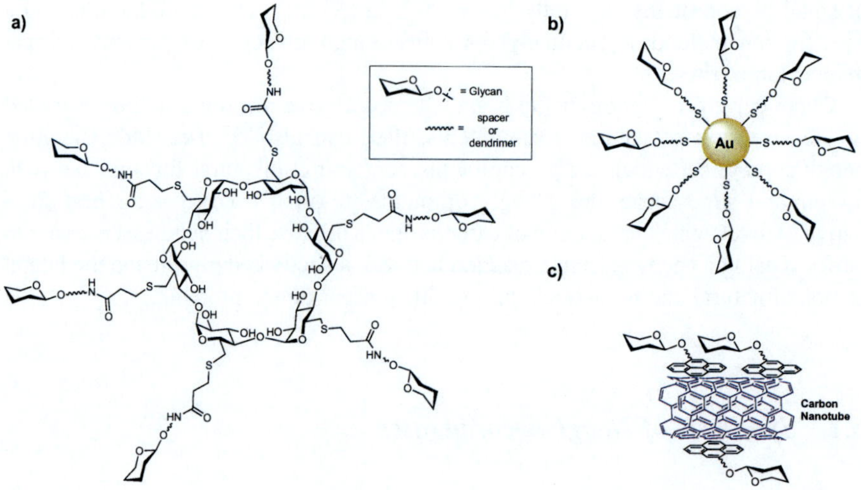


**Fig. 13** Synthetic glycoconjugates and glyconanoparticles. (**a**) Cyclodextrins conjugates, (**b**) Gold glyconanoparticles, and (**c**) glyconanotubes

use glycoproteins as novel protein-based therapeutics, carbohydrate antigen derivatives and immunogens [63].

Specific carbohydrate structures are present on the surface of pathogens. During infections, these structures activate the host immune system inducing the expression of cytokines and the production of anti-glycan antibodies. Small oligosaccharides are poor immunogens. However, a conjugation to carrier proteins can enhance the immune response against glycans. Some active groups present on glycans like alkenes, thiols, or activated esters introduced as part of a linker at the reducing end, can be used for linking glycans to proteins. The conjugation usually involves a chemoselective reaction of activated glycan and amines, carboxylates, or thiols present on the protein, or with function introduced by site-specific modification of the protein [64]. Alternatively, the so-called cross-linkers, di-functionalized spacers having two active moieties, are reacted from one side with carboxylates, amine, and thiols from the protein and with an amine or thiol on the carbohydrate from the other side (Fig. 13b) [43].

Glycoconjugates are also accessible by attaching glycan structures to clustering scaffolds such as dendrimers, cyclodextrins, gold nanoparticles, and carbon nanotubes, among others (Fig. 13) [65–67]. Similar to the preparation of protein glycoconjugates, the glycans are modified with an active functionality that undergoes a chemoselective reaction with a reactive function on the scaffold-like dendrimers or with the surface of a metal. An example of the conjugation to metals is the attachment of sugar-thiols to gold nanoparticles forming gold glyconanoconjugates. All these conjugates have gained special attention in studies to investigate carbohydrate–protein interactions, increasing the binding affinity of carbohydrates to proteins through a multivalent presentation of the carbohydrate [65]. Glycoconjugates are also used for the elaboration of microarrays and their application as a high-throughput platform to evaluate the binding of one or multiple glycoconjugates with different samples in one single experiment [68].

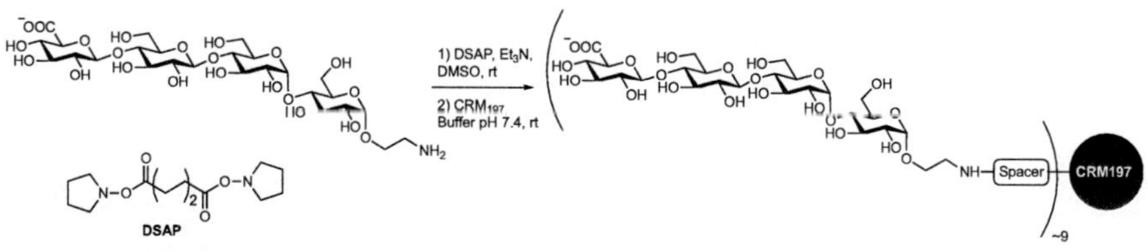


**Fig. 14** Synthetic glycoconjugates vaccine for *S. pneumonia* (ST8, [69])

Immunogens based on protein neoglycoconjugate behave similarly to natural glycoconjugates and require the presence of multiple copies of the carbohydrate to favor the interaction with the immune system. Neoglycoconjugates have found application in the development of carbohydrate-based vaccines using immunogenic active carrier proteins such as tetanus toxoid or its nontoxic variant CRM197. These proteins increase and facilitate the presentation of glycans and the production of a specific immune response to the glycan (Fig. 14). Using these strategies diverse carbohydrate-based vaccines have been designed and are under development [69, 70].

The formation of a complex mixture of products with a variable number of glycans at different positions on the sequence is the main limitation of neoglycoconjugates obtained by reaction with the side chain of amino acids. This heterogeneous modification of activated proteins results from the distinctive accessibility of the active groups on the protein surface for the reaction with the synthetic glycans. The lack of homogeneity hinders the determination of thermodynamic and kinetic parameters of binding events between glycoconjugates and proteins by SPR, ITC or any other method [71]. A requirement that certainly has to be fulfilled to conduct the aforementioned experiments is access to pure and defined glycoconjugates and their corresponding derivatives.

The production of a multivalent system based on the assembly of protein monomers carrying a polyvalent glycan motif was recently introduced as a strategy to produce glycoconjugates. In this strategy, an alkyne is introduced as a site-specific modification of an expressed protein. The capsid protein of the bacteriophage Qβ was used as a carrier protein. This protein assembles into a 180-copy virus-like multimer. The alkynes of the protein nanoparticle can react with a glycodendrimer azide using a cycloaddition reaction, forming a homogeneous protein-glycodendrimer. This well-defined polyvalent glycoprotein assemblies forming a virus-like glycodendrinanoparticle are assemblies presenting on their surface up to 1,620 copies of a glycan (Fig. 15) [72].

## 5.2 Synthesis of Natural Glycoconjugates

The synthesis of glycoconjugates having natural bonds and site-specific modifications is demanding and has required the development of multiple strategies that are specific for the formation of the linkage between glycan and protein for

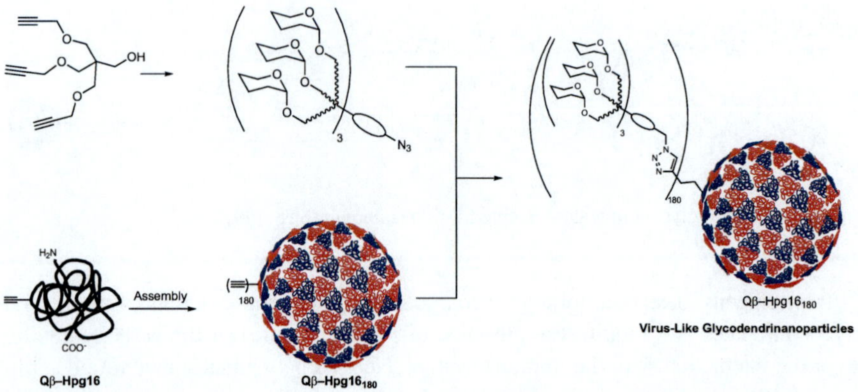


**Fig. 15** Synthesis of a virus-like glycodendrinanoparticle. This glycoconjugate contains up to 1,620 glycans exposed on the surface [72].

delivering *O*- and *N*-glycoproteins and to the lipids by forming glycolipids Whereas in *O*-glycoproteins glycans are attached to serine or threonine residues, in *N*-glycoproteins the glycans are attached to asparagine residues in the consensus sequence Asn-Xxx-Ser/Thr, Xxx being any amino acid different than proline.

*N*-glycoproteins are accessible via expression systems in cell lines such as Chinese Hamster Ovary cells (CHO), Human Embryonic Kidney (HEK) cells or other human cell lines. However, due to the difficult control of the cell glycosylation machinery, which is out of genetic control, the expressed proteins are generally isolated as a mixture of glycoforms. Therefore, investigations aimed to investigate the role of single glycan structure on proteins, require additional strategies [53]. Proteins synthesis and semi-synthesis have emerged as suitable strategies to obtain well-defined glycoproteins [53]. In these strategies, synthetic peptides or expressed protein fragments are ligated with synthetic glycopeptides having a defined glycan structure [73, 74].

Chemical Synthesis of glycoproteins has been accomplished using sequential ligation of active peptide and glycopeptide fragments [53]. Among the different ligation reactions, native chemical ligation with peptide thioesters has been the preferred strategy [75]. The peptides are obtained directly as thioester or as peptide thioester precursors that are activated in situ. The thioesters undergo a two-step chemoselective reaction with peptides having an *N*-terminal cysteine residue ending with the formation of a native peptide bond between the *C*-terminal amino acid of one peptide and the cysteine of the other. The synthesis of the glycoprotein using more than two peptides, which is generally the case, requires protection on cysteine at the *N*-terminus of the internal peptides or glycopeptides and it is only released after the first ligation is completed (Fig. 16a). Recently, a strategy involving the use of thioester precursors that are activated after a completed ligation has been established [76].

In addition to glycoprotein semi-synthesis, the use of enzymes for glycan remodeling and for transferring glycans to a protein having a monosaccharide precursor, trans-glycosylation, are becoming useful methods (Fig. 16b) [77]. The

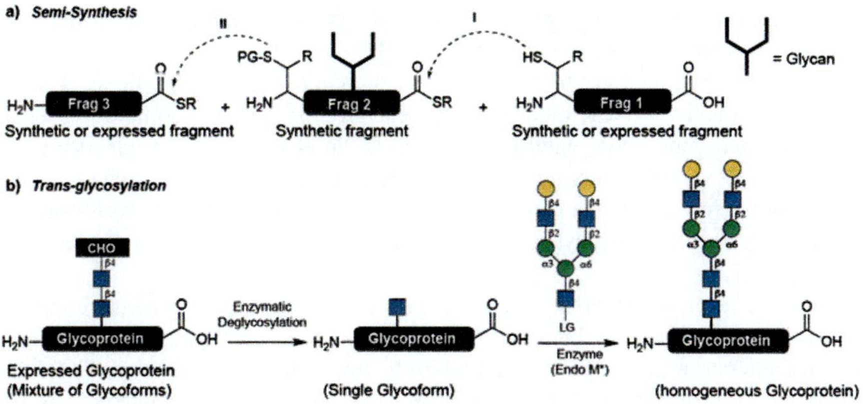


**Fig. 16** Strategies for the semi-synthesis of homogeneous glycoproteins. CHO: mixtures of structures

glycoproteins are generally expressed on cell lines and are treated with an endoglycosidase to leave only a glucosamine unit attached to the protein. This glucosamine is used as an acceptor for the transfer of an oxazoline activated synthetic glycan [78]. Noteworthy here is the use of glycan endoglycosidase that has been mutated to act as a glycosyltransferase [79].

# 6 Glycolipids and Amphiphilic Glycoconjugates

Carbohydrates are also conjugated to lipids forming amphiphilic glycolipids that are localized at the interface between the cell membrane and the extracellular matrix of cells. The hydrophobic part of glycolipids attaches these molecules to the membrane bilayer and participates in the formation of supramolecular assemblies such as lipid rafts or membrane microdomains [80]. Glycolipids also contribute to the structural and morphological changes of the membrane and participate in cell processes through interactions of the glycan part with other molecules present on the extracellular environment, the membrane of the same cell, or from other cells [81].

There are different types of glycolipids and high variation in both, the glycan and lipid structure. In bacteria and plants, glycans are mainly attached to diacylglycerol having lipid chains of different lengths and degrees of saturation; however, they attach also to sterols, alcohols, and aminoalcohols, although in minor extend. In animals and humans, the main part of glycolipids contains ceramide as a lipid moiety forming sphingolipids [82].

Glycolipids can be synthesized as single chains or as glycodendrimers having a lipid tail for its interaction with a hydrophobic environment [83, 84]. Most of the strategies developed to synthesize natural and non-natural glycolipids include a key step for the attachment of the glycan and the lipid [85]. This step can be a glycosylation of the hydroxyl group of glycerolipids and ceramides or the formation of a phosphodiester with a phosphoglycan [81].

To investigate the effect of the structure and composition of glycolipids in their behavior and activity, these molecules are generally inserted into complex systems containing an amphiphilic environment such as liposomes [83], vesicles [86], or monolayers as model membranes [85]. In all these systems, the hydrophobic part of the glycolipid inserts into the membrane and interacts with the alkyl chains forming a stabilized system that displays the glycan on the hydrophilic surface to the aqueous phase (Fig. 17).

Contrary to pure carbohydrates, glycolipids are strong activators of the immune system in mammals. They bind with lectins and other carbohydrate-binding receptors and interact with receptors that bind to the hydrophobic part, such as Toll-like receptors TLR, inducing a strong cell-mediated response. These interactions of glycolipids convert them in good adjuvants and modulators of the immune system and are the motivation for their application in the development of vaccines, i.e. Globo H is a glycolipid used as vaccine candidates against cancer (Fig. 17f) [87]. To illustrate the potency of glycolipid adjuvants, a fully synthetic strategy for the development of vaccines has been introduced. This strategy involves a synthetic conjugate having a glycolipid adjuvant connected to the desired glycan antigen (Fig. 17e) [88].

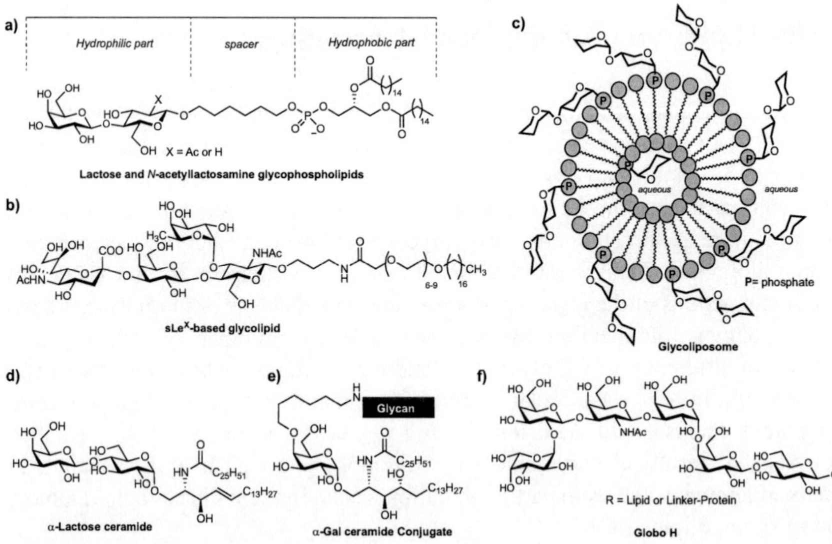


**Fig. 17** Structure of glycolipids. (**a**) Lactose containing phospholipids; (**b**) Structure of a sialyl-Lewis X (sLe$^{X}$)-based glycolipid; (**c**) representation of a glycoliposome; (**d**) alpha lactose ceramide; (**e**) Representation of a synthetic vaccine glycoconjugate containing an alpha Gal-Cer; (**f**) structure of Globo H glycan

## 7 Conclusions and Outlook

Carbohydrates are essential biomolecules participating directly or as glycocojugates in a plethora of cell functions and processes. Research to determine the biological activity and potential applications of carbohydrates using high-throughput methods have as main limitation the access to libraries of these molecules in good purity and amounts. Oligosaccharide synthesis and the development of automated platforms are emerging as tools to provide complex structures and overcome the challenging isolation of carbohydrates from natural sources. Synthetic and semi-synthetic strategies provide carbohydrates and glycoconjugates in pure form and are an essential technology for the application of carbohydrates in material and biomedical research. However, the synthesis of oligosaccharides still requires new protecting groups and high stereoselective glycosylations methods.

Glycoconjugates are available by chemical synthesis in different forms and levels of complexity. However, the methods used to get these conjugates still have limitations to provide natural glycoproteins. The synthesis of homogeneous glycoproteins involving a combination of chemical and enzymatic methods is becoming the standard method to access small to medium size glycoproteins and for supporting investigations to understand the interplay between glycan structure and glycoprotein function. These strategies do not provide long and multi-glycosylated glycoproteins. Thus, the development of methods for introducing complex carbohydrates into large proteins such as trans-glycosylation with enzymes and methods involving multiple ligations are gaining interest and are under development for application in large-scale processes with high efficiency.

## References

1. Varki A (1993) Biological roles of oligosaccharides: all of the theories are correct. Glycobiology 3(2):97–130. https://doi.org/10.1093/glycob/3.2.97
2. Seeberger PH, Werz DB (2007) Synthesis and medical applications of oligosaccharides. Nature 446:1046. https://doi.org/10.1038/nature05819
3. Wang Z, Chinoy ZS, Ambre SG, Peng WJ, McBride R, de Vries RP, Glushka J, Paulson JC, Boons GJ (2013) A general strategy for the chemoenzymatic synthesis of asymmetrically branched N-glycans. Science 341(6144):379–383. https://doi.org/10.1126/science.1236231
4. Koizumi A, Matsuo I, Takatani M, Seko A, Hachisu M, Takeda Y, Ito Y (2013) Top-down chemoenzymatic approach to a high-mannose-type glycan library: synthesis of a common precursor and its enzymatic trimming. Angew Chem Int Ed 52(29):7426–7431. https://doi.org/10.1002/anie.201301613
5. Tsai TI, Lee HY, Chang SH, Wang CH, Tu YC, Lin YC, Hwang DR, Wu CY, Wong CH (2013) Effective sugar nucleotide regeneration for the large-scale enzymatic synthesis of Globo H and SSEA4. J Am Chem Soc 135(39):14831–14839. https://doi.org/10.1021/ja4075584
6. Kiessling LL, Splain RA (2010) Chemical approaches to glycobiology. Annu Rev Biochem 79 (1):619–653. https://doi.org/10.1146/annurev.biochem.77.070606.100917
7. Rillahan CD, Schwartz E, McBride R, Fokin VV, Paulson JC (2012) Click and pick: identification of sialoside analogues for siglec-based cell targeting. Angew Chem Int Ed 51 (44):11014–11018. https://doi.org/10.1002/anie.201205831

8. Johnson MA, Bundle DR (2013) Designing a new antifungal glycoconjugate vaccine. Chem Soc Rev 42(10):4327–4344. https://doi.org/10.1039/c2cs35382b
9. Zeng Y, Ramya TNC, Dirksen A, Dawson PE, Paulson JC (2009) High-efficiency labeling of sialylated glycoproteins on living cells. Nat Methods 6(3):207–209. https://doi.org/10.1038/nmeth.1305
10. Volbeda AG, van der Marel GA, Codee JDC (2019) Protecting group strategies in carbohydrate chemistry. In: Sebastien V (ed) Protecting groups. Wiley-VCH Verlag, Manheim, pp 1–27. https://doi.org/10.1002/9783527697014.ch1
11. Polyakova SM, Nizovtsev AV, Kunetskiy RA, Bovin NV (2015) New protecting groups in the synthesis of oligosaccharides. Russ Chem B 64(5):973–989. https://doi.org/10.1007/s11172-015-0968-5
12. Bols M (1996) Synthesis of Kojitriose using silicon-tethered glycosidation. Acta Chem Scand 50(10):931–937. https://doi.org/10.3891/acta.chem.scand.50-0931
13. Kim JH, Yang H, Boons GJ (2005) Stereoselective glycosylation reactions with chiral auxiliaries. Angew Chem Int Ed 44(6):947–949. https://doi.org/10.1002/anie.200461745
14. Wang H-Y, Blaszczyk SA, Xiao G, Tang W (2018) Chiral reagents in glycosylation and modification of carbohydrates. Chem Soc Rev 47(3):681–701. https://doi.org/10.1039/C7CS00432J
15. Tsai Y-H, Gotze S, Vilotijevic I, Grube M, Varon Silva D, Seeberger PH (2013) A general and convergent synthesis of diverse glycosylphosphatidylinositol glycolipids. Chem Sci 4 (1):468–481
16. Swarts BM, Guo ZW (2010) Synthesis of a glycosylphosphatidylinositol anchor bearing unsaturated lipid chains. J Am Chem Soc 132(19):6648–6650. https://doi.org/10.1021/ja1009037
17. Lee BY, Seeberger PH, Varon Silva D (2016) Synthesis of glycosylphosphatidylinositol (GPI)-anchor glycolipids bearing unsaturated lipids. Chem Commun (Camb) 52(8):1586–1589. https://doi.org/10.1039/c5cc07694c
18. Baeschlin DK, Chaperon AR, Green LG, Hahn MG, Ince SJ, Ley SV (2000) 1,2-Diacetals in synthesis: total synthesis of a glycosylphosphatidylinositol anchor of Trypanosoma brucei. Chem A Eur J 6(1):172–186. https://doi.org/10.1002/(SICI)1521-3765(20000103)6:1<172::AID-CHEM172>3.0.CO;2-5
19. Wang CC, Lee JC, Luo SY, Kulkarni SS, Huang YW, Lee CC, Chang KL, Hung SC (2007) Regioselective one-pot protection of carbohydrates. Nature 446(7138):896–899. https://doi.org/10.1038/nature05730
20. Demchenko AV (2008) General aspects of the glycosidic bond formation. In: Demchenko AV (ed) Handbook of chemical glycosylation. Wiley-VCH, Weinheim, pp 1–27. https://doi.org/10.1002/9783527621644.ch1
21. McKay MJ, Nguyen HM (2012) Recent advances in transition metal-catalyzed glycosylation. ACS Catal 2(8):1563–1595. https://doi.org/10.1021/cs3002513
22. Nukada T, Berces A, Zgierski MZ, Whitfield DM (1998) Exploring the mechanism of neighboring group assisted glycosylation reactions. J Am Chem Soc 120(51):13291–13295. https://doi.org/10.1021/ja981041m
23. Nigudkar SS, Demchenko AV (2015) Stereocontrolled 1,2-cis glycosylation as the driving force of progress in synthetic carbohydrate chemistry. Chem Sci 6(5):2687–2704. https://doi.org/10.1039/C5SC00280J
24. Murakata C, Ogawa T (1992) Stereoselective synthesis of glycobiosyl phosphatidylinositol, a part structure of the glycosyl-phosphatidylinositol (GPI) anchor of Trypanosoma brucei. Carbohydr Res 234:75–91
25. Hahm HS, Hurevich M, Seeberger PH (2016) Automated assembly of oligosaccharides containing multiple cis-glycosidic linkages. Nat Commun 7:12482. https://doi.org/10.1038/ncomms12482
26. Wulff G, Rohle G (1974) Results and problems of O-glycoside synthesis. Angew Chem Int Ed 13(3):157–170. https://doi.org/10.1002/anie.197401571

27. Schmidt RR, Behrendt M, Toepfer A (1990) Nitriles as solvents in glycosylation reactions – highly selective beta-glycoside synthesis. Synlett (11):694–696
28. Satoh H, Hansen HS, Manabe S, van Gunsteren WF, Hünenberger PH (2010) Theoretical investigation of solvent effects on glycosylation reactions: stereoselectivity controlled by preferential conformations of the intermediate oxacarbenium-counterion complex. J Chem Theory Comput 6(6):1783–1797. https://doi.org/10.1021/ct1001347
29. Bartetzko MP, Schuhmacher F, Hahm HS, Seeberger PH, Pfrengle F (2015) Automated glycan assembly of oligosaccharides related to arabinogalactan proteins. Org Lett 17(17):4344–4347. https://doi.org/10.1021/acs.orglett.5b02185
30. Plante OJ, Palmacci ER, Seeberger PH (2001) Automated solid-phase synthesis of oligosaccharides. Science 291(5508):1523–1527. https://doi.org/10.1126/science.1057324
31. Seeberger PH (2015) The logic of automated glycan assembly. Acc Chem Res 48 (5):1450–1463. https://doi.org/10.1021/ar5004362
32. Bennett CS (2014) Principles of modern solid-phase oligosaccharide synthesis. Org Biomol Chem 12(11):1686–1698. https://doi.org/10.1039/C3OB42343C
33. Jaipuri FA, Pohl NL (2008) Toward solution-phase automated iterative synthesis: fluorous-tag assisted solution-phase synthesis of linear and branched mannose oligomers. Org Biomol Chem 6(15):2686–2691. https://doi.org/10.1039/B803451F
34. Weiss H, Unverzagt C (2003) Highly branched oligosaccharides: a general strategy for the synthesis of multiantennary N-glycans with a bisected motif. Angew Chem Int Ed 42 (35):4261–4263. https://doi.org/10.1002/anie.200351625
35. Walvoort MTC, van den Elst H, Plante OJ, Krock L, Seeberger PH, Overkleeft HS, van der Marel GA, Codee JDC (2012) Automated solid-phase synthesis of beta-Mannuronic acid alginates. Angew Chem Int Ed 51(18):4393–4396. https://doi.org/10.1002/anie.201108744
36. Zhang Z, Ollmann IR, Ye X-S, Wischnat R, Baasov T, Wong C-H (1999) Programmable one-pot oligosaccharide synthesis. J Am Chem Soc 121(4):734–753. https://doi.org/10.1021/ja982232s
37. Lu J, Jayaprakash KN, Schlueter U, Fraser-Reid B (2004) Synthesis of a malaria candidate glycosylphosphatidylinositol (GPI) structure: a strategy for fully inositol acylated and phosphorylated GPIs. J Am Chem Soc 126(24):7540–7547. https://doi.org/10.1021/ja038807p
38. Muthana SM, Campbell CT, Gildersleeve JC (2012) Modifications of glycans: biological significance and therapeutic opportunities. ACS Chem Biol 7(1):31–43. https://doi.org/10.1021/cb2004466
39. Yu H, Chen X (2007) Carbohydrate post-glycosylational modifications. Org Biomol Chem 5 (6):865–872. https://doi.org/10.1039/B700034K
40. Mende M, Bednarek C, Wawryszyn M, Sauter P, Biskup MB, Schepers U, Brase S (2016) Chemical synthesis of glycosaminoglycans. Chem Rev 116(14):8193–8255. https://doi.org/10.1021/acs.chemrev.6b00010
41. Birch AJ, Smith H (1958) Reduction by metal-amine solutions: applications in synthesis and determination of structure. Q Rev Chem Soc 12(1):17–33. https://doi.org/10.1039/QR9581200017
42. Niemietz M, Perkams L, Hoffman J, Eller S, Unverzagt C (2011) Selective oxidative debenzylation of mono- and oligosaccharides in the presence of azides. Chem Commun 47 (37):10485–10487. https://doi.org/10.1039/c1cc13884g
43. Ding N, Li X, Chinoy ZS, Boons G-J (2017) Synthesis of a glycosylphosphatidylinositol anchor derived from Leishmania donovani that can be functionalized by Cu-catalyzed azide–alkyne cycloadditions. Org Lett 19(14):3827–3830. https://doi.org/10.1021/acs.orglett.7b01703
44. Manabe S, Ito Y, Ogawa T (1998) Solvent effect in glycosylation reaction on polymer support. Synlett (6):628
45. Plante OJ, Andrade RB, Seeberger PH (1999) Synthesis and use of glycosyl phosphates as glycosyl donors. Org Lett 1(2):211–214. https://doi.org/10.1021/ol9905452

46. Shimizu H, Sakamoto M, Nagahori N, Nishimura SI (2007) A new glycosylation method. Part 2: study of carbohydrate elongation onto the gold nanoparticles in a colloidal phase. Tetrahedron 63(11):2418–2425. https://doi.org/10.1016/j.tet.2007.01.011
47. Pornsuriyasak P, Ranade SC, Li AX, Parlato MC, Sims CR, Shulga OV, Stine KJ, Demchenko AV (2009) STICS: surface-tethered iterative carbohydrate synthesis. Chem Commun (14):1834–1836. https://doi.org/10.1039/b817684a
48. Le Mai Hoang K, Pardo-Vargas A, Zhu Y, Yu Y, Loria M, Delbianco M, Seeberger PH (2019) Traceless photolabile linker expedites the chemical synthesis of complex oligosaccharides by automated glycan assembly. J Am Chem Soc 141(22):9079–9086. https://doi.org/10.1021/jacs.9b03769
49. Andrade RB, Plante OJ, Melean LG, Seeberger PH (1999) Solid-phase oligosaccharide synthesis: preparation of complex structures using a novel linker and different glycosylating agents. Org Lett 1(11):1811–1814. https://doi.org/10.1021/ol991071+
50. Nguyen SH, Trotta AH, Cao J, Straub TJ, Bennett CS (2012) Direct O-glycosidation of resin bound thioglycosides. Org Biomol Chem 10(12):2373–2376. https://doi.org/10.1039/c2ob06883d
51. Eller S, Collot M, Yin J, Hahm HS, Seeberger PH (2013) Automated solid-phase synthesis of chondroitin sulfate glycosaminoglycans. Angew Chem Int Ed 52(22):5858–5861. https://doi.org/10.1002/anie.201210132
52. Yu Y, Kononov A, Delbianco M, Seeberger PH (2018) A capping step during automated glycan assembly enables access to complex glycans in high yield. Chem A Eur J 24(23):6075–6078. https://doi.org/10.1002/chem.201801023
53. Unverzagt C, Kajihara Y (2013) Chemical assembly of N-glycoproteins: a refined toolbox to address a ubiquitous posttranslational modification. Chem Soc Rev 42(10):4408–4420. https://doi.org/10.1039/c3cs35485g
54. Rudd PM, Joao HC, Coghill E, Fiten P, Saunders MR, Opdenakker G, Dwek RA (1994) Glycoforms modify the dynamic stability and functional activity of an enzyme. Biochemistry 33 (1):17–22
55. Yamaguchi Y, Nishimura M, Nagano M, Yagi H, Sasakawa H, Uchida K, Shitara K, Kato K (2006) Glycoform-dependent conformational alteration of the fc region of human immunoglobulin G1 as revealed by NMR spectroscopy. Biochim Biophys Acta 1760(4):693–700
56. Taylor CM (1998) Glycopeptides and glycoproteins: focus on the glycosidic linkage. Tetrahedron 54(38):11317–11362
57. Wang L-X, Lomino JV (2011) Emerging technologies for making glycan-defined glycoproteins. ACS Chem Biol 7(1):110–122. https://doi.org/10.1021/cb200429n
58. Pratt MR, Bertozzi CR (2005) Synthetic glycopeptides and glycoproteins as tools for biology. Chem Soc Rev 34(1):58–68
59. Liu L, Bennett CS, Wong C-H (2006) Advances in glycoprotein synthesis. Chem Commun (1):21–33
60. Grogan MJ, Pratt MR, Marcaurelle LA, Bertozzi CR (2002) Homogeneous glycopeptides and glycoproteins for biological investigation. Annu Rev Biochem 71:593–634
61. Davis BG (2002) Synthesis of glycoproteins. Chem Rev 102:579–601
62. Bill RM, Flitsch S (1996) Chemical and biological approaches to glycoprotein synthesis. Chem Biol 3(3):145–149
63. Wiederschain GY (2013) Glycobiology: progress, problems, and perspectives. Biochemistry (Mosc) 78(7):679–696. https://doi.org/10.1134/s0006297913070018
64. van Hest JCM, Kiick KL, Tirrell DA (2000) Efficient incorporation of unsaturated methionine analogues into proteins in vivo. J Am Chem Soc 122(7):1282–1288. https://doi.org/10.1021/ja992749j
65. Solís D, Bovin NV, Davis AP, Jiménez-Barbero J, Romero A, Roy R, Smetana K, Gabius H-J (2015) A guide into glycosciences: how chemistry, biochemistry and biology cooperate to crack the sugar code. Biochim Biophys Acta 1850(1):186–235. https://doi.org/10.1016/j.bbagen.2014.03.016

66. Baoyong L, Xiaowei L, Hong R, Jiwei C, Hongguang L (2016) Synthesis and applications of glyconanoparticles. Curr Org Chem 20(14):1502–1511. https://doi.org/10.2174/1385272820666151207194453
67. Kottari N, Chabre YM, Sharma R, Roy R (2013) Applications of glyconanoparticles as "Sweet" glycobiological therapeutics and diagnostics. In: Dutta PK, Dutta J (eds) Multifaceted development and application of biopolymers for biology, biomedicine and nanotechnology. Advances in polymer science, vol 254. pp 297–341. https://doi.org/10.1007/12_2012_208
68. Horlacher T, Seeberger PH (2008) Carbohydrate arrays as tools for research and diagnostics. Chem Soc Rev 37(7):1414–1422. https://doi.org/10.1039/b708016f
69. Schumann B, Hahm HS, Parameswarappa SG, Reppe K, Wahlbrink A, Govindan S, Kaplonek P, L-a P, Witzenrath M, Anish C, Pereira CL, Seeberger PH (2017) A semisynthetic *Streptococcus pneumoniae* serotype 8 glycoconjugate vaccine. Sci Transl Med 9:380
70. Parameswarappa SG, Reppe K, Geissner A, Menova P, Govindan S, Calow ADJ, Wahlbrink A, Weishaupt MW, Monnanda BP, Bell RL, Pirofski LA, Suttorp N, Sander LE, Witzenrath M, Pereira CL, Anish C, Seeberger PH (2016) A semi-synthetic oligosaccharide conjugate vaccine candidate confers protection against *Streptococcus pneumoniae* serotype 3 infection. Cell Chem Biol 23(11):1407–1416. https://doi.org/10.1016/j.chembiol.2016.09.016
71. Velazquez-Campoy A, Ohtaka H, Nezami A, Muzammil S, Freire E (2004) Isothermal titration calorimetry. Curr Protoc Cell Biol. Chapter 17:Unit 17 18. https://doi.org/10.1002/0471143030.cb1708s23
72. Ribeiro-Viana R, Sánchez-Navarro M, Luczkowiak J, Koeppe JR, Delgado R, Rojo J, Davis BG (2012) Virus-like glycodendrinanoparticles displaying quasi-equivalent nested polyvalency upon glycoprotein platforms potently block viral infection. Nat Commun 3:1303. https://doi.org/10.1038/ncomms2302. https://www.nature.com/articles/ncomms2302#supplementary-information
73. Piontek C, Ring P, Harjes O, Heinlein C, Mezzato S, Lombana N, Pöhner C, Püttner M, Varón-Silva D, Martin A, Schmid F-X, Unverzagt C (2009) Semisynthese eines homogenen glycoprotein-enzyms: ribonuclease-C (Teil-1). Angew Chem 121(11):1968–1973
74. Piontek C, Ring P, Harjes O, Heinlein C, Mezzato S, Lombana N, Pohner C, Puettner M, Varon-Silva D, Martin A, Schmid FX, Unverzagt C (2009) Semisynthesis of a homogeneous glycoprotein enzyme: ribonuclease C: part 1. Angew Chem Int Ed 48(11):1936–1940. https://doi.org/10.1002/anie.200804734
75. Dawson PE, Muir TW, Clark-Lewis I, Kent SB (1994) Synthesis of proteins by native chemical ligation. Science 266(5186):776–779
76. Reif A, Siebenhaar S, Troster A, Schmalzlein M, Lechner C, Velisetty P, Gottwald K, Pohner C, Boos I, Schubert V, Rose-John S, Unverzagt C (2014) Semisynthesis of biologically active glycoforms of the human cytokine interleukin 6. Angew Chem Int Ed 53(45):12125–12131. https://doi.org/10.1002/anie.201407160
77. Wang LX, Amin MN (2014) Chemical and chemoenzymatic synthesis of glycoproteins for deciphering functions. Chem Biol 21(1):51–66. https://doi.org/10.1016/j.chembiol.2014.01.001
78. Ochiai H, Huang W, Wang L-X (2008) Expeditious chemoenzymatic synthesis of homogeneous N-glycoproteins carrying defined oligosaccharide ligands. J Am Chem Soc 130(41):13790–13803. https://doi.org/10.1021/ja805044x
79. Umekawa M, Huang W, Li B, Fujita K, Ashida H, Wang LX, Yamamoto K (2008) Mutants of Mucor hiemalis endo-beta-N-acetylglucosaminidase show enhanced transglycosylation and glycosynthase-like activities. J Biol Chem 283(8):4469–4479. https://doi.org/10.1074/jbc.M707137200
80. Lingwood D, Simons K (2010) Lipid rafts as a membrane-organizing principle. Science 327(5961):46. https://doi.org/10.1126/science.1174621
81. Latxague L, Gaubert A, Barthélémy P (2018) Recent advances in the chemistry of glycoconjugate amphiphiles. Molecules 23(1):89

82. Kopitz J (2017) Lipid glycosylation: a primer for histochemists and cell biologists. Histochem Cell Biol 147(2):175–198. https://doi.org/10.1007/s00418-016-1518-4
83. Harada Y, Murata T, Totani K, Kajimoto T, Masum SM, Tamba Y, Yamazaki M, Usui T (2005) Design and facile synthesis of neoglycolipids as lactosylceramide mimetics and their transformation into glycoliposomes. Biosci Biotechnol Biochem 69(1):166–178. https://doi.org/10.1271/bbb.69.166
84. René R, Mi KJ (1999) Amphiphilic p-tert-butylcalix[4]arene scaffolds containing exposed carbohydrate dendrons. Angew Chem Int Ed 38(3):369–372. https://doi.org/10.1002/(SICI)1521-3773(19990201)38:3<369::AID-ANIE369>3.0.CO;2-1
85. Jayaraman N, Maiti K, Naresh K (2013) Multivalent glycoliposomes and micelles to study carbohydrate-protein and carbohydrate-carbohydrate interactions. Chem Soc Rev 42(11):4640–4656. https://doi.org/10.1039/C3CS00001J
86. Kameta N, Matsuzawa T, Yaoi K, Fukuda J, Masuda M (2017) Glycolipid-based nanostructures with thermal-phase transition behavior functioning as solubilizers and refolding accelerators for protein aggregates. Soft Matter 13(17):3084–3090. https://doi.org/10.1039/C7SM00310B
87. Danishefsky SJ, Shue Y-K, Chang MN, Wong C-H (2015) Development of globo-H cancer vaccine. Acc Chem Res 48(3):643–652. https://doi.org/10.1021/ar5004187
88. Cavallari M, Stallforth P, Kalinichenko A, Rathwell DCK, Gronewold TMA, Adibekian A, Mori L, Landmann R, Seeberger PH, De Libero G (2014) A semisynthetic carbohydrate-lipid vaccine that protects against *S. pneumoniae* in mice. Nat Chem Biol 10:950. https://doi.org/10.1038/nchembio.1650. https://www.nature.com/articles/nchembio.1650#supplementary-information

# Enzymatic Synthesis of Glycans and Glycoconjugates

**Thomas Rexer, Dominic Laaf, Johannes Gottschalk, Hannes Frohnmeyer, Erdmann Rapp, and Lothar Elling**

## Contents



---

Thomas Rexer and Dominic Laaf contributed equally to this work.

The original version of this chapter was revised: A correction to this chapter can be found at https://doi.org/10.1007/10_2020_159

---

T. Rexer
Max Planck Institute for Dynamics of Complex Technical Systems Magdeburg, Department of Bioprocess Engineering, Magdeburg, Germany
e-mail: rexer@mpi-magdeburg.mpg.de

D. Laaf, J. Gottschalk, H. Frohnmeyer, and L. Elling (✉)
Laboratory for Biomaterials, Institute for Biotechnology and Helmholtz-Institute for Biomedical Engineering, RWTH Aachen University, Aachen, Germany
e-mail: d.laaf@biotec.rwth-aachen.de; j.gottschalk@biotec.rwth-aachen.de; h.frohnmeyer@biotec.rwth-aachen.de; l.elling@biotec.rwth-aachen.de

E. Rapp
Max Planck Institute for Dynamics of Complex Technical Systems, Magdeburg, Germany

glyXera GmbH, Magdeburg, Germany
e-mail: rapp@mpi-magdeburg.mpg.de; e.rapp@glyxera.com

**Abstract** Glycoconjugates have great potential to improve human health in a multitude of different ways and fields. Prominent examples are human milk oligosaccharides and glycosaminoglycans. The typical choice for the production of homogeneous glycoconjugates is enzymatic synthesis. Through the availability of expression and purification protocols, recombinant Leloir glycosyltransferases are widely applied as catalysts for the synthesis of a wide range of glycoconjugates. Extensive utilization of these enzymes also depends on the availability of activated sugars as building blocks. Multi-enzyme cascades have proven a versatile technique to synthesize and in situ regenerate nucleotide sugar.

In this chapter, the functions and mechanisms of Leloir glycosyltransferases are revisited, and the advantage of prokaryotic sources and production systems is discussed. Moreover, in vivo and in vitro pathways for the synthesis of nucleotide sugar are reviewed. In the second part, recent and prominent examples of the application of Leloir glycosyltransferase are given, i.e., the synthesis of glycosaminoglycans, glycoconjugate vaccines, and human milk oligosaccharides as well as the re-glycosylation of biopharmaceuticals, and the status of automated glycan assembly is revisited.

**Graphical Abstract**

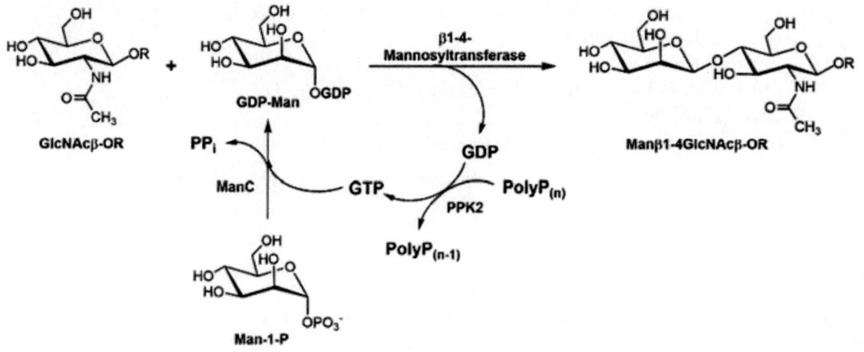


**Keywords** Biocatalysis, Glycoconjugates, Glycoengineering, Glycoproteins, Glycosaminoglycans, Glycosyltransferases, Microreactors, Milk Oligosaccharides, Nucleotide sugars

## Abbreviations

| | |
|---|---|
| ADP | Adenosine diphosphate |
| Asp | Asparagine |
| CHO | Chinese hamster ovary |
| CMP | Cytidine monophosphate |
| CS | Chondroitin sulfate |
| CTP | Cytidine triphosphate |

| | |
|---|---|
| DS | Dermatan sulfate |
| DSP | Downstream processing |
| dTDP | Deoxythymidine diphosphate |
| FucT | Fucosyltransferase |
| GalT | Galactosyltransferase |
| GDP | Guanosine diphosphate |
| Glc | Glucose |
| GlcA | Glucuronic acid |
| GlcNAc | *N*-acetylglucosamine |
| GRAS | Generally recognized as safe |
| GT | Glycosyltransferase |
| HMO | Human milk oligosaccharide |
| HMW | High-molecular-weight |
| HNK | Human natural killer cell |
| Hp | Heparin sulfate |
| HS | Heparan sulfate |
| IdoA | Iduronic acid |
| IgG | Immunoglobulin G |
| LacNAc | *N*-Acetyl-D-lactosamine |
| LMW | Low-molecular-weight |
| LNT II | Lacto-*N*-triose |
| Man | Mannose |
| MBP | Maltose-binding protein |
| MP-CE | Multiplexed capillary electrophoresis |
| NDP | Nucleoside diphosphate |
| Neu5Ac | *N*-Acetylneuraminic acid |
| NMPK | Nucleoside monophosphate kinase |
| NMP | Nucleoside monophosphate |
| OPME | One-pot multi-enzyme |
| OST | Oligosaccharyltransferase |
| PEP | Phosphoenolpyruvate |
| PG | Proteoglycans |
| PGCS | Proteoglycan carrying chondroitin sulfate |
| PGDS | Proteoglycan carrying dermatan sulfate |
| PGHS | Proteoglycan carrying heparan sulfate |
| PK | Pyruvate kinase |
| PPK | Polyphosphate kinase |
| PolyP | Polyphosphate |
| Ser | Serine |
| SiaT | Sialyltransferase |
| STY | Space-time yield |
| SuSy | Sucrose synthase |
| TTN | Total turnover numbers |
| UDP | Uridine diphosphate |
| UTP | Uridine triphosphate |
| Xyl | Xylose |

# 1 Introduction

This chapter covers the most recent advances in enzymatic synthesis of glycans and glycoconjugates. The term "glycoconjugates" refers to glycoproteins, glycolipids, carbohydrate-based polymers, and glycosides of natural products. We will focus on novel biocatalytic in vitro cascades. Advances in carbohydrate-active enzyme engineering based on their reaction mechanisms and protein structure have significantly contributed to expanding their substrate acceptance. We will highlight the scope of Leloir glycosyltransferase and provide some examples where the strength of Leloir glycosyltransferases in glycan and glycoconjugate synthesis is demonstrated.

This chapter should serve as a guide in enzymatic carbohydrate synthesis utilizing Leloir glycosyltransferases. Databanks such as CAZy and BRENDA are very helpful for planning a retrosynthetic biocatalytic route for the synthesis of glycans and finding the right biocatalysts. Elucidation of reaction mechanisms as well as protein engineering and directed evolution of carbohydrate-active enzymes opens the cross-talk between chemistry and biocatalysis and expands the spectrum of reachable glycan products. Glycosyltransferases are indispensable biocatalysts for large-scale glycan and glycopolymer (including biopolymers such as glycosaminoglycans) synthesis. Enzymatic synthesis is often the method of choice to obtain homogeneous glycosylation patterns on glycoconjugates, e.g., glycoproteins, glycolipids, or glycosides of natural products. In this context, numerous examples testify the strength of biocatalysts in glycoconjugate synthesis. We like to refer our readers to excellent reviews and book chapters summarizing the topics that are not fully covered by this chapter (Table 1).

**Table 1** List of reviews covering the indicated topics of enzymes in glycan synthesis

| |
|---|
| *Data mining for carbohydrate-active enzymes:* |
| CAZy (www.cazy.org/) and CAZypedia (www.cazypedia.org) |
| Computational biology methods: [1] |
| *Reaction mechanisms and protein structures:* |
| Reaction mechanisms: [2–4] |
| Protein structures: [5–7] |
| *Protein engineering/directed evolution:* |
| Screening for novel enzymes: [8, 9] |
| Glycosyltransferases: [10, 11] |
| Non-Leloir glycosyltransferases: [12, 13] |
| Glycosidases/glycosynthases: [11, 14, 15] |
| *Glycan, oligosaccharide, and glyco(bio)polymer synthesis:* |
| HMOS: [16–18] |
| Food additives and oligosaccharides: [13, 19–21] |
| Glycosaminoglycans: [22, 23] |
| *Glycoconjugate synthesis:* |
| Glycoproteins: [24–27] |
| Glycolipids: [28, 29] |
| Vaccines: [30] |
| Glycosides of natural products: [31–33] |

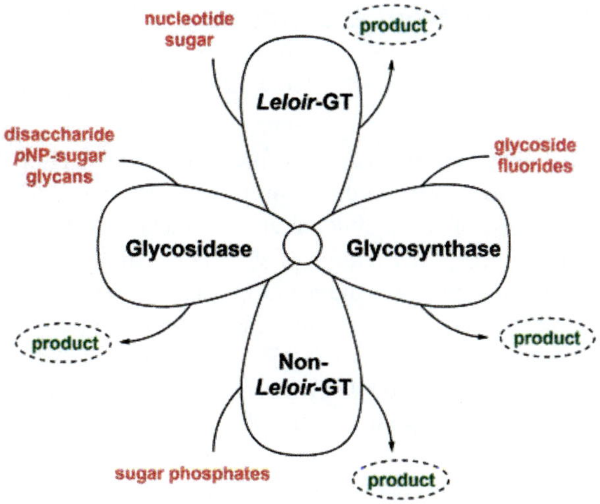


**Fig. 1** Characteristics of the enzymes used in carbohydrate synthesis

What are the challenges? The main challenge is enabling non-experts to synthesize carbohydrate/glycan structures. In terms of enzymatic synthesis, platform technologies and a platform of chemicals and enzymes should be available. This could be synthetic kits or microreactors as small unit operation modules. Even automation of enzymatic synthesis is realistic. Multiple use of stable immobilized biocatalysts, not yet fully exploited in enzymatic glycan and glycoconjugate synthesis, opens the door to obtain larger product amounts. Most importantly, the integration of chemical and enzymatic carbohydrate synthesis into multistep synthesis should offer valuable and novel synthetic pathways. We will present and discuss examples that address challenges for applications of Leloir glycosyltransferases.

Figure 1 depicts the most important characteristics of their use in carbohydrate synthesis. Leloir glycosyltransferases utilize nucleotide sugars as donor substrates and are classified by their reaction mechanisms of retaining or inverting the anomeric configuration of the transferred sugar in the newly formed glycosidic bond. Being dependent on nucleotide sugars, we will first discuss the pros and cons of this enzyme class before we highlight recent advances for their use in cascade reactions. Similarly, classified as retaining and inverting enzymes are exo-glycosidases and endoglycosidases. Substrates for their kinetically driven transglycosylation activity are disaccharides and aryl glycosides, respectively. Their corresponding engineered non-hydrolytic active pendants are glycosynthases. Non-Leloir glycosyltransferases include most often sugar phosphorylases as well as glycosyltransferases, which use monosaccharide-1-phosphates or "energy-rich" disaccharides (e.g., sucrose), respectively, for their synthetic transfer activity.

# 2 Glycan Synthesis with Leloir Glycosyltransferase

## 2.1 *The Choice of Leloir Glycosyltransferases*

Availability and production – "many microbial GTs do the job"

The enzymatic synthesis of glycans follows a retrosynthetic strategy for the connection of the individual sugar moieties by regio- and stereoselective bonds. Leloir glycosyltransferases (Leloir GTs) from biosynthetic pathways of natural glycan structures are often the first choice. In terms of synthesis strategy, these Leloir GTs are combined in different ways. Figure 2 depicts the general reaction equation for a Leloir GT (GT1) transferring sugar B onto the acceptor sugar A. Practically, Leloir GTs assemble four potential glycan structures. For linear oligosaccharides with a sugar sequence D-C-B (**1**), cascades of GT reactions offer theoretically a synthetic one-pot approach since each GT depends on the action of the preceding, which delivers the acceptor substrate for the next glycosylation step. Sugar units are incorporated in linear glycan sequence C-B-C-B (**2**) by sequential reactions, e.g., sugar unit D is introduced at each B unit introducing multiple branching points in a linear glycan structure. Glycan structures with more than one branch at one sugar unit (**3**) are synthesized by a sequential reaction mode where the branching point (GT2 or GT2') is critical for the order of GTs in the reaction sequence. Finally, repeating glycan units $[C\text{-}B]_n$ (**4**) are assembled by the alternate order of the appropriate GT.

Structural and mechanistic features of GTs have been reviewed in detail elsewhere [6, 34, 35]. Two different mechanistic courses are discussed for retaining

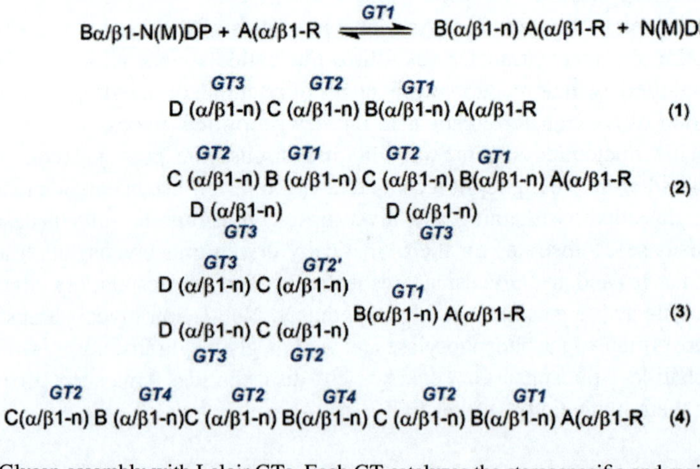


**Fig. 2** Glycan assembly with Leloir GTs. Each GT catalyzes the stereospecific and regioselective formation of glycosidic bonds between its specific donor substrate (nucleotide sugar) and sugar acceptor substrate. The exemplary GT reaction is depicted for the transfer of sugar B onto the acceptor sugar. Glycan building blocks are assembled by a combination of Leloir GTs: linear (**1**), linear with multiple branching points (**2**), branched (**3**), and linear with repeating sugar units (**4**)

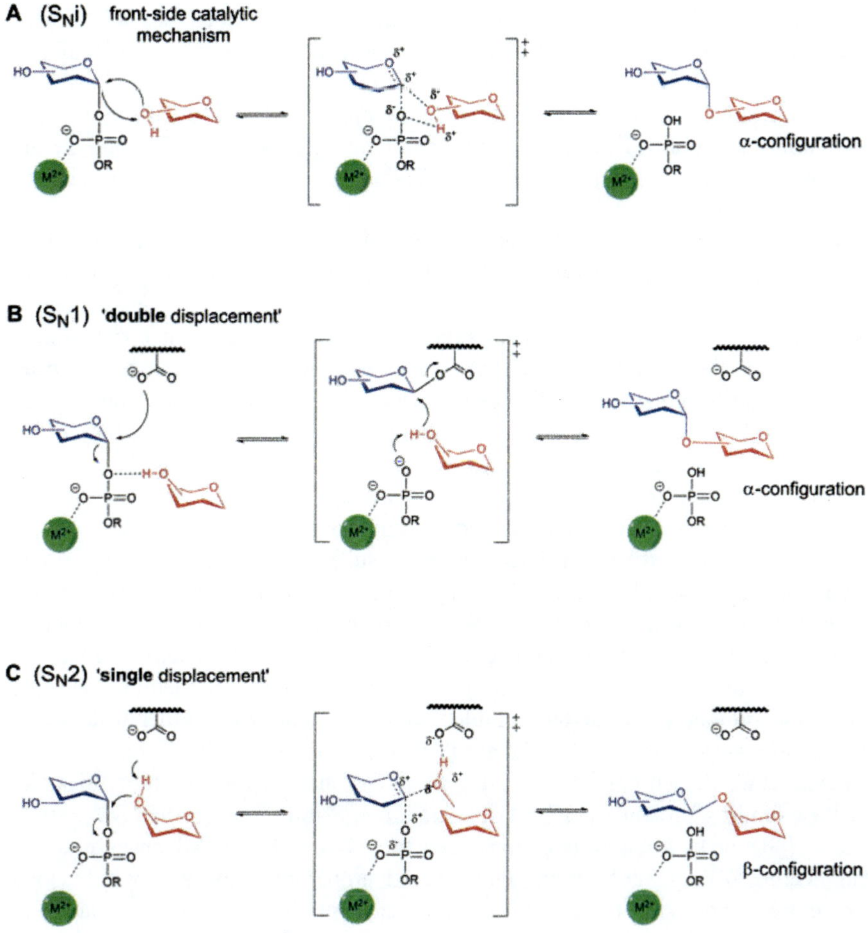


**Fig. 3** Mechanistic features of retaining (**a**, **b**) and inverting (**c**) Leloir glycosyltransferases [2, 3]

GTs: the front-face ($S_Ni$) and double-displacement ($S_N1$) catalyzed reaction (Fig. 3a, b). The latter one has been solved for enzymes belonging to GT family 6, e.g., α3-galactosyltransferase from *Bos taurus* [36]. The exact retaining mechanisms of the front-face catalyzed $S_Ni$-like mechanisms remain elusive because of the open question of whether the leaving phosphate is capable of acting as a base. On the other hand, the inverting GTs reverse the anomeric configuration determined by the nucleotide sugar by a single bimolecular nucleophilic attack (Fig. 3c), described as $S_N2$ mechanism. During this direct displacement, a negatively charged amino acid, predominantly aspartic acid (Asp) or glutamic acid (Glu), stabilizes the oxocarbenium ion-like intermediate [2, 3].

The respective GT-protein fold family (GT-A and GT-B and GT-C fold) is also helpful for setting biochemical parameters in enzymatic reactions. Leloir GTs belonging to the GT-A fold family are dependent on divalent metal ions ($Mn^{2+}$, $Mg^{2+}$) essential for the binding of the nucleotide sugar donor substrate at a typical DXD (H) motif. GT-B fold GTs are more or less independent from metal ions, and nucleotide sugar binding is mediated by basic amino acids. Reaction mechanisms are similar as described in Fig. 3. Some retaining and inverting GTs catalyze the hydrolysis of their donor substrates (α3GalT, FucTs, SiaTs), also in the presence of an acceptor substrate.

In conclusion, the type of sugar and its glycosidic linkage to the acceptor substrate determine the choice of GT. Biosynthetic pathways of glycans and databanks (CAZy, BRENDA) assist to find the right or at least putative biocatalyst. Although CAZy classifies more than 100,000 putative GTs, more than 95% remain to be characterized. Table 2 summarizes microbial Leloir glycosyltransferases used for the synthesis of mammalian-like glycan structures. The list includes retaining and inverting GTs concerning the anomeric configuration of the donor sugar before and after the transfer reaction. Recent reviews cover expression systems for human Leloir glycosyltransferases [37] as well as plant and microbial Leloir GTs for the synthesis of glycosylated natural secondary products [38, 39].

Leloir GTs from prokaryotic hosts lack any disulfide bridges and posttranslational modifications. Production is, therefore, most often achieved as $His_6$-tagged proteins by *Escherichia coli* (*E. coli*), known as the simplest of all recombinant production systems (Table 2). Reaction parameter engineering (amount of inductor, compatible solutes, decreased temperature during induction, fermentation medium) ensures satisfying protein yields in the soluble fraction. *E. coli* was further tested as a production system for more advanced viral and/or eukaryotic Leloir GTs by co-expressing chaperones [40] or fusing the GT of interest with solubility tags like maltose-binding protein (MBP) [41], glutathione S-transferase (GST) [42], small lectins [43], or hydrophilic peptide sequences [44, 45]. Recent advances in strain engineering offer *E. coli* strains with altered properties (tRNAs, disulfide bond isomerases, expression in periplasm) for efficient production of mammalian GTs [46, 47] that were previously declared as "difficult-to-express." However, one major disadvantage of the *E. coli* expression system is the fact that cell lysis is required to purify the protein from the cytoplasmic or periplasmic crude extract.

A more elegant but rather less explored field is the use of *Bacillus subtilis* (*B. subtilis*) and the methylotrophic yeast *Pichia pastoris* [48–50], which are recorded as GRAS (generally recognized as safe) organism owing to the lack of endotoxins. Therefore, they are suitable alternatives to *E. coli* when aiming at an application of GT in the biomedical context. The major advantage is the secretion of the target proteins into the medium leading to less laborious downstream processing and high titers/yields. In this context, insect cell cultures [51] and Chinese hamster ovary (CHO) cells [52] or COS cells [53] are common alternatives for producing GTs that need posttranslational modifications. However, the laborious and cost-intensive processes may restrict large-scale protein production.

A promising upcoming field for the production of GTs is cell-free protein synthesis [54]. Cell extracts from microbes, animal cell lines, and plants are utilized for the production of soluble and transmembrane proteins [55]. However, being in

**Table 2** Microbial Leloir glycosyltransferases for the synthesis of mammalian glycoconjugates

| Enzyme name (CAZy) | Origin | Production[a] | Synthesized glycoconjugates | References |
|---|---|---|---|---|
| *GT family/donor substrate: β3Galactosyltransferases/UDP-Gal* | | | | |
| CgtB | *Campylobacter jejuni* | *E. coli* | Glycosphingolipids | [56] |
| WbuP | *E. coli* O114 | *E. coli* | *O*-Antigen precursor | [57] |
| WbgO | *E. coli* O55:H7 | *E. coli* | Type 1 LacNAc, poly-LacNAc, HMOS | [58–61] |
| Cvβ3GalT | *C. violaceum* | *E. coli* | HMOS | [62] |
| *GT family/donor substrate: β4Galactosyltransferases/UDP-Gal* | | | | |
| NgLgtB | *N. gonorrhoeae* | CFPS/ *E. coli* | Glycoproteins | [54] |
| NmLgtB | *N. meningitidis* | CFPS/ *E. coli* | Glycoproteins | [54, 63] |
| Hpβ4GalT | *H. pylori* | CFPS/ *E. coli* | Glycoproteins; HMOS, poly-LacNAc | [54, 63] |
| SpWchK | *S. pleuropnemoniae* | CFPS | Glycoproteins | [54] |
| HP0826 | *H. pylori* | *E. coli* | Thioglycosides | [64] |
| *GT family/donor substrate: α3Galactosyltransferases/UDP-Gal* | | | | |
| BgtB | *Escherichia coli* O86 | *E. coli* | Blood group B | [65] |
| *GT family/donor substrate: α4Galactosyltransferases/UDP-Gal* | | | | |
| LgtC | *Neisseria meningitidis* | *E. coli* | Glycosphingolipids | [66] |
| *GT family/donor substrate: β3N-Acetyl-Glucosaminyltransferases/UDP-GlcNAc* | | | | |
| Hpβ3GlcNAcT | *Helicobacter pylori* | *E. coli* | HMOS; poly-LacNAc | [63, 67] |
| NmLgtA | *Neisseria meningitidis* | *E. coli* | HMOS; poly-LacNAc | [63] |
| HP1105 | *H. pylori* strain 26695 | *E. coli* | HMOS; poly-LacNAc | [68] |
| *GT family: a4N-Acetyl-Glucosaminyltransferases/UDP-GlcNAc* | | | | |
| PmHS2 | *Pasteurella multocida* | *E. coli* | Heparosan | [22, 69, 70] |
| KfiA | *E. coli* K5 | *E. coli* | Heparosan | [22, 69, 71] |

(continued)

**Table 2** (continued)

| Enzyme name (CAZy) | Origin | Production[a] | Synthesized glycoconjugates | References |
|---|---|---|---|---|
| *GT family/donor substrate: α3N-Acetyl-Galactosaminyltransferases/UDP-GalNAc* | | | | |
| BgtA | *Helicobacter mustelae* | *E. coli* | Blood group A | [72] |
| *GT family/donor substrate: β4N-Acetyl-Galactosaminyltransferases/UDP-GalNAc* | | | | |
| CgtA | *Campylobacter jejuni* | *E. coli* | Glycosphingolipids | [56] |
| KfoC | *Escherichia coli* K4 | *E. coli* | Chondroitin | [73, 74] |
| CpCS | *Chlorobium phaeobacteroides* | *E. coli* | Chondroitin | [75] |
| *GT family/donor substrate: β3Glucuronosyltransferases/UDP-GlcA* | | | | |
| PmCS | *Pasteurella multocida* type F | *E. coli* | Chondroitin | [22] |
| KfoC | *Escherichia coli* K4 | *E. coli* | Chondroitin | [73, 74] |
| CpCS | *Chlorobium phaeobacteroides* | *E. coli* | Chondroitin | [75] |
| *GT family/donor substrate: β4Glucuronosyltransferases/UDP-GlcA* | | | | |
| PmHS2 | *Pasteurella multocida* | *E. coli* | Heparosan | [22, 70, 71, 76] |
| *GT family/donor substrate: β4Mannosyltransferases/GDP-Man* | | | | |
| Alg1 | *Saccharomyces cerevisiae* | *E. coli* | Phytanyl-PP-$(GlcNAc)_2$-$Man_1$ | [77] |
| *GT family/donor substrate: α2Fucosyltransferases/GDP-Fuc* | | | | |
| FutC | *H. pylori* strain 26695 | *E. coli* | HMOS | [68] |
| WbgL | *E. coli* O126 | *E. coli* | HMOS | [45] |
| Te2FT | *Thermosynechococcus elongatus* | *E. coli* | HMOS | [78] |
| WbsJ | *E. coli* O128:B12 | *E. coli* | Blood group antigens; HMOS | [68, 79] |
| WbiQ | *E. coli* O127:K63(B8) | *E. coli* | Blood group antigens | [80] |
| WbwK | *E. coli* O86:B7 | *E. coli* | Blood group antigens | [79, 81] |
| WbnK | *E. coli* O86:K62:H2 | *E. coli* | Blood group antigens | [82] |
| *GT family/donor substrate: α3Fucosyltransferases/GDP-Fuc* | | | | |
| α1,3/4-Fucosyltransferase (FucTIII) | *Helicobacter pylori* DSM6709 | *E. coli* | HMOS | [83] |
| α1−3/4-Fucosyltransferase (Hp3/4FT) | *Helicobacter pylori* UA948 | *E. coli* | HMOS | [84, 85] |

| | | | | |
|---|---|---|---|---|
| C-terminal 66-amino acid truncated version (Hp3FT) | *Helicobacter pylori* | *E. coli* | HMOS | [86] |
| FutA-delta52/triple mutant A128N/H129E/Y132I | *Helicobacter pylori* | *E. coli* | HMOS | [85, 87] |
| HhFT2 | *Helicobacter hepaticus* | *E. coli* | HMOS | [88] |
| *GT family/donor substrate: α4Fucosyltransferases/GDP-Fuc* | | | | |
| α1,3/4-Fucosyltransferase (FucTIII) | *Helicobacter pylori* DSM6709 | *E. coli* | HMOS | [83] |
| α1−3/4-Fucosyltransferase (Hp3/4FT) | *Helicobacter pylori* UA948 | | HMOS; Lewis blood groups | [84, 85, 89] |
| *GT family/donor substrate: a3Sialyltransferases/CMP-Neu5Ac* | | | | |
| Cst-I | *Campylobacter jejuni* | *E. coli* | Sialylated glycans | [90, 91] |
| Cst-II | *Campylobacter jejuni* | *E. coli* | Glycosphingolipids | [56] |
| PmST E271F/R313Y | *Pasteurella multocida* | *E. coli* | Sialylated glycans | [92] |
| PmST1 and mutant PmST1 M144D | *Pasteurella multocida* | *E. coli* | Lewis blood group | [89] |
| PmST3 | *Pasteurella multocida* | *E. coli* | *O*-Glycopeptides | [93] |
| SiaT | *Photobacterium phosphoreum* T-ISH-467 | *E. coli* | Sialylated glycans/lactose | [94] |
| PdST | *Pasteurella dagmatis* | *E. coli* | Sialylated lactose | [95] |
| *GT family/donor substrate: a6Sialyltransferases/CMP-Neu5Ac* | | | | |
| Mutant PmST P34H | *Pasteurella multocida* | *E. coli* | Sialylated glycans | [92] |
| Pd2,6ST | *Photobacterium damselae* | *E. coli* | HMOS | [84, 85, 96] |
| PlSt6 | *Photobacterium leiognathi* JT-SHIZ-145 | *E. coli* | Sialylated glycans | [97] |
| Psp2,6ST | *Photobacterium* sp. JH-ISH-224 | *E. coli* | Sialylated Tn antigen | [98] |
| *GT family/donor substrate: a8Sialyltransferases/CMP-Neu5Ac* | | | | |
| Cst-II | *Campylobacter jejuni* | *E. coli* | Glycosphingolipids | [56] |
| α2,8-Polysialyltransferase | *Neisseria meningitidis* | *E. coli* | Poly-sialylated glycoproteins | [99, 100] |

[a]*CFPS* Cell-free protein synthesis

the infancy for Leloir-GT production, only a few examples are documented up to date (Table 2).

This background is crucial when aiming at an application of recombinant Leloir GTs for glycan assembly. Depending on the desired glycan sequence and its glycosidic linkages, one has to overthink carefully what GTs are most suitable for the job. Picking the right biocatalysts is often dependent on the chemical nature of the acceptor molecule. The identification of functional GTs is therefore laborious. However, the enzymatic toolbox is continuously increasing by the combination of a rapid expression system with high-throughput microarray screening technologies and sophisticated analytics. For example, combining GTs' reactions on microarrays of self-assembled monolayers of sugar substrates with mass spectrometry analysis (SAMDI technology) resulted in the annotation of 4 novel GTs out of more than 14,000 combinations of putative GTs, donors, acceptors, and buffers [101]. SAMDI technology was used for the characterization of GTs produced by *E. coli*-based cell-free protein synthesis [54, 102]. Over 3,000 peptide substrates were screened in more than 13,000 reactions [102], and a modular platform was established for rapid prototyping of protein glycosylation pathways [54]. In a different approach, Tritium-labeled donor substrates were used to image the resulting glycan products on a microarray of immobilized acceptor substrates [103]. A more glycan structure related source of putative GTs is the *E. coli O*-antigen database (ECODAB) [104, 105], which has been established to connect respective genes encoding for certain GTs to known *O*-antigen structures.

In summary, with new technologies in hand, more versatile Leloir-GT tools for glycoconjugate synthesis are expected. However, translation into bioeconomic processes for glycoconjugate production is still challenging.

## 2.2 *Nucleotide Sugars*

Observing the basic components of glycan chains reveals an order of a variety of monosaccharides raising the question for the biocatalytic machinery by which organisms can produce those specified structures. The answer to this question is the coordinated supply of nucleotide-activated monosaccharides, known as the nucleotide sugars. The conjugation of a nucleotide as a high energetic compound to a monosaccharide leads to the donor substrate of Leloir glycosyltransferases, which provides the specificity and energy for the specific and effective glycosylation reaction [106].

### 2.2.1 Biosynthesis of Nucleotide Sugars

Nine nucleotide sugars are common for the biosynthesis of glycoproteins, glycosphingolipids, and glycosaminoglycans [107, 108], namely, uridine 5′-diphospho-α-D-glucose (UDP-Glc), uridine 5′-diphospho-α-D-galactose

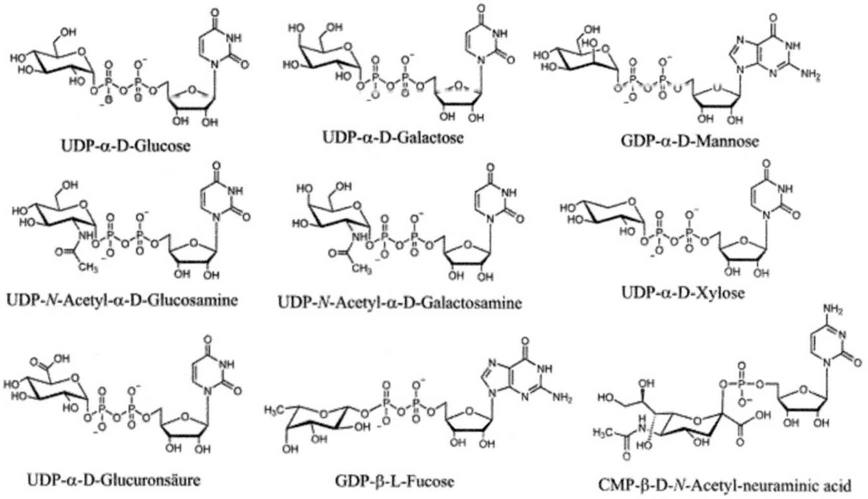


**Fig. 4** Nucleotide sugars for the synthesis of glycoproteins, glycosphingolipids, and glycosaminoglycans

(UDP-Gal), uridine 5′-diphospho-α-D-glucuronic acid (UDP-GlcA), uridine 5′-diphospho-α-D-xylose (UDP-Xyl), uridine 5′-diphospho-*N*-acetyl-α-D-glucosamine (UDP-GlcNAc), uridine 5′-diphospho-*N*-acetyl-α-D-galactosamine (UDP-GalNAc), guanidine 5′-diphospho-α-D-mannose (GDP-Man), guanidine 5′-diphospho-β-L-fucose (GDP-Fuc), and cytidine 5′-monophospho-*N*-acetyl-β-D-neuraminic acid (CMP-Neu5Ac) (Fig. 4).

The biosynthesis of nucleotide sugars is complex, and the understanding of the pathways is essential for their production. In de novo pathways, sugar-1-phosphates originate from metabolic sugar pathways, and primary nucleotide sugars generate from the reaction of nucleoside triphosphates (NTPs) with sugar-1-phosphates, e.g., nucleoside diphosphate (NDP)-hexoses and NDP–*N*-acetylhexosamines (HexNAc). Further modifications of NDP-sugars, by, e.g., epimerization, deoxygenation, reduction, oxidation, and decarboxylation, yield secondary nucleotide sugars (UDP-Gal, UDP-GlcA, UDP-Xyl, GDP-Fuc). For example, the de novo pathway of GDP-Fuc starts by dehydration of GDP-Man yielding GDP-6-deoxy-4-keto-D-mannose (Fig. 5). Further 3,5-epimerization gives the intermediate GDP-6-deoxy-4-keto-L-galactose, which is reduced at the 4-keto position to the final product. CMP-Neu5Ac is the exception since it is synthesized from UDP-GlcNAc or directly from Neu5Ac. In salvage pathways, sugar-1-phosphates of Gal, GlcNAc, GalNAc, GlcA, and Fuc are formed by phosphorylation of the anomeric C atom [108] and then converted with the corresponding NTP. Overviews of biosynthetic pathways of nucleotide sugars in bacteria, plants, and mammals are given in further reviews [111–113].

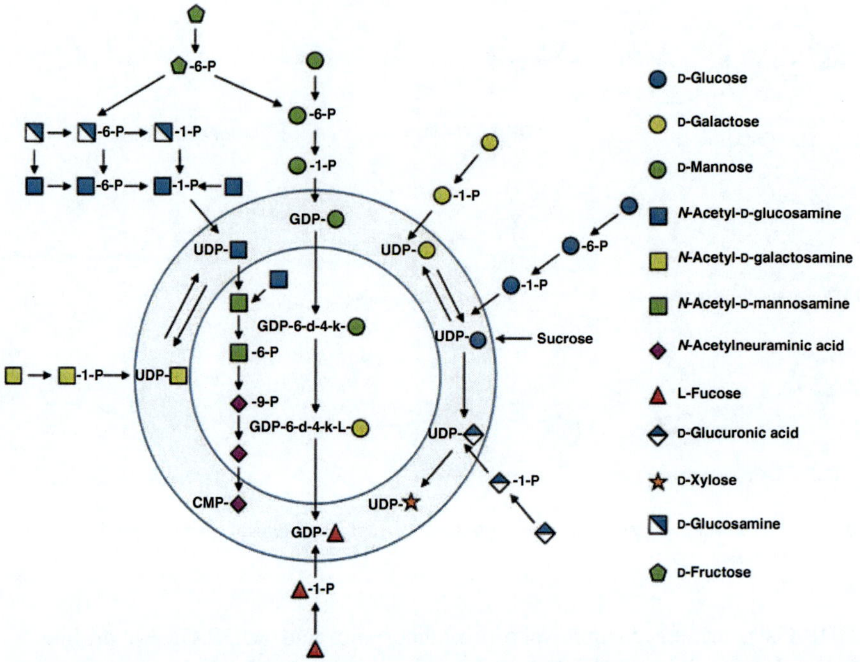


**Fig. 5** Overview of mammalian, microbial, and plant biosynthetic pathways for the nine most important nucleotide sugars. Detailed pathways are outlined in *Essentials of Glycobiology* [109] (https://www.ncbi.nlm.nih.gov/books/NBK453043/figure/ch5.f1/?report=objectonly). Symbol nomenclature follows the Consortium for Functional Glycomics [110] (https://www.ncbi.nlm.nih.gov/glycans/snfg.html)

### 2.2.2 Synthesis of Nucleotide Sugars

Enzymatic synthesis of glycoconjugates is mostly propagated as a more reliable and less wasteful approach; however, usage of nucleotide sugars is still uncommon in the broader field of industrial glycan production. This is owed to the general consideration that nucleotide sugars are costly in production and not available in larger amounts. Large-scale production of nucleotide sugars is scarce, and approaches have been realized by in vivo biotransformation and in vitro biocatalysis.

Fermentation processes yielded nucleotide sugars up to the kg scale. Biosynthetic pathways were tailored in single *E. coli* strains and *Corynebacterium ammoniagenes* for high product titers and space-time yields (STY) of UDP-Gal (44 g $L^{-1}$, STY 2.1 g $L^{-1}$ $h^{-1}$) [114], UDP-GlcNAc (7.4 g $L^{-1}$, STY 0.93 g $L^{-1}$ $h^{-1}$) [115], and GDP-Fuc (18.4 g $L^{-1}$, STY 0.84 g $L^{-1}$ $h^{-1}$) [116]. CMP-Neu5Ac was produced in a single *E. coli* strain with a product titer of 53 g $L^{-1}$ and STY of 2.4 g $L^{-1}$ $h^{-1}$ [117]. A whole-cell biotransformation process for UDP-Glc synthesis with *E. coli* yielded 0.7 kg isolated product per L bioreactor volume with a STY of 10 g $L^{-1}$ $h^{-1}$ [118].

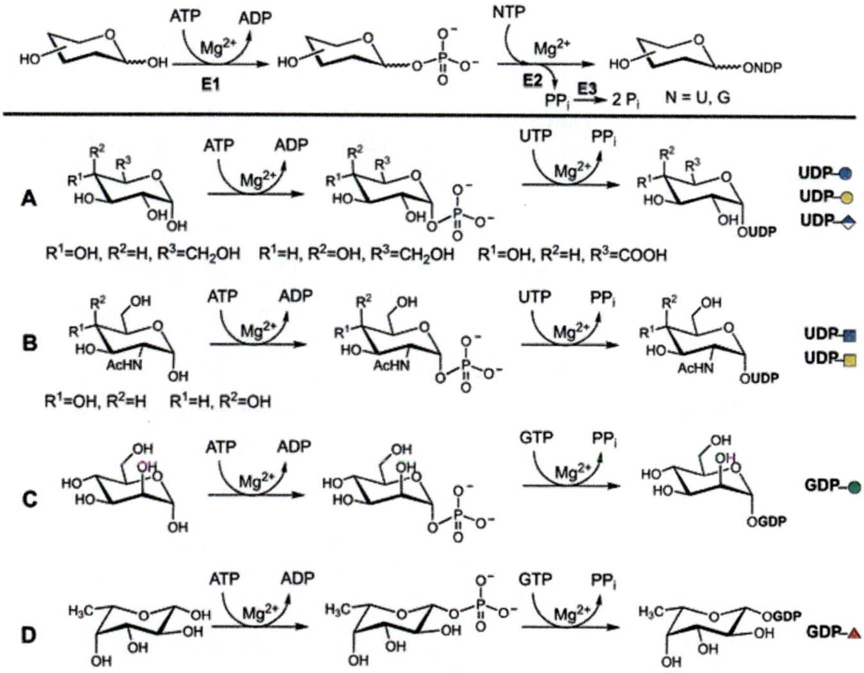


**Fig. 6** Enzyme modules for the synthesis of nucleotide sugars by salvage pathway enzymes. Enzyme cascades for NDP-sugar synthesis consist of sugar-1-phosphate kinases (**E1**), NDP-sugar pyrophosphorylases (**E2**), and pyrophosphatase (**E3**). (**a**) UDP-α-D-glucose (UDP-Glc), UDP-α-D-galactose (UDP-Gal), UDP-α-D-glucuronic acid (UDP-GlcA); (**b**) UDP-α-D-*N*-acetyl-glucosamine (UDP-GlcNAc), UDP-α-D-*N*-acetyl-galactosamine (UDP-GalNAc); (**c**) GDP-α-D-mannose (GDP-Man); (**d**) GDP-β-L-fucose (GDP-Fuc)

The elucidation of salvage pathways enabled the development of enzyme modules for the in vitro biocatalytic production of nucleotide sugars. Enzyme cascades with sugar-1-phosphate kinases and pyrophosphorylases were created to generate sugar-1-phosphates and nucleotide sugars (Fig. 6) [108, 119]. The modular enzyme systems can be generally recognized as very robust and reliable to produce specific nucleotide sugar products.

UDP-Xyl [120] and CMP-Neu5Ac [121] are synthesized by enzyme cascades from their de novo biosynthetic pathways as depicted in Fig. 7. A one-pot multi-enzyme (OPME) system was propagated by Chen and Wang [119] for in situ generation of nucleotide sugars and concomitant synthesis of glycans [84, 122, 123]. Substrate promiscuity of the involved enzymes also enables the chemoenzymatic synthesis of modified nucleotide sugars [119].

The synthesis of nucleotide sugars is closely connected to sugar phosphates as precursors. An alternative comes from plant metabolism using the high-energy substrate sucrose as an economically attractive substrate. Sucrose synthase (SuSy) is a glycosyltransferase and catalyzes the synthesis of nucleotide-activated Glc

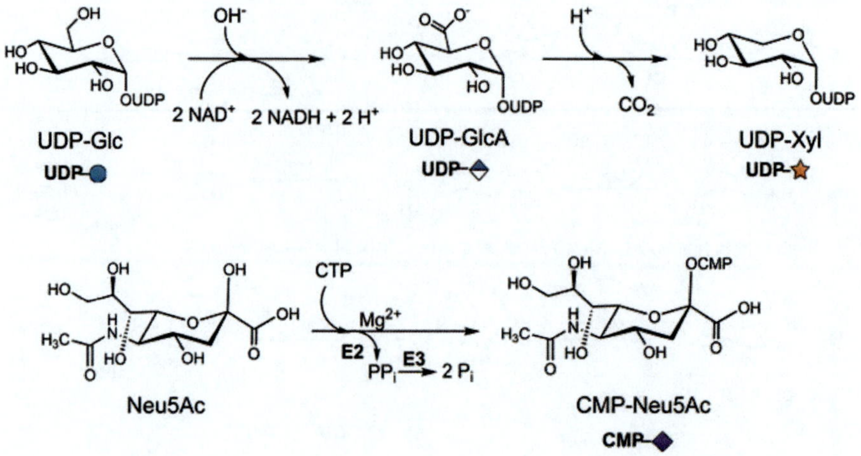


**Fig. 7** Synthesis of UDP-Xyl and CMP-Neu5Ac by de novo biosynthetic enzyme cascades [120, 121]

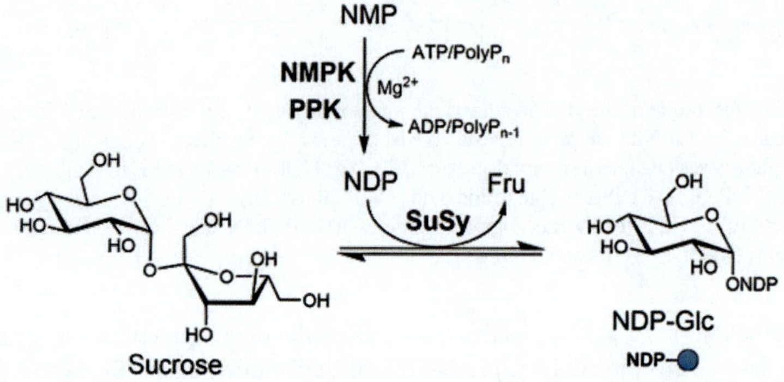


**Fig. 8** Synthesis of NDP-Glc from sucrose in enzyme cascades of sucrose synthases and nucleoside monophosphate kinases (NMPK) or polyphosphate kinase (PPK). N: uridine, deoxythymidine, cytidine, adenine [131, 136]

(NDP-Glc) from sucrose and NDPs (Fig. 8). The concept of utilizing plant SuSys for nucleotide sugar synthesis was introduced by the Elling group [118, 124–127]. In vitro synthesis of NDP-Glc is preferred at lower pH values between 5 and 7. The broad substrate spectrum of SuSy for UDP, dTDP, CDP, and ADP was exploited in enzyme cascades for the synthesis of UDP-Gal [128], UDP-GlcA [129], CDP-Glc [130], ADP-Glc [131], dTDP-Glc [132], and dTDP-deoxysugars [133]. Recently, the SuSy concept was extended by enzymes from non-photosynthetic bacteria [134] for the synthesis of UDP-Glc [135] and ADP-Glc [136].

### 2.2.3 Production of Nucleotide Sugars

Standard operation protocols were recently developed for the production and downstream processing of nucleotide sugars. In the Golgi Glycan Factory (GGF) project, the Elling group introduced high-throughput screening of one-pot enzymatic reactions by multiplexed capillary electrophoresis (MP-CE) to optimize system parameters for high space-time yields (STY; g $L^{-1}$ $h^{-1}$) and total turnover numbers (TTN; g enzyme $g^{-1}$ $product^{-1}$) of enzyme cascades [137]. MP-CE reaction screening was instrumental for the thorough characterization of novel enzyme cascades [138–140] and the synthesis of $^{13}C$- and $^{15}N$-labeled UDP-Gal and UDP-GlcNAc [61]. With optimized parameters and high stability of enzyme cascades, multi-g scale synthesis was run in repetitive batch mode (RBM), as depicted in Fig. 9 [141]. This principle utilizes enzyme cascades multiple times, by splitting enzymes and products to prevent the accumulation of side products, which destabilize the synthesis reaction. The high product yield facilitates further use of the nucleotide sugar solutions in one-pot glycosylation steps with Leloir glycosyltransferases [60, 61].

Repetitive batch mode was also applied for the g-scale synthesis of ADP-Glc with potato SuSy [131]. In combination with an appropriate nucleoside monophosphate (NMP) kinase, the enzyme cascades start from NMP [130] (Fig. 9). This enzyme cascade was recently utilized for the efficient g-scale synthesis of ADP-Glc using a bacterial SuSy [135, 136]. Table 3 summarizes the production of nucleotide sugars obtained on the multi-g scale.

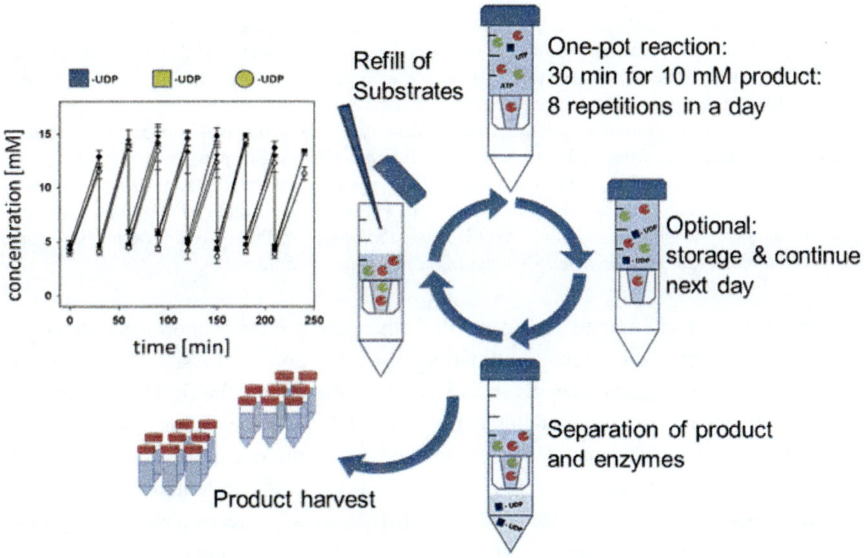


**Fig. 9** General principle of repetitive-batch-mode synthesis for the multi-gram scale production of UDP-GlcNAc, UDP-GalNAc, and UDP-Gal [141]

**Table 3** Biocatalytic production of nucleotide sugars

| Nucleotide sugar | Enzyme module[a] | Process results | References |
|---|---|---|---|
| **UDP-Glc** | *Ac*SuSy<br>*Ne*SuSy (immobilized) 3 cycles<br>*Ac*SuSy (immobilized) 5 cycles | **STY:**<br>25 g $L^{-1}$ $h^{-1}$<br>**TTN:** 1440<br>**STY:**<br>20 g $L^{-1}$ $h^{-1}$<br>**TTN:** 500<br>**STY:**<br>25 g $L^{-1}$ $h^{-1}$<br>**TTN:** 833 | [149]<br>[145, 146] |
| **UDP-Gal** | RBM/A: *Ec*GalK/*Hv*USP/PPiase 40 cycles | **STY:**<br>10.7 g $L^{-1}$ $h^{-1}$<br>**TTN:** 494 | [141] |
| **UDP-GlcA** | A: *At*GlcAK, *At*USP, *Pm*PpA, *Hs*UGDH | **STY:**<br>9.3 g $L^{-1}$ $h^{-1}$<br>**TTN:** 1950 | [139]<br>[120] |
| **UDP-GlcNAc** | RBM/B: BlNahK/*Hs*AGX1/PPiase 40 cycles | **STY:**<br>9.9 g $L^{-1}$ $h^{-1}$<br>**TTN:** 522 | [141] |
| **UDP-GalNAc** | RBM/B: BlNahK/*Hs*AGX1/PPiase 20 cycles | **STY:**<br>19.4 g $L^{-1}$ $h^{-1}$<br>**TTN:** 103 | [141] |
| **UDP-Xyl** | *Hs*UGDH/*Hs*UXS | **TTN:** 650 | [120] |
| **CMP-Neu5Ac** | CSTR: CMP-Neu5Ac synthetase (*E. coli* K-235/CS1) | **STY:**<br>10.6 g $L^{-1}$ $h^{-1}$ | [150] |

See Figs. 7, 8, and 9 for enzyme modules. RBM, Repetitive batch mode; CSTR, continuous stirred tank reactor, enzyme-membrane reactor; STY, space-time yield (g product $L^{-1}$ $h^{-1}$); TTN, mass-based turnover number (g product per g enzyme)

[a]*Ac*SuSy, SuSy from *Acidithiobacillus caldus*; *Ne*SuSy, SuSy from *Nitrosomonas europaea*; *Ec*GalK, galactokinase from *Escherichia coli*; *Hv*USP, UDP-sugar pyrophosphorylase from *Hordeum vulgare*; *Hs*AGX1, UDP-GlcNAc/GalNAc pyrophosphorylase from *Homo sapiens*; PPiase, pyrophosphatase; *At*GlcAK, glucuronic acid kinase from *Arabidopsis thaliana*; AtUSP, UDP-sugar pyrophosphorylase from *Arabidopsis thaliana*; PmPpA, pyrophosphatase from *Pasteurella multocida*; *Hs*UXS, UDP-Xyl synthase from *Homo sapiens*

Immobilization of enzyme cascades will further advance the economic synthesis of nucleotide sugars and the automation of glycan synthesis with Leloir glycosyltransferases. Examples have been demonstrated by the Wang group and used for in situ regeneration of nucleotide sugars [142]. The Elling and Franzreb group developed an automated compartmented flow microreactor system (CFMS) where single nucleotide sugars are synthesized by compartmented immobilized enzyme modules [143, 144] (see Chap. 3.3.). Enzymes are recovered by magnetic separation, and nucleotide sugars are delivered to the next compartment for glycan synthesis by Leloir glycosyltransferases. The non-sulfated human natural killer cell HNK-1 glycan epitope, a trisaccharide, was obtained with 96% yield in 210 min starting from *N*-acetyl-glucosamine [143]. Multi-point immobilization was demonstrated for bacterial SuSys [145, 146]. The immobilized enzymes could be reused with high substrate conversion in 3–5 cycles for the synthesis of UDP-Glc [145, 146].

Downstream processing (DSP) protocols for nucleotide sugars include two chromatographic steps (anion-exchange chromatography, AEC, and size-exclusion chromatography, SEC) [147], or AEC combined with desalting by EtOH precipitation of nucleotide sugars [131, 148]. An efficient chromatography-free DSP protocol was recently developed including treatment with alkaline phosphatase and EtOH precipitation of nucleotide sugars, which reduce significantly the solvent volume [135].

### 2.2.4 In Situ Regeneration of Nucleotides and Nucleotide Sugars

Although the g-scale production of some nucleotide sugars has been demonstrated, their broader application in glycobiotechnology still faces the problem of high costs (e.g., GDP-Fuc) and availability of rare nucleotide sugars (e.g., NDP-deoxysugars). High costs for nucleotide sugars are directly related to the use of nucleotides as substrates. ATP plays an important role in the nucleotide sugar synthesis steps and is considered as a cost factor. Some nucleotides such as UTP or CTP are less available and therefore more expensive for a nucleotide sugar-based glycosylation reaction. Nucleotides are also by-products and often inhibit nucleotide sugar enzymes and Leloir-glycosyltransferase reactions. In conclusion, glycobiotechnology is facing the challenge to find efficient ways for the economic generation of nucleoside triphosphates (NTPs) from abundant mono- or diphosphate nucleosides (NMP or NDPs). Nucleotide sugar regeneration focuses on the recycling of NMP/NDP generated by the Leloir-glycosylation reaction. Thus, the NDP-sugars are generated in situ avoiding their complex product isolation. Different enzyme cascades have been developed (Fig. 10).

Nucleotide sugar regeneration has been pioneered by C.H. Wong [151–153] developing and advancing system A (Fig. 10) by utilization salvage pathway enzymes for nucleotide sugar synthesis. Regeneration of NTPs relies on pyruvate kinase using phosphoenolpyruvate (PEP) as a high-energy substrate. Phosphorylated sugars in de novo and salvage pathways are built by sugar kinases using ATP as a cofactor. Conversion of sugar-1-P or a free sugar (e.g., Neu5Ac) with nucleoside triphosphates (NTPs) yields NDP-/NMP-sugars. System A is very robust and effective. However, since PEP is a costly substrate and regeneration of one nucleotide sugar from free sugar needs two PEP molecules, it appears not as an economic process (Fig. 10, system A). Nevertheless, it has been applied for the g-scale synthesis of the glycosphingolipid glycans Globo H and SSEA4 [154].

Recent studies focused on the use of polyphosphate (polyP) for nucleotide regeneration by polyphosphate kinases (PPKs) as an economically attractive alternative to the PEP/PK system (Fig. 10, system B) [155–158]. PolyP consists of long phosphate chains, which are used to regenerate single ATP molecules multiple times. The polyP/PPK system showed to be very efficient regarding turnover numbers and space-time yields. The Reichl/Rapp group demonstrated the first example for nucleotide sugar regeneration employing the polyP/PPK system (Fig. 11) [77].

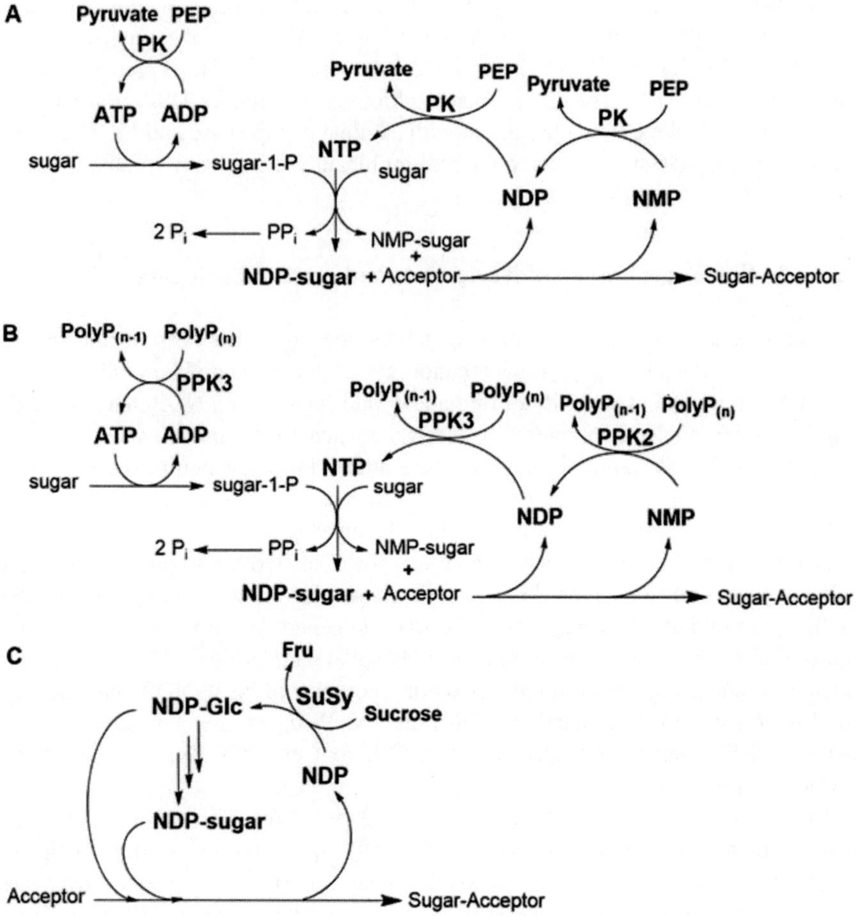


**Fig. 10** In situ nucleotide and nucleotide sugar regeneration systems. (**a**) Regeneration by the PEP/PK module: phosphorylation of nucleoside mono- and diphosphates (NMP/NDP) with phosphoenolpyruvate (PEP) by pyruvate kinase (PK) releasing pyruvate. (**b**) Regeneration by polyphosphate (polyP). Polyphosphate kinases use polyP for the phosphorylation of NMP/NDP. In A and B, the regenerated nucleoside triphosphate (NTP) is used for nucleotide sugar synthesis and recycled in glycosyltransferase reactions. (**c**) Nucleotide sugar regeneration by sucrose synthase (SuSy) using sucrose and NDP to form NDP-Glc and releasing fructose. NDP-Glc can be used or processed to secondary nucleotide sugars in Leloir-glycosyltransferase reactions

PolyP consists of long phosphate chains, which are used by PPKs to regenerate ATP multiple times. The systems were shown to be very efficient regarding turnover numbers and space-time yields. Although very potent, polyP is not well characterized as a commercially available molecule, since the chain length of polyphosphate is often not well defined and lays between two to several thousand phosphate subunits [159]. This can lead to deviations of the regeneration efficiency and is therefore often given in high concentrations to the synthesis reaction. Furthermore,

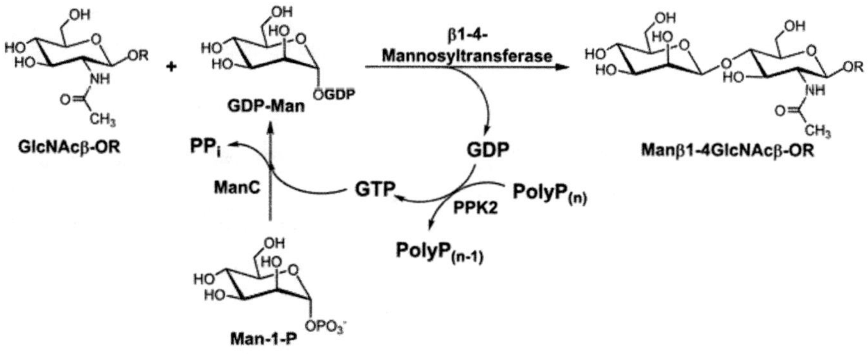


**Fig. 11** In situ regeneration of GTP and GDP-Man by polyP/PPK2. *ManC* Mannose-1-phosphate guanylyltransferase, *R* GlcNAcα1-lipid linker

polyP cannot be entirely degraded causing the problem of high phosphate impurities for the product [159].

A strategy for nucleotide sugar regeneration from sucrose was introduced by the Elling group (Fig. 10, system C) [124]. NDP-Glc is regenerated by the conversion of NDP with sucrose synthase (SuSy). Sucrose is a cheap high-energy source to fuel regeneration cycles of nucleotide sugars. In comparison to the other described systems, it is phosphate-free. However, NDP-sugar regeneration depends on the substrate specificity of plant and bacterial SuSys and covers UDP-, dTDP-, CDP-, and ADP-activated glucose [127, 134, 160]. Further conversion of UDP-Glc by UDP-Glc 4′-epimerase or UDP-Glc dehydrogenase yields UDP-Gal and UDP-GlcA, respectively, for the synthesis of glycan epitopes [128, 129, 161] and hyaluronic acid [138]. Furthermore, dTDP-deoxysugars, e.g., dTDP-L-rhamnose, are regenerated from dTDP-Glc in combination with pathway enzymes [111]. More recently, the SuSy system has been applied in natural product glycosylation for the regeneration of UDP-Glc [160, 162–165], UDP-Gal [166], and UDP-L-Rha [167].

# 3 Practical Application of Enzymatic Glycosylation Reactions

## 3.1 *Glycosaminoglycans: Hyaluronic Acid, Heparan Sulfate, Heparin, Chondroitin Sulfate, Dermatan Sulfate*

### 3.1.1 Structure and Synthesis of GAGs

Glycosaminoglycans (GAGs) are complex, long-chain polysaccharides of disaccharide building blocks consisting of a hexosamine (*N*-acetylglucosamine (GlcNAc), *N*-acetylgalactosamine (GalNAc)) and a uronic acid (glucuronic acid (GlcA), iduronic

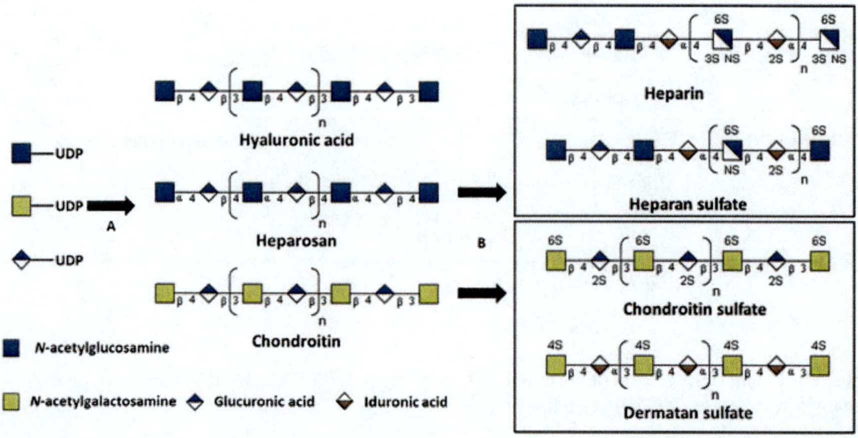


**Fig. 12** Enzymatic synthesis of heparosan (HP) and chondroitin (CH) and subsequent epimerization and sulfation. Precursors are the activated monosaccharides UDP-GlcNAc, UDP-GalNAc, and UDP-GlcA. (**a**) GAG glycosyltransferases; (**b**) 5-epimerase and sulfotransferases. Disaccharide repeating units are for **HP** [4GlcNAcα1-4GlcAβ1-]$_n$ and **CH** [3GalNAcβ1-4GlcAβ1-]$_n$. 5-Epimerase converts GlcA to IdoA; sulfotransferases are dependent on the substrate PAPS (3′-phosphoadenosine-5′-phosphosulfate)

acid (IdoA)) moiety (Fig. 12). Depending on the hexosamine saccharide unit, GAGs have different backbones and are therefore divided into glucosamino- and galactosaminoglycan. The best-studied GAGs are the glucosaminoglycans heparin/heparan sulfate (Hp/HS) and hyaluronan and the galactosaminoglycans chondroitin/dermatan sulfate (CS/DS) [168–171]. The precursor for Hp/HS is heparosan (HP) (β4GlcA-α1-4GlcNAc1-)$_n$, and the precursor for CS/DS is chondroitin (CH) (β4GlcA-β1-3GalNAc1-)$_n$. Hyaluronan (β4GlcA-β3GlcNAc1-)$_n$ is not further processed and is the only known unsulfated GAG and not bound to a protein backbone [172]. GAGs normally have a size that ranges between 1.5 and $2 \times 10^4$ Da. An exception is hyaluronan with a size of up to $10^7$ Da [173].

While the synthesis of the polymeric hyaluronan happens at the inner face of the plasma membrane, the syntheses of Hp/HS and CS/DS as parts of proteoglycans take place in the endoplasmic reticulum and Golgi apparatus [174, 175]. The enzymes are either membrane-bound or transmembrane proteins [176, 177].

The enzymatic synthesis of GAGs starts with the polymerization of the activated monosaccharides UDP-GlcNAc and UDP-GlcA or UDP-GalNAc and UDP-GlcA, respectively, by glycosyltransferases [178]. Glycosyltransferases for hyaluronan, heparosan, and chondroitin synthesis are often bifunctional enzymes extending the GAG chain with both saccharides [172, 176, 177, 179]. In biosynthetic GAG pathways, enzymatic sulfation and epimerization occur after polymer formation. In the case of Hp, HS, and DS, a C5-epimerase is transforming GlcA into IdoA [23]. *O*-Sulfation of the 2-hydroxyl groups of IdoA and GlcA, as well as the 3-hydroxyl and 6-hydroxyl groups and *N*-sulfation of the 2-amino group of the glucosamine residue, is accomplished by deacetylases/sulfotransferases [23]. Sulfotransferases depend on a nucleotide-activated sulfate PAPS (3′-phosphoadenosine-5′-phosphosulfate) as a donor substrate. The pattern of sulfation varies between the GAGs. For example, Hp is nearly fully *N*-sulfated, while HS shows mixed patterns of *N*-sulfated and *N*-acetylated regions [180]. It is also reported that desulfation regulates these patterns [181]. Due to the various carboxyl and/or sulfate groups, GAGs are polyanions, which influence the interaction with other ions and biomolecules [182]. For example, HA binds huge amounts of water, which results in a viscoelastic gel [168]. In proteoglycans (PGs), Hp, HS, CS, and DS are covalently linked to a protein core via serine (Ser) and a linker composed of galactose (Gal) and xylose (Xyl) (GlcA*β*1–3-Galβ1–3Gal*β*1–4Xylβ1–*O*-Ser) [170, 183, 184]. More than half of the molecular weight is the GAG chain of these PGs and is pivotal for physiological activity, e.g., binding of growth factors [185]. The production of PGs can be strongly regulated by the first attachment of Xyl to Ser [170]. With a different expression of genes encoding core proteins, various lengths and types of GAGs, variable sulfation, and attachment of different GAG chains to various Ser residues, there is an enormous diversity of PGs [173].

### 3.1.2 Biology of GAGs

Because of the diversity, GAGs and PGs play an enormous role in different biological processes [186, 187]. GAGs occur in the extracellular matrix (ECM) and pericellular coating and therefore maintain the structural integrity of cells and tissues [188]. Due to the negatively charged sulfate and carboxyl groups, GAGs can bind electrostatically to a variety of proteins. There are several evidences that the sulfation pattern of GAG chains (sulfation code) encodes biological information leading to a physiological or pathological state [189]. Thus, the GAG sulfation code is important for coagulation, inflammation, cell adhesion, metastasis, cell growth, tissue differentiation, and pathogen/viral defense [188, 190–196].

Hyaluronan functions in two mechanisms: first as a passive structure and second as a signaling molecule [168]. The passive structure performs because of its viscoelastic behavior as a lubricant in joint fluids and as a moisturizer in skins and eye fluid [197–200]. Hyaluronan binds also to receptors like CD44, which activates a lymphatic response. Interestingly the size of hyaluronan plays an important role. The longer the hyaluronan chain, the more proteins can bind to that hyaluronan chain, building a complex signaling structure [168]. Small and long hyaluronan chains can have even contrary effects. For example, low-molecular-weight hyaluronan can have pro-inflammatory effects, while high-molecular-weight chains can have anti-inflammatory effects [201].

Similar to hyaluronan, CS is found in joint fluid, where it is responsible for anti-inflammation and enhanced syntheses of PGs and hyaluronan [169, 202]. Recently there are more studies, which indicate that proteoglycans carrying chondroitin sulfate (PGCS) are involved in the neuronal outgrowth [203, 204]. There are hints that the expression of PGCS can regulate the amount of transient receptor potential cation channels in the astrocytes. A dysfunction could be the trigger of multiple sclerosis [205].

DS chains regulate and localize transglutaminases in the ECM and are therefore associated with wound healing, fibrosis, and vascular remodeling [206]. A defective biosynthesis of proteoglycans carrying dermatan sulfate (PGDS) is also responsible for human skin and skeleton disorder. The disorder comes probably with a change of matrix proteins like collagen and cell signaling proteins [207].

Hp's major role lays in the control of blood coagulation. Via a specific pentasaccharide sequence, Hp binds to antithrombin, which triggers inhibition of the coagulation cascade by inactivation of the serine proteases Factor X and thrombin [208, 209]. It is also reported that DS interacts with a Hp cofactor, which inhibits thrombin [210]. HS is also involved in coagulation but with less effectiveness [208]. However, proteoglycans carrying heparan sulfate (PGHS) can bind to many ligands, for example, the growth factor Wnt, which is involved in cell proliferation and the early stage of embryo development [211, 212].

### 3.1.3 Biomedical Application of GAGs

Because of their variable biological functionality, GAGs are used in many medical applications and already sold as drugs [187]. Through the ability to bind several proteins, GAGs are used as a drug delivery system. One example is Solaraze®, a crème with 2.5% HA, which is bound to diclofenac. The crème is used in the treatment of actinic keratosis [213].

Hyaluronan as an active ingredient is used to treat wounds and skin irritation (Connettivina®) [214, 215], and already in 1980 Healon® containing hyaluronan was sold for ophthalmic surgery [216]. Hyaluronan is also used in the cosmetic field as moisture in crèmes and serums and as a soft tissue regeneration agent [168]. The most often medical application of hyaluronan is to treat arthritis and reduce pain by injection into joints. Examples are Supartz FX® and Synvisc® [217, 218]. Because of its anti-inflammatory feature, drugs based on CS, for example, Condrosulf® or Theraflex®, were developed to treat osteoarthritis [219, 220].

Hp is since 1930 a long known agent for anti-thrombosis, but normal Hp had many side effects – because of its length, the chance that heparin can interact with other receptors was increased. Therefore low-molecular-weight LMW Hp was developed. Clexane®, which consists of LMW Hp, is a medication for thrombosis and sold for decades [221]. DS and CS are also handled as an alternative for heparin with fewer side effects [222]. A mixture from LMW Hp, DS, and CS (danaparoid, Orgaran®) already has been developed and successfully applied as an anti-thrombosis agent [223].

There is no specific drug including HS on the market, but it is a promising target for antiviral resistance. Viruses often use HS as an attachment receptor. The idea is either to block human HS with HS antagonists or to use HS mimics to block the protein receptors of the virus. Through this interaction, the virus cannot attach to a cell [224]. An HS mimic was a successful treatment of the dengue virus [225].

Another example of an HS mimic is ReGeneraTing Agents (RGTA®), which help in tissue recovery. The mimic is fulfilling the role of degraded HS and therefore stabilizes and reconstructs damaged ECM [226].

GAGs are also involved in tissue engineering. Biomaterials are combined with differentiating cells and bioactive factors to produce functional tissues and organs. For example, hydrogels are giving an aqueous environment, which eases the nutrient and signaling transfer to the cells [227, 228]. A hydrogel composed of hyaluronan showed a better transition of transforming growth factor beta-3 to mesenchymal stem cells [229]. By comparison of hydrogels either consisting of hyaluronan/Hp or only hyaluronan, the hyaluronan/Hp hydrogels showed a delayed release of an osteoinductive factor, which could improve bone formation [230]. Cai et al. showed that Hp is an important part of vascular scaffolds, which inhibits thrombogenesis and therefore could improve the treatment of patients with cardiovascular disease [231]. Also, CS, often in combination with chitosan, is used in hydrogels and was successfully applied to grow multipotent bone marrow-derived stromal cells [232].

### 3.1.4 Industrial Production of GAGs

Due to the huge number of applications of GAGs in biomedicine, biomaterial research, and cosmetics, the demand for GAGs is increasing. For hyaluronan it is expected to rise to 16.6 billion dollars by 2027 which is a growth rate of about 8.1% [233]. Hp market is expected to grow to 14.6 billion dollars by 2027 with a rise to 4.3% [234]. LMW heparin is derived from chemical depolymerization of extracted tissue and dominates the US market with 55% together with high-molecular-weight (HMW) heparin (40%) [23]. In 2018 the market for CS is evaluated to 1.1 billion dollars and is estimated to grow with 3.2% between 2019 and 2025 [235]. Till today, processes for the production of Hp, hyaluronan, and CS currently rely on the extraction from animal tissues (Hp, porcine intestines; CS, bovine or porcine trachea and shark fins; hyaluronan, rooster combs) [236, 237]. The extraction comes with inherent product variability and impurities as well as poor control of source material [236]. Therefore, naturally occurring GAGs are extremely heterogeneous regarding dispersity and sulfuration. This limits the further development of these compounds. Contaminations with animal proteins, sugars, and viruses can cause tremendous effects. For example, oversulfated CS in lots of pharmaceutical porcine Hp caused several deaths in the United States and hundreds of adverse reactions worldwide in 2007 and 2008 [238, 239]. The contaminant CS showed a disaccharide repeat unit with an unusual sulfation pattern (GlcA2S3S(β1-)3GalNAc4S6S) with anti-coagulant activity and could not be traced by the required drug safety tests at this time. However, several patients suffered from a rapid and severe anaphylactic response with deadly outcome. As a consequence, new regulations have been implemented by the regulatory authorities to ensure the integrity and quality for naturally sourced biologic drugs. Out of these reasons, new methods were developed like fermentation or chemical/enzymatic synthesis [189, 240]. Classical organic de novo synthesis of GAGs is a multistep process with low product yields due to the

**Table 4** Glycosyltransferases used for the enzymatic in vitro synthesis of hyaluronan, heparosan, and chondroitin

| GAG | Enzyme/source | Function | References |
|---|---|---|---|
| Hyaluronan | PmHAS/*P. multocida* | β4GlcNAc-T/β3GlcA-T | [138, 177, 179, 252–254] |
| Heparosan (**HP**) | KfiA and KfiC/*E. coli* K5<br>PmHS2/*P. multocida*<br>PmHS1/*P. multocida* | α4GlcNAc-T and β4GlcA-T<br>α4GlcNAc-T/β4GlcA-T<br>α4GlcNAc-T/β4GlcA-T | [69]<br>[70, 76, 139, 255] |
| Chondroitin (**CH**) | KfoC/*E. coli* K4 | β4GalNAc-T/β3GlcA-T | [74, 256, 257] |

complex modification patterns including sulfation and epimerization [189]. This affects the prize. For example, the chemically synthesized ultra-LMW heparin product fondaparinux (Arixtra®) is a sulfated pentasaccharide with a narrowed medical indication being 1,000-fold more expensive than HMW heparin [236].

*Streptococcus* strains, as natural hyaluronan producer, are used for fermentation reaching product titers of 6–7 g/L [241]. Due to toxin formation by *Streptococcus*, GRAS (generally recognized as safe) organisms like *Bacillus* or *Lactococcus* were metabolically engineered using recombinant hyaluronan synthases [168, 237]. The fermentation of sulfated GAGs is more difficult to accomplish a certain sulfate pattern for a desired biological function. Therefore, the unsulfated precursors, HP and CH, are produced and purified for subsequent chemoenzymatic sulfation [240, 242]. Processes including recombinant GAG synthases (glycosyltransferases) and enzymes for the synthesis of nucleotide sugar and sulfate donor substrates are highly attractive to obtain homogeneous GAG polymers from well-defined biobased resources [23]. Metabolic engineered microbial hosts (*E. coli* and *B. subtilis*) have been constructed to implement pathways for the precursor nucleotide sugars UDP-GlcNAc/UDP-GalNAc and UDP-GlcA and the respective GAG synthases. Product yields for HP and CH of up to 5 g/L with a molecular weight distribution between 30 and 110 kDa were obtained [243–246]. Unfortunately, a low dispersity of HA and the right sulfate pattern of HP and CS are still challenges and deeply depending on the culture conditions and used enzymes [237, 247].

A new approach is enzymatic in vitro syntheses of GAGs. Soluble glucosyltransferases have been characterized and applied as isolated/immobilized enzymes to produce HA, HP, and CH (Table 4) [22, 23, 76, 173, 176, 248–251].

Moreover, one-pot syntheses were established to provide the activated UDP-saccharides, which are further processed to hyaluronan or HP, respectively [138, 139, 254, 258]. The approaches to produce HA showed a better dispersity and control of the size compared to the common production processes [138, 139]. There are even examples, where the sulfated Hp/HS are directly produced with an enzymatic one-pot synthesis [259]. However, the next step for the enzymatic syntheses of GAGs needs the establishment of a scale-up to further compete with fermentation or extraction from animal tissue.

## 3.2 Enzymatic Synthesis of Human Milk Oligosaccharides

Human milk oligosaccharides (HMOs) are a major solid constituent of human milk in addition to protein, lactose, and fat [260]. Overall more than 100 different oligosaccharide structures from human milk have been elucidated. Various studies over the last years showed health benefits associated with the consumption of HMOs such as reduced infant morbidity and enhanced brain development [261]. For this reason and the fact that they do not naturally occur in other mammals, interest in the synthesis of HMOs has almost exponentially increased over the last two decades. Eventually, the successful commercial synthesis of two abundant and simple structures, 2′-fucosyllactose and lacto-*N*-neotetraose, has led to their addition to infant formulas as novel food ingredients [262]. More HMO mixes are currently under development by various companies. All of these HMOs are exclusively produced by the fermentation of genetically *E. coli* strains [262]. However, a wide range of HMOs can also be produced by enzymatic synthesis [263, 264]. Through the establishment of the recombinant synthesis of a wide range of glycosyltransferases in bacterial production systems, pathways for the synthesis of a multitude of HMOs can be synthesized in multi-enzyme cascade reactions [264]. Typically, HMOs prepared by enzymatic synthesis are used for initial testing of oligosaccharide function, e.g., in animal models [263]. However, due to the high cost of substrates such as nucleotide sugars, enzymatic synthesis is at present not able to compete with fermentation even though much higher space-time yields can be achieved [265]. Moreover, little effort has been made to date to transfer lab-scale enzymatic synthesis into viable, robust, and scalable processes. An exception in this respect is the work by Nidetzky and co-workers who established a packed bed reactor containing an engineered glycosynthases for the synthesis of lacto-*N*-triose (LNT II) [266]. By optimizing the residence time, a stable, quantitative, and continuous LNT II synthesis was achieved. In addition, engineering glycosynthases for the exploitation of readily available HMOs as building block donors, e.g., as fucose and sialic acid donors, viable industrial synthesis of specific HMOs might become feasible [267].

## 3.3 Microreactors for Automated Enzymatic Glycan Synthesis

To elucidate the function of specific glycans, pure standards need to be accessible and affordable to the glycoscience community [268–272]. The synthesis of glycan standards in milligram to gram quantities has been identified as a milestone for carbohydrate research by the US National Research Council in 2012 [270]. Due to a wide range of possible structures of glycans of all categories, i.e., *N*-glycans, human milk oligosaccharides, poly-*N*-acetyllactosamine derivatives, and gangliosides, only the establishment of efficient automated synthesis can warrant achieving this objective [269]. The development of automated glycan synthesis can be divided primarily into chemical and enzymatic synthesis while also studies in combining both have been undertaken [272, 273]. The first efforts to tackle automated synthesis were based on chemical glycosylation, and automated systems have been commercialized [274]. The drawbacks of this approach compared to enzymatic synthesis are low yields of complex glycans [143, 272, 275–277]. The key development for the automation of enzymatic synthesis of complex glycan structures was the emerging accessibility of a wide range of recombinant bacterial and mammalian glycosyltransferases and the establishment of operationally simple glycan purification processes [269, 272, 278]. Through the most advanced development to date concerning the number of synthesized glycans by Boons and co-workers, a wide range of glycans can be synthesized in milligram quantities in up to 15 subsequent reaction cycles. The technique relies on the solid-phase extraction for intermediate product purification using a sulfonate tag, which circumvents lyophilization or other tedious purification steps of intermediates and, thus, enables automation. To drive the glycosyltransferase reaction to completion, product inhibition by nucleotides is prevented by using phosphatases [269].

In a proof-of-concept study, a peptide synthesizer has been adopted for the automated synthesis of gangliosides and poly-LacNAc derivatives. For intermediate glycan purification in between glycosylation reactions, a poly(*N*-isopropyl-acrylamide) polymer which can be precipitated and filtered at elevated temperatures is used [275]. Contrary to these approaches where the solid-phase synthesis has been applied, Elling and co-workers have established the fundamentals for a microreactor using immobilized enzymes [143]. To screen for suitable solid supports as well as reaction conditions is a labor-intensive task – especially when complex structures are synthesized and a multitude of enzymes are required. However, once cascades of immobilized enzymes are established and integrated into a reactor such as a compartmented flow microreactor, the system has major advantages. Intermediate glycan purification is not required. Coupling glycan synthesis to nucleotide sugar synthesis reduces costs for substrates, especially when synthesizing larger amounts. Most notably, immobilized enzymes can be removed for product separation and reused for multiple reactions reducing enzyme production and downstream processing. Immobilized enzyme cascades can also be used to develop reactors that can synthesize gram amounts of glycans. However, one of the drawbacks of using immobilized cascades is that subsequent product purification is still required. An overview of the three discussed setups for automated synthesis is given in Table 5.

### 3.4 Glycoconjugate Vaccines

Glycoconjugate vaccines directed against infectious diseases are typically composed of a bacterial polysaccharide linked to a protein. There are numerous examples of licensed vaccines or vaccines in clinical trials [279]. The polysaccharides component

**Table 5** Most notable automated enzymatic glycan synthesis developed and published so far

| System | Method | Structures synthesized | Scale and proposed scalability |
|---|---|---|---|
| "Automated platform for the enzyme-mediated assembly of complex oligosaccharides" [269] | Exploiting a sulfonate tag for solid-phase extraction, automation using a robotic workstation | Poly-LacNAc derivatives, human milk oligosaccharides, gangliosides, and *N*-glycans | Milligram quantities; up to 100 mg possible |
| "Machine-driven enzymatic oligosaccharide synthesis" [275] | Poly(*N*-isopropyl-acrylamide) polymer for intermediate purification, automation using a peptide synthesizer | Ganglioside, poly-LacNAc derivatives | Milligram quantities (8–20 mg) |
| "Automated enzymatic glycan synthesis in a compartmented flow microreactor system" [143] | Compartments with immobilized enzymes, sugar nucleotide synthesis modules, automation in a flow microreactor | HNK-1 glycan epitope | 40 mg; the concept is scalable to > gram amounts |

is conventionally obtained through microbial cultivation, harvesting, and purification [280–283]. This is associated with multiple drawbacks such as the occurrence of pathogenic contaminants and slow development processes [284]. Various other production methods have been developed of which especially full glycoprotein synthesis by *E. coli* equipped with an in vivo glycosylation machinery has been used to produce several vaccine candidates that are currently in different phases of clinical trials [279]. Enzymatic polysaccharides synthesis, especially for automated synthesis, is mostly seen as a fast-track approach to synthesize a wide range of polysaccharides for screening for effective vaccines [279]. However, recently a route toward an effective enzymatic production process has been shown for *N. meningitidis* serogroup X (MenX) polysaccharide fragments [284]. For elongation of MenX polysaccharide, an engineered MenX capsular polymerase was immobilized through metal affinity chromatography. By pumping a reaction mixture composed of the sugar donor UDP-GlcNAc and acceptor trisaccharide through the column under optimized conditions, the polysaccharides with defined average length are obtained. The process is pathogen-free and has advantages over chemical polysaccharides synthesis, i.e., the stereoselective synthesis resulting in high yields in addition to the reduction of steps needed for the synthesis. However, high-cost sugar donor UDP-GlcNAc is excessively needed as a building block, and for these types of enzymatic processes to become viable at large scales, either sugar nucleotides need to become available to reduced prices in large amounts or the synthesis needs to be in situ coupled to sugar nucleotide synthesis cascades as described in Chap. 2.2.4 [285].

### 3.5 *In Vitro Glycoengineering of Pharma Glycoproteins*

Due to their integral commercial relevance in the (bio)pharmaceutical industry, the enzymatic in vitro glycoengineering of recombinant monoclonal antibodies is especially well-studied [286–290]. mAbs are most notably used as anti-cancer drugs and anti-inflammatory medication. Immunoglobulin G (IgG) antibodies are glycosylated at the Asn residue number 297 in the crystallizable region (Fc) of both heavy chains. Excellent reviews on the importance of Fc glycosylation and its impact on IgG conformation, stability, and binding to Fc receptors have been published [286, 288, 290, 291]. Current mAb glycoengineering strategies are mostly based on Chinese hamster ovary (CHO) cell-line engineering, e.g., mogamulizumab [288, 292, 293]. However, current manufacturing processes of mAbs still offer insufficient control over Fc-glycosylation and batch-to-batch variation thereof [294]. By using in vitro glycan remodeling as an intermediate process step between upstream and downstream processing or as part of the downstream processing, full control over the glycosylation can be exerted [295]. The associated advantages are increasing the IgG efficacy and safety by generating tailored homogenous glycoforms and eliminating batch-to-batch variations in glycosylation [294]. Moreover, the in vitro glycoengineering will increase process flexibility concerning the choice of expression host and facilitate the process development by evading both, cell-line engineering and the adjustment of fermentation conditions to achieve high fractions of the desired glycoform [294].

Recent enzymatic in vitro glycoengineering strategies of glycoproteins can be divided into three different approaches (Table 6 and Fig. 13):

(a) Re-glycosylation using Leloir glycosyltransferases and sugar nucleotides to stepwise shorten and extend glycans on glycoproteins. Optionally, glycosidases are used for prior glycan trimming.
(b) Transglycosylation using glycosynthases for the en block transfer of glycans to GlcNAc or glucose residues on proteins.
(c) Full *N*-glycosylation of aglycosylated proteins, i.e., in vitro modeling of the ER-glycosylation machinery by using oligosaccharyltransferases (OSTs) and lipid-linked oligosaccharides as substrates [302]. With this approach aglycosylated proteins, e.g., recombinant proteins, expressed in bacterial production systems can be glycosylated. However, elaborated chemical and enzymatic synthesis of eukaryotic-type lipid-linked oligosaccharide is a major challenge to overcome before wider application in higher scales is possible.

The transglycosylation approach involves splitting off the *N*-glycan using an endoglycosidase and the transglycosylation step with an (engineered) glycosynthase and glycan oxazoline as substrate [297]. The great advantage of this approach is the effective removal of the core fucose from core GlcNAc residues of IgG using fucosidase that is mostly inactive toward complex glycans of commercial monoclonal antibodies [303, 304]. Moreover, using substrate-site-selective fucosidases and glycosynthases also offers a route toward site-selective glycoengineering. However, only a few glycans can be readily purified from natural resources, and further large-

**Table 6** The three main approaches for the enzymatic in vitro glycoengineering of proteins

| Approach | Method | Advantage | Challenges |
|---|---|---|---|
| In vitro re-glycosylation [296] | Building of homogeneous glycoforms using Leloir glycosyltransferases, optionally prior trimming of glycans by glycosidases | Wide range of enzymes and substrates are commercially available | Large-scale expression of glycosyltransferases and sugar nucleotides; enzymatic defucosylation |
| In vitro transglycosylation [297–299] | Cleavage of the core GlcNAc residues and subsequent en bloc transfer of glycan oxazolines by glycosynthases | Enzymatic cleavage of core fucose by fucosidases possible; potential for site-specific glycoengineering | Large-scale synthesis of glycan oxazolines |
| In vitro glycosylation [300, 301] | In vitro modeling of the ER glycosylation machinery: using OSTs to transfer glycans from lipid-linked oligosaccharides to proteins | Glycosylation of aglycosylated proteins, i.e., glycosylation of "empty" consensus sequences | Effective (chemo)-enzymatic synthesis of eukaryotic-type lipid-linked oligosaccharides |

The major challenges for large-scale application are listed

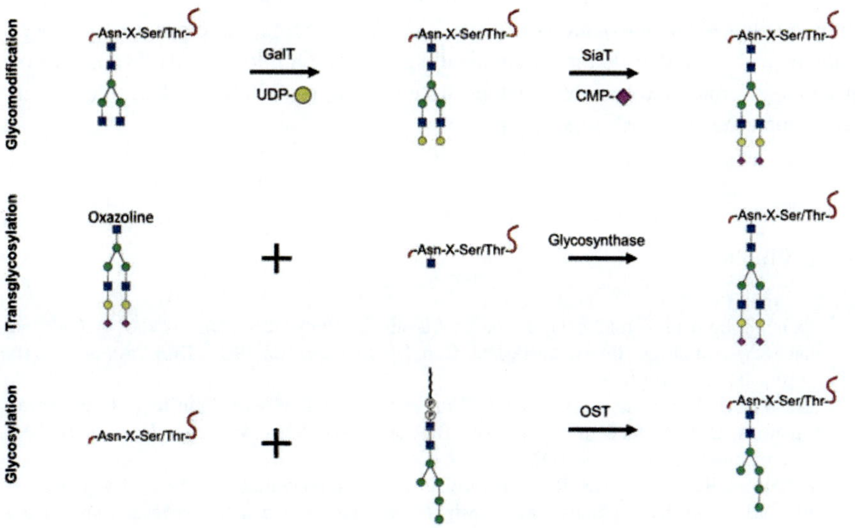


**Fig. 13** The three most commonly used strategies for manipulating the glycosylation of proteins: in vitro glycomodification, transglycosylation, and in vitro glycosylation

scale application of the transglycosylation of mAbs will hinge on the effective synthesis of glycans oxazolines [305, 306]. Excellent reviews on engineered glycosynthases and their applications can be found elsewhere and are not reviewed here [297].

The most advanced in vitro glycoengineering approach is the re-glycosylation of glycans on proteins using Leloir glycosyltransferases [296, 307]. For the generation of homogeneous glycoforms, this approach involves the optional trimming of glycans by glycosidases down to (mammalian-type) core structures such as $(GlcNAc)_2Man_3GlcNAc_2$ and stepwise building homogenous glycoforms using Leloir glycosyltransferases and sugar nucleotides. In the most comprehensive studies on IgG1, aglycosylated, fully galactosylated, and mono- and di-sialylated variants were produced using commercially available galactosidase, galactosyltransferase, and two variants of sialyltransferases, together with the sugar nucleotides UDP-Gal and CMP-Neu5Ac [296]. In vitro re-glycosylation can remodel the glycan of not only isolated IgG but also of IgG in cell-culture supernatant as well as semi-purified IgG [295]. Moreover, in vitro glycomodification can be facilitated as a one-pot process, and, thus, it could be integrated into existing processing steps without additional downstream processing [295]. Alternatively, antibodies can also be in vitro modified as part of the downstream processing, when the antibody is immobilized on a protein A or G column [307]. Fully galactosylated glycoforms were produced in one-stage and sialylated glycoform in a two-stage conversion process when antibodies were immobilized using commercially available enzymes and substrates [307]. While remarkable progress has been made over the last decade on enzymatically producing homogeneous glycoforms on proteins, it remains to be seen whether these techniques will be applied to commercial biopharmaceuticals in the future.

## References

1. André I, Potocki-Véronèse G, Barbe S, Moulis C, Remaud-Siméon M (2014) CAZyme discovery and design for sweet dreams. Curr Opin Chem Biol 19:17–24. https://doi.org/10.1016/j.cbpa.2013.11.014
2. Lairson LL, Henrissat B, Davies GJ, Withers SG (2008) Glycosyltransferases: structures, functions, and mechanisms. Annu Rev Biochem 77(1):521–555. https://doi.org/10.1146/annurev.biochem.76.061005.092322
3. Ardèvol A, Rovira C (2015) Reaction mechanisms in carbohydrate-active enzymes: glycoside hydrolases and glycosyltransferases. Insights from ab initio quantum mechanics/molecular mechanics dynamic simulations. J Am Chem Soc 137(24):7528–7547. https://doi.org/10.1021/jacs.5b01156
4. Tvaroška I (2015) Atomistic insight into the catalytic mechanism of glycosyltransferases by combined quantum mechanics/molecular mechanics (QM/MM) methods. Carbohydr Res 403:38–47. https://doi.org/10.1016/j.carres.2014.06.017
5. Breton C, Snajdrova L, Jeanneau C, Koca J, Imberty A (2006) Structures and mechanisms of glycosyltransferases. Glycobiology 16(2):29–37

6. Gloster TM (2014) Advances in understanding glycosyltransferases from a structural perspective. Curr Opin Struct Biol 28:131–141. https://doi.org/10.1016/j.sbi.2014.08.012
7. Teppa R, Petit D, Plechakova O, Cogez V, Harduin-Lepers A (2016) Phylogenetic-derived insights into the evolution of Sialylation in eukaryotes: comprehensive analysis of vertebrate β-galactoside α2,3/6-Sialyltransferases (ST3Gal and ST6Gal). Int J Mol Sci 17(8):1286
8. Chao L, Jongkees S (2019) High-throughput approaches in carbohydrate-active enzymology: glycosidase and glycosyl transferase inhibitors, evolution, and discovery. Angew Chem 131 (37):12880–12890. https://doi.org/10.1002/ange.201900055
9. Benkoulouche M, Fauré R, Remaud-Siméon M, Moulis C, André I (2019) Harnessing glycoenzyme engineering for synthesis of bioactive oligosaccharides. Interface Focus 9 (2):20180069. https://doi.org/10.1098/rsfs.2018.0069
10. McArthur John B, Chen X (2016) Glycosyltransferase engineering for carbohydrate synthesis. Biochem Soc Trans 44(1):129–142. https://doi.org/10.1042/bst20150200
11. Hancock SM, Vaughan MD, Withers SG (2006) Engineering of glycosidases and glycosyltransferases. Curr Opin Chem Biol 10(5):509–519
12. Daude D, Andre I, Monsan P, Remaud-Simeon M (2014) Chapter 28 successes in engineering glucansucrases to enhance glycodiversification. In: Carbohydrate chemistry, vol 40. The Royal Society of Chemistry, London, pp 624–645. https://doi.org/10.1039/9781849739986-00624
13. O'Neill EC, Field RA (2015) Enzymatic synthesis using glycoside phosphorylases. Carbohydr Res 403:23–37. https://doi.org/10.1016/j.carres.2014.06.010
14. Slámová K, Bojarová P (2017) Engineered N-acetylhexosamine-active enzymes in glycoscience. BBA 1861(8):2070–2087. https://doi.org/10.1016/j.bbagen.2017.03.019
15. Cobucci-Ponzano B, Moracci M (2012) Glycosynthases as tools for the production of glycan analogs of natural products. Nat Prod Rep 29(6):697–709
16. Sprenger GA, Baumgärtner F, Albermann C (2017) Production of human milk oligosaccharides by enzymatic and whole-cell microbial biotransformations. J Biotechnol 258(Suppl C):79–91. https://doi.org/10.1016/j.jbiotec.2017.07.030
17. Han NS, Kim T-J, Park Y-C, Kim J, Seo J-H (2012) Biotechnological production of human milk oligosaccharides. Biotechnol Adv 30(6):1268–1278. https://doi.org/10.1016/j.biotechadv.2011.11.003
18. Petschacher B, Nidetzky B (2016) Biotechnological production of fucosylated human milk oligosaccharides: prokaryotic fucosyltransferases and their use in biocatalytic cascades or whole cell conversion systems. J Biotechnol 235:61–83. https://doi.org/10.1016/j.jbiotec.2016.03.052
19. Seibel J, Buchholz K, Derek H (2010) Tools in oligosaccharide synthesis: current research and application. In: Advances in carbohydrate chemistry and biochemistry, vol 63. Academic Press, Cambridge, pp 101–138
20. Daude D, Remaud-Simeon M, Andre I (2012) Sucrose analogs: an attractive (bio)source for glycodiversification. Nat Prod Rep 29(9):945–960. https://doi.org/10.1039/C2NP20054F
21. Nidetzky B, Gutmann A, Zhong C (2018) Leloir glycosyltransferases as biocatalysts for chemical production. ACS Catal:6283–6300. https://doi.org/10.1021/acscatal.8b00710
22. DeAngelis PL, Liu J, Linhardt RJ (2013) Chemoenzymatic synthesis of glycosaminoglycans: re-creating, re-modeling and re-designing nature's longest or most complex carbohydrate chains. Glycobiology 23(7):764–777. https://doi.org/10.1093/glycob/cwt016
23. Suflita M, Fu L, He W, Koffas M, Linhardt RJ (2015) Heparin and related polysaccharides: synthesis using recombinant enzymes and metabolic engineering. Appl Microbiol Biotechnol 99(18):7465–7479. https://doi.org/10.1007/s00253-015-6821-9
24. Danby PM, Withers SG (2016) Advances in enzymatic glycoside synthesis. ACS Chem Biol 11(7):1784–1794. https://doi.org/10.1021/acschembio.6b00340
25. Rich JR, Withers SG (2009) Emerging methods for the production of homogeneous human glycoproteins. Nat Chem Biol 5(4):206–215

26. Pratta MR, Bertozzi CR (2005) Synthetic glycopeptides and glycoproteins as tools for biology. Chem Soc Rev 34:58–68
27. Krasnova L, Wong C-H (2019) Oligosaccharide synthesis and translational innovation. J Am Chem Soc. https://doi.org/10.1021/jacs.8b11005
28. Kittl R, Withers SG (2010) New approaches to enzymatic glycoside synthesis through directed evolution. Carbohydr Res 345(10):1272–1279. https://doi.org/10.1016/j.carres.2010.04.002
29. Armstrong Z, Withers SG (2013) Synthesis of glycans and glycopolymers through engineered enzymes. Biopolymers 99(10):666–674. https://doi.org/10.1002/bip.22335
30. Harding CM, Feldman MF (2019) Glycoengineering bioconjugate vaccines, therapeutics, and diagnostics in E. coli. Glycobiology 29(7):519–529. https://doi.org/10.1093/glycob/cwz031
31. Kim BG, Yang SM, Kim SY, Cha MN, Ahn JH (2015) Biosynthesis and production of glycosylated flavonoids in Escherichia coli: current state and perspectives. Appl Microbiol Biotechnol 99(7):2979–2988. https://doi.org/10.1007/s00253-015-6504-6
32. Marienhagen J, Bott M (2013) Metabolic engineering of microorganisms for the synthesis of plant natural products. J Biotechnol 163(2):166–178. https://doi.org/10.1016/j.jbiotec.2012.06.001
33. Zhou J, Du G, Chen J (2014) Novel fermentation processes for manufacturing plant natural products. Curr Opin Biotechnol 25(0):17–23. https://doi.org/10.1016/j.copbio.2013.08.009
34. Chang A, Singh S, Phillips Jr GN, Thorson JS (2011) Glycosyltransferase structural biology and its role in the design of catalysts for glycosylation. Curr Opin Biotechnol 22(6):800–808. https://doi.org/10.1016/j.copbio.2011.04.013
35. Breton C, Fournel-Gigleux S, Palcic MM (2012) Recent structures, evolution and mechanisms of glycosyltransferases. Curr Opin Struct Biol 22(5):540–549. https://doi.org/10.1016/j.sbi.2012.06.007
36. Albesa-Jove D, Sainz-Polo MA, Marina A, Guerin ME (2017) Structural snapshots of alpha-1,3-Galactosyltransferase with native substrates: insight into the catalytic mechanism of retaining glycosyltransferases. Angew Chem Int Ed Engl 56(47):14853–14857. https://doi.org/10.1002/anie.201707922
37. Moremen KW, Ramiah A, Stuart M, Steel J, Meng L, Forouhar F, Moniz HA, Gahlay G, Gao Z, Chapla D, Wang S, Yang J-Y, Prabhakar PK, Johnson R, Rosa MD, Geisler C, Nairn AV, Seetharaman J, Wu S-C, Tong L, Gilbert HJ, LaBaer J, Jarvis DL (2017) Expression system for structural and functional studies of human glycosylation enzymes. Nat Chem Biol 14:156. https://doi.org/10.1038/nchembio.2539
38. Tiwari P, Sangwan RS, Sangwan NS (2016) Plant secondary metabolism linked glycosyltransferases: an update on expanding knowledge and scopes. Biotechnol Adv 34 (5):714–739. https://doi.org/10.1016/j.biotechadv.2016.03.006
39. Liang D-M, Liu J-H, Wu H, Wang B-B, Zhu H-J, Qiao J-J (2015) Glycosyltransferases: mechanisms and applications in natural product development. Chem Soc Rev 44 (22):8350–8374. https://doi.org/10.1039/C5CS00600G
40. Skretas G, Carroll S, DeFrees S, Schwartz MF, Johnson KF, Georgiou G (2009) Expression of active human sialyltransferase ST6GalNAcI in *Escherichia coli*. Microb Cell Factories 8 (1):50. https://doi.org/10.1186/1475-2859-8-50
41. Bernatchez S, Gilbert M, Blanchard M-C, Karwaski M-F, Li J, DeFrees S, Wakarchuk WW (2007) Variants of the β1,3-Galactosyltransferase CgtB from the bacterium *Campylobacter jejuni* have distinct acceptor specificities. Glycobiology 17(12):1333–1343. https://doi.org/10.1093/glycob/cwm090
42. Li M, Shen J, Liu X, Shao J, Yi W, Chow CS, Wang PG (2008) Identification of a new a1,2-Fucosyltransferase involved in O-antigen biosynthesis of Escherichia coli O86:B7 and formation of H-type 3 blood group antigen. Biochemistry 47(44):11590–11597
43. Pasek M, Boeggeman E, Ramakrishnan B, Qasba PK (2010) Galectin-1 as a fusion partner for the production of soluble and folded human β-1,4-galactosyltransferase-T7 in *E. coli*. Biochem Biophys Res Commun 394(3):679–684

44. Sauerzapfe B, Namdjou DJ, Schumacher T, Linden N, Křenek K, Křen V, Elling L (2008) Characterization of recombinant fusion constructs of human β1,4-galactosyltransferase 1 and the lipase pre-propeptide from *Staphylococcus hyicus*. J Mol Catal B Enzym 50(2):128–140. https://doi.org/10.1016/j.molcatb.2007.09.009
45. Engels L, Elling L (2014) WbgL: a novel bacterial α1,2-fucosyltransferase for the synthesis of 2′-fucosyllactose. Glycobiology 24(2):170–178. https://doi.org/10.1093/glycob/cwt096
46. Lauber J, Handrick R, Leptihn S, Dürre P, Gaisser S (2015) Expression of the functional recombinant human glycosyltransferase GalNAcT2 in *Escherichia coli*. Microb Cell Factories 14:3. https://doi.org/10.1186/s12934-014-0186-0
47. Lobstein J, Emrich CA, Jeans C, Faulkner M, Riggs P, Berkmen M (2012) SHuffle, a novel *Escherichia coli* protein expression strain capable of correctly folding disulfide bonded proteins in its cytoplasm. Microb Cell Factories 11(1):753. https://doi.org/10.1186/1475-2859-11-56
48. Luley-Goedl C, Czabany T, Longus K, Schmölzer K, Zitzenbacher S, Ribitsch D, Schwab H, Nidetzky B (2016) Combining expression and process engineering for high-quality production of human sialyltransferase in Pichia pastoris. J Biotechnol 235:54–60. https://doi.org/10.1016/j.jbiotec.2016.03.046
49. Bencúrová M, Rendić D, Fabini G, Kopecky E-M, Altmann F, Wilson IBH (2003) Expression of eukaryotic glycosyltransferases in the yeast *Pichia pastoris*. Biochimie 85(3):413–422. https://doi.org/10.1016/S0300-9084(03)00072-5
50. Malissard M, Zeng S, Berger EG (1999) The yeast expression system for recombinant glycosyltransferases. Glycoconj J 16(2):125–139
51. Kim HG, Yang SM, Lee YC, Do SI, Chung IS, Yang JM (2003) High-level expression of human glycosyltransferases in insect cells as biochemically active form. Biochem Biophys Res Commun 305(3):488–493. https://doi.org/10.1016/S0006-291X(03)00795-2
52. Umana P, Jean-Mairet J, Bailey JE (1999) Tetracycline-regulated overexpression of glycosyltransferases in Chinese hamster ovary cells. Biotechnol Bioeng 65(5):542–549
53. De Vries T, Knegtel RMA, Holmes EH, Macher BA (2001) Fucosyltransferases: structure/function studies. Glycobiology 11(10):119–128
54. Kightlinger W, Duncker KE, Ramesh A, Thames AH, Natarajan A, Stark JC, Yang A, Lin L, Mrksich M, DeLisa MP, Jewett MC (2019) A cell-free biosynthesis platform for modular construction of protein glycosylation pathways. Nat Commun 10(1):5404. https://doi.org/10.1038/s41467-019-12024-9
55. Dondapati SK, Stech M, Zemella A, Kubick S (2020) Cell-free protein synthesis: a promising option for future drug development. BioDrugs. https://doi.org/10.1007/s40259-020-00417-y
56. Blixt O, Vasiliu D, Allin K, Jacobsen N, Warnock D, Razi N, Paulson JC, Bernatchez S, Gilbert M, Wakarchuk W (2005) Chemoenzymatic synthesis of 2-azidoethyl-ganglio-oligosaccharides GD3, GT3, GM2, GD2, GT2, GM1, and GD1a. Carbohydr Res 340(12):1963–1972. https://doi.org/10.1016/j.carres.2005.06.008
57. Zhou D, Utkina N, Li D, Dong C, Druzhinina T, Veselovsky V, Liu B (2013) Biochemical characterization of a new β-1,3-galactosyltransferase WbuP from Escherichia coli O114 that catalyzes the second step in O-antigen repeating-unit. Carbohydr Res 381:43–50. https://doi.org/10.1016/j.carres.2013.08.021
58. X-w L, Xia C, Li L, Guan W-Y, Pettit N, Zhang H-C, Chen M, Wang PG (2009) Characterization and synthetic application of a novel β1,3-galactosyltransferase from Escherichia coli O55:H7. Bioorg Med Chem 17(14):4910–4915. https://doi.org/10.1016/j.bmc.2009.06.005
59. Fischöder T, Laaf D, Dey C, Elling L (2017) Enzymatic synthesis of N-Acetyllactosamine (LacNAc) type 1 oligomers and characterization as multivalent galectin ligands. Molecules 22(8):1320
60. Fischöder T, Cajic S, Reichl U, Rapp E, Elling L (2019) Enzymatic cascade synthesis provides novel linear human Milk oligosaccharides as reference standards for xCGE-LIF based high-throughput analysis. Biotechnol J 14(3):e1800305. https://doi.org/10.1002/biot.201800305

61. Fischöder T, Cajic S, Grote V, Heinzler R, Reichl U, Franzreb M, Rapp E, Elling L (2019) Enzymatic cascades for tailored $^{13}C_6$ and $^{15}N$ enriched human Milk oligosaccharides. Molecules 24(19):3482
62. McArthur JB, Yu H, Chen X (2019) A bacterial β1–3-galactosyltransferase enables multigram-scale synthesis of human milk lacto-N-tetraose (LNT) and its fucosides. ACS Catal 9(12):10721–10726. https://doi.org/10.1021/acscatal.9b03990
63. Li Y, Xue M, Sheng X, Yu H, Zeng J, Thon V, Chen Y, Muthana MM, Wang PG, Chen X (2016) Donor substrate promiscuity of bacterial β1–3-N-acetylglucosaminyltransferases and acceptor substrate flexibility of β1–4-galactosyltransferases. Bioorg Med Chem 24 (8):1696–1705. https://doi.org/10.1016/j.bmc.2016.02.043
64. Namdjou D-J, Chen H-M, Vinogradov E, Brochu D, Withers SG, Wakarchuk WW (2008) A beta-1,4-Galactosyltransferase from *Helicobacter pylori* is an efficient and versatile biocatalyst displaying a novel activity for thioglycoside synthesis. Chembiochem 9(10):1632–1640
65. Yi W, Shao J, Zhu L, Li M, Singh M, Lu Y, Lin S, Li H, Ryu K, Shen J, Guo H, Yao Q, Bush CA, Wang PG (2005) Escherichia coli O86 O-antigen biosynthetic gene cluster and stepwise enzymatic synthesis of human blood group B antigen tetrasaccharide. J Am Chem Soc 127 (7):2040–2041. https://doi.org/10.1021/ja045021y
66. Lairson LL, Watts AG, Wakarchuk WW, Withers SG (2006) Using substrate engineering to harness enzymatic promiscuity and expand biological catalysis. Nat Chem Biol 2 (12):724–728. https://doi.org/10.1038/nchembio828
67. Rech C, Rosencrantz RR, Křenek K, Pelantová H, Bojarová P, Römer CE, Hanisch F-G, Křen V, Elling L (2011) Combinatorial one-pot synthesis of poly-*N*-acetyllactosamine oligosaccharides with *Leloir*-glycosyltransferases. Adv Synth Catal 353(13):2492–2500. https://doi.org/10.1002/adsc.201100375
68. Fang J-L, Tsai T-W, Liang C-Y, Li J-Y, Yu C-C (2018) Enzymatic synthesis of human Milk Fucosides α1,2-Fucosyl Para-lacto-N-Hexaose and its isomeric derivatives. Adv Synth Catal 360(17):3213–3219. https://doi.org/10.1002/adsc.201800518
69. Li Y, Yu H, Thon V, Chen Y, Muthana MM, Qu J, Hie L, Chen X (2014) Donor substrate promiscuity of the N-acetylglucosaminyltransferase activities of Pasteurella multocida heparosan synthase 2 (PmHS2) and Escherichia coli K5 KfiA. Appl Microbiol Biotechnol 98(3):1127–1134. https://doi.org/10.1007/s00253-013-4947-1
70. Chavaroche AAE, van den Broek LAM, Boeriu C, Eggink G (2012) Synthesis of heparosan oligosaccharides by *Pasteurella multocida* PmHS2 single-action transferases. Appl Microbiol Biotechnol 95(5):1199–1210. https://doi.org/10.1007/s00253-011-3813-2
71. Cai C, Edgar K, Liu J, Linhardt RJ (2013) Preparation and application of a 'clickable' acceptor for enzymatic synthesis of heparin oligosaccharides. Carbohydr Res 372(0):30–34. https://doi.org/10.1016/j.carres.2013.02.010
72. Yi W, Shen J, Zhou G, Li J, Wang PG (2008) Bacterial homologue of human blood group A transferase. J Am Chem Soc 130(44):14420–14421. https://doi.org/10.1021/ja805844y
73. Li J, Su G, Liu J (2017) Enzymatic synthesis of homogeneous chondroitin sulfate oligosaccharides. Angew Chem Int Ed 56(39):11784–11787. https://doi.org/10.1002/anie.201705638
74. Sugiura N, Shimokata S, Minamisawa T, Hirabayashi J, Kimata K, Watanabe H (2008) Sequential synthesis of chondroitin oligosaccharides by immobilized chondroitin polymerase mutants. Glycoconj J 25(6):521–530. https://doi.org/10.1007/s10719-008-9105-0
75. Green DE, DeAngelis PL (2017) Identification of a chondroitin synthase from an unexpected source, the green sulfur bacterium chlorobium phaeobacteroides. Glycobiology 27 (5):469–476. https://doi.org/10.1093/glycob/cwx008
76. Chavaroche AAE, Springer J, Kooy F, Boeriu C, Eggink G (2010) In vitro synthesis of heparosan using recombinant *Pasteurella multocida* heparosan synthase PmHS2. Appl Microbiol Biotechnol 85(6):1881–1891. https://doi.org/10.1007/s00253-009-2214-2
77. Rexer TFT, Schildbach A, Klapproth J, Schierhorn A, Mahour R, Pietzsch M, Rapp E, Reichl U (2018) One pot synthesis of GDP-mannose by a multi-enzyme cascade for enzymatic assembly of lipid-linked oligosaccharides. Biotechnol Bioeng 115(1):192–205. https://doi.org/10.1002/bit.26454

78. Zhao C, Wu Y, Yu H, Shah IM, Li Y, Zeng J, Liu B, Mills DA, Chen X (2016) One-pot multienzyme (OPME) synthesis of human blood group H antigens and a human milk oligosaccharide (HMOS) with highly active Thermosynechococcus elongatus [small alpha]1-2-fucosyltransferase. Chem Commun. https://doi.org/10.1039/C5CC10646J
79. Li M, Liu X-W, Shao J, Shen J, Jia Q, Yi W, Song JK, Woodward R, Chow CS, Wang PG (2008) Characterization of a novel a1,2-Fucosyltransferase of *Escherichia coli* O128:B12 and functional investigation of its common motif. Biochemistry 47(1):378–387
80. Pettit N, Styslinger T, Mei Z, Han W, Zhao G, Wang PG (2010) Characterization of WbiQ: an [alpha]1,2-fucosyltransferase from Escherichia coli O127:K63(B8), and synthesis of H-type 3 blood group antigen. Biochem Biophys Res Commun 402(2):190–195
81. Li Q, Li Z, Duan X, Yi W (2014) A tandem enzymatic approach for detecting and imaging tumor-associated Thomsen–Friedenreich antigen disaccharide. J Am Chem Soc 136 (36):12536–12539. https://doi.org/10.1021/ja5054225
82. Yi W, Liu X, Li Y, Li J, Xia C, Zhou G, Zhang W, Zhao W, Chen X, Wang PG (2009) Remodeling bacterial polysaccharides by metabolic pathway engineering. Proc Natl Acad Sci USA 106(11):4207–4212. https://doi.org/10.1073/pnas.0812432106
83. Tsai T-W, Fang J-L, Liang C-Y, Wang C-J, Huang Y-T, Wang Y-J, Li J-Y, Yu C-C (2019) Exploring the synthetic application of helicobacter pylori α1,3/4-fucosyltransferase FucTIII toward the syntheses of fucosylated human milk glycans and Lewis antigens. ACS Catal. https://doi.org/10.1021/acscatal.9b03752
84. Yu H, Yan X, Autran CA, Li Y, Etzold S, Latasiewicz J, Robertson BM, Li J, Bode L, Chen X (2017) Enzymatic and chemoenzymatic syntheses of disialyl glycans and their necrotizing enterocolitis preventing effects. J Org Chem 82(24):13152–13160. https://doi.org/10.1021/acs.joc.7b02167
85. Ye J, Xia H, Sun N, Liu C-C, Sheng A, Chi L, Liu X-W, Gu G, Wang S-Q, Zhao J, Wang P, Xiao M, Wang F, Cao H (2019) Reprogramming the enzymatic assembly line for site-specific fucosylation. Nat Catal. https://doi.org/10.1038/s41929-019-0281-z
86. Bai J, Wu Z, Sugiarto G, Gadi MR, Yu H, Li Y, Xiao C, Ngo A, Zhao B, Chen X, Guan W (2019) Biochemical characterization of helicobacter pylori α1–3-fucosyltransferase and its application in the synthesis of fucosylated human milk oligosaccharides. Carbohydr Res 480:1–6. https://doi.org/10.1016/j.carres.2019.05.007
87. Choi YH, Kim JH, Park BS, Kim B-G (2016) Solubilization and iterative saturation mutagenesis of α1,3-fucosyltransferase from helicobacter pylori to enhance its catalytic efficiency. Biotechnol Bioeng 113(8):1666–1675. https://doi.org/10.1002/bit.25944
88. Zhang L, Lau K, Cheng J, Yu H, Li Y, Sugiarto G, Huang S, Ding L, Thon V, Wang PG, Chen X (2010) *Helicobacter hepaticus* Hh0072 gene encodes a novel Î±1-3-fucosyltransferase belonging to CAZy GT11 family. Glycobiology 20(9):1077–1088. https://doi.org/10.1093/glycob/cwq068
89. Tasnima N, Yu H, Yan X, Li W, Xiao A, Chen X (2019) Facile chemoenzymatic synthesis of Lewis a (lea) antigen in gram-scale and sialyl Lewis a (sLea) antigens containing diverse sialic acid forms. Carbohydr Res 472:115–121. https://doi.org/10.1016/j.carres.2018.12.004
90. Gilbert M, Brisson J-R, Karwaski M-F, Michniewicz J, Cunningham A-M, Wu Y, Young NM, Wakarchuk WW (2000) Biosynthesis of ganglioside mimics in campylobacter jejuni OH4384: identification of the glycosyltransferase genes, enzymatic synthesis of model compounds, and characterization of nanomole amounts by 600-MHz 1H and 13C NMR analysis. J Biol Chem 275(6):3896–3906. https://doi.org/10.1074/jbc.275.6.3896
91. Morley TJ, Withers SG (2010) Chemoenzymatic synthesis and enzymatic analysis of 8-modified Cytidine monophosphate-Sialic acid and Sialyl lactose derivatives. J Am Chem Soc 132(27):9430–9437. https://doi.org/10.1021/ja102644a
92. Guo Y, Jers C, Meyer AS, Li H, Kirpekar F, Mikkelsen JD (2015) Modulating the regioselectivity of a pasteurella multocida sialyltransferase for biocatalytic production of 3′- and 6′-sialyllactose. Enzym Microb Technol 78:54–62. https://doi.org/10.1016/j.enzmictec.2015.06.012

93. Malekan H, Fung G, Thon V, Khedri Z, Yu H, Qu J, Li Y, Ding L, Lam KS, Chen X (2013) One-pot multi-enzyme (OPME) chemoenzymatic synthesis of sialyl-Tn-MUC1 and sialyl-T-MUC1 glycopeptides containing natural or non-natural sialic acid. Bioorg Med Chem 21 (16):4778–4785. https://doi.org/10.1016/j.bmc.2013.02.040
94. Tsukamoto H, Takakura Y, Yamamoto T (2007) Purification, cloning, and expression of an α/β-galactoside α-2,3-sialyltransferase from a luminous marine bacterium, photobacterium phosphoreum. J Biol Chem 282(41):29794–29802. https://doi.org/10.1074/jbc.M701907200
95. Schmölzer K, Eibinger M, Nidetzky B (2017) Active-site His85 of Pasteurella dagmatis sialyltransferase facilitates productive sialyl transfer and so prevents futile hydrolysis of CMP-Neu5Ac. ChemBioChem. https://doi.org/10.1002/cbic.201700113
96. Chokhawala HA, Huang S, Lau K, Yu H, Cheng J, Thon V, Hurtado-Ziola N, Guerrero JA, Varki A, Chen X (2008) Combinatorial chemoenzymatic synthesis and high-throughput screening of sialosides. ACS Chem Biol 3(9):567–576
97. Yamamoto T, Hamada Y, Ichikawa M, Kajiwara H, Mine T, Tsukamoto H, Takakura Y (2007) A {beta}-galactoside {alpha}2,6-sialyltransferase produced by a marine bacterium, photobacterium leiognathi JT-SHIZ-145, is active at pH 8. Glycobiology 17(11):1167–1174. https://doi.org/10.1093/glycob/cwm086
98. Ding L, Yu H, Lau K, Li Y, Muthana S, Wang J, Chen X (2011) Efficient chemoenzymatic synthesis of sialyl Tn-antigens and derivatives. Chem Commun 47(30):8691–8693. https://doi.org/10.1039/C1CC12732B
99. Willis LM, Gilbert M, Karwaski MF, Blanchard MC, Wakarchuk WW (2008) Characterization of the alpha-2,8-polysialyltransferase from Neisseria meningitidis with synthetic acceptors, and the development of a self-priming polysialyltransferase fusion enzyme. Glycobiology 18(2):177–186. https://doi.org/10.1093/glycob/cwm126
100. Lindhout T, Iqbal U, Willis LM, Reid AN, Li J, Liu X, Moreno M, Wakarchuk WW (2011) Site-specific enzymatic polysialylation of therapeutic proteins using bacterial enzymes. Proc Natl Acad Sci 108(18):7397–7402. https://doi.org/10.1073/pnas.1019266108
101. Ban L, Pettit N, Li L, Stuparu AD, Cai L, Chen W, Guan W, Han W, Wang PG, Mrksich M (2012) Discovery of glycosyltransferases using carbohydrate arrays and mass spectrometry. Nat Chem Biol 8(9):769–773. https://doi.org/10.1038/nchembio.1022
102. Kightlinger W, Lin L, Rosztoczy M, Li W, DeLisa MP, Mrksich M, Jewett MC (2018) Design of glycosylation sites by rapid synthesis and analysis of glycosyltransferases. Nat Chem Biol 14(6):627–635. https://doi.org/10.1038/s41589-018-0051-2
103. Serna S, Hokke CH, Weissenborn M, Flitsch S, Martin-Lomas M, Reichardt NC (2013) Profiling glycosyltransferase activities by tritium imaging of glycan microarrays. Chembiochem 14(7):862–869. https://doi.org/10.1002/cbic.201300051
104. Lundborg M, Modhukur V, Widmalm G (2010) Glycosyltransferase functions of *E. coli* O-antigens. Glycobiology 20(3):366–368. https://doi.org/10.1093/glycob/cwp185
105. Rojas-Macias MA, Ståhle J, Lütteke T, Widmalm G (2015) Development of the ECODAB into a relational database for *Escherichia coli* O-antigens and other bacterial polysaccharides. Glycobiology 25(3):341–347. https://doi.org/10.1093/glycob/cwu116
106. Elling L (1997) Glycobiotechnology: enzymes for the synthesis of nucleotide sugars. In: Scheper T (ed) Advances in biochemical engineering/biotechnology, vol 58. Springer, Berlin, pp 89–144
107. Bülter T, Elling L (1999) Enzymatic synthesis of nucleotide sugars. Glycoconj J 16 (2):147–159
108. Engels L, Elling L (2016) Enzymatic and chemoenzymatic synthesis of nucleotide sugars: novel enzymes, novel substrates, novel products, and novel routes. In: Grunwald P (ed) Handbook of carbohydrate-modifying biocatalysts. Stanford Publishing, Stanford, pp 297–320. https://doi.org/10.4032/9789814303484
109. Freeze HH, Hart GW, Schnaar RL (2017) Chapter 5: Glycosylation precursors. Essentials of glycobiology.3rd edn. Cold Spring Harbor Laboratory Press, New York

110. Varki A (2017) New and updated glycoscience-related resources at NCBI. Glycobiology 27 (11):993–993. https://doi.org/10.1093/glycob/cwx077
111. Rupprath C, Kopp M, Hirtz D, Müller R, Elling L (2007) An enzyme module system for *in situ* regeneration of dTDP-activated deoxysugars. Adv Synth Catal 349(8–9):1489–1496
112. Bar-Peled M, O'Neill MA (2011) Plant nucleotide sugar formation, interconversion, and salvage by sugar recycling. Annu Rev Plant Biol 62:127–155. https://doi.org/10.1146/annurev-arplant-042110-103918
113. Cai L (2012) Recent Progress in enzymatic synthesis of sugar nucleotides. J Carbohydr Chem 31(7):535–552. https://doi.org/10.1080/07328303.2012.687059
114. Koizumi S, Endo T, Tabata K, Ozaki A (1998) Large-scale production of UDP-galactose and globotriose by coupling metabolically engineered bacteria. Nat Biotechnol 16(9):847–850. https://doi.org/10.1038/nbt0998-847
115. Tabata K, Koizumi S, Endo T, Ozaki A (2000) Production of UDP-N-acetylglucosamine by coupling metabolically engineered bacteria. Biotechnol Lett 22(6):479–483
116. Koizumi S, Endo T, Tabata K, Nagano H, Ohnishi J, Ozaki A (2000) Large-scale production of GDP-fucose and Lewis X by bacterial coupling. J Ind Microbiol Biotech 25(4):213–217. https://doi.org/10.1038/sj.jim.7000055
117. Ishikawa M, Koizumi S (2010) Microbial production of N-acetylneuraminic acid by genetically engineered Escherichia coli. Carbohydr Res 345(18):2605–2609
118. Schmolzer K, Gutmann A, Diricks M, Desmet T, Nidetzky B (2016) Sucrose synthase: a unique glycosyltransferase for biocatalytic glycosylation process development. Biotechnol Adv 34(2):88–111. https://doi.org/10.1016/j.biotechadv.2015.11.003
119. Yu H, Chen X (2016) One-pot multienzyme (OPME) systems for chemoenzymatic synthesis of carbohydrates. Org Biomol Chem 14(10):2809–2818. https://doi.org/10.1039/C6OB00058D
120. Eixelsberger T, Nidetzky B (2014) Enzymatic redox cascade for one-pot synthesis of uridine 5′-diphosphate xylose from uridine 5′-diphosphate glucose. Adv Synth Catal 356 (17):3575–3584. https://doi.org/10.1002/adsc.201400766
121. Gilormini P-A, Lion C, Noel M, Krzewinski-Recchi M-A, Harduin-Lepers A, Guérardel Y, Biot C (2016) Improved workflow for the efficient preparation of ready to use CMP-activated sialic acids. Glycobiology 26(11):1151–1156. https://doi.org/10.1093/glycob/cww084
122. Ye J, X-w L, Peng P, Yi W, Chen X, Wang F, Cao H (2016) Diversity-oriented enzymatic modular assembly of ABO Histo-blood group antigens. ACS Catal 6(12):8140–8144. https://doi.org/10.1021/acscatal.6b02755
123. Yu H, Li Y, Zeng J, Thon V, Nguyen DM, Ly T, Kuang HY, Ngo A, Chen X (2016) Sequential one-pot multienzyme chemoenzymatic synthesis of glycosphingolipid glycans. J Org Chem. https://doi.org/10.1021/acs.joc.6b01905
124. Elling L, Grothus M, Kula M-R (1993) Investigation of sucrose synthase from rice for the synthesis of various nucleotide sugars and saccharides. Glycobiology 3:349–355
125. Elling L, Güldenberg B, Grothus M, Zervosen A, Peus M, Helfer A, Stein A, Adrian H, Kula M-R (1995) Isolation of sucrose synthase from rice (Oryza sativa) grains in pilot scale for application in carbohydrate synthesis. Biotechnol Appl Biochem 21(1):29–37
126. Sauerzapfe B, Engels L, Elling L (2008) Broadening the biocatalytic properties of recombinant sucrose synthase 1 from potato (Solanum tuberosum L.) by expression in Escherichia coli and Saccharomyces cerevisiae. Enzym Microb Technol 43(3):289–296. https://doi.org/10.1016/j.enzmictec.2008.04.001
127. Römer U, Schrader H, Gunther N, Nettelstroth N, Frommer WB, Elling L (2004) Expression, purification and characterization of recombinant sucrose synthase 1 from *Solanum tuberosum* L. for carbohydrate engineering. J Biotechnol 107(2):135–149. https://doi.org/10.1016/j.jbiotec.2003.10.017
128. Zervosen A, Elling L (1996) A novel three-enzyme reaction cycle for the synthesis of *N*-acetyllactosamine with in situ regeneration of uridine 5′- diphosphate glucose and uridine 5′-diphosphate galactose. J Am Chem Soc 118(8):1836–1840

129. Engels L, Henze M, Hummel W, Elling L (2015) Enzyme module systems for the synthesis of uridine 5′-diphospho-α-D-glucuronic acid and non-sulfated human natural killer cell-1 (HNK-1) epitope. Adv Synth Catal 357(8):1751–1762. https://doi.org/10.1002/adsc.201500180
130. Zervosen A, Stein A, Adrian H, Elling L (1996) Combined enzymatic synthesis of nucleotide (deoxy) sugars from sucrose and nucleoside monophosphates. Tetrahedron 52(7):2395–2404
131. Zervosen A, Römer U, Elling L (1998) Application of recombinant sucrose synthase large scale synthesis of ADP-glucose. J Mol Catal B Enzym 5(1–4):25–28. https://doi.org/10.1016/s1381-1177(98)00040-x
132. Zervosen A, Elling L, Kula MR (1994) Continuous enzymatic synthesis of 2-′-deoxythymidine-5′(alpha-D-glucopyranosyl)diphosphate. Angew Chem Int Ed 33 (5):571–572. https://doi.org/10.1002/anie.199405711
133. Elling L, Rupprath C, Günther N, Römer U, Verseck S, Weingarten P, Dräger G, Kirschning A, Piepersberg W (2005) An enzyme module system for the synthesis of dTDP-activated Deoxysugars from dTMP and sucrose. Chembiochem 6:1423–1430
134. Diricks M, De Bruyn F, Van Daele P, Walmagh M, Desmet T (2015) Identification of sucrose synthase in nonphotosynthetic bacteria and characterization of the recombinant enzymes. Appl Microbiol Biotechnol 99(20):8465–8474. https://doi.org/10.1007/s00253-015-6548-7
135. Lemmerer M, Schmolzer K, Gutmann A, Nidetzky B (2016) Downstream processing of nucleoside-Diphospho-sugars from sucrose synthase reaction mixtures at decreased solvent consumption. Adv Synth Catal 358(19):3113–3122. https://doi.org/10.1002/adsc.201600540
136. Kulmer ST, Gutmann A, Lemmerer M, Nidetzky B (2017) Biocatalytic cascade of polyphosphate kinase and sucrose synthase for synthesis of nucleotide-activated derivatives of glucose. Adv Synth Catal 359(2):292–301. https://doi.org/10.1002/adsc.201601078
137. Wahl C, Hirtz D, Elling L (2016) Multiplexed capillary electrophoresis as analytical tool for fast optimization of multi-enzyme Cascade reactions – synthesis of nucleotide sugars. Biotechnol J 11(10):1298–1308. https://doi.org/10.1002/biot.201600265
138. Eisele A, Zaun H, Kuballa J, Elling L (2018) In vitro one-pot enzymatic synthesis of hyaluronic acid from sucrose and *N*-acetylglucosamine: optimization of the enzyme module system and nucleotide sugar regeneration. ChemCatChem 10(14):2969–2981. https://doi.org/10.1002/cctc.201800370
139. Gottschalk J, Zaun H, Eisele A, Kuballa J, Elling L (2019) Key factors for a one-pot enzyme cascade synthesis of high molecular weight hyaluronic acid. Int J Mol Sci 20(22):5664
140. Wahl C, Spiertz M, Elling L (2017) Characterization of a new UDP-sugar pyrophosphorylase from *Hordeum vulgare* (barley). J Biotechnol 258(Supplement C):51–55. https://doi.org/10.1016/j.jbiotec.2017.03.025
141. Fischöder T, Wahl C, Zerhusen C, Elling L (2019) Repetitive batch mode facilitates enzymatic synthesis of the nucleotide sugars UDP-gal, UDP-GlcNAc, and UDP-GalNAc on a multi-gram scale. Biotechnol J 14(4). https://doi.org/10.1002/biot.201800386
142. Nahalka J, Liu Z, Chen X, Wang PG (2003) Superbeads: immobilization in sweet chemistry. Chem Eur J 9(2):372–377
143. Heinzler R, Fischöder T, Elling L, Franzreb M (2019) Toward automated enzymatic glycan synthesis in a compartmented flow microreactor system. Adv Synth Catal 361(19):4506–4516. https://doi.org/10.1002/adsc.201900709
144. Heinzler R, Hubner J, Fischöder T, Elling L, Franzreb M (2018) A compartmented flow microreactor system for automated optimization of bioprocesses applying immobilized enzymes. Front Bioeng Biotechnol 6(189):189. https://doi.org/10.3389/fbioe.2018.00189
145. Orrego AH, Trobo-Maseda L, Rocha-Martin J, Guisan JM (2017) Immobilization-stabilization of a complex multimeric sucrose synthase from *Nitrosomonas europaea*. Synthesis of UDP-glucose. Enzym Microb Technol 105(Supplement C):51–58. https://doi.org/10.1016/j.enzmictec.2017.06.008
146. Trobo-Maseda L, Orrego AH, Moreno-Pérez S, Fernández-Lorente G, Guisan JM, Rocha-Martin J (2017) Stabilization of multimeric sucrose synthase from *Acidithiobacillus caldus* via immobilization and post-immobilization techniques for synthesis of UDP-glucose. Appl Microbiol Biotechnol. https://doi.org/10.1007/s00253-017-8649-y

147. Zhao G, Guan W, Cai L, Wang PG (2010) Enzymatic route to preparative-scale synthesis of UDP-GlcNAc/GalNAc, their analogues and GDP-fucose. Nat Protocols 5(4):636–646
148. Ohashi H, Wahl C, Ohashi T, Elling L, Fujiyama K (2017) Effective synthesis of guanosine 5 '-diphospho-beta-L-galactose using bacterial L-Fucokinase/Guanosine 5 '-Diphosphate-L-fucose pyrophosphorylase. Adv Synth Catal 359(23):4227–4234. https://doi.org/10.1002/adsc.201700901
149. Gutmann A, Nidetzky B (2016) Unlocking the potential of Leloir glycosyltransferases for applied biocatalysis: efficient synthesis of uridine 5 '-diphosphate-glucose by sucrose synthase. Adv Synth Catal 358(22):3600–3609. https://doi.org/10.1002/adsc.201600754
150. Kragl U, Klein T, Vasic-Racki D, Kittelmann M, Ghisalba O, Wandrey C (1996) Reaction engineering aspects of activated sugar production. CMP-Neu5Ac as an example. Ann N Y Acad Sci 799(1):577–583. https://doi.org/10.1111/j.1749-6632.1996.tb33260.x
151. Schmaltz RM, Hanson SR, Wong C-H (2011) Enzymes in the synthesis of Glycoconjugates. Chem Rev 111(7):4259–4307
152. Wong C-H, Halcomb RL, Ichikawa Y, Kajimoto T (1995) Enzymes in organic synthesis: application to the problems of carbohydrate recognition (part 1). Angew Chem Int Ed Engl 34 (4):412–432
153. Chi-Huey Wong RLH, Ichikawa Y, Kajimoto T (1995) Enzymes in organic synthesis: application to the problems of carbohydrate recognition (part 2). Angew Chem Int Ed Engl 34(5):521–546
154. Tsai T-I, Lee H-Y, Chang S-H, Wang C-H, Tu Y-C, Lin Y-C, Hwang D-R, Wu C-Y, Wong C-H (2013) Effective sugar nucleotide regeneration for the large-scale enzymatic synthesis of Globo H and SSEA4. J Am Chem Soc 135(39):14831–14839. https://doi.org/10.1021/ja4075584
155. Andexer JN, Richter M (2015) Emerging enzymes for ATP regeneration in biocatalytic processes. Chembiochem 16(3):380–386. https://doi.org/10.1002/cbic.201402550
156. Mordhorst S, Singh J, Mohr MKF, Hinkelmann R, Keppler M, Jessen HJ, Andexer JN (2019) Several polyphosphate kinase 2 enzymes catalyse the production of adenosine 5-′-polyphosphates. Chembiochem 20(8):1019–1022. https://doi.org/10.1002/cbic.201800704
157. Mordhorst S, Maurer A, Popadić D, Brech J, Andexer JN (2017) A flexible polyphosphate-driven regeneration system for coenzyme A dependent catalysis. ChemCatChem 9 (22):4164–4168. https://doi.org/10.1002/cctc.201700848
158. Nocek B, Kochinyan S, Proudfoot M, Brown G, Evdokimova E, Osipiuk J, Edwards AM, Savchenko A, Joachimiak A, Yakunin AF (2008) Polyphosphate-dependent synthesis of ATP and ADP by the family-2 polyphosphate kinases in bacteria. Proc Natl Acad Sci 105 (46):17730–17735. https://doi.org/10.1073/pnas.0807563105
159. Christ JJ, Blank LM (2018) Enzymatic quantification and length determination of polyphosphate down to a chain length of two. Anal Biochem 548:82–90. https://doi.org/10.1016/j.ab.2018.02.018
160. Gutmann A, Lepak A, Diricks M, Desmet T, Nidetzky B (2017) Glycosyltransferase cascades for natural product glycosylation: use of plant instead of bacterial sucrose synthases improves the UDP-glucose recycling from sucrose and UDP. Biotechnol J 12(7):1600557. https://doi.org/10.1002/biot.201600557
161. Hokke CH, Zervosen A, Elling L, Joziasse DH, van den Eijnden DH (1996) One-pot enzymatic synthesis of the gal(a1-3)gal(b1-4)GlcNAc sequence with *in situ* UDP-gal regeneration. Glycoconj J 13(4):687–692
162. Gutmann A, Bungaruang L, Weber H, Leypold M, Breinbauer R, Nidetzky B (2014) Towards the synthesis of glycosylated dihydrochalcone natural products using glycosyltransferase-catalysed cascade reactions. Green Chem 16(9):4417–4425. https://doi.org/10.1039/C4GC00960F
163. Schmölzer K, Lemmerer M, Nidetzky B (2018) Glycosyltransferase cascades made fit for chemical production: integrated biocatalytic process for the natural polyphenol C-glucoside nothofagin. Biotechnol Bioeng 115(3):545–556. https://doi.org/10.1002/bit.26491

164. Zhang L, Gao Y, Liu X, Guo F, Ma C, Liang J, Feng X, Li C (2019) Mining of sucrose synthases from *Glycyrrhiza uralensis* and their application in the construction of an efficient UDP-recycling system. J Agric Food Chem 67(42):11694–11702. https://doi.org/10.1021/acs.jafc.9b05178
165. Sun P, Cai R, Chen L, Li Y, Jia H, Yan M, Chen K (2020) Natural product glycosylation: biocatalytic synthesis of Quercetin-3,4′-O-diglucoside. Appl Biochem Biotechnol 190 (2):464–474. https://doi.org/10.1007/s12010-019-03103-0
166. Pei J, Chen A, Zhao L, Cao F, Ding G, Xiao W (2017) One-pot synthesis of hyperoside by a three-enzyme cascade using a UDP-galactose regeneration system. J Agric Food Chem 65 (29):6042–6048. https://doi.org/10.1021/acs.jafc.7b02320
167. Pei J, Chen A, Sun Q, Zhao L, Cao F, Tang F (2018) Construction of a novel UDP-rhamnose regeneration system by a two-enzyme reaction system and application in glycosylation of flavonoid. Biochem Eng J 139:33–42. https://doi.org/10.1016/j.bej.2018.08.007
168. Fallacara A, Baldini E, Manfredini S, Vertuani S (2018) Hyaluronic acid in the third millennium. Polymers 10(7):701–701. https://doi.org/10.3390/polym10070701
169. Bishnoi M, Jain A, Hurkat P, Jain SK (2016) Chondroitin sulphate: a focus on osteoarthritis. Glycoconj J 33(5):693–705. https://doi.org/10.1007/s10719-016-9665-3
170. Mikami T, Kitagawa H (2017) Sulfated glycosaminoglycans: their distinct roles in stem cell biology. Glycoconj J 34(6):725–735. https://doi.org/10.1007/s10719-016-9732-9
171. Pomin VH (2015) Sulfated glycans in inflammation. Eur J Med Chem 92:353–369. https://doi.org/10.1016/j.ejmech.2015.01.002
172. DeAngelis PL, White CL (2002) Identification and molecular cloning of a Heparosan synthase from pasteurella multocida type D. J Biol Chem 277(9):7209–7213. https://doi.org/10.1074/jbc.M112130200
173. Gandhi NS, Mancera RL (2008) The structure of glycosaminoglycans and their interactions with proteins. Chem Biol Drug Des 72(6):455–482. https://doi.org/10.1111/j.1747-0285.2008.00741.x
174. Akintayo A, Stanley P (2019) Roles for golgi glycans in oogenesis and spermatogenesis. Front Cell Dev Biol 7(JUN):1–9. https://doi.org/10.3389/fcell.2019.00098
175. Knopf-Marques H, Pravda M, Wolfova L, Velebny V, Schaaf P, Vrana NE, Lavalle P (2016) Hyaluronic acid and its derivatives in coating and delivery systems: applications in tissue engineering, regenerative medicine and immunomodulation. Adv Healthc Mater 5 (22):2841–2855. https://doi.org/10.1002/adhm.201600316
176. DeAngelis PL, Padgett-McCue AJ (2000) Identification and molecular cloning of a chondroitin synthase from Pasteurella multocida type F. J Biol Chem 275(31):24124–24129. https://doi.org/10.1074/jbc.M003385200
177. Weigel PH, DeAngelis PL (2007) Hyaluronan synthases: a decade-plus of novel glycosyltransferases. J Biol Chem 282(51):36777–36781
178. Zhang X, Lin L, Huang H, Linhardt RJ (2020) Chemoenzymatic synthesis of glycosaminoglycans. Acc Chem Res 53(2):335–346. https://doi.org/10.1021/acs.accounts.9b00420
179. Jing W, DeAngelis PL (2003) Analysis of the two active sites of the hyaluronan synthase and the chondroitin synthase of Pasteurella multocida. Glycobiology 13(10):661–671. https://doi.org/10.1093/glycob/cwg085
180. Lindahl U, Kusche-Gullberg M, Kjellén L (1998) Regulated diversity of Heparan sulfate. J Biol Chem 273(39):24979–24982. https://doi.org/10.1074/jbc.273.39.24979
181. Dhoot GK (2001) Regulation of Wnt signaling and embryo patterning by an extracellular sulfatase. Science 293(5535):1663–1666. https://doi.org/10.1126/science.293.5535.1663
182. Lamberg SI, Stoolmiller AC (1974) Glycosaminoglycans. A biochemical and clinical review. J Investig Dermatol 63(6):433–449. https://doi.org/10.1111/1523-1747.ep12680346
183. Silbert JE, Sugumaran G (2002) Biosynthesis of chondroitin/Dermatan sulfate. IUBMB Life (Int Union Biochem Mol Biol Life) 54(4):177–186. https://doi.org/10.1080/15216540214923
184. Sugahara K, Kitagawa H (2002) Heparin and Heparan sulfate biosynthesis. IUBMB Life (Int Union Biochem Mol Biol Life) 54(4):163–175. https://doi.org/10.1080/15216540214928

185. Victor XV, Nguyen TKN, Ethirajan M, Tran VM, Nguyen KV, Kuberan B (2009) Investigating the elusive mechanism of glycosaminoglycan biosynthesis. J Biol Chem 284 (38):25842–25853. https://doi.org/10.1074/jbc.M109.043208
186. Bülow HE, Hobert O (2006) The molecular diversity of glycosaminoglycans shapes animal development. Annu Rev Cell Dev Biol 22(1):375–407. https://doi.org/10.1146/annurev.cellbio.22.010605.093433
187. Köwitsch A, Zhou G, Groth T (2018) Medical application of glycosaminoglycans: a review. J Tissue Eng Regen Med 12(1):e23–e41. https://doi.org/10.1002/term.2398
188. Morla S (2019) Glycosaminoglycans and glycosaminoglycan mimetics in cancer and inflammation. Int J Mol Sci 20(8):1963–1963. https://doi.org/10.3390/ijms20081963
189. Bedini E, Laezza A, Iadonisi A (2016) Chemical derivatization of sulfated glycosaminoglycans. Eur J Org Chem 2016(18):3018–3042. https://doi.org/10.1002/ejoc.201600108
190. Gatto F, Volpi N, Nilsson H, Nookaew I, Maruzzo M, Roma A, Johansson ME, Stierner U, Lundstam S, Basso U, Nielsen J (2016) Glycosaminoglycan profiling in patients' plasma and urine predicts the occurrence of metastatic clear cell renal cell carcinoma. Cell Rep 15 (8):1822–1836. https://doi.org/10.1016/j.celrep.2016.04.056
191. Graham GJ, Handel TM, Proudfoot AEI (2019) Leukocyte adhesion: reconceptualizing chemokine presentation by glycosaminoglycans. Trends Immunol 40(6):472–481. https://doi.org/10.1016/j.it.2019.03.009
192. Gschwandtner M, Strutzmann E, Teixeira MM, Anders HJ, Diedrichs-Möhring M, Gerlza T, Wildner G, Russo RC, Adage T, Kungl AJ (2017) Glycosaminoglycans are important mediators of neutrophilic inflammation in vivo. Cytokine 91:65–73. https://doi.org/10.1016/j.cyto.2016.12.008
193. Honarpardaz A, Irani S, Pezeshki-Modaress M, Zandi M, Sadeghi A (2019) Enhanced chondrogenic differentiation of bone marrow mesenchymal stem cells on gelatin/glycosaminoglycan electrospun nanofibers with different amount of glycosaminoglycan. J Biomed Mater Res A 107(1):38–48. https://doi.org/10.1002/jbm.a.36501
194. Mhanna R, Becher J, Schnabelrauch M, Reis RL, Pashkuleva I (2017) Sulfated alginate as a mimic of sulfated glycosaminoglycans: binding of growth factors and effect on stem cell behavior. Adv Biosyst 1(7):1700043–1700043. https://doi.org/10.1002/adbi.201700043
195. Park PW (2016) Glycosaminoglycans and infection. Front Biosci 21(6):4455–4455. https://doi.org/10.2741/4455
196. Sobczak AIS, Pitt SJ, Stewart AJ (2018) Glycosaminoglycan neutralization in coagulation control. Arterioscler Thromb Vasc Biol 38(6):1258–1270. https://doi.org/10.1161/ATVBAHA.118.311102
197. Cooper C, Rannou F, Richette P, Bruyère O, Al-Daghri N, Altman RD, Brandi ML, Collaud Basset S, Herrero-Beaumont G, Migliore A, Pavelka K, Uebelhart D, Reginster JY (2017) Use of intraarticular hyaluronic acid in the management of knee osteoarthritis in clinical practice. Arthritis Care Res 69(9):1287–1296. https://doi.org/10.1002/acr.23204
198. Fallacara A, Vertuani S, Panozzo G, Pecorelli A, Valacchi G, Manfredini S (2017) Novel artificial tears containing cross-linked hyaluronic acid: an in vitro re-epithelialization study. Molecules 22(12):1–13. https://doi.org/10.3390/molecules22122104
199. Kawada C, Yoshida T, Yoshida H, Sakamoto W, Odanaka W, Sato T, Yamasaki T, Kanemitsu T, Masuda Y, Urushibata O (2015) Ingestion of hyaluronans (molecular weights 800 k and 300 k) improves dry skin conditions: a randomized, double blind, controlled study. J Clin Biochem Nutr 56(1):66–73. https://doi.org/10.3164/jcbn.14-81
200. Kawada C, Yoshida T, Yoshida H, Matsuoka R, Sakamoto W, Odanaka W, Sato T, Yamasaki T, Kanemitsu T, Masuda Y, Urushibata O (2014) Ingested hyaluronan moisturizes dry skin. Nutr J 13(1):70–70. https://doi.org/10.1186/1475-2891-13-70
201. Cyphert JM, Trempus CS, Garantziotis S (2015) Size matters: molecular weight specificity of Hyaluronan effects in cell biology. Int J Cell Biol 2015:1–8. https://doi.org/10.1155/2015/563818

202. Simental-Mendía M, Sánchez-García A, Vilchez-Cavazos F, Acosta-Olivo CA, Peña-Martínez VM, Simental-Mendía LE (2018) Effect of glucosamine and chondroitin sulfate in symptomatic knee osteoarthritis: a systematic review and meta-analysis of randomized placebo-controlled trials. Rheumatol Int 38 (8):1413–1428. doi:https://doi.org/10.1007/s00296-018-4077-2
203. Jin J, Tilve S, Huang Z, Zhou L, Geller H, Yu P (2018) Effect of chondroitin sulfate proteoglycans on neuronal cell adhesion, spreading and neurite growth in culture. Neural Regen Res 13(2):289–297. https://doi.org/10.4103/1673-5374.226398
204. Shida M, Mikami T, Tamura J-I, Kitagawa H (2017) A characteristic chondroitin sulfate trisaccharide unit with a sulfated fucose branch exhibits neurite outgrowth-promoting activity: novel biological roles of fucosylated chondroitin sulfates isolated from the sea cucumber Apostichopus japonicus. Biochem Biophys Res Commun 487(3):678–683. https://doi.org/10.1016/j.bbrc.2017.04.114
205. Kamermans A, Planting KE, Jalink K, van Horssen J, de Vries HE (2019) Reactive astrocytes in multiple sclerosis impair neuronal outgrowth through TRPM7-mediated chondroitin sulfate proteoglycan production. Glia 67(1):68–77. https://doi.org/10.1002/glia.23526
206. Wisowski G, Koźma EM, Bielecki T, Pudełko A, Olczyk K (2017) Dermatan sulfate is a player in the transglutaminase 2 interaction network. PLoS One 12(2):e0172263. https://doi.org/10.1371/journal.pone.0172263
207. Mizumoto S, Kosho T, Yamada S, Sugahara K (2017) Pathophysiological significance of dermatan sulfate proteoglycans revealed by human genetic disorders. Pharmaceuticals 10 (4):34–34. https://doi.org/10.3390/ph10020034
208. Biran R, Pond D (2017) Heparin coatings for improving blood compatibility of medical devices. Adv Drug Deliv Rev 112:12–23. https://doi.org/10.1016/j.addr.2016.12.002
209. Linhardt RJ (2016) Heparin and anticoagulation. Front Biosci 21(7):4462–4462. https://doi.org/10.2741/4462
210. Tykesson E, Maccarana M, Thorsson H, Liu J, Malmström A, Ellervik U, Westergren-Thorsson G (2019) Recombinant dermatan sulfate is a potent activator of heparin cofactor II-dependent inhibition of thrombin. Glycobiology 29(6):446–451. https://doi.org/10.1093/glycob/cwz019
211. Gao W, Xu Y, Liu J, Ho M (2016) Epitope mapping by a Wnt-blocking antibody: evidence of the Wnt binding domain in heparan sulfate. Sci Rep 6(1):26245–26245. https://doi.org/10.1038/srep26245
212. Yin Y, Wang A, Feng L, Wang Y, Zhang H, Zhang I, Bany BM, Ma L (2018) Heparan sulfate proteoglycan Sulfation regulates uterine differentiation and signaling during embryo implantation. Endocrinology 159(6):2459–2472. https://doi.org/10.1210/en.2018-00105
213. Pirard D, Vereecken P, Mélot C, Heenen M (2005) Three percent diclofenac in 2.5% hyaluronan gel in the treatment of actinic keratoses: a meta-analysis of the recent studies. Arch Dermatol Res 297(5):185–189. https://doi.org/10.1007/s00403-005-0601-9
214. Aya KL, Stern R (2014) Hyaluronan in wound healing: rediscovering a major player. Wound Repair Regen 22(5):579–593. https://doi.org/10.1111/wrr.12214
215. Tagliagambe M, Elstrom TA, Ward DB (2017) Hyaluronic acid sodium salt 0.2% gel in the treatment of a recalcitrant distal leg ulcer: a case report. J Clin Aesthetic Dermatol 10 (11):49–51
216. Liesegang TJ (1990) Viscoelastic substances in ophthalmology. Surv Ophthalmol 34 (4):268–293. https://doi.org/10.1016/0039-6257(90)90027-S
217. Bowman S, Awad ME, Hamrick MW, Hunter M, Fulzele S (2018) Recent advances in hyaluronic acid based therapy for osteoarthritis. Clin Transl Med 7(1). https://doi.org/10.1186/s40169-017-0180-3
218. Sun SF, Hsu CW, Lin HS, Liou IH, Chen YH, Hung CL (2017) Comparison of single intra-articular injection of novel Hyaluronan (HYA-JOINT plus) with synvisc-one for knee osteoarthritis: a randomized, controlled, double-blind trial of efficacy and safety. J Bone Joint Surg (Am Vol) 99(6):462–471. https://doi.org/10.2106/JBJS.16.00469

219. Singh JA, Noorbaloochi S, MacDonald R, Maxwell LJ (2015) Singh JA (ed) Chondroitin for osteoarthritis, vol 176. Wiley, Chichester, pp 139–148. https://doi.org/10.1002/14651858.CD005614.pub2
220. Volpi N (2002) Oral bioavailability of chondroitin sulfate (Condrosulf®) and its constituents in healthy male volunteers. Osteoarthr Cartil 10(10):768–777. https://doi.org/10.1053/joca.2002.0824
221. Saltissi D, Morgan C, Westhuyzen J, Healy H (1999) Comparison of low-molecular-weight heparin (enoxaparin sodium) and standard unfractionated heparin for haemodialysis anticoagulation. Nephrol Dial Transplant 14(11):2698–2703. https://doi.org/10.1093/ndt/14.11.2698
222. Vitale C, Berutti S, Bagnis C, Soragna G, Gabella P, Fruttero C, Marangella M (2013) Dermatan sulfate: an alternative to unfractionated heparin for anticoagulation in hemodialysis patients. J Nephrol 26(1):158–163. https://doi.org/10.5301/jn.5000105
223. Hayashi T, Takatori H, Horii R, Nio K, Terashima T, Iida N, Kitahara M, Shimakami T, Arai K, Kitamura K, Kawaguchi K, Yamashita T, Sakai Y, Yamashita T, Mizukoshi E, Honda M, Toyama T, Okumura K, Kozaka K, Kaneko S (2019) Danaparoid sodium-based anticoagulation therapy for portal vein thrombosis in cirrhosis patients. BMC Gastroenterol 19(1):217–217. https://doi.org/10.1186/s12876-019-1140-8
224. Rusnati M, Lembo D (2016) Heparan sulfate proteoglycans: a multifaceted target for novel approaches in antiviral drug discovery. J Bioeng Biomed Sci 6(2):6–8. https://doi.org/10.4172/2155-9538.1000177
225. Modhiran N, Gandhi NS, Wimmer N, Cheung S, Stacey K, Young PR, Ferro V, Watterson D (2019) Dual targeting of dengue virus virions and NS1 protein with the heparan sulfate mimic PG545. Antivir Res 168(April):121–127. https://doi.org/10.1016/j.antiviral.2019.05.004
226. Barritault D, Gilbert-Sirieix M, Rice KL, Siñeriz F, Papy-Garcia D, Baudouin C, Desgranges P, Zakine G, Saffar J-L, van Neck J (2017) RGTA® or ReGeneraTing agents mimic heparan sulfate in regenerative medicine: from concept to curing patients. Glycoconj J 34(3):325–338. https://doi.org/10.1007/s10719-016-9744-5
227. Ayerst BI, Merry CLR, Day AJ (2017) The good the bad and the ugly of Glycosaminoglycans in tissue engineering applications. Pharmaceuticals 10(4):54–54. https://doi.org/10.3390/ph10020054
228. Celikkin N, Rinoldi C, Costantini M, Trombetta M, Rainer A, Święszkowski W (2017) Naturally derived proteins and glycosaminoglycan scaffolds for tissue engineering applications. Mater Sci Eng C 78:1277–1299. https://doi.org/10.1016/j.msec.2017.04.016
229. Kim M, Erickson IE, Choudhury M, Pleshko N, Mauck RL (2012) Transient exposure to TGF-β3 improves the functional chondrogenesis of MSC-laden hyaluronic acid hydrogels. J Mech Behav Biomed Mater 11:92–101. https://doi.org/10.1016/j.jmbbm.2012.03.006
230. Bhakta G, Rai B, Lim ZXH, Hui JH, Stein GS, van Wijnen AJ, Nurcombe V, Prestwich GD, Cool SM (2012) Hyaluronic acid-based hydrogels functionalized with heparin that support controlled release of bioactive BMP-2. Biomaterials 33(26):6113–6122. https://doi.org/10.1016/j.biomaterials.2012.05.030
231. Cai Z, Gu Y, Cheng J, Li J, Xu Z, Xing Y, Wang C, Wang Z (2019) Decellularization, cross-linking and heparin immobilization of porcine carotid arteries for tissue engineering vascular grafts. Cell Tissue Bank 20(4):569–578. https://doi.org/10.1007/s10561-019-09792-5
232. Silva JM, Georgi N, Costa R, Sher P, Reis RL, van Blitterswijk CA, Karperien M, Mano JF (2013) Nanostructured 3D constructs based on chitosan and chondroitin Sulphate multilayers for cartilage tissue engineering. PLoS One 8(2). https://doi.org/10.1371/journal.pone.0055451
233. Grand View R (2020) Hyaluronic acid market size worth $16.6 billion by 2027. CAGR: 8.1%
234. iHealthcareAnaylyst (2020) Global heparin anticoagulant market $14.6 billion by 2027
235. Grand View R (2019) Chondroitin sulfate market size, share & trends analysis report by source (synthetic, bovine, swine, poultry, shark), by application (Nutraceuticals, pharmaceuticals, animal feed, personal care), and segment forecasts, 2019–2025

236. Badri A, Williams A, Linhardt RJ, Koffas MAG (2018) The road to animal-free glycosaminoglycan production: current efforts and bottlenecks. Curr Opin Biotechnol 53:85–92. https://doi.org/10.1016/j.copbio.2017.12.018
237. Sze JH, Brownlie JC, Love CA (2016) Biotechnological production of hyaluronic acid: a mini review. 3 Biotech 6(1):1–9. https://doi.org/10.1007/s13205-016-0379-9
238. Guerrini M, Beccati D, Shriver Z, Naggi A, Viswanathan K, Bisio A, Capila I, Lansing JC, Guglieri S, Fraser B, Al-Hakim A, Gunay NS, Zhang Z, Robinson L, Buhse L, Nasr M, Woodcock J, Langer R, Venkataraman G, Linhardt RJ, Casu B, Torri G, Sasisekharan R (2008) Oversulfated chondroitin sulfate is a contaminant in heparin associated with adverse clinical events. Nat Biotechnol 26(6):669–675. https://doi.org/10.1038/nbt1407
239. Szajek AY, Chess E, Johansen K, Gratzl G, Gray E, Keire D, Linhardt RJ, Liu J, Morris T, Mulloy B, Nasr M, Shriver Z, Torralba P, Viskov C, Williams R, Woodcock J, Workman W, Al-Hakim A (2016) The US regulatory and pharmacopeia response to the global heparin contamination crisis. Nat Biotechnol 34(6):625–630. https://doi.org/10.1038/nbt.3606
240. Datta P, Linhardt RJ, Sharfstein ST (2019) Industrial production of glycosaminoglycans. Encycl microbiol:681–690. https://doi.org/10.1016/B978-0-12-809633-8.12224-1
241. Zhang J, Ding X, Yang L, Kong Z (2006) A serum-free medium for colony growth and hyaluronic acid production by Streptococcus zooepidemicus NJUST01. Appl Microbiol Biotechnol 72(1):168–172. https://doi.org/10.1007/s00253-005-0253-x
242. Xiong J, Bhaskar U, Li G, Fu L, Li L, Zhang F, Dordick JS, Linhardt RJ (2013) Immobilized enzymes to convert N-sulfo, N-acetyl heparosan to a critical intermediate in the production of bioengineered heparin. J Biotechnol 167(3):241–247. https://doi.org/10.1016/j.jbiotec.2013.06.018
243. Chen X, Chen R, Yu X, Tang D, Yao W, Gao X (2017) Metabolic engineering of Bacillus subtilis for biosynthesis of heparosan using heparosan synthase from Pasteurella multocida, PmHS1. Bioprocess Biosyst Eng 40(5):675–681. https://doi.org/10.1007/s00449-016-1732-4
244. He W, Fu L, Li G, Andrew Jones J, Linhardt RJ, Koffas M (2015) Production of chondroitin in metabolically engineered E. coli. Metab Eng 27:92–100. https://doi.org/10.1016/j.ymben.2014.11.003
245. Jin P, Zhang L, Yuan P, Kang Z, Du G, Chen J (2016) Efficient biosynthesis of polysaccharides chondroitin and heparosan by metabolically engineered Bacillus subtilis. Carbohydr Polym 140:424–432. https://doi.org/10.1016/j.carbpol.2015.12.065
246. Zhou Z, Li Q, Huang H, Wang H, Wang Y, Du G, Chen J, Kang Z (2018) A microbial-enzymatic strategy for producing chondroitin sulfate glycosaminoglycans. Biotechnol Bioeng 115(6):1561–1570. https://doi.org/10.1002/bit.26577
247. DeAngelis PL (2012) Glycosaminoglycan polysaccharide biosynthesis and production: today and tomorrow. Appl Microbiol Biotechnol 94(2):295–305. https://doi.org/10.1007/s00253-011-3801-6
248. DeAngelis PL, Jing W, Drake RR, Achyuthan AM (1998) Identification and molecular cloning of a unique hyaluronan synthase from Pasteurella multocida. J Biol Chem 273(14):8454–8458. https://doi.org/10.1074/jbc.273.14.8454
249. Kobayashi S, Fujikawa S-I, Ohmae M (2003) Enzymatic synthesis of chondroitin and its derivatives catalyzed by Hyaluronidase. J Am Chem Soc 125(47):14357–14369. https://doi.org/10.1021/ja036584x
250. Lane RS, Ange KS, Zolghadr B, Liu X, Schäffer C, Linhardt RJ, DeAngelis PL (2017) Expanding glycosaminoglycan chemical space: towards the creation of sulfated analogs, novel polymers and chimeric constructs. Glycobiology 27(7):646–656. https://doi.org/10.1093/glycob/cwx021
251. Wang Y, Li S, Xu X, Tan Y, Liu X-W, Fang J (2020) Chemoenzymatic synthesis of homogeneous chondroitin polymers and its derivatives. Carbohydr Polym 232 (2019):115822–115822. https://doi.org/10.1016/j.carbpol.2019.115822

252. Zhang L, Huang H, Wang H, Chen J, Du G, Kang Z (2016) Rapid evolution of hyaluronan synthase to improve hyaluronan production and molecular mass in Bacillus subtilis. Biotechnol Lett 38(12):2103–2108. https://doi.org/10.1007/s10529-016-2193-1
253. Fu X, Shang W, Wang S, Liu Y, Qu J, Chen X, Wang PG, Fang J (2017) A general strategy for the synthesis of homogeneous hyaluronan conjugates and their biological applications. Chem Commun 53(25):3555–3558. https://doi.org/10.1039/C6CC09431G
254. Li S, Wang S, Fu X, Liu XW, Wang PG, Fang J (2017) Sequential one-pot multienzyme synthesis of hyaluronan and its derivative. Carbohydr Polym 178:221–227. https://doi.org/10.1016/j.carbpol.2017.09.041
255. Chavaroche AAE, van den Broek LAM, Springer J, Boeriu C, Eggink G (2011) Analysis of the polymerization initiation and activity of *Pasteurella multocida* Heparosan synthase PmHS2, an enzyme with Glycosyltransferase and UDP-sugar hydrolase activity. J Biol Chem 286(3):1777–1785. https://doi.org/10.1074/jbc.M110.136754
256. Ninomiya T, Sugiura N, Tawada A, Sugimoto K, Watanabe H, Kimata K (2002) Molecular cloning and characterization of chondroitin polymerase from *Escherichia coli* strain K4. J Biol Chem 277(24):21567–21575. https://doi.org/10.1074/jbc.M201719200
257. Osawa T, Sugiura N, Shimada H, Hirooka R, Tsuji A, Shirakawa T, Fukuyama K, Kimura M, Kimata K, Kakuta Y (2009) Crystal structure of chondroitin polymerase from *Escherichia coli* K4. Biochem Biophys Res Commun 378(1):10–14
258. Chen Y, Li Y, Yu H, Sugiarto G, Thon V, Hwang J, Ding L, Hie L, Chen X (2013) Tailored design and synthesis of Heparan sulfate oligosaccharide analogues using sequential one-pot multienzyme systems. Angew Chem Int Ed 52(45):11852–11856. https://doi.org/10.1002/anie.201305667
259. Bhaskar U, Li G, Fu L, Onishi A, Suflita M, Dordick JS, Linhardt RJ (2015) Combinatorial one-pot chemoenzymatic synthesis of heparin. Carbohydr Polym 122:399–407. https://doi.org/10.1016/j.carbpol.2014.10.054
260. Bode L (2012) Human milk oligosaccharides: every baby needs a sugar mama. Glycobiology 22(9):1147–1162. https://doi.org/10.1093/glycob/cws074
261. Morrow AL, Yu Y (2017) Chapter 7 – potential public health impact of human Milk oligosaccharides. In: McGuire MK, McGuire MA, Bode L (eds) Prebiotics and probiotics in human Milk. Academic Press, San Diego, pp 207–222. https://doi.org/10.1016/B978-0-12-802725-7.00007-5
262. Bych K, Mikš MH, Johanson T, Hederos MJ, Vigsnæs LK, Becker P (2019) Production of HMOs using microbial hosts — from cell engineering to large scale production. Curr Opin Biotechnol 56:130–137. https://doi.org/10.1016/j.copbio.2018.11.003
263. Faijes M, Castejón-Vilatersana M, Val-Cid C, Planas A (2019) Enzymatic and cell factory approaches to the production of human milk oligosaccharides. Biotechnol Adv 37 (5):667–697. https://doi.org/10.1016/j.biotechadv.2019.03.014
264. Bode L, Campbell S, Furneaux R, Beauprez J, Muscroft-Taylor A (2017) Chapter 9 – making human Milk oligosaccharides available for research and application – approaches, challenges, and future opportunities. In: McGuire MK, McGuire MA, Bode L (eds) Prebiotics and probiotics in human Milk. Academic Press, San Diego, pp 251–293. https://doi.org/10.1016/B978-0-12-802725-7.00009-9
265. Ruzic L, Bolivar JM, Nidetzky B (2020) Glycosynthase reaction meets the flow: continuous synthesis of lacto-N-triose II by engineered β-hexosaminidase immobilized on solid support. Biotechnol Bioeng 117(5):1597–1602. https://doi.org/10.1002/bit.27293
266. Schmölzer K, Weingarten M, Baldenius K, Nidetzky B (2019) Glycosynthase principle transformed into biocatalytic process technology: lacto- N-triose II production with engineered exo-Hexosaminidase. ACS Catal 9(6):5503–5514. https://doi.org/10.1021/acscatal.9b01288
267. Industrial Enzyme Applications (2019) Industrial enzyme applications. Wiley-V C H Verlag Gmbh, Weinheim. https://doi.org/10.1002/9783527813780
268. Seeberger PH (2015) The logic of automated glycan assembly. Acc Chem Res 48 (5):1450–1463. https://doi.org/10.1021/ar5004362

269. Li T, Liu L, Wei N, Yang J-Y, Chapla DG, Moremen KW, Boons G-J (2019) An automated platform for the enzyme-mediated assembly of complex oligosaccharides. Nat Chem 11 (3):229–236. https://doi.org/10.1038/s41557-019-0219-8
270. Council NR (2012) Transforming Glycoscience: a roadmap for the future. The National Academies Press, Washington. https://doi.org/10.17226/13446
271. Panza M, Pistorio SG, Stine KJ, Demchenko AV (2018) Automated chemical oligosaccharide synthesis: novel approach to traditional challenges. Chem Rev 118(17):8105–8150. https://doi.org/10.1021/acs.chemrev.8b00051
272. Wen L, Edmunds G, Gibbons C, Zhang J, Gadi MR, Zhu H, Fang J, Liu X, Kong Y, Wang PG (2018) Toward automated enzymatic synthesis of oligosaccharides. Chem Rev 118 (17):8151–8187. https://doi.org/10.1021/acs.chemrev.8b00066
273. Fair RJ, Hahm HS, Seeberger PH (2015) Combination of automated solid-phase and enzymatic oligosaccharide synthesis provides access to α(2,3)-sialylated glycans. Chem Commun 51(28):6183–6185. https://doi.org/10.1039/C5CC01368B
274. Hahm HS, Schlegel MK, Hurevich M, Eller S, Schuhmacher F, Hofmann J, Pagel K, Seeberger PH (2017) Automated glycan assembly using the Glyconeer 2.1 synthesizer. Proc Natl Acad Sci 114(17):E3385–E3389. https://doi.org/10.1073/pnas.1700141114
275. Zhang J, Chen C, Gadi MR, Gibbons C, Guo Y, Cao X, Edmunds G, Wang S, Liu D, Yu J, Wen L, Wang PG (2018) Machine-driven enzymatic oligosaccharide synthesis by using a peptide synthesizer. Angew Chem Int Ed 57(51):16638–16642. https://doi.org/10.1002/anie.201810661
276. Guberman M, Seeberger PH (2019) Automated glycan assembly: a perspective. J Am Chem Soc 141(14):5581–5592. https://doi.org/10.1021/jacs.9b00638
277. Hahm HS, Broecker F, Kawasaki F, Mietzsch M, Heilbronn R, Fukuda M, Seeberger PH (2017) Automated glycan assembly of Oligo-N-acetyllactosamine and Keratan sulfate probes to study virus-glycan interactions. Chem 2(1):114–124. https://doi.org/10.1016/j.chempr.2016.12.004
278. Moremen KW, Ramiah A, Stuart M, Steel J, Meng L, Forouhar F, Moniz HA, Gahlay G, Gao Z, Chapla D, Wang S, Yang J-Y, Prabhakar PK, Johnson R, Rosa MD, Geisler C, Nairn AV, Seetharaman J, Wu S-C, Tong L, Gilbert HJ, LaBaer J, Jarvis DL (2018) Expression system for structural and functional studies of human glycosylation enzymes. Nat Chem Biol 14(2):156–162. https://doi.org/10.1038/nchembio.2539
279. Micoli F, Del Bino L, Alfini R, Carboni F, Romano MR, Adamo R (2019) Glycoconjugate vaccines: current approaches towards faster vaccine design. Expert Rev Vaccines 18 (9):881–895. https://doi.org/10.1080/14760584.2019.1657012
280. Rappuoli R (2018) Glycoconjugate vaccines: principles and mechanisms. Sci Transl Med 10 (456):eaat4615. https://doi.org/10.1126/scitranslmed.aat4615
281. Costantino P, Rappuoli R, Berti F (2011) The design of semi-synthetic and synthetic glycoconjugate vaccines. Expert Opin Drug Discovery 6(10):1045–1066. https://doi.org/10.1517/17460441.2011.609554
282. Kay E, Cuccui J, Wren BW (2019) Recent advances in the production of recombinant glycoconjugate vaccines. NPJ Vaccines 4(1):16. https://doi.org/10.1038/s41541-019-0110-z
283. MacCalman TE, Phillips-Jones MK, Harding SE (2019) Glycoconjugate vaccines: some observations on carrier and production methods. Biotechnol Genet Eng Rev 35(2):93–125. https://doi.org/10.1080/02648725.2019.1703614
284. Oldrini D, Fiebig T, Romano MR, Proietti D, Berger M, Tontini M, De Ricco R, Santini L, Morelli L, Lay L, Gerardy-Schahn R, Berti F, Adamo R (2018) Combined chemical synthesis and tailored enzymatic elongation provide fully synthetic and conjugation-ready Neisseria meningitidis Serogroup X vaccine antigens. ACS Chem Biol 13(4):984–994. https://doi.org/10.1021/acschembio.7b01057
285. Mahour R, Klapproth J, Rexer TFT, Schildbach A, Klamt S, Pietzsch M, Rapp E, Reichl U (2018) Establishment of a five-enzyme cell-free cascade for the synthesis of uridine diphosphate N-acetylglucosamine. J Biotechnol 283:120–129. https://doi.org/10.1016/j.jbiotec.2018.07.027

286. Cymer F, Beck H, Rohde A, Reusch D (2018) Therapeutic monoclonal antibody N-glycosylation – structure, function and therapeutic potential. Biologicals 52:1–11. https://doi.org/10.1016/j.biologicals.2017.11.001
287. Van Landuyt L, Lonigro C, Meuris L, Callewaert N (2019) Customized protein glycosylation to improve biopharmaceutical function and targeting. Curr Opin Biotechnol 60:17–28. https://doi.org/10.1016/j.copbio.2018.11.017
288. Wang Z, Zhu J, Lu H (2020) Antibody glycosylation: impact on antibody drug characteristics and quality control. Appl Microbiol Biotechnol 104(5):1905–1914. https://doi.org/10.1007/s00253-020-10368-7
289. Jefferis R (2016) Posttranslational modifications and the immunogenicity of biotherapeutics. J Immunol Res 2016. https://doi.org/10.1155/2016/5358272
290. Alter G, Ottenhoff THM, Joosten SA (2018) Antibody glycosylation in inflammation, disease and vaccination. Semin Immunol 39:102–110. https://doi.org/10.1016/j.smim.2018.05.003
291. Liu L (2015) Antibody glycosylation and its impact on the pharmacokinetics and pharmacodynamics of monoclonal antibodies and fc-fusion proteins. J Pharm Sci 104(6):1866–1884. https://doi.org/10.1002/jps.24444
292. Beck A, Reichert JM (2012) Marketing approval of mogamulizumab: a triumph for glycoengineering. MAbs 4(4):419–425. https://doi.org/10.4161/mabs.20996
293. Wang Q, Chung C-Y, Chough S, Betenbaugh MJ (2018) Antibody glycoengineering strategies in mammalian cells. Biotechnol Bioeng 115(6):1378–1393. https://doi.org/10.1002/bit.26567
294. Mastrangeli R, Palinsky W, Bierau H (2018) Glycoengineered antibodies: towards the next-generation of immunotherapeutics. Glycobiology 29(3):199–210. https://doi.org/10.1093/glycob/cwy092
295. Malik S, Thomann M (2016) In vitro glycoengineering – suitability for BioPharmamanufacturing. Application Note p 8
296. Thomann M, Schlothauer T, Dashivets T, Malik S, Avenal C, Bulau P, Rüger P, Reusch D (2015) In vitro glycoengineering of IgG1 and its effect on fc receptor binding and ADCC activity. PLoS One 10(8). https://doi.org/10.1371/journal.pone.0134949
297. Li C, Wang L-X (2018) Chemoenzymatic methods for the synthesis of glycoproteins. Chem Rev 118(17):8359–8413. https://doi.org/10.1021/acs.chemrev.8b00238
298. Wang L-X, Tong X, Li C, Giddens JP, Li T (2019) Glycoengineering of antibodies for modulating functions. Annu Rev Biochem 88(1):433–459. https://doi.org/10.1146/annurev-biochem-062917-012911
299. Giddens JP, Lomino JV, DiLillo DJ, Ravetch JV, Wang LX (2018) Site-selective chemoenzymatic glycoengineering of fab and fc glycans of a therapeutic antibody. Proc Natl Acad Sci U S A 115(47):12023–12027. https://doi.org/10.1073/pnas.1812833115
300. Ramírez AS, Boilevin J, Biswas R, Gan BH, Janser D, Aebi M, Darbre T, Reymond J-L, Locher KP (2017) Characterization of the single-subunit oligosaccharyltransferase STT3A from Trypanosoma brucei using synthetic peptides and lipid-linked oligosaccharide analogs. Glycobiology 27(6):525–535. https://doi.org/10.1093/glycob/cwx017
301. Jaffee MB, Imperiali B (2013) Optimized protocol for expression and purification of membrane-bound PglB, a bacterial oligosaccharyl transferase. Protein Expr Purif 89 (2):241–250. https://doi.org/10.1016/j.pep.2013.04.001
302. Boilevin JM, Reymond JL (2018) Synthesis of lipid-linked oligosaccharides (LLOs) and their Phosphonate analogues as probes to study protein glycosylation enzymes. Synthesis 50 (14):2631–2654. https://doi.org/10.1055/s-0037-1609735
303. Tsai TI, Li ST, Liu CP, Chen KY, Shivatare SS, Lin CW, Liao SF, Lin CW, Hsu TL, Wu YT, Tsai MH, Lai MY, Lin NH, Wu CY, Wong CH (2017) An effective bacterial fucosidase for glycoprotein remodeling. ACS Chem Biol 12(1):63–72. https://doi.org/10.1021/acschembio.6b00821

304. Li C, Li T, Wang LX (2018) Chemoenzymatic defucosylation of therapeutic antibodies for enhanced effector functions using bacterial α-fucosidases. Methods Mol Biol 1827. https://doi.org/10.1007/978-1-4939-8648-4_19
305. Sun B, Bao W, Tian X, Li M, Liu H, Dong J, Huang W (2014) A simplified procedure for gram-scale production of sialylglycopeptide (SGP) from egg yolks and subsequent semi-synthesis of Man3GlcNAc oxazoline. Carbohydr Res 396:62–69. https://doi.org/10.1016/j.carres.2014.07.013
306. Liu L, Prudden AR, Bosman GP, Boons GJ (2017) Improved isolation and characterization procedure of sialylglycopeptide from egg yolk powder. Carbohydr Res 452:122–128. https://doi.org/10.1016/j.carres.2017.10.001
307. Tayi VS, Butler M (2018) Solid-phase enzymatic remodeling produces high yields of single Glycoform antibodies. Biotechnol J 13(4). https://doi.org/10.1002/biot.201700381

# Recombinant Proteins and Monoclonal Antibodies

**Roy Jefferis**

## Contents



**Abstract** The human genome has become a subject of public interest, whilst the proteome remains the province of specialists. Less appreciated is the human glycoprotein (GP) repertoire (proteoglycome!); however, some 50% of open reading frame genes encode for proteins (P) that may accept the addition of *N*-linked

The original version of this chapter was revised: Graphical Abstract included as an additional correction in this Chapter.

R. Jefferis (✉)
Institute of Immunology and Immunotherapy, College of Medical and Dental Sciences, University of Birmingham, Birmingham B15 2TT, UK
e-mail: R.Jefferis@bham.ac.uk

and/or *O*-linked sugar chains (oligosaccharides). It is established that the attachment of defined oligosaccharide structures impacts mechanisms of action (MoAs), pharmacokinetics, pharmacodynamics, etc., and is a critical quality attribute (CQA) for recombinant GP therapeutics. The oligosaccharide structure attached at a given site may exhibit structural heterogeneity, and individual structures (glycoforms) may modulate MoAs. The biopharmaceutical industry is challenged, therefore, to produce recombinant GP therapeutics that have structural fidelity to the natural (endogenous) molecule, in non-human cells. Multiple production platforms have been developed that, in addition to the natural glycoform, may produce unnatural glycoforms, including sugar residues that can be immunogenic in human subjects. Following a general introduction to the field, this review discusses glycosylation of recombinant monoclonal antibodies (mAbs), the contribution of glycoforms to MoAs and the development of customised mAb therapeutic glycoforms to optimise MoAs for individual disease indications.

## Graphical Abstract

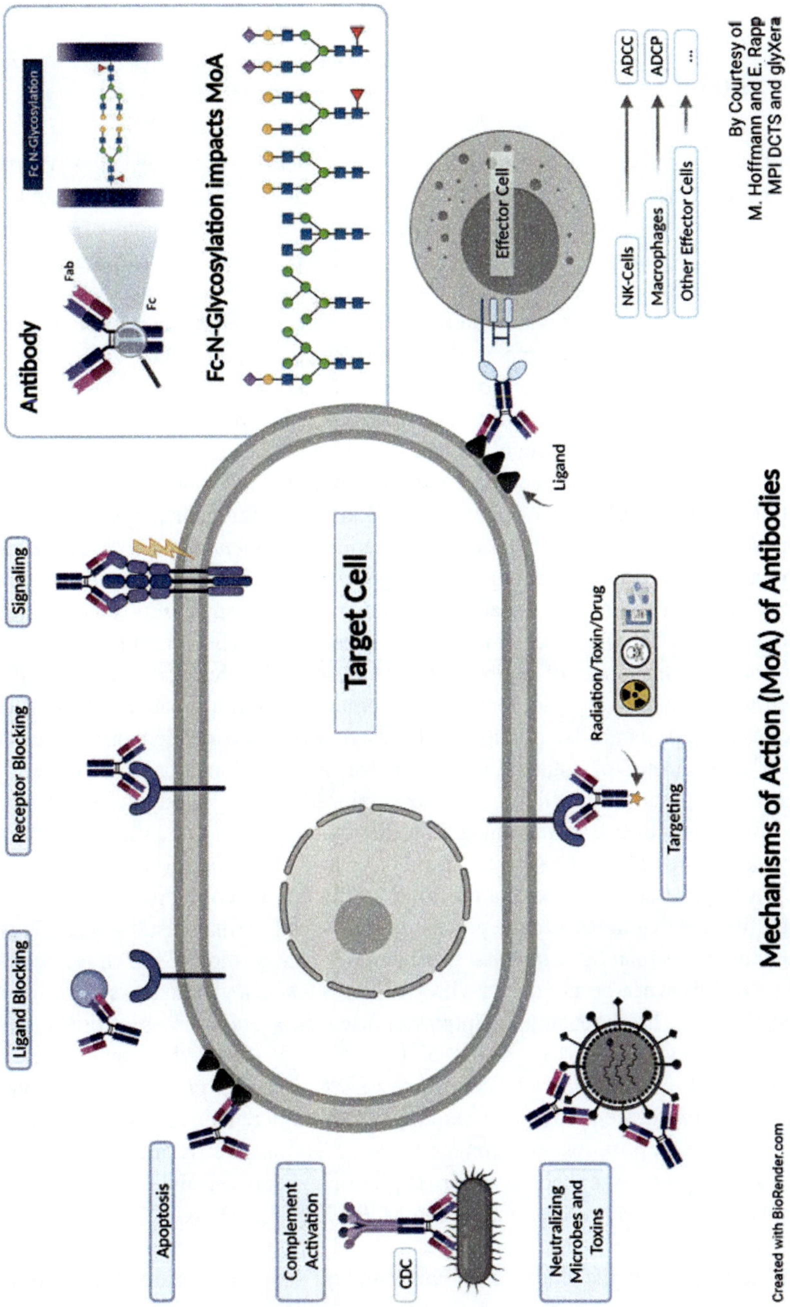

**Keywords** Critical quality attributes, Glycoforms, Glycoproteins, IgG subclasses, Mechanisms of action, Oligosaccharides, Recombinant antibody therapeutics

## 1 Introduction

The moment when life begins can be defined in various ways, depending on physiological evidence and/or spiritual conviction; however, conception (i.e., fusion of a sperm with an oocyte) is a prerequisite. The initial event leading to fusion is recognition of glycoproteins (GPs; proteins with attached chains of sugars, oligosaccharides) present on the surface of the oocyte by receptors expressed on the head (acrosome) of the sperm. The oocyte of metazoans is surrounded by a translucent matrix, the *zona pellucida* (ZP), composed of four glycoproteins designated ZP1, ZP2, ZP3, and ZP4. Interactions between receptors on the acrosome and the ZP GPs activate the release of enzymes that break down the matrix, allowing passage of the sperm nucleus into the oocyte [1]. An oligosaccharide ("oligo" meaning "few" and "saccharide" meaning a "chain of sugars") present on a GP can be linked to a nitrogen atom of an asparagine residue (*N*-linked oligosaccharide) or an oxygen atom of serine, threonine, or tyrosine (*O*-linked oligosaccharide). Carbohydrates/oligosaccharides are essential macromolecules for the growth and survival of living organisms, together with lipids, proteins, and nucleic acids.

Protein receptors that selectively bind individual sugar molecules, expressed within oligosaccharides, are collectively termed lectins (from Latin *legere*, meaning "to select"). One family of lectins is characterized by the presence of a $Ca^+$ ion (C-type lectins) in the carbohydrate recognition domain (CRD); a broader family of lectins express C-type lectin-like domains (CTLDs) that are not dependent on the presence of a $Ca^+$ ion for binding sugars. Lectin–oligosaccharide interactions contribute to cell–cell interactions, cell trafficking, glycoprotein turnover, etc. Endogenous lectins are essential components of the innate immune system and specifically bind exogenous glycans expressed on the surface of infective microorganisms (bacteria, yeasts, etc.) [2]. It follows that absence of a machinery effecting glycosylation is not compatible with life and that defects in the process of glycosylation may result in pathology. For humans, more than 80 congenital disorders of glycosylation (CDG) have been identified and shown to be associated with symptoms that can vary in severity from mild to disabling or life-threatening (http://rarediseases.org/rare-diseases/congenital-disorders-of-glycosylation/) [3]. About 500 genes (0.5–1% of the transcribed human genome) have been shown to contribute to glycosylation processes; therefore, it is likely that further genetic defects leading to pathology remain to be discovered (http://rarediseases.org/rare-diseases/congenital-disorders-of-glycosylation/) [3].

Of the proteins encoded within the human genome, about 50% include the sequence asparagine–X–serine/threonine (N-X-S/T), where X is any amino acid other than proline. The sequence is termed the glycosylation sequon and is a potential site for the addition of an *N*-linked oligosaccharide. Occupancy of a potential site varies according to the local secondary structure formed as the polypeptide is extruded from the ribosome channel. The addition of *O*-linked sugars/oligosaccharides to the hydroxyl groups of serine, threonine, and tyrosine residues takes place as the polypeptide traverses the Golgi apparatus; potential sites for the addition of *O*-linked sugars cannot be predicted from amino acid sequence. Humans utilize nine basic monosaccharides and their derivatives in stereospecific linkages to generate libraries of oligosaccharides. Stereospecificity allows the generation of an estimated repertoire of around $10^{12}$ unique hexasaccharides [4]. It is common for the oligosaccharide attached at a given site to exhibit a degree of structural heterogeneity that varies with cell type, gender, or species in which it is expressed [5]. Thus, the capacity to attach sugars and oligosaccharides to proteins, lipids, etc. extends the diversity of the proteome, generating the proteoglycome and, hence, the complexity and individuality of an organism. The machinery that generates this complexity can be subverted by pathogens. Thus, a virus can exploit the glycosylation machinery of its "host" to disguise itself through the expression of host oligosaccharides. For example, the HIV-1 envelope is covered by a glycan shield of about 90 *N*-linked oligosaccharides, comprising half of its mass, which is a key component of HIV evasion from humoral immunity [6, 7]. Some DNA viruses encode glycosyltransferases that exploit the Golgi apparatus to synthesize and attach unique (non-self) oligosaccharides [8, 9].

## 2 Impact of Glycosylation on Structure and Function

Development of each recombinant GP therapeutic presents a unique challenge because, unlike transcription and translation, glycosylation is a nontemplated process and endogenous GPs may express a heterogeneous glycoform profile that can vary over time and with health or disease. The consensus protein and glycoform structure of an endogenous protein defines critical quality attributes (CQAs) that should be mirrored by a potential recombinant GP therapeutic. A further challenge arises from the necessity to express a potential protein or GP therapeutic within a production platform employing nonhuman cell lines. Such platforms can result in the production of nonhuman glycoforms that can be immunogenic and lead to the generation of antidrug antibodies (ADA). The first recombinant protein therapeutics approved by the US Food and Drug Administration (FDA) were insulin (1982) and interferon 2α (Roferon; 1986), each produced in *Escherichia coli*. Endogenous insulin is a small, 51 amino acid residue (aar), protein that is not glycosylated; however, endogenous interferon 2α (166 aar) bears one *O*-linked oligosaccharide. The absence of the *O*-linked oligosaccharide from this recombinant protein does not appear to compromise its activity, although it may be more susceptible to enzymatic degradation in vivo [9, 10]. Similarly, recombinant forms of granulocyte-colony stimulating factor (G-CSF; 174–177 aar) that naturally bears a single *O*-linked oligosaccharide have been approved both as glycosylated (Lenograstim) and aglycosylated (Filgrastrim) products; the former is produced in CHO cells and the latter in *E. coli* [11, 12]. The related cytokine granulocyte-macrophage colony stimulating factor (GM-CSF) presents a different challenge because, although comprising only ~127 aar, it expresses two potential *N*-linked glycosylation sites and one *O*-linked sugar [13]. The FDA-approved recombinant therapeutics Sargramostim (produced in *Pichia pastoris* yeast cells) and Regramostim (produced in CHO cells) are each composed of a complex mixture of glycoforms. This glycan heterogeneity reflects a lack of specificity in post-translational glycosylation, which has been reported to affect the in vivo properties of the therapeutics [14]. Molgramostim, an aglycosylated form produced in *E. coli*, is approved in Europe, but has been associated with increased adverse side-effects, perhaps caused by its enhanced susceptibility to truncation [14, 15]. A graphic illustration of the impact of glycosylation on function is provided by glycodelin-A, glycodelin-S, glycodelin-F, and glycodelin-C [16, 17]. Glycodelin-S is present in seminal plasma and is essential for sperm capacitation; glycodelins A, F, and C are present in the female reproductive tract and are protective of sperm while attaching to the ovum. Each glycodelin has an identical aar sequence but bears a different glycoform at three potential *N*-linked glycosylation sites [13, 14]. Glycodelins are pleomorphic and exhibit hormonal activity in addition to influencing reproduction [18, 19].

The importance of glycoform fidelity between natural and recombinant GPs was demonstrated during the development of recombinant erythropoietin (EPO). This protein comprises 165 aar and bears one *O*-linked and three *N*-linked oligosaccharides, which account for ~40% of its mass [20, 21]. The principal function of EPO is to promote red cell production, meaning that it is an erythropoiesis stimulating agent (ESA) [22]. The EPO produced in CHO cells was initially shown to exhibit enhanced functional activity relative to the natural product, in vitro. However, trials in vivo demonstrated a lack of functional activity because of rapid degradation and a short half-life. Fractionation of bulk product allowed the isolation of a minor component (epoetin) that proved to be efficacious in vivo and received regulatory approval in 1989.

Glycoform identity between endogenous and recombinant GPs cannot always be achieved; however, in the absence of an approved therapeutic, a product demonstrated to have clinical efficacy may be approved, even without strict comparability. Thus, recombinant antithrombin (ATryn) produced in transgenic goats was approved although the glycoform profile differs from that of the natural product [23, 24]. At the time of its approval, this was the only effective therapeutic available. A different regulatory decision is exemplified for recombinant forms of aglucosidase alpha in the treatment of Pompe disease, a lysozyme storage disease. A recombinant form (Myozyme), produced in a small scale bioreactor (160 L), was approved in 2006 and its clinical success led to a demand that exceeded production capacity. Production was scaled-up to 2,000 L; however, the FDA declined approval for the product to be marketed as Myozyme because of a difference in glycoform profile. A new BLA (Biologics License Application) was submitted and approved, but the product had to be marketed under a different brand name (Lumizyme) [25–27]. The mechanism of action (MoA) requires that these drugs express terminal mannose residues to enable entry into macrophages via the mannose receptor. The primary drug substance does not express terminal mannose residues; therefore, it is exposed to glucosidases (neuraminidase, β galactosidase, and β hexosaminidase) in vitro to generate product bearing exposed mannose residues [28].

Each endogenous protein or GP may be assigned a dominant physiological role; however, its structure also determines its pharmacokinetic and pharmacodynamic profile (e.g., absorption, distribution, metabolism, catabolism, elimination/excretion). The liver has a major role in catabolism and the turnover of both proteins and GPs. Liver resident asialoglycoprotein receptor (ASR) and the mannose receptor (MR) lectins bind, ingest, and catabolize GPs expressing terminal galactose or mannose sugar residues, respectively [29–31]. Terminal sialic acid residues are naturally subject to loss in vivo, resulting in the exposure of a terminal galactose residue recognized by the ASR. The physiological function and half-life of EPO is dependent on its glycoform and the affinity of binding to the EPO receptor (EOPR) on red blood cells. The short half-life of the original EPO product was primarily a result of the absence of terminal sialic acid residues and, hence, accelerated clearance. The introduction of two additional glycosylation sequons into the EPO gene results in expression of a product (Darbepoeitin) that bears additional highly sialylated oligosaccharides. Reduced affinity for EPOR and increased sialic acid content result in enhanced biologic activity [22]. Thus, lectin receptors can be exploited to target appropriately glycosylated drugs for cellular uptake [32]. Similarly, recombinant coagulation factor VIII (FVIII), gonadotrophin, and tissue plasminogen activator (tPA) exhibit differing catabolic rates depending on the product glycoform profile [33–35]. In the case of tPA, the 570 aar protein has three *N*-linked glycosylation sites at residues N-117, N-184, and N-448; type I and type II tPA are characterized by differences in oligosaccharides expressed at N-117 that influence enzymatic and catabolic activities [35]. Control and/or manipulation of the glycoform profile of recombinant GPs can be achieved by protein and/or glycosylation engineering, selection of the producer cell line, or fine tuning of the culture conditions [36]. Interestingly, the catabolic half-lives of the two proteins found at the highest concentrations in serum, albumin and IgG, are independent of glycoform, albumin being a nonglycosylated protein. They are protected from enzymatic degradation in intracellular vacuoles by binding to the neonatal Fc receptor (FcRn) [37, 38].

With the exception of IgG, the structure and function(s) of recombinant GPs can be compared with those determined for the purified endogenous GPs; however, each monoclonal antibody (mAb) therapeutic has to be independently assessed because each has a unique sequence and specificity for a unique target. The MoA of a mAb depends on the activation of effector activities, which vary with isotype and glycoform. This difficulty is compensated by the opportunity to select and customize each mAb to deliver maximum therapeutic efficacy for a given disease indication. Accordingly, mAb therapeutics are the main focus of the remainder of this review.

## 3 Humoral Immune Response and Recombinant Antibody Therapeutics

The defining property of a protective humoral immune response is its specificity for a given target. This is achieved by the generation and production of antibodies of unique sequence that express a unique antigen binding site (paratope) complementary to a unique structure expressed on the antigen (antigenic determinant, epitope). The human antibody response comprises one or a mixture of nine immunoglobulin (Ig) isotypes, namely the IgM, IgD, and IgE classes together with the four subclasses of IgG (IgG1, IgG2, IgG3 and IgG4) and two of IgA (gA1 and IgA2). Each isotype exhibits unique structural and functional properties. In addition, the genes encoding the IgG and IgA isotypes are polymorphic and inherited as a haplotype [39–42]. The separation of populations over the course of human evolution has resulted in a characteristic distribution of haplotypes among racial groups [42]. The biologic effector mechanisms activated within a protective, polyclonal antibody response differ according to the isotype, or mixture of isotypes, of antibody forming immune complexes (IC). The unique properties of each isotype can be exploited in the development and clinical application of a recombinant mAb therapeutic. Because antibodies are, minimally, divalent and an antigen can express multiple identical epitopes, the structure and size of the IC formed varies according to the antigen/antibody ratio. Although the formation of an IC can immobilize and neutralize an offending "foreign body" (antigen), protection requires that it be removed and destroyed. This is achieved when the IC interacts with soluble and/or cell-borne effector ligands to initiate downstream biologic activities. The IgG antibody class predominates in human blood, equilibrates with the extravascular space, and activates a wide range of effector activities that can result in the killing, removal, and/or destruction of specifically targeted pathogens. To date, all approved recombinant antibody therapeutics have been based on the IgG format.

## 4 Polypeptide Structure of Human IgG

The characteristic $H_2L_2$ (two heavy and two light) four-chain homodimeric structure of IgG antibodies was established in the 1950s and the contributions of Rodney Porter (UK) and Gerald Edelman (USA) recognized with the Nobel Prize in 1972. The Edelman laboratory was the first to publish the complete covalent structure of a monoclonal human IgG1 subclass protein (Eu, IgG1K), isolated from the serum of a patient with multiple myeloma [43]. This protein defines the sequence and enumeration of amino acid residues in both the heavy and light chains for all IgG molecules; for example, asparagine 297 (N-297) is the attachment site for oligosaccharides. The actual residue number of this asparagine varies for each mAb, depending on the length of the heavy chain variable region. At the protein sequence level, the light (~25 kDa) and heavy (~50 kDa) chains are composed of two and four sequence homology regions, respectively, of ~110 amino acid residues (Fig. 1a).

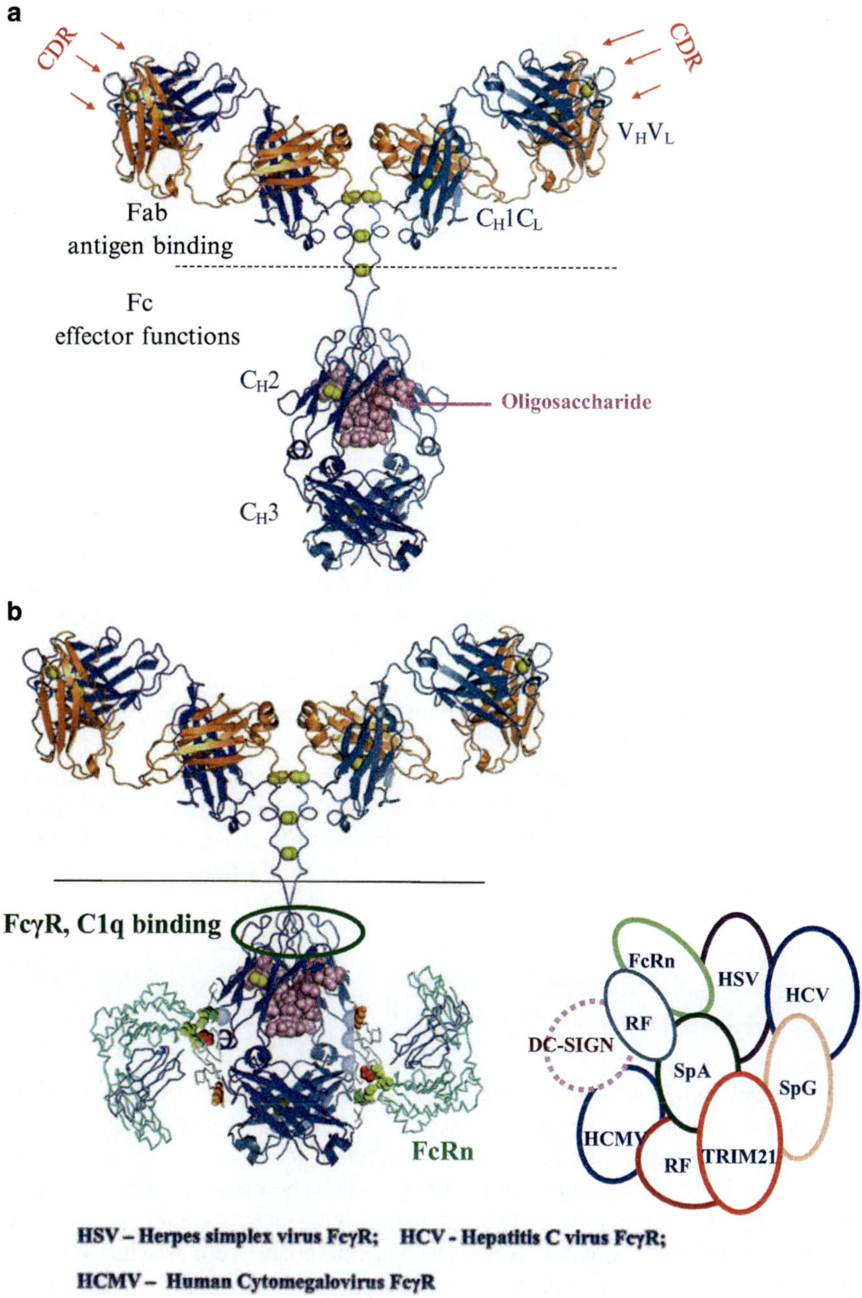


**Fig. 1** (**a**) Alpha carbon backbone structure of an IgG1 molecule. Digestion within the hinge region, by papain, releases the Fab (fragment antigen binding) and Fc (fragment crystallizable) fragments. (**b**) Alpha carbon backbone structure of an IgG1 molecule illustrating ligands binding to overlapping sites at the $C_H2$–$C_H3$ interface. Structures generated by Peter Artymiuk (University of Sheffield, UK) using PyMOL (http://www.pymol.sourceforge.net)

At the gene level, each homology region is encoded within an exon separated by intervening introns. Each homology region folds to form a β-barrel structure composed of two antiparallel β-pleated sheets connected through β-bends and bridged by an intrachain disulfide bond. Hydrophobic side chains are orientated toward the interior, whereas hydrophilic side chains are exposed to solvent [39–44]. This stable protein "scaffold" is referred to as the immunoglobulin fold or domain: It is widely used within the proteome and allows virtually unlimited sequence variation (particularly within the β-bends) and the generation of unique interaction/receptor sites [39–44].

The *N*-terminal variable regions of the light ($L_V$) and heavy ($H_V$) chains differ in length between antibodies, and the unique sequence determines epitope specificity. Maximum sequence diversity is localized within three hypervariable or complementarity-determining regions (CDRs), formed at β-bends, of both the heavy and light chains. The six CDRs are brought into spatial proximity by the immunoglobulin fold to form a unique epitope-binding paratope [39–41]. Humans express two light chain isotypes, kappa (κ) and lambda (λ), and four gamma (γ) IgG heavy chain isotypes or subclasses (γ1, γ2, γ3, γ4), encoded by genes on chromosomes 2, 22, and 14. Each light chain is characterized by one constant homology domain, $C_\kappa$ or $C_\lambda$, and each heavy chain by three constant homology regions, $C_H1$, $C_H2$, and $C_H3$. The $C_\kappa$ and $C_\lambda$ domains each bind with the heavy chain $C_H1$ domain through multiple noncovalent interactions and a single interchain disulfide bridge. Plasma cells express only one heavy chain and one light chain gene to secrete antibodies that are either $H_2\kappa_2$ or $H_2\lambda_2$ homodimers, comprising $[V_H/V_L\text{-}C_K/C_H1\text{-}h\text{-}C_H2\text{-}C_H3]2$ or $[V_H/V_L\text{-}C_\lambda/C_H1\text{-}h\text{-}C_H2\text{-}C_H3]2$ (where h indicates a hinge region) homology regions. Formation of the $H_2L_2$ homodimer is dependent on formation of a single disulfide bridge between the heavy and light chains, multiple interheavy chain disulfide bridges within the hinge region, multiple noncovalent interactions between the $C_H3$ domains, and lateral noncovalent interactions at the $C_H2$–$C_H3$ interface.

## 5 IgG Subclasses

The four human IgG subclasses are enumerated according to their relative concentrations in normal human serum; thus, IgG1, IgG2, IgG3, and IgG4 account for ~60, 25, 10, and 5% of total serum IgG, respectively. Each IgG subclass exhibits a unique profile of biologic effector activities in vitro [39–42]. Therefore, when developing a mAb therapeutic, the choice of IgG subclass is guided by the anticipated MoA in vivo, although the presumption that one can extrapolate from activities demonstrated in vitro to function realized in vivo may be naive. The broad generalization can be made that protein antigens provoke predominantly IgG1 and IgG3 responses, carbohydrate antigens an IgG2 response, and IgG4 responses predominate as a consequence of chronic antigen stimulation [39–42]. Attachment of oligosaccharide at N-297 of the IgG-Fc is essential for full expression of effector functions, and the glycoform profile is a CQA for each therapeutic IgG mAb. The production process that delivers mAbs having a consistent glycoform profile is achieved by the development and practice of quality-by-design (QbD) parameters that are the intellectual property of the innovator company. It is established that ~30% of serum polyclonal IgGs bear *N*-linked oligosaccharides within their V-regions, the glycosylation sequon primarily resulting from somatic hypermutation and selection. The presence of oligosaccharides attached to V-regions can impact paratope specificity and affinity; it can also contribute to the solubility and stability of drug substance and drug product.

## 6 Antigens

Pathogens and self-macromolecules are complex in structure and can present hundreds, if not thousands, of overlapping, nonidentical epitopes to the immune system. The protective human antibody response produces a similarly diverse library of paratopes. Hence, the structure and "architecture" of ICs formed are diverse and influence the MoA. Parameters that contribute to the size/architecture of the ICs formed include: (1) antibody isotype, (2) epitope specificity, (3) Fc glycoform profile, (4) antibody/antigen ratio, (5) valency of the antibody, (6) affinity/avidity of the antibody population, (7) valency or epitope density of the antigen, (8) access and density of effector ligands, (9) cumulative valency when multiple ligands are engaged, and (10) proportions of each antibody isotype present within a polyclonal response [39–45]. This is exemplified by IgG1 subclass anti-CD20 antibody therapeutics having differing epitope specificities that exhibit differing MoAs [46]. Thus, paratope and isotype selection can be used to generate mAbs expressing MoAs deemed appropriate for treatment of given disease manifestations [46–48].

## 7 IgG-Fc Glycosylation Is Essential for Effector Function Activation

The first therapeutic mAb approved by the FDA was Rituximab (Rituxan) in 1998. Rituximab is a chimeric mAb with specificity for the CD20 molecule expressed on normal B cells, but may be overexpressed on the B cells of patients with non-Hodgkin's lymphoma. On administration of Rituximab, the B cells become highly sensitized (opsonized) with the mAb and are targets for IgG-Fc receptor (Fc$\gamma$R)-expressing effector cells and/or the classical complement pathway is activated, with consequent lysis. This "blockbuster" drug has served as a model for glycosylation and protein engineering studies to elucidate structure–function relationships. The understanding achieved is being exploited for the generation of biosimilar and/or "biobetter" analogs. Biobetters can cause either attenuation or reduction in MoAs, depending on the disease indication. A further avenue to improved efficacy of a mAb drug is to extend its half-life by genetic engineering of the IgG-Fc sequence to manipulate the binding affinity for FcRn between pH values of 7.2 and 6.5.

Humans express three classes and six isotypes of FcγR that are coexpressed and/or differentially expressed on multiple leukocyte cell types [39–41, 49–53]. The FcγR types and subtypes are structurally homologous and their engagement by ICs results in activation of one or more MoAs, including antibody-dependent cellular cytotoxicity (ADCC), antibody-dependent cell-mediated phagocytosis (ADCP), release of inflammatory mediators, induction of cellular apoptosis, and regulation of immune function [49–53]. Early studies demonstrating the binding of monomeric IgG or IgG-Fc to the cell surface of leukocytes (monocytes) led to the identification of a receptor referred to as the high-affinity Fc gamma receptor (FcγRI) [39–41]. Subsequent studies identified two low-affinity classes (FcγRII and FcγRIII) and five subtypes (FcγRIIa, FcγRIIb, FcγRIIc FcγRIIIa, and FcγRIIIb); the FcγR gene locus is at chromosome 1q23.3. Polymorphisms of FcγR exist within and between populations [49–52]. Engagement of the FcγR results in positive cellular activation, mediated through the immune-tyrosine activating motif (ITAM). The FcγRIIb receptor is an exception as it delivers an inhibitory activity mediated through the immune-tyrosine inhibitory motif (ITIM) [49–54]. All FcγR, except FcγRIIIb, are transmembrane GPs and the glycoform profile of the ectodomain modulates their activity. There is also evidence that the glycoform profile of each expressed FcγR differs between cell types; FcγRIIIb is a glycosphingolipid membrane-bound molecule.

Although IgG-Fc glycosylation, at N-297, is essential for full effector activity [39–41, 52–56], residual activity can be detected for ICs composed of multiple aglycosylated IgG mAb complexes [45, 56]; thus, cumulative avidity can compensate for low affinity. Comparison of IgG binding (or not binding) to FcγRI in human and other animal species suggested that the IgG1/IgG3 sequence -$_{234}$L-L-G-G$_{237}$- proximal to the hinge region is associated with FcγRI binding [39, 49–52]. Human IgG2 that does not bind FcγRI has the sequence -V$_{234}$-A-G-, with a deletion at 237, whereas IgG4 binds with lower affinity because of a leucine/phenylalanine (L/F) replacement giving the sequence -$_{234}$F-L-G-G$_{237}$- [39–41, 52–56]. Subsequently, extensive protein engineering has been applied in attempts to generate panels of IgG1 proteins exhibiting increased, decreased, and/or selective binding to each of the FcγR types [48–55].

Immune complexes of glycosylated, but not aglycosylated, IgG1 and IgG3 subclass antibodies bind and activate the C1q component of the classical complement system [39–41, 55–57]. Binding triggers a cascade of enzyme cleavage events, with the addition of some complement component breakdown products to the IC. Leucocytes express receptors having specificity for these breakdown products, and their engagement enhances opsonization and phagocytosis or lysis, following the formation of a "membrane attack complex" (MAC). The hydrophobic MAC mediates CDC by insertion into target cellular membranes to form pores that allow ingress and egress of water and small molecules, with consequent loss of integrity and osmotic control. The epitope specificity of a mAb determines the morphology (architecture) of the IC formed and the ability to activate CDC [56, 57].

An important property of mAb drugs, in contrast with small molecule drugs, is their long half-lives in vivo: about 21 days for IgG1, IgG2, and IgG4 and 7 days for IgG3 [37–41, 58–61]. This offers protection over an extended time period, limiting the frequency of attendance at the clinic and reducing the cost of treatment. Catabolism of IgG is mediated through FcRn, which is expressed on the membrane of many cell types. The natural process of pinocytosis results in the uptake of extracellular fluid and the formation of a vacuole lined with membrane-bound FcRn. Subsequent acidification to pH 6.5 promotes the binding of IgG and albumin (present in the ingested fluid) to FcRn and protection from cleavage by enzymes released into the vacuole; unbound IgG and albumin are degraded [58, 59]. When the membrane of the vacuole is re-cycled to the external cellular surface, the IgG/FcRn complex is exposed to extravascular fluid, at pH 7.2, and the IgG is released. Protein engineering has been applied to increase the affinity of a mAb for FcRn at pH 6.5, but not change its release at pH 7.2, to provide preferential protection of mAb relative to the normal IgG present and extension of the half-life [58, 61]. This further enhances therapeutic efficacy and reduces cost, particularly for self-treatment with mAbs formulated at high concentrations. As the name implies, FcRn functions in the transport of IgG from mother to fetus. Transport is initiated in the third trimester; at term, IgG levels in cord blood and the blood of the newborn exceed that of maternal blood [62].

Despite the diversity of the immune response, humans remain subject to infection and consequent disease. This reflects the long coevolution history of humankind within a hostile environment that is constantly changing, sometimes precipitately and at other times over millennia. Chance mutations result in the emergence of structurally altered pathogens that may escape or frustrate immune protection [39–41, 44, 48, 63]. Familiar examples are the production of staphylococcal protein A (SpA) by *Staphylococcus aureus*, and streptococcal protein G (SpG) by streptococcal strains C and G. A simplistic explanation for their MoA is that these bacterial proteins bind nonspecifically to the IgG-Fc of serum polyclonal IgG to masquerade as self. In practice, pathogen–host interactions are more complex; for example, SpA is also a polyclonal B cell activator. The biopharmaceutical industry exploits these bacterial proteins for industrial-scale purification of mAb drug substances. Some viruses have been shown to carry genes that encode proteins that, when expressed on the surface of infected cells, bind the Fc region of serum IgG (i.e., function as pseudo-FcγR). It is posited that the binding of serum IgG to virus-encoded pseudo-FcγRs blocks binding to effector cell FcγR and/or the C1 component of complement. In concert, these interactions frustrate immune clearance. Interestingly, to date, all non-self ligands have been shown to bind IgG-Fc at the $C_H2$–$C_H3$ interface at sites overlapping but not identical to FcRn (Fig. 1b) [39–41, 63].

A continuing problem associated with mAb therapy is the potential for immunogenicity and the development of ADA, which can be neutralizing and/or give rise to adverse reactions on re-exposure to the therapeutic [64–66]. These responses are mostly limited to epitopes expressed by the unique variable region sequences (idiotypes), but attempts to modulate the MoA by protein engineering may create new non-self structures (epitopes) and enhance immunogenicity. It could also compromise relationships between coevolved human pathogens and protective innate and adaptive immune responses. Ideally, a holistic approach should be adopted and any IgG sequence mutant should be evaluated for interactions with all currently identified endogenous and exogenous ligands. Selection between the natural glycoforms of IgG-Fc can impact the MoA but not immunogenicity.

## 8 Glycosylation of IgG-Fc, Derived from Polyclonal Human Serum IgG

Although neutralization of a toxin can provide immediate protection, resolution of an infection requires that the invading organism is removed and destroyed. This is achieved through IC activation of a cascade of downstream biologic mechanisms that constitute the MoA [39–41, 55–58]. *N*-Linked glycosylation of the IgG-Fc is essential for optimal effector ligand binding and activation. Analysis of oligosaccharides released from normal polyclonal human IgG and monoclonal human IgG proteins produced by neoplastic plasma cells (multiple myeloma) reveals a heterogeneous population of diantennary structures. However, each paraprotein analyzed exhibits a unique glycoform profile that appears to be a "signature" of the neoplastic clone; in addition, the profile for each patient can vary between samples analyzed at diagnosis, remission, and relapse [67–69]. Approved mAb drugs are produced in mammalian [CHO (hamster), NS0/Sp2/0 (murine)] cell lines that produce mAb with a restricted IgG-Fc glycoform profile; however, they may also add nonhuman glycoforms. Because glycosylation is essential for expression of the full range of effector functions, efficacy can also vary between different glycoforms. Structural studies have shown that IgG-Fc oligosaccharide (s) impact the tertiary/quaternary conformation of a mAb and that an attached fucose residue inhibits interactions between the IgG-Fc and the ectodomain of FcγRIIIa [39–41].

The oligosaccharide released from normal human serum IgG-Fc is essentially composed of a core heptasaccharide with variable addition of fucose, galactose, bisecting *N*-acetylglucosamine, and sialic acid residues (Fig. 2) [39–41, 67–70].

Carbohydrate chemists, glycobiologists, and mass spectrometry specialists have developed different systems of nomenclature to represent oligosaccharide structures [39–41, 71, 72]. Antibody "practitioners" use a shorthand nomenclature to represent the oligosaccharides released from normal serum polyclonal IgG. In Fig. 2, the core heptasaccharide highlighted in blue is designated G0 (zero galactose); the core bearing one or two galactose residues is designated G1 or G2, respectively. The core + fucose is designated G0F and the core + fucose + galactose

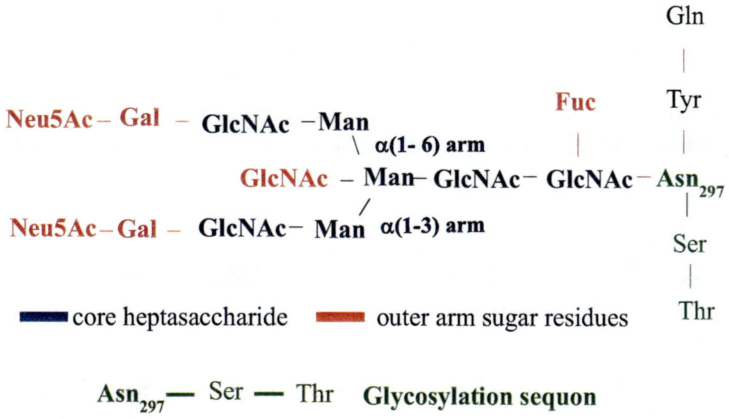


**Fig. 2** Representative IgG complex diantennnary oligosaccharide. The "core" heptasaccharide residues, (GlcNAc)2-Man3-(GlcNAc)2, are shown in *blue*; other sugar residues that may be present are in *red*. *GlcNAc N*-acetylglucosamine, *Neu5Ac N*-acetylneuraminic acid

is G1F, G2F, etc. When a bisecting *N*-acetylglucosamine is present, "B" is added (e.g., G0B, G0BF, G1BF, etc.). Sialylation of the galactose residues is designated by G1FS, G2FBS, etc. The approximate composition of neutral oligosaccharides released from normal polyclonal human IgG-Fc is G0 3%, G1 3%, G2 6%, G0F 23%, G1F 30%, G2F 24%, G0BF 3%, G1BF 4%, and G2BF 7% [73–75]. It is important to define the glycoform of the intact IgG molecule (e.g., [G0/G1F], [G1F/G2BF]) because individual IgG molecules can be composed of symmetrical or asymmetrical heavy chain glycoform pairs [76–78]. This has important consequences for the engagement and activation of FcγRIIIa-mediated ADCC, which requires that only one heavy chain bears an oligosaccharide devoid of fucose; thus, the [G0/G0F] glycoform could be as potent in ADCC as the [G0/G0] glycoform.

Minor oligosaccharide structures present in polyclonal IgG-Fc may be functionally significant because each could be the predominant glycoform of an individual antibody secreted from a single plasma cell. Although analysis of monoclonal myeloma IgG has shown that the IgG-Fc glycoform profile of each paraprotein (patient) is essentially unique, subtle differences in oligosaccharide processing between subclasses and allotypes were also observed, such as a preference for addition of galactose to the α(1–6) arm of IgG1-Fc and the α(1–3) arm of IgG2-Fc. The arm preference for IgG3 proteins correlated with allotype [67–69]. These data suggest that critical conformations of the IgG-Fc are necessary to accommodate the steric requirements for glycosyltranferase-mediated sugar additions. Such conformations may be sensitive to niche environments because the GP transits the Golgi apparatus.

The glycoform profile of polyclonal serum-derived IgG can vary significantly in health and disease, particularly in autoimmune and inflammatory diseases [39–41, 78–82]. Methods have been developed that allow the glycoform profile of antigen-specific polyclonal IgG autoantibodies to be determined. Significant differences in the glycoform profiles of IgG autoantibodies and the bulk IgG have been reported

[79–82]. The [G0F/G0F] oligosaccharide glycoform predominates for mAb produced in mammalian cells but can vary according to producer cell type, the production platform, and the precise culture conditions employed. Under conditions of stress (e.g., nutrient depletion, acid pH), deviant glycosylation may be observed, as shown by the presence of high mannose forms and/or incomplete site occupancy [83–88].

## 9 IgG-Fc Glycoform Profiles of Recombinant IgG Antibody Therapeutics

The glycoform profile is a CQA for each approved mAb therapeutic. The glycoform profile may be selected to optimize effector functions, depending on the required or presumed MoA. The first criterion, therefore, is either 100 or 0% oligosaccharide occupancy. Although CHO, NS0, and Sp2/0 cell lines deliver essentially 100% occupancy, they produce mAbs bearing predominantly G0F heavy chain glycoforms with relatively low levels of galactosylated and nonfucosylated glycoforms, relative to normal polyclonal IgG-Fc. Control of culture conditions during a production run allows minor changes in glycoform profile and maintenance of product fidelity [83–88]. Producer cell lines may also add sugars that are not expressed on human glycoproteins and can be immunogenic in human recipients. Thus, although CHO cell lines add *N*-acetylneuraminic acid residues, they do so in α(2,3) linkage rather than the α(2,6) linkage present in human IgG-Fc. A particular concern is the addition of galactose in α(1,3) linkage to galactose linked β(1,4) to the *N*-acetylglucosamine residues by NS0 and Sp2/0 cells [89–91]. Humans and higher primates do not have a functional gene encoding the transferase that adds galactose in α(1,3) linkage. However, as a result of environmental exposure to the gal-α(1,3)-gal epitope (e.g., in red meat), humans can develop IgG antibodies specific to this antigen. The gal-α(1,3)-gal epitope is widely expressed on hamster cells in vivo but rarely encountered on CHO-expressed mAbs, although some CHO cell lines have been shown to revert to expression of the gal-α(1,3)-gal epitope [88]. Similarly, CH0, NS0, and Sp2/0 cells may add *N*-glycolylneuraminic acid in α(2,3) linkage that may be immunogenic in humans [89–91]. A significant population of normal human IgG-Fc bears a bisecting *N*-acetylglucosamine residue that is absent from IgG-Fc produced in CHO, NS0, or Sp2/0 cells. Studies of homogeneous IgG-Fc glycoforms, generated in vitro, have shown qualitative and quantitative differences in effector function activities between the IgG subclasses and for differing glycoforms within each subclass [39–41]. To date, it has not been possible to manipulate culture medium conditions to generate mAbs expressing a predetermined homogeneous glycoform profile.

# 10 Impact of IgG-Fc Glycoform on Downstream Effector Functions

Homogeneous IgG-Fc glycoforms have been generated in vivo using glycosidases and/or glycotransferases and their functional properties probed [39, 87, 92, 93]. An alternative approach has been to engineer cell lines by "knocking-in" or "knocking-out" glycosyl-transferase genes or blocking selected stages of maturation during passage though the Golgi apparatus [67, 94–98]. The demonstration of radical functional differences between glycoforms suggests that the immune system responds to pathogens by production of an antibody response composed of antibody isotype(s) and glycoform(s) optimal for immune protection. Most structure/function studies have employed intact IgG1 antibodies or the IgG1-Fc fragment; similar results may be anticipated for IgG3 antibodies but caution should be exercised in extending these observations to IgG2 and IgG4 antibodies. Differences in IgG-Fc-mediated functions have also been reported between intact IgG1 and its Fc fragment, suggesting that the presence of the Fab modulates structure and function [99–103]. There is an emerging consensus for effector ligand engagement and activation of IgG mAbs, but quantitative discrepancies have been reported due to differences in the assay systems employed, such as binding to recombinant FcγR immobilized on a matrix or in free solution, and binding FcγR expressed on effector cells harvested from fresh blood or immortal cell lines rendered transgenic for FcγR expression [48–55]. Current analytical protocols allow accurate and reproducible determination of the glycoform profile of each IgG subclass contributing to specific autoantibody responses (e.g., citrullinated peptides, platelets, the PR3 antigen, and antivirus antibodies) [79–82]. Nevertheless, it remains to be determined whether these differences relate to disease activity and/or resolution. In the following section, the impact of individual IgG-Fc glycoforms on function are summarized prior to attempting a structural rationale.

## *10.1 Influence of Fucose and Bisecting N-Acetylglucosamine on IgG-Fc Activity*

The influence of recombinant protein glycoforms on biologic activity has been explored through their production in mutant CHO cells lacking the ability to add one or more sugar residues [104]. The cell line Lec 13 lacks the ability to add fucose to the primary *N*-acetylglucosamine residue; antibodies of the IgG1 subclass produced in this cell line exhibit enhanced ability to kill cancer cells by natural killer (NK) cell-mediated ADCC [94]. This finding was confirmed and extended to all IgG subclasses when antibodies were produced in a α(1,6)-fucosyltransferase knockout CHO cell line or alternative platforms generating substantially nonfucosylated IgG [98, 105–107]; the α(1,6)-fucosyltransferase knockout CHO cell line is available commercially and provides access to the "Potelligent" production platform [108]. A nonfucosylated anti-CCR4 antibody (Mogamulizumab) expressed in this cell line has been approved in Japan for the treatment of patients with relapsed or refractory CCR4-positive adult T-cell leukemia-lymphoma (ATL) [108, 109] and is in phase III trials in Europe and the USA. A similar improvement in ADCC was reported for IgG1 antibody produced in a knock-in CHO cell line transfected with human β-1,4-*N*-acetylglucosaminyltransferase III (GnTIII) gene, resulting in the addition of bisecting *N*-acetylglucosamine residues [110, 111]. The early addition of bisecting *N*-acetylglucosamine during passage through the Golgi apparatus was shown to inhibit the addition of fucose by endogenous α(1,6)-fucosyltransferase [111]. It was posited, therefore, that the absence of fucose is the main factor determining increased NK cell-mediated ADCC for these glycoforms. The latter platform has been employed by Glycart-Roche for production of the biobetter anti-CD20 antibody Obinutuzumab, which was approved for previously untreated chronic lymphocytic leukemia (CLL) in 2013; approval was extended to follicular lymphoma in 2016 [112, 113]. Multiple technologies are being developed in attempts to generate mAbs expressing a single glycoform, selected to activate downstream biologic activities appropriate to specific disease indications [114]. These IgG glycoforms may be minor components of the oligosaccharides present in normal polyclonal human IgG-Fc; however, because they are normal (self) structures they do not present immunogenicity issues [39–41, 67, 73–75]. By contrast, some glycoforms produced by nonhuman (mammalian) cell lines may be immunogenic [89, 90].

The above discussion was centered on ADCC mediated by peripheral blood mononuclear leucocytes; however, the impact of fucosylation is different for polymorphonuclear cells [115–118]. A study employing batches of an IgG mAb with high and low fucose contents reported that a higher fucose content resulted in more active neutrophil-mediated ADCC, whereas a lower content resulted in higher neutrophil-mediated phagocytosis and apoptosis [115]. Results for ADCC studies employing cell lines expressing cellular receptors in vitro can vary because the glycoform of the receptor is also a critical parameter and can differ between effector cell lines [52, 110]. The presence or absence of fucose has not been reported to impact CDC, but an IgG1/IgG3 hybrid molecule exhibited enhanced CDC for both fucosylated and nonfucosylated IgG-Fc glycoforms [119].

The enhanced ADCC mediated by nonfucosylated antibodies has led academic and commercial laboratories to explore alternative routes for the generation of nonfucosylated glycoproteins. Engineering CHO cells to generate homogeneous Man5/Man6 glycoforms results in lack of addition of fucose [120–124]. Similarly, inhibitors targeting enzymes within the Golgi apparatus enable production of nonfucosylated molecules; for example, kifunensine has been employed by several groups for the generation of nonfucosylated high mannose (Man6–Man9) glycoforms [121, 122]. Other platforms include GlymaxX, which engineers mammalian cells to express a bacterial enzyme that inhibits the pathway leading to the addition of fucose [114], and the addition to the culture medium of sugar analogs that inhibit incorporation of the natural sugar [123, 124].

The influence of fucose on FcγRIIIa-mediated ADCC is also dependent on the glycoform of the receptor. The FcγRIIIa receptor expresses five *N*-linked glycosylation sites, and the glycoform attached at N-162 is expressed at the interface of the FcγRIIIa/IgG-Fc interaction site. Enhanced FcγRIIIa/IgG-Fc binding affinity and ADCC has been demonstrated for afucosylated IgG; aglycosylated FcγRIIIa has the same binding affinity for fucosylated and afucosylated IgG-Fc [110, 125]. The presence of a further *N*-linked oligosaccharide at N-45 has a negative impact on FcγRIIIa binding [126].

## *10.2 Influence of Galactosylation on IgG-Fc Activity*

The extent of IgG-Fc galactosylation is a major source of glycoform heterogeneity, in both health and disease. Accepting the levels of galactosylation observed for young adults as the norm, a decline is observed with ageing [62, 127–129]. Levels of IgG-Fc galactosylation increase over the course of normal pregnancy but return to the adult norm following parturition [62, 130]. Hypogalactosylation of IgG-Fc is reported for a number of inflammatory states associated with autoimmune disease [79–82, 131–133]. The extent of IgG-Fc galactosylation observed between monoclonal myeloma IgG proteins is highly variable, indicating that the level of IgG-Fc galactosylation is a clonal property [73, 74, 134]. The antibody products of CHO, Sp2/0, and NS0 cell lines used in commercial production of recombinant antibody are generally highly fucosylated, but hypogalactosylated relative to polyclonal human IgG [135–137]; it is necessary therefore, to consider the possible impact of differential IgG-Fc galactosylation on functional activity.

The variations in galactosylation observed in health and disease suggest that it is either of functional significance or an epiphenomenon. The increase in galactosylation in pregnancy is particularly intriguing because it coincides with FcRn-mediated transcytosis of maternal IgG to the fetus in the third trimester. It follows, therefore, that IgG present in neonatal blood is similarly highly galactosylated [58, 138, 139]. Studies of the binding affinity of the human IgG for FcRn have not revealed differences between the various natural glycoforms; however, oligosaccharide present at the single glycosylation site in FcRn does influence IgG-Fc binding affinity [58]. The possible impact of the level of galactosylation of recombinant mAbs on in vivo activity has been extrapolated from in vitro cell-based assays and animal experiments. Removal of terminal galactose residues from Campath-1H reduced classical complement activation but had no effect on FcγR-mediated functions [140]. Similarly, the ability of Rituximab to kill tumor cells via the classical complement route was maximal for the [G2F]2 glycoform, in comparison with the [G0F]2 glycoform [141]. The product that gained licensing approval contained of ~25% galactosylated oligosaccharides; therefore, this proportion must be maintained over the life span of the drug. The level of galactosylation of an approved drug substance is identified as a CQA and its maintenance can serve as a measure of control over the production process. In the absence of galactose, the terminal sugar residue is *N*-acetylglucosamine, which may be accessible to bind the mannose receptors expressed on many cell types, including antigen-presenting dendritic cells. ICs formed with agalactosylated IgG can bind the mannan-binding lectin (MBL) to activate the lectin complement pathway [39–41].

## 10.3 Sialylation of IgG-Fc Oligosaccharides

Although reports of the impact of fucosylation and galactosylation on the MoAs are relatively consistent, as determined by in vitro assays, reports of the impact of sialylation vary considerably. Less than 10% of oligosaccharides released from polyclonal IgG-Fc bear terminal α(2–6) *N*-acetylneuraminic acid residues [39, 64, 67–70, 74, 75]. Given the observed asymmetry of heavy chain glycoforms, a maximum of 5% of molecules can bear sialylated oligosaccharides on both heavy chains and 10% on one heavy chain only. The paucity of sialylation may reflect the absence of galactosylation and/or restricted access of the α(2–6) *N*-acetylneuraminic transferase enzyme to terminal galactose residues, rather than an inherent deficit in the sialylation machinery. This conclusion is supported by the finding that when oligosaccharides are present in both IgG-Fc and IgG-Fab the latter bears highly galactosylated and sialylated structures, demonstrating that the glycosylation machinery is fully functional [69, 73, 74, 76, 142–144]. In contrast to most serum proteins, the presence or absence of terminal galactose and/or sialic acid residues does not influence IgG half-life because it is not catabolized via the asialo-glycoprotein receptor (ASGPR) in the liver but in multiple cell types expressing FcRn. The impact of IgG-Fc structure on glycoform profile was demonstrated for a panel of IgG1 antibodies in which amino acid residues known to interact with oligosaccharide residues were sequentially replaced by alanine. In each case, hypergalactosylated and highly sialylated glycoforms resulted, suggesting some relaxation of structure that allowed access to glycosyl transferases [142–144].

The early demonstration of increased levels of serum (G0F)2 IgG-Fc glycoforms associated with inflammatory autoimmune disease led to this glycoform being regarded as a possible mediator of inflammation; by contrast, galactosylated and sialylated glycoforms are considered relatively anti-inflammatory. Similarly, the dramatic impact of the absence or presence of fucosylated oligosaccharides on IgG-Fc MoA (e.g., ADCC) could be equated with inflammatory versus anti-inflammatory antibody glycoforms. Therefore, association of the term "anti-inflammatory" to sialylated IgG-Fc glycoforms alone may overemphasize its significance. Activation of complement by ICs is also an inflammatory cascade, for which (G2F)2 glycoforms of Rituximab and Campath-1H are increased relative to (G0F)2 glycoforms [140, 141]. The focus on sialylation emerged with attempts to elucidate the mechanism(s) by which intravenous IgG (IVIG) mediates an anti-inflammatory activity in some autoimmune diseases [81, 133, 145, 146]. Multiple MoAs have been proposed and one "school" consistently reports that the α 2–6 *N*-acetylneuraminic acid IgG-Fc glycoform is essential for the anti-inflammatory activity and is mediated by engagement of the DC-SIGN lectin receptor, a "knock-on" effect being upregulation of inhibitory FγRIIb receptor expression, resulting in attenuation of autoantibody-mediated inflammation [81, 144–152].

Initially, attempts to further investigate the functional activity of sialylated antibodies were hampered by the low levels of sialylation present in serum IgG and mAbs produced in CHO cells. In consequence, protein and glycosylation engineering have been employed to generate IgG antibodies expressing elevated levels of sialylated IgG-Fc [69, 144, 153]. Some studies have consistently reported an anti-inflammatory role for IgG-Fc sialylated antibodies [81, 144–152]; in other studies, anti-inflammatory activity has either not been observed or claimed for sialylated IgG-(Fab′)2 fragments [150, 154–161]. These discrepancies have been addressed in numerous review articles but currently are unresolved [150, 153, 161].

## 11 Recombinant Glycoproteins Bearing High Mannose Oligosaccharides

Although the presence of high mannose (Man5–Man9) glycoforms has not been reported for normal human serum IgG-Fc, they are usually present at low levels in mAbs. There has been a concern that this glycoform can compromise the efficacy of a mAb therapeutics and/or result in more rapid clearance. This question has been investigated for mAb produced in CHO-Lec3.2.8.1 or human embryonic kidney (HEK)293S cells that lack GnT1 activity, restricting maturation at the Man5 glycoform [31, 85, 104, 162–165]. The Man5 oligosaccharide is normally an intermediate in GP processing and is rarely present on mature human GP products. When present on recombinant glycoproteins, Man5 may be regarded as an artefact of the cell line and/or the production platform employed. However, for some recombinant glycoprotein therapeutics the presence of terminal mannose residues may be beneficial or essential. The GnT1-deficient cell lines have been exploited to produce homogeneous Man5 glycoforms that can target cells bearing mannose receptors. In addition, being structurally homogeneous, the proteins are more amenable to crystallization and subsequent x-ray crystallographic studies. Although multiple parameters impact Golgi-mediated glycoprotein processing, some control of Man5 levels by manipulation of cell culture conditions has been reported [162, 163]. Inhibition of enzymes within the Golgi apparatus provides another avenue for the production of high mannose glycoforms. Thus, kifunensine inhibits the mannosidase I enzyme, resulting in production of Man6–Man9 glycoforms [162, 165]. It has recently been demonstrated that incomplete processing in vivo, with consequent generation of truncated mannose oligosaccharides, can result from restricted access for mannose transferases Thus, although the surface of recombinant HIV GP120 glycoprotein is almost entirely covered by *N*-linked high mannose oligosaccharide structures, native GP120, expressed on HIV virus isolates, bears a number of truncated oligomannose structures. It appears that the density of the early oligomannose structures limits enzyme processing [166].

Glycoproteins bearing exposed mannose residues can be internalized by cells expressing mannose receptor(s) and/or activate multiple biologic pathways in vivo (e.g., the lectin pathway of complement activation) [167–169]. Exposed terminal mannose residues are required for some GPs to facilitate cellular internalization via the mannose receptor. An interesting example is the approved biologic Cerezyme as enzyme replacement therapy for patients with Gaucher's disease. This lysosome storage disease results from deficient production of the enzyme β-glucocerebrosidase within macrophage lysosomes [167]. The product produced by CHO cells can express terminal *N*-acetylglucosamine, galactose, or sialic acid sugar residues that are not bound by the mannose receptor. Consequently, the CHO cell product is exposed to sialidase, galactosidase, and *N*-acetlyglucosaminidase to remove these sugar residues and expose the terminal trimannose core. The recently developed CHO-gmt4 cell line harbors a dysfunctional *N*-acetylglucosaminyltransferase 1 (GnT-1) gene; therefore recombinant glucocerebrosidase produced by these cells does not require further processing. Macrophage uptake did not differ significantly between Man2–Man9 glycoforms, but the high mannose products were shown to bind to MBL, with possible unwanted lectin pathway activation of the complement cascade [168]. A comprehensive review by Jaumouillé and Grinstein of receptors mediating phagocytosis, protection, and the initiation of immune responses is recommended [169].

## 12 IgG-Fc Glycoform–Ligand Interactions: An Attempt to Rationalize

As previously commented, the structure of the IgG molecule allows each Fab moiety to bind to spatially distinct epitopes while the IgG-Fc remains available for interaction with one or more effector ligand. The necessary mobility for the Fab and Fc regions is provided by the intervening hinge region, which differs significantly in length and flexibility between the IgG subclasses. Each IgG subclass protein expresses a unique ligand binding profile and, consequently, potentially differing MoA profiles. It is not possible to offer a comprehensive review of the structure–function relationships for each of the IgG isotypes because most studies have probed these relationships for IgG1 subclass proteins only. Multiple orthogonal techniques have been applied for structural characterization of IgG proteins and relating structural parameters to in vitro biologic activities. Such studies have been conducted under widely differing conditions of temperature, but rarely at body temperature. Similarly, binding and biologic activities have employed various individually unique assay protocols at "room temperature!" or 37°C; not infrequently, they generate conflicting data and conclusions. That being said, a consensus is emerging, although extrapolation to MoAs in vivo remains challenging.

The IgG-Fc X-ray crystal structure reported by Deisenhofer in 1981 was generated by papain cleavage of polyclonal IgG at the $Lys_{222}$–$Thr_{223}$ peptide bond, within the hinge region, and extending to a C-terminus residue at

446 [44, 170]. Data was collected at ~ 100 K (−173°C). At this temperature, vibrational mobility of the molecules is limited and weak intermolecular interactions establish a relatively stable three-dimensional structure. Interpretable electron density could be resolved for residues 238–443 but not for residues 223–237 (which comprise the core hinge sequence and the hinge proximal region of the $C_H2$ domain) or C-terminal residues 444–446. Unexpectedly, a defined structure for the diantennary oligosaccharide was obtained, showing it to be "sequestered" within the internal "horseshoe" structure of the IgG-Fc. Thus, the conformation of the protein and oligosaccharide moieties were shown to be interdependent, with multiple noncovalent interactions between constituent sugar residues with amino acid side chains and main chain atoms of the $C_H2$ domain, in addition to the covalent protein–oligosaccharide bond at N-297. These interactions substitute for the domain pairing observed for the $V_H/V_L$, $C_H1/C_L$, and $C_H3/C_H3$ regions. These structural characteristics have been confirmed and extended for crystal structures obtained for human IgG-Fc alone or in complex with SpA [170, 171], SpG [172], rheumatoid factor (RF) [173], and recombinant soluble ectodomains of human FcγRIIa [174], FcγRIIIb [175], and FcγRIIIa [176, 177]. There are several common structural features reported for IgG-Fc, as follows:

1. The $C_H3$ domains are well defined because of noncovalent pairing, involving ~2,000 Å$^2$ of accessible surface area in the $(C_H3)2$ module.
2. The area of noncovalent contact between the $C_H2$ and $C_H3$ domains is ~800 Å. This suggests that the $C_H2$–$C_H3$ contact contributes to the relative stability observed for the C-terminal proximal region of $C_H2$ domains, as opposed to the "softness" of the $C_H2$ domain proximal to the hinge region.
3. The hydrophobic surface of each $C_H2$ domain is "overlaid" by the carbohydrate. Hydrophobic and polar interactions between the oligosaccharide and the $C_H2$ domain surface occupy ~500 Å$^2$ and substitute for domain pairing [170, 171].
4. One $C_H2$ domain is less ordered than the other as a result of crystal contact with a neighboring $C_H2$ domain.
5. The more disordered structure for the hinge proximal region of the $C_H2$ domain is reflected in higher temperature factors (i.e., unfolding at relatively low temperatures).
6. The intrinsic stability of the immunoglobulin fold is reflected in higher structural resolution of β-sheets regions compared with β-bends.

The disorder reported for the hinge proximal regions of the $C_H2$ domains reflects mobility, which can be significantly enhanced at body temperature and result in the generation of dynamic equilibrium of high-order structural conformers. Each ligand (e.g., one of the three homologous Fcγ receptors or the C1q component of complement) may bind a unique IgG-Fc conformer [53, 178]. Presumably, this is a reciprocal property, such that each effector ligand can exist as an equilibrium of conformers (e.g., the FcγR family of receptors each binds a unique IgG-Fc conformer). This idea is supported by the demonstration that residues of the lower hinge region that cannot be resolved for the IgG-Fc crystals are ordered in the IgG1-

Fc/FcγR complexes and directly involved in receptor binding [174–178]. Some amino acid residue side chains and/or main chain atoms may contribute to the binding of different ligands [39–41, 170–177], as shown by the presence of a "proline sandwich" as a common structural feature for each IgG-Fc–FcγR interaction [174–177]. The binding sites for soluble recombinant FcγRIIa, FcγRIIIa, and FcγRIIIb are asymmetric, with each heavy chain engaging distinct regions of the receptor. Consequently, monomeric IgG is univalent for Fcγ receptors and the C1 component of complement. By contrast, IgG-Fc is functionally divalent for ligands binding at the $C_H2$–$C_H3$ interface (e.g., FcRn, RF, SpA, and SpG). Because of the symmetry of the IgG-Fc, these two interaction sites are opposed at ~180° and each is accessible to bind macromolecular ligands to form multimeric complexes.

It is important to consider IgG-Fc glycoform symmetry/asymmetry when attempting to optimize the IgG-Fc glycoform for a selected MoA. Fucosylation of (G0)2 glycoforms during passage from the medial to the *trans*-Golgi region of the endoplasmic reticulum can result in generation of asymmetric (G0F/G0) and symmetric (G0/G0) or (G0F/G0F) IgG-Fc glycoforms. As previously stated, a (G0F/G0) IgG in which only one heavy chain is devoid of fucose may express the same level of FcγRIIIa-mediated ADCC as a (G0F)2 molecule [76, 175–178]. Increased FcγRIIIa-mediated ADCC, independent of glycoform, has also been achieved for protein engineered IgG-Fc. Because each heavy chain of the IgG molecule binds a distinct region of the FcγRIIIa receptor, the optimal IgG-Fc structure requires generation of a molecule in which the two heavy chains have different sequences. This objective has been realized employing the "knobs-into-holes" approach to generate an IgG molecule with asymmetric heavy chain amino acid sequences [179].

Submission for regulatory approval of a mAb therapeutic requires comprehensive structural characterization employing multiple orthogonal techniques. A plethora of techniques are available and a consensus view of the most relevant techniques and protocols is sought. This challenge has been addressed by a study emanating from the US National Institute for Standards and Techniques (NIST). An IgG1 protein molecule was structurally characterized by major biopharmaceutical companies, employing all currently available state of the art techniques. This allowed insight into the selection of appropriate techniques and the availability of a proposed reference material that can be employed to standardize performance across laboratories. The fruits of this exercise have been published in a three-volume series [180–182].

## 13 IgG-Fab Glycosylation

It has been established that about 30% of polyclonal human IgG molecules bear *N*-linked oligosaccharides within the variable regions of the kappa (Vκ), lambda ($V_\lambda$), or heavy ($V_H$) chains, and sometimes both [39–41, 76, 81, 134, 141–144]. In the immunoglobulin sequence database, about 20% of expressed IgG variable regions have *N*-linked glycosylation consensus sequences. Interestingly, these consensus sequences are mostly not germline encoded but result from somatic hypermutation, which is suggestive of positive selection for improved antigen binding. Analysis of oligosaccharides released from polyclonal human serum-derived IgG-Fab fragments revealed the presence of diantennary oligosaccharides with high levels of G2F and substantial levels of G2FS oligosaccharides, in contrast to the diantennary oligosaccharides released from IgG-Fc [39–41, 81, 134, 142–144]. This pattern was maintained for IgG-Fab prepared from IgG isolated from the sera of patients with Wegner's granulomatosis or microscopic polyangiitis, which expressed hypogalactosylated Fc glycans [143]. Thus, the in vivo environment of IgG-producing plasma cells influences the efficacy of glycoprocessing of IgG-Fc but not IgG-Fab during passage through the Golgi apparatus. The functional significance for IgG-Fab glycosylation of polyclonal IgG has not been fully determined, but data emerging for mAbs suggest that Vκ, $V_\lambda$, or $V_H$ glycosylation can have a neutral, positive, or negative influence on antigen binding [183, 184]. The differences observed for polyclonal IgG-Fc and IgG-Fab glycoforms has been maintained for mAbs produced in CHO cells and monoclonal human myeloma IgG proteins [81, 142–144].

The oligosaccharide present in GPs and IgG-Fc, in particular, has been shown to contribute positively to solubility and stability and it is possible that IgG-Fab glycosylation confers similar benefits [170–177]. Thus, IgG-Fab glycosylation may contribute to mAb formulation at concentrations of >100 mg/mL [145–150, 185, 186], levels required to allow the development of self-administration protocols. These concentrations result in longer dosing intervals, reducing the necessity for attendance at the clinic and, consequently, reducing the cost of treatment. The demand for control of glycoform fidelity at both Fab and Fc sites is a further challenge for the biopharmaceutical industry.

The licensed mAb Erbitux (cetuximab), expressed in Sp2/0 cells, bears an *N*-linked oligosaccharide at N-88 of the $V_H$ region; interestingly there is an unoccupied glycosylation sequon within the light chain at N-41 [187, 188]. Analysis of the oligosaccharides released from the IgG-Fc and IgG-Fab fragments of Erbitux revealed highly significant differences in composition. Although the IgG-Fc oligosaccharides were typical (i.e., composed predominantly of diantennary G0F oligosaccharides), the IgG-Fab oligosaccharides were extremely heterogeneous and included complex diantennary, triantennary, and hybrid oligosaccharides. Nonhuman oligosaccharides such as galactose in α(1,3) linkage to galactose and *N*-glycylneuraminic acid residues were also present.

Severe adverse reactions to cetuximab therapy have been reported. In a study of 76 patients treated with Erbitux, 25 experienced hypersensitivity reactions due to the presence of IgE antibodies targeting gal-α(1,3)-gal. Interestingly, environmental factors appeared to influence the development of IgE anti-gal-α(1,3)-gal responses and IgE antibodies were detected in pretreatment samples from 17 of the patients [189–192]. The incidence of hypersensitivity varied significantly between treatment centers and could be linked to differences in predominant infectious agents present in local environments. Subsequently, it has been demonstrated that most individuals that consume meat (beef, lamb, pork, etc.) have IgG anti-gal-α(1,3)-gal antibodies and a minority have IgE anti-gal-α(1,3)-gal antibodies. It is becoming routine, therefore, to monitor patients for the presence of IgE anti-gal-α(1,3)-gal antibodies prior to exposure to Erbitux [193, 194].

A detailed analysis of the glycoforms of a humanized IgG anti-amyloid-β mAb, also expressed in Sp2/0 cells, reveals the expected IgG-Fc glycoform profile of predominantly G0F oligosaccharides, but an additional oligosaccharide at N-56 of the $V_H$. Eleven oligosaccharides were released from the IgG-Fab, including diantennary and triantennary oligosaccharides bearing gal-α(1,3)-gal, *N*-glycylneuraminic acid, and *N*-acetyl galactosamine residues [195]. The consistent observation of higher levels of galactosylation and sialylation for IgG-Fab *N*-linked oligosaccharides, in comparison to IgG-Fc, is thought to reflect its attachment at the surface of the molecule, thus providing accessibility to glycosyltransferases. In view of these experiences, the perceived virtues of the NS0 and Sp2/0 cells might best be pursued by knocking out or otherwise inactivating the gal-α(1,3) and *N*-glycylneuraminic acid transferases.

The challenge of controlling the glycoform profile of mAbs in both IgG-Fc and IgG-Fab has generally led companies to remove $V_H$ or $V_L$ glycosylation sequons (e.g., by substitution of asparagine residues by alanine). In contrast, recent reports suggest that mAbs expressed in CHO cells can generate $V_H$ and/or $V_L$ glycoforms similar to those present in normal polyclonal IgG [185, 196, 197]. Because oligosaccharides are hydrophilic, the addition of glycans within $V_H$ and/or $V_L$ regions could impact the physicochemical properties of an antibody molecule and affect its pharmacokinetics [196, 197], solubility [185], aggregation, etc. A $V_H$ glycosylated human IgG mAb was shown to have the same pharmacokinetics as the $V_H$ deglycosylated molecule in a mouse model [196]; however, introduction of a glycosylation site within bispecific single-chain diabodies resulted in a significant increase in serum half-lives [185]. Studies of the solubility of an anti-IL-13 mAb are revealing. The clone selected for development included a glycosylation sequon (53NSS55) within the heavy chain CDR2 [185]. Initially, this site was engineered out by replacing N-53 by an aspartic acid residue; however, the product exhibited very limited solubility (~13 mg/mL) and high levels of aggregation. Reverting to development of the original N-53 molecule, with limited engineering of the $V_L$, generated a $V_H$ glycosylated mAb with a solubility $>$110 mg/mL [185].

## 14 Concluding Remarks

It is important to emphasize that the structural studies discussed here mostly employed natural or glycosylation engineered IgG-Fc fragments, alone or in complex with a recombinant form of a natural ligand such as SpA. There is a paucity of data for full-length IgG molecules or full-length IgG antibodies in complex with their target ligand. By contrast, many X-ray crystal structures of Fab fragments in complex with their target antigens have been solved. The challenge remains to solve the structure of full-length IgG mAb/antigen complexes binding to a membrane-bound effector ligand. Currently, we only have an indication that IgG-Fc–ligand interactions are favored when the $C_H2$ domains assume a relatively open structure. However, the impact of single and multiple amino acid replacements on structure and effector ligand binding/activation suggests that more sophisticated approaches are required, particularly for understanding how a single amino acid residue replacement within the $C_H3$ domain impacts FcγR binding at the lower hinge region. An increasing number of studies have reported Fab–Fc interactions within intact IgG mAbs that modulate functional activity [99–103]. Therefore, the conformation of the IgG molecule is a CQA that may undergo subtle dynamic changes in vivo and within experimental protocols. This could account for the tendency of monomeric mAb molecules to form aggregates in the absence of antigen, a property that could result in enhanced immunogenicity and the production of ADA. It is essential, therefore, that multiple orthogonal physicochemical techniques should be employed to characterize a potential mAb therapeutic as drug substance or drug product, and following exposure to accelerated storage conditions. Industry and academia will be best served by having access to a reference material that has been comprehensively characterized using state of the art techniques [160]. A consensus view may emerge enumerating the techniques considered essential and that could become mandatory within QbD protocols. It is interesting to note that different ligands bind to the IgG-Fc through the same amino acid residues within the hinge proximal region for FcγR and C1q and at the $C_H2$–$C_H3$ interface for FcRn, SpA, SpG, RFs, and IgG-Fc-like receptors encoded within the genomes of some viruses. The presence of sialic acid might further influence Fc–ligand interactions. The topography of FcγR and C1q ligand binding sites could be a functional necessity for circulating IgG to be monovalent for these ligands, to prevent continuous cellular activation. However, the significance of ligand binding divalency at the $C_H2$–$C_H3$ interface is not immediately evident. The influence of the IgG-Fc glycoform on functional activity can be exploited to generate homogeneous glycoforms selected for a predetermined functional profile considered optimal for a given disease indication. It is important to note that this can be achieved for each glycoform present within normal polyclonal IgG-Fc; therefore, they do not have the potential to be immunogenic. Many innovative studies have explored engineering of the protein moiety for selective enhancement of biologic activities; however, these are mutant forms of IgG (i.e., non-self) that might enhance immunogenicity. This is probably not an issue when treating patients for cancer because they may be receiving chemotherapy, with consequent immune suppression. However, it is a concern in treatment of chronic diseases that require long-term and/or interrupted exposure to mAbs. The reductionist approach of studying interactions of individual mAb molecules with a defined target antigen or effector ligand has provided a rationale for the development of mAb therapeutics; however, we must be aware of its limitations when attempting to predict outcomes in vivo, when different MoAs may be activated simultaneously or trigger unexpected outcomes or unintended consequences.

## References


1. Gupta SK, Bhandari B, Shrestha A, Biswal BK, Palaniappan C, Malhotra SS, Gupta N (2012) Mammalian zona pellucida glycoproteins: structure and function during fertilization. Cell Tissue Res 349(3):665–678
2. Drickamer K, Taylor ME (2015) Recent insights into structures and functions of C-type lectins in the immune system. Curr Opin Struct Biol 34:26–34
3. Monticelli M, Ferro T, Jaeken J, Dos Reis Ferreira V, Videira PA (2016) Immunological aspects of congenital disorders of glycosylation (CDG). J Inherit Metab Dis 39:765–780
4. Laine RA (1994). Glycobiology 4(6):759–767
5. Higel F, Seidl A, Sörgel F, Friess W (2016) N-glycosylation heterogeneity and the influence on structure, function and pharmacokinetics of monoclonal antibodies and Fc fusion proteins. Eur J Pharm Biopharm 100:94–100
6. Walsh D, Matthews MB, Mohr I (2013) Tinkering with translation: protein synthesis in virus-infected cells. Cold Spring Harb Perspect Biol 5(1):a012351
7. Stewart-Jones GB (2016) Trimeric HIV-1-Env structures define glycan shields from clades A, B, and G. Cell 165(4):813–826
8. Piacente F, Gaglianone M, Laugieri ME, Tonetti MG (2015) The autonomous glycosylation of large DNA viruses. Int J Mol Sci 16(12):29315–29328
9. Adolf GR, Kalsner I, Ahorn H, Maurer-Fogy I, Cantell K (1991) Natural human interferon-alpha 2 is O-glycosylated. Biochem J 276(Pt 2):511–518
10. Jonasch E, Haluska FG (2001) Interferon in oncological practice: review of interferon biology, clinical applications, and toxicities. Oncologist 6(1):34–55
11. Hogland M (1998) Glycosylated and non-glycosylated recombinant human granulocyte colony-stimulating factor (rhG-CSF) – what is the difference? Med Oncol 15(4):229–233
12. Welte K (2014) G-CSF: filgrastim, lenograstim and biosimilars. Expert Opin Biol Ther 14 (7):983–993
13. Okamoto M, Nakai M, Nakayama C, Yanagi H, Matsui H, Noguchi H, Namiki M, Sakai J, Kadota K, Fukui M, Hara H (1991) Purification and characterization of three forms of differently glycosylated recombinant human granulocyte–macrophage colony-stimulating factor. Arch Biochem Biophys 286:562–568
14. Zhang Q, Johnston EV, Shieh J-H, Moore MAS, Danishefsky SJ (2014) Synthesis of granulocyte–macrophage colony-stimulating factor as homogeneous glycoforms and early comparisons with yeast cell-derived material. Proc Natl Acad Sci U S A 111(8):2885–2890
15. Palash Bhatacharya AE, Gaurav Pandey AE, Mukherjee KJ (2007) Production and purification of recombinant human granulocyte–macrophage colony stimulating factor (GM-CSF) from high cell density cultures of Pichia pastoris. Bioprocess Biosyst Eng 30:305–312
16. Seppälä M, Koistinen H, Koistinen R, Chiu PC, Yeung WS (2007) Glycosylation related actions of glycodelin: gamete, cumulus cell, immune cell and clinical associations. Hum Reprod Update 13(3):275–287
17. Yeung WS, Lee KF, Koistinen R, Koistinen H, Seppälä M, Chiu PC (2009) Effects of glycodelins on functional competence of spermatozoa. J Reprod Immunol 83(1–2):26–30
18. Lee CL, Pang PC, Yeung WS, Tissot B, Panico M, Lao TT, Chu IK, Lee KF, Chung MK, Lam KK, Koistinen R, Koistinen H, Seppälä M, Morris HR, Dell A, Chiu PC (2009) Effects of differential glycosylation of glycodelins on lymphocyte survival. J Biol Chem 284(22):15084–15096
19. Seppälä M, Koistinen H, Koistinen R, Hautala L, Chiu PC, Yeung WS (2009) Glycodelin in reproductive endocrinology and hormone-related cancer. Eur J Endocrinol 160(2):121–133. https://doi.org/10.1530/EJE-08-0756
20. Jelkmann W (1992) Erythropoietin: structure, control of production, and function. Physiol Rev 72(2):449–489
21. Gong B, Burnina I, Stadheim TA, Li H (2013) Glycosylation characterization of recombinant human erythropoietin produced in glycoengineered Pichia pastoris by mass spectrometry. J Mass Spectrom 48(12):1308–1317

22. Jelkmann W (2013) Physiology and pharmacology of erythropoietin. Transfus Med Hemother 40(5):302–309
23. Bertolini LR, Meade H, Lazzarotto CR, Martins LT, Tavares KC, Bertolini M, Murray JD (2016) The transgenic animal platform for biopharmaceutical production. Transgenic Res 25 (3):329–343
24. Zhou Q, Kyazike J, Echelard Y, Meade HM, Higgins E, Cole ES, Edmunds T (2005) Effect of genetic background on glycosylation heterogeneity in human antithrombin produced in the mammary gland of transgenic goats. J Biotechnol 117(1):57–72
25. Guo J, Kelton CM, Guo JJ (2012) Recent developments, utilization, and spending trends for Pompe disease therapies. Am Health Drug Benefits 5(3):182–189
26. Schoser B, Stewart A, Kanters S, Hamed A, Jansen J, Chan K, Karamouzian M, Toscano A (2016) Survival and long-term outcomes in late-onset Pompe disease following alglucosidase alfa treatment: a systematic review and meta-analysis. J Neurol 264:621–630
27. Ratner M (2009) Genzyme's Lumizyme clears bioequivalence hurdles. Nat Biotechnol 27:685
28. Grinnell BW, Yan SB, Macias WL (2006) Activated protein C. In: McGrath B, Walsh G (eds) Directory of therapeutic enzymes. CRC, Boca Raton, pp 69–95
29. Ashwell G, Harford J (1982) Carbohydrate-specific receptors of the liver. Annu Rev Biochem 51:531–554
30. Lee SJ, Zheng NY, Clavijo M, Nussenzweig MC (2003) Normal host defense during systemic candidiasis in mannose receptor-deficient mice. Infect Immun 71(1):437–445
31. Goh JS, Liu Y, Chan KF, Wan C, Teo G, Zhang P, Zhang Y, Song Z (2014) Producing recombinant therapeutic glycoproteins with enhanced sialylation using CHO-gmt4 glycosylation mutant cells. Bioengineered 5(4):269–273
32. Ahmed M, Narain R (2015) Carbohydrate-based materials for targeted delivery of drugs and genes to the liver. Nanomedicine 10(14):2263–2288
33. Kessler C, Oldenburg J, Ettingshausen CE, Tiede A, Khair K, Négrier C, Klamroth R (2015) Spotlight on the human factor: building a foundation for the future of haemophilia A management: report from a symposium on human recombinant FVIII at the World Federation of Hemophilia World Congress, Melbourne, Australia on 12 May 2014. Haemophilia 21 (Suppl 1):1–12. https://doi.org/10.1111/hae.12582
34. Bousfield GR, Dias JA (2011) Synthesis and secretion of gonadotropins including structure-function correlates. Rev Endocr Metab Disord 12(4):289–302
35. Howard SC, Wittwer AJ, Welply JK (1991) Oligosaccharides at each glycosylation site make structure-dependent contributions to biological properties of human tissue plasminogen activator. Glycobiology 1(4):411–418
36. Solá RJ, Griebenow K (2010) Glycosylation of therapeutic proteins: an effective strategy to optimize efficacy. BioDrugs 24(1):9–21
37. Sockolosky JT, Szoka FC (2015) The neonatal Fc receptor, FcRn, as a target for drug delivery and therapy. Adv Drug Deliv Rev 91:109–124
38. Ward ES, Devanaboyina SC, Ober RJ (2015) Targeting FcRn for the modulation of antibody dynamics. Mol Immunol 67(2 Pt A):131–141
39. Jefferis R (2012) Isotype and glycoform selection for antibody therapeutics. Arch Biochem Biophys 526(2):159–166
40. Irani V, Guy AJ, Andrew D, Beeson JG, Ramsland PA, Richards JS (2015) Molecular properties of human IgG subclasses and their implications for designing therapeutic monoclonal antibodies against infectious diseases. Mol Immunol 67(2 Pt A):171–182
41. Vidarsson G, Dekkers G, Rispens T (2014) IgG subclasses and allotypes: from structure to effector functions. Front Immunol 5:520
42. Jefferis R, Lefranc M-P (2009) Human immunoglobulin allotypes: possible implications for immunogenicity. MAbs 1:332–338
43. Edelman GM, Cunningham BA, Gall WE et al (2004) The covalent structure of an entire gamma G immunoglobulin molecule. J Immunol 173(9):5335–5342

44. Deisenhofer J (1981) Crystallographic refinement and atomic models of a human Fc fragment and its complex with fragment B of protein A from *Staphylococcus aureus* at 2.9- and 2.8-A resolution. Biochemistry 20(9):2361–2370
45. Jefferis R (2011) Aggregation, immune complexes and immunogenicity. MAbs 3:503–504
46. Okroj M, Österborg A, Blom AM (2013) Effector mechanisms of anti-CD20 monoclonal antibodies in B cell malignancies. Cancer Treat Rev 39(6):632–639
47. Tipton TR, Roghanian A, Oldham RJ, Carter MJ, Cox KL, Ian Mockridge C, French RR, Dahal LN, Duriez PJ, Hargreaves PG, Cragg MS, Beers SA (2015) Antigenic modulation limits the effector cell mechanisms employed by type I anti-CD20 monoclonal antibodies. Blood 125:1901–1909
48. Narciso JE, Uy ID, Cabang AB et al (2011) Analysis of the antibody structure based on high-resolution crystallographic studies. Nat Biotechnol 28(5):435–447
49. Bruhns P, Jonsson F (2015) Mouse and human FcR effector functions. Immunol Rev 268 (1):25–51
50. Caaveiro JM, Kiyoshi M, Tsumoto K (2015) Structural analysis of Fc/FcγR complexes: a blueprint for antibody design. Immunol Rev 268(1):201–221
51. Hargreaves CE, Rose-Zerilli MJ, Machado LR, Iriyama C, Hollox EJ, Cragg MS, Strefford JC (2015) Fcγ receptors: genetic variation, function, and disease. Immunol Rev 268(1):6–24
52. Hayes JM, Cosgrave EF, Struwe WB, Wormald M, Davey GP, Jefferis R, Rudd PM (2014) Glycosylation and Fc receptors. Curr Top Microbiol Immunol 382:165–199
53. Subedi GP, Barb AW (2016) The immunoglobulin G1 N-glycan composition affects binding to each low affinity Fc γ receptor. MAbs 8:1512–1524
54. Shields RL, Namenuk AK, Hong K, Meng YG, Rae J, Briggs J, Xie D, Lai J, Stadlen A, Li B, Fox JA, Presta LG (2001) High resolution mapping of the binding site on human IgG1 for Fc gamma RI, Fc gamma RII, Fc gamma RIII, and FcRn and design of IgG1 variants with improved binding to the Fc gamma R. J Biol Chem 276(9):6591–6604
55. Kellner C, Derer S, Valerius T, Peipp M (2014) Boosting ADCC and CDC activity by Fc engineering and evaluation of antibody effector functions. Methods 65(1):105–113
56. Rojko JL, Evans MG, Price SA, Han B, Waine G, DeWitte M, Haynes J, Freimark B, Martin P, Raymond JT, Evering W, Rebelatto MC, Schenck E, Horvath C (2014) Formation, clearance, deposition, pathogenicity, and identification of biopharmaceutical-related immune complexes: review and case studies. Toxicol Pathol 42(4):725–764
57. Taylor RP, Lindorfer MA (2016) Cytotoxic mechanisms of immunotherapy: harnessing complement in the action of anti-tumor monoclonal antibodies. Semin Immunol 28 (3):309–316
58. Pyzik M, Rath T, Lencer WI, Baker K, Blumberg RS (2015) FcRn: the architect behind the immune and nonimmune functions of IgG and albumin. J Immunol 194(10):4595–4603
59. Stapleton NM, Einarsdóttir HK, Stemerding AM, Vidarsson G (2015) The multiple facets of FcRn in immunity. Immunol Rev 268(1):253–268
60. Strohl WR (2015) Fusion proteins for half-life extension of biologics as a strategy to make biobetters. BioDrugs 29(4):215–239
61. Monnet C, Jorieux S, Urbain R, Fournier N, Bouayadi K, De Romeuf C, Behrens CK, Fontayne A, Mondon P (2015) Selection of IgG variants with increased FcRn binding using random and directed mutagenesis: impact on effector functions. Front Immunol 6:39
62. de Haan N, Reiding KR, Driessen G, van der Burg M, Wuhrer M (2016) Changes in healthy human IgG Fc-glycosylation after birth and during early childhood. J Proteome Res 15 (6):1853–1861
63. Tong HF, Lin DQ, Zhang QL, Wang RZ, Yao SJ (2014) Molecular recognition of Fc-specific ligands binding onto the consensus binding site of IgG: insights from molecular simulation. J Mol Recognit 27(8):501–509
64. Jefferis R (2016) Post-translational modifications and the immunogenicity of biotherapeutics. J Immunol Res 2016:5358272

65. Moss AC, Brinks V, Carpenter JF (2013) Review article: immunogenicity of anti-TNF biologics in IBD – the role of patient, product and prescriber factors. Aliment Pharmacol Ther 38(10):1188–1197
66. Filipe V, Jiskoot W, Basmeleh AH, Halim A, Schellekens H, Brinks V (2012) Immunogenicity of different stressed IgG monoclonal antibody formulations in immune tolerant transgenic mice. MAbs 4(6):740–752
67. Masuda K, Kubota T, Kaneko E et al (2007) Enhanced binding affinity for FcgammaRIIIa of fucose-negative antibody is sufficient to induce maximal antibody-dependent cellular cytotoxicity. Mol Immunol 44(12):3122–3131
68. Sorensen M, Harmes DC, Stoll DR, Staples GO, Fekete S, Guillarme D, Beck A (2016) Comparison of originator and biosimilar therapeutic monoclonal antibodies using comprehensive two-dimensional liquid chromatography coupled with time-of-flight mass spectrometry. MAbs 8:1224–1234
69. Mimura Y, Kelly RM, Unwin L, Albrecht S, Jefferis R, Goodall M, Mizukami Y, Mimura-Kimura Y, Matsumoto T, Ueoka H, Rudd PM (2016) Enhanced sialylation of a human chimeric IgG1 variant produced in human and rodent cell lines. J Immunol Methods 428:30–36
70. Reusch D, Haberger M, Maier B, Maier M, Kloseck R, Zimmermann B, Hook M, Szabo Z, Tep S, Wegstein J, Alt N, Bulau P, Wuhrer M (2015) Comparison of methods for the analysis of therapeutic immunoglobulin G Fc-glycosylation profiles–part 1: separation-based methods. MAbs 7(1):167–179
71. Consortium for Functional Glycomics (2016) Symbol and text nomenclature for representation of glycan stucture. http://glycomics.scripps.edu/CFGnomenclature.pdf. Accessed 8 Sept 2016
72. NIBRT (2016) Glycobase 3.2.4. http://glycobase.nibrt.ie/glycobase/about.action. Accessed 8 Sept 2016
73. Jefferis R, Lund J, Mizutani H et al (1990) A comparative study of the N-linked oligosaccharide structures of human IgG subclass proteins. Biochem J 268(3):529–537
74. Kobata A (2008) The N-linked sugar chains of human immunoglobulin G: their unique pattern, and their functional roles. Biochim Biophys Acta 1780(3):472–478
75. Farooq M, Takahashi N, Arrol H et al (1997) Glycosylation of polyclonal and paraprotein IgG in multiple myeloma. Glycoconj J 14(4):489–492
76. Mimura Y, Ashton PR, Takahashi N et al (2007). J Immunol Methods 326(1–2):116–126
77. Xue J, Zhu LP, Wei Q (2013) IgG-Fc N-glycosylation at Asn297 and IgA O-glycosylation in the hinge region in health and disease. Glycoconj J 30(8):735–745. https://doi.org/10.1007/s10719-013-9481-y
78. Lauc G, Huffman JE, Pučić M, Zgaga L, Adamczyk B, Mužinić A, Novokmet M, Polašek O, Gornik O, Krištić J, Keser T, Vitart V, Scheijen B, Uh HW, Molokhia M, Patrick AL, McKeigue P, Kolčić I, Lukić IK, Swann O, van Leeuwen FN, Ruhaak LR, Houwing-Duistermaat JJ, Slagboom PE, Beekman M, de Craen AJ, Deelder AM, Zeng Q, Wang W, Hastie ND, Gyllensten U, Wilson JF, Wuhrer M, Wright AF, Rudd PM, Hayward C, Aulchenko Y, Campbell H, Rudan I (2013) Loci associated with N-glycosylation of human immunoglobulin G show pleiotropy with autoimmune diseases and haematological cancers. PLoS Genet 9(1):e1003225
79. Sonneveld ME, Koelewijn J, de Haas M, Admiraal J, Plomp R, Koeleman CA, Hipgrave Ederveen AL, Ligthart P, Wuhrer M, van der Schoot CE, Vidarsson G (2016) Antigen specificity determines anti-red blood cell IgG-Fc alloantibody glycosylation and thereby severity of haemolytic disease of the fetus and newborn. Br J Haematol 176(4):651–660. https://doi.org/10.1111/bjh.14438
80. Wuhrer M, Stavenhagen K, Koeleman CAM, Selman MHJ, Harper L, Jacobs BJ, Savage COS, Jefferis R, Deelder AM, Morgan M (2015) Skewed Fc glycosylation profiles of anti-proteinase 3 immunoglobulin G1 autoantibodies from granulomatosis with polyangiitis patients show low levels of bisection, galactosylation and sialylation. J Proteome Res 14 (4):1657–1665

81. Hafkenscheid L, Bondt A, Scherer HU, Huizinga TW, Wuhrer M, Toes RE, Rombouts Y (2016) Structural analysis of variable domain glycosylation of anti-citrullinated protein antibodies in rheumatoid arthritis reveals the presence of highly sialylated glycans. Mol Cell Proteomics 16:278–287
82. Ackerman ME, Crispin M, Yu X, Baruah K, Boesch AW, Harvey DJ, Dugast A-S, Heizen EL, Ercan A, Choi I, Streeck H, Nigrovic PA, Bailey-Kellogg C, Scanlan C, Alter G (2013) Natural variation in Fc glycosylation of HIV-specific antibodies impacts antiviral activity. J Clin Invest 123(5):2183–2192
83. Blondeel EJ, Braasch K, McGill T, Chang D, Engel C, Spearman M, Butler M, Aucoin MG (2015) Tuning a MAb glycan profile in cell culture: supplementing N-acetylglucosamine to favour G0 glycans without compromising productivity and cell growth. J Biotechnol 214:105–112
84. Sha S, Agarabi C, Brorson K, Lee DY, Yoon S (2016) N-Glycosylation design and control of therapeutic monoclonal antibodies. Trends Biotechnol 34(10):835–846
85. Hossler P (2012) Protein glycosylation control in mammalian cell culture: past precedents and contemporary prospects. Adv Biochem Eng Biotechnol 127:187–219
86. Jimenez del Val I, Nagy JM, Kontoravdi C (2011) A dynamic mathematical model for monoclonal antibody N-linked glycosylation and nucleotide sugar donor transport within a maturing Golgi apparatus. Biotechnol Prog 27(6):1730–1743
87. Lin CW, Tsai MH, Li ST, Tsai TI, Chu KC, Liu YC, Lai MY, Wu CY, Tseng YC, Shivatare SS, Wang CH, Chao P, Wang SY, Shih HW, Zeng YF, You TH, Liao JY, Tu YC, Lin YS, Chuang HY, Chen CL, Tsai CS, Huang CC, Lin NH, Ma C, Wu CY, Wong CH (2015) A common glycan structure on immunoglobulin G for enhancement of effector functions. Proc Natl Acad Sci U S A 112(34):10611–10616
88. Andersen DC, Bridges T, Grawlitzek M, Hoy C (2000) Multiple cell culture factors can affect the glycosylation of Asn-184 in CHO-produced tissue-type plasminogen activator. Biotechnol Bioeng 70(1):25–31
89. Bosques CJ, Collins BE, Meador 3rd JW, Sarvaiya H, Murphy JL, Dellorusso G, Bulik DA, Hsu IH, Washburn N, Sipsey SF, Myette JR, Raman R, Shriver Z, Sasisekharan R, Venkataraman G (2010) Chinese hamster ovary cells can produce galactose–α1,3-galactose antigens on proteins. Nat Biotechnol 28(11):1153–1156
90. Galili U (2016) Natural anticarbohydrate antibodies contributing to evolutionary survival of primates in viral epidemics? Glycobiology 26:1140–1150
91. Ghaderi D, Zhang M, Hurtado-Ziola N, Varki A (2012) Production platforms for biotherapeutic glycoproteins. Occurrence, impact, and challenges of non-human sialylation. Biotechnol Genet Eng Rev 28:147–176
92. Mimura Y, Church S, Ghirlando R et al (2000) The influence of glycosylation on the thermal stability and effector function expression of human IgG1-Fc: properties of a series of truncated glycoforms. Mol Immunol 37:697–706
93. Krapp S, Mimura Y, Jefferis R et al (2003) Structural analysis of human IgG glycoforms reveals a correlation between oligosaccharide content, structural integrity and Fc-receptor affinity. J Mol Biol 325:979–989
94. Shields RL, Lai J, Keck R et al (2002) Lack of fucose on human IgG1 N-Linked oligosaccharide improves binding to human FcγRIII and antibody-dependent cellular toxicity. J Biol Chem 277:26733–26740
95. Davies J, Jiang L, Labarre MJ et al (2001) Expression of GTIII in a recombinant anti-CD20 CHO production cell line: expression of antibodies of altered glycoforms leads to an increase in ADCC thro' higher affinity for FcRIII. Biotechnol Bioeng 74:288–294
96. Ferrara C, Brünker P, Suter T, Moser S, Püntener U, Umaña P (2006) Modulation of therapeutic antibody effector functions by glycosylation engineering: influence of Golgi enzyme localization domain and co-expression of heterologous beta1, 4-N-acetylglucosaminyltransferase III and Golgi alpha-mannosidase II. Biotechnol Bioeng 93(5):851–861

97. Shinkawa T, Nakamura K, Yamane N, Shoji-Hosaka E, Hanai N, Kanda Y, Sakurada M, Uchida K, Anazawa H, Satoh M, Yamasaki M, Hanai N, Shitara K (2003) The absence of fucose but not the presence of galactose or bisecting *N*-acetylglucosamine of human IgG1 complex-type oligosaccharides shows the critical role of enhancing antibody-dependent cellular cytotoxicity. J Biol Chem 278:3466–3473
98. Liu SD, Chalouni C, Young JC, Junttila TT, Sliwkowski MX, Lowe JB (2015) Afucosylated antibodies increase activation of FcγRIIIa-dependent signaling components to intensify processes promoting ADCC. Cancer Immunol Res 3(2):173–183
99. Sagawa T, Oda M, Morii H, Takizawa H, Kozono H, Azuma T (2005) Conformational changes in the antibody constant domains upon hapten-binding. Mol Immunol 42(1):9–18
100. Dall'Acqua WF, Cook KE, Damschroder MM, Woods RM, Herren W (2006) Modulation of the effector functions of a human IgG1 through engineering of its hinge region. J Immunol 177(2):1129–1138
101. Xia Y, Pawar RD, Nakouzi AS, Herlitz L, Broder A, Liu K et al (2012) The constant region contributes to the antigenic specificity and renal pathogenicity of murine anti-DNA antibodies. J Autoimmun 39:398–411
102. Crespillo S, Casares S, Mateo PL, Conejero-Lara F (2014) Thermodynamic analysis of the binding of 2F5 (Fab and immunoglobulin G forms) to its gp41 epitope reveals a strong influence of the immunoglobulin Fc region on affinity. J Biol Chem 289:594–599
103. Janda A, Bowen A, Greenspan NS, Casadevall A (2016) Ig constant region effects on variable region structure and function. Front Microbiol 7:22
104. Stanley P (2011) Golgi glycosylation. Cold Spring Harb Perspect Biol 3(4):a005199
105. Gomathinayagam S, Laface D, Houston-Cummings NR, Mangadu R, Moore R, Shandil I, Sharkey N, Li H, Stadheim TA, Zha D (2015) In vivo anti-tumor efficacy of afucosylated anti-CS1 monoclonal antibody produced in glycoengineered Pichia pastoris. J Biotechnol 208:13–21
106. Shibata-Koyama M, Iida S, Misaka H, Mori K, Yano K, Shitara K, Satoh M (2009) Nonfucosylated rituximab potentiates human neutrophil phagocytosis through its high binding for FcgammaRIIIb and MHC class II expression on the phagocytic neutrophils. Exp Hematol 37:309–321
107. Hmiel LK, Brorson KA (2015) Boyne MT 2nd post-translational structural modifications of immunoglobulin G and their effect on biological activity. Anal Bioanal Chem 407(1):79–94. https://doi.org/10.1007/s00216-014-8108-x
108. Yamane-Ohnuki NM, Satoh M (2009) Production of therapeutic antibodies with controlled fucosylation. MAbs 1:230–236
109. Subramaniam JM, Whiteside G, McKeage K, Croxtall JC (2012) Mogamulizumab: first global approval. Drugs 72:1293–1298
110. Ferrara C, Grau S, Jäger C, Sondermann P, Brünker P, Waldhauer I, Hennig M, Ruf A, Rufer AC, Stihle M, Umaña P, Benz J (2011) Unique carbohydrate-carbohydrate interactions are required for high affinity binding between FcgammaRIII and antibodies lacking core fucose. Proc Natl Acad Sci U S A 108(31):12669–12674
111. Golay J, Da Roit F, Bologna L, Ferrara C, Leusen JH, Rambaldi A, Klein C, Introna M (2013) Glycoengineered CD20 antibody obinutuzumab activates neutrophils and mediates phagocytosis through CD16B more efficiently than rituximab. Blood 122(20):3482–3491
112. Shah A (2015) New developments in the treatment of chronic lymphocytic leukemia: role of obinutuzumab. Ther Clin Risk Manage 11:1113–1122
113. Reddy V, Dahal LN, Cragg MS, Leandro M (2016) Optimising B-cell depletion in autoimmune disease: is obinutuzumab the answer? Drug Discov Today 21(8):1330–1338. https://doi.org/10.1016/j.drudis.2016.06.009
114. Ogorek C, Jordan I, Sandig V, von Horsten HH (2012) Fucose-targeted glycoengineering of pharmaceutical cell lines. Methods Mol Biol 907:507–517
115. Peipp M, van Bueren JJL, Schneider-Merck T, Bleeker WW, Dechant M, Beyer T, Repp R, van Berkel PH, Vink T, van de Winkel JG, Parren PW, Valerius T (2008) Antibody

fucosylation differentially impacts cytotoxicity mediated by NK and PMN effector cells. Blood 112(6):2390–2399
116. Derer S, Kellner C, Berger S, Valerius T, Peipp M (2012) Fc engineering: design, expression, and functional characterization of antibody variants with improved effector function. Methods Mol Biol 907:519–536
117. Nakagawa T, Natsume A, Satoh M, Niwa R (2010) Non-fucosylated anti-CD20 antibody potentially induces apoptosis in lymphoma cells through enhanced interaction with FcgammaRIIIb on neutrophils. Leuk Res 34:666–671
118. Derer S, Glorius P, Schlaeth M, Lohse S, Klausz K, Muchhal U, Desjarlais JR, Humpe A, Valerius T, Peipp M (2014) Increasing FcγRIIa affinity of an FcγRIII-optimized anti-EGFR antibody restores neutrophil-mediated cytotoxicity. MAbs 6(2):409–421
119. Natsume A, In M, Takamura H, Nakagawa T, Shimizu Y, Kitajima K, Wakitani M, Ohta S, Satoh M, Shitara K, Niwa R (2008) Engineered antibodies of IgG1/IgG3 mixed isotype with enhanced cytotoxic activities. Cancer Res 68:3863–3872
120. Le NPL, Bowden TA, Struwe WB, Crispin M (2016) Immune recruitment or suppression by glycan engineering of endogenous and therapeutic antibodies. Biochim Biophys Acta 1860(8):1655–1668
121. Yu C, Crispin M, Sonnen A, Harvey DJ, Chang VT, Evans EJ, Scanlan CJ, Stuart DI, Gilbert RJC, Davis SJ (2011) Use of the α-mannosidase I inhibitor kifunensine allows the crystallization of apo CTLA-4 homodimer produced in long-term cultures of Chinese hamster ovary cells. Acta Crystallogr Sect F Struct Biol Cryst Commun 67:785–789
122. Gloster TM, Vocadlo DJ (2012) Developing inhibitors of glycan processing enzymes as tools for enabling glycobiology. Nat Chem Biol 8:683–694
123. Sealover NR, Davis AM, Brooks JK, George HJ, Kayser KJ, Lin N (2013) Engineering Chinese hamster ovary (CHO) cells for producing recombinant proteins with simple glycoforms by zinc-finger nuclease (ZFN)-mediated gene knockout of mannosyl (alpha-1,3-)-glycoprotein beta-1,2-N-acetylglucosaminyltransferase (Mgat1). J Biotechnol 167:24–32
124. Okeley NM, Alley SC, Anderson ME, Boursalian TE, Burke PJ, Emmerton KM, Jeffrey SC, Klussman K, Law CL, Sussman D, Toki BE, Westendorf L, Zeng W, Zhang X, Benjamin DR, Senter PD (2013) Development of orally active inhibitors of protein and cellular fucosylation. Proc Natl Acad Sci U S A 110(14):5404–5409
125. Mizushima T, Yagi H, Takemoto E, Shibata-Koyama M, Isoda Y, Iida S, Masuda K, Satoh M, Kato K (2011) Structural basis for improved efficacy of therapeutic antibodies on defucosylation of their Fc glycans. Genes Cells 16(11):1071–1080
126. Shibata-Koyama M, Iida S, Okazaki A, Mori K, Kitajima-Miyama K, Saitou S, Kakita S, Kanda Y, Shitara K, Kato K et al (2009) The N-linked oligosaccharide at Fc gamma RIIIa Asn-45: an inhibitory element for high Fc gamma RIIIa binding affinity to IgG glycoforms lacking core fucosylation. Glycobiology 19:126–134
127. Pucić M, Knezević A, Vidic J, Adamczyk B, Novokmet M, Polasek O, Gornik O, Supraha-Goreta S, Wormald MR, Redzić I, Campbell H, Wright A, Hastie ND, Wilson JF, Rudan I, Wuhrer M, Rudd PM, Josić D, Lauc G (2011) High throughput isolation and glycosylation analysis of IgG-variability and heritability of the IgG glycome in three isolated human populations. Mol Cell Proteomics 10(10):M111.010090
128. de Jong SE, Selman MH, Adegnika AA, Amoah AS, van Riet E, Kruize YC, Raynes JG, Rodriguez A, Boakye D, von Mutius E, Knulst AC, Genuneit J, Cooper PJ, Hokke CH, Wuhrer M, Yazdanbakhsh M (2016) IgG1 Fc N-glycan galactosylation as a biomarker for immune activation. Sci Rep 6:28207
129. Einarsdottir HK, Selman MH, Kapur R, Scherjon S, Koeleman CA, Deelder AM, van der Schoot CE, Vidarsson G, Wuhrer M (2013) Comparison of the Fc glycosylation of fetal and maternal immunoglobulin G. Glycoconj J 30(2):147–157
130. Bondt A, Selman MH, Deelder AM et al (2013) Association between galactosylation of immunoglobulin G and improvement of rheumatoid arthritis during pregnancy is independent of sialylation. J Proteome Res 12(10):4522–4531

131. Plomp R, Bondt A, de Haan N, Rombouts Y, Wuhrer M (2016) Recent advances in clinical glycoproteomics of immunoglobulins (Igs). Mol Cell Proteomics 15(7):2217–2228
132. Miyoshi E, Shinzaki S, Fujii H, Iijima H, Kamada Y, Takehara T (2016) Role of aberrant IgG glycosylation in the pathogenesis of inflammatory bowel disease. Proteomics Clin Appl 10:384–390
133. Maverakis E, Kim K, Shimoda M, Gershwin ME, Patel F, Wilken R, Raychaudhuri S, Ruhaak LR, Lebrilla CB (2015) Glycans in the immune system and the altered glycan theory of autoimmunity: a critical review. J Autoimmun 57:1–13
134. Farooq M, Takahashi N, Drayson M, Lund J, Jefferis R (1998) A longitudinal study of glycosylation of a human IgG3 paraprotein in a patient with multiple myeloma. Adv Exp Med Biol 435:95–103
135. Butler M, Spearman M (2014) The choice of mammalian cell host and possibilities for glycosylation engineering. Curr Opin Biotechnol 30:107–112
136. Liu L (2015) Antibody glycosylation and its impact on the pharmacokinetics and pharmacodynamics of monoclonal antibodies and Fc-fusion proteins. J Pharm Sci 104:1866–1884
137. Dumont J, Euwart D, Mei B, Estes S, Kshirsagar R (2015) Human cell lines for biopharmaceutical manufacturing: history, status, and future perspectives. Crit Rev Biotechnol 18:1–13
138. Kuo TT, Baker K, Yoshida M, Qiao SW, Aveson VG, Lencer WI, Blumberg RS (2010) Neonatal Fc receptor: from immunity to therapeutics. J Clin Immunol 30(6):777–789
139. Sand KM, Bern M, Nilsen J, Noordzij HT, Sandlie I, Andersen JT (2015) Unraveling the interaction between FcRn and albumin: opportunities for design of albumin-based therapeutics. Front Immunol 5:682
140. Boyd PN, Lines AC, Patel AK (1995) The effect of the removal of sialic acid, galactose and total carbohydrate on the functional activity of Campath-1H. Mol Immunol 32:1311–1318
141. Raju TS, Jordan R (2012) Galactosylation variations in marketed therapeutic antibodies. MAbs 4(3):385–391
142. van de Bovenkamp FS, Hafkenscheid L, Rispens T, Rombouts Y (2016) The emerging importance of IgG fab glycosylation in immunity. J Immunol 196(4):1435–1441
143. Holland M, Yagi H, Takahashi N, Kato K, Savage CO, Goodall DM, Jefferis R (2006) Differential glycosylation of polyclonal IgG, IgG-Fc and IgG-Fab isolated from the sera of patients with ANCA-associated systemic vasculitis. Biochim Biophys Acta 1760(4):669–677
144. Ahmed AA, Giddens J, Pincetic A, Lomino JV, Ravetch JV, Wang LX, Bjorkman PJ (2014) Structural characterization of anti-inflammatory immunoglobulin G Fc proteins. J Mol Biol 426(18):3166–3179
145. Nagelkerke SQ, Kuijpers TW (2015) Immunomodulation by IVIg and the role of Fc-gamma receptors: classic mechanisms of action after all? Front Immunol 5:674
146. Biermann MH, Griffante G, Podolska MJ, Boeltz S, Stürmer J, Muñoz LE, Bilyy R, Herrmann M (2016) Sweet but dangerous – the role of immunoglobulin G glycosylation in autoimmunity and inflammation. Lupus 25(8):934–942
147. Nimmerjahn F, Ravetch JV (2007) The antiinflammatory activity of IgG: the intravenous IgG paradox. J Exp Med 204:11–15
148. Magorivska I, Muñoz LE, Janko C, Dumych T, Rech J, Schett G, Nimmerjahn F, Bilyy R, Herrmann M (2016) Sialylation of anti-histone immunoglobulin G autoantibodies determines their capabilities to participate in the clearance of late apoptotic cells. Clin Exp Immunol 184 (1):110–117
149. Schwab I, Lux A, Nimmerjahn F (2015) Pathways responsible for human autoantibody and therapeutic intravenous igg activity in humanized mice. Cell Rep 13(3):610–620
150. Schwab I, Nimmerjahn F (2014) Role of sialylation in the anti-inflammatory activity of intravenous immunoglobulin – $F(ab')_2$ versus Fc sialylation. Clin Exp Immunol 178(Suppl 1):97–99
151. Ohmi Y, Ise W, Harazono A, Takakura D, Fukuyama H, Baba Y, Narazaki M, Shoda H, Takahashi N, Ohkawa Y, Ji S, Sugiyama F, Fujio K, Kumanogoh A, Yamamoto K,

Kawasaki N, Kurosaki T, Takahashi Y, Furukawa K (2016) Sialylation converts arthritogenic IgG into inhibitors of collagen-induced arthritis. Nat Commun 7:11205
152. Wong AH, Fukami Y, Sudo M, Kokubun N, Hamada S, Yuki N (2016) Sialylated IgG-Fc: a novel biomarker of chronic inflammatory demyelinating polyneuropathy. J Neurol Neurosurg Psychiatry 87(3):275–279
153. Raymond C, Robotham A, Spearman M, Butler M, Kelly J, Durocher Y (2015) Production of α2,6-sialylated IgG1 in CHO cells. MAbs 7(3):571–583
154. Bayry J, Bansal K, Kazatchkine MD et al (2009) DC-SIGN and alpha2,6-sialylated IgG Fc interaction is dispensable for the anti-inflammatory activity of IVIg on human dendritic cells. Proc Natl Acad Sci U S A 106:E24
155. Campbell IK, Miescher S, Branch DR, Mott PJ, Lazarus AH, Han D, Maraskovsky E, Zuercher AW, Neschadim A, Leontyev D, McKenzie BS, Käsermann F (2014) Therapeutic effect of IVIG on inflammatory arthritis in mice is dependent on the Fc portion and independent of sialylation or basophils. J Immunol 192(11):5031–5038
156. Quast I, Peschke B, Lünemann JD (2016) Regulation of antibody effector functions through IgG Fc N-glycosylation. Cell Mol Life Sci 75:837–847
157. Bouhlal H, Martinvalet D, Teillaud JL, Fridman C, Kazatchkine MD, Bayry J, Lacroix-Desmazes S, Kaveri SV (2014) Natural autoantibodies to Fcγ receptors in intravenous immunoglobulins. J Clin Immunol 34(Suppl 1):S4–11
158. Nagelkerke SQ, Dekkers G, Kustiawan I, van de Bovenkamp FS, Geissler J, Plomp R, Wuhrer M, Vidarsson G, Rispens T, van den Berg TK, Kuijpers TW (2014) Inhibition of FcγR-mediated phagocytosis by IVIg is independent of IgG-Fc sialylation and FcγRIIb in human macrophages. Blood 124(25):3709–3718
159. Thomann M, Schlothauer T, Dashivets T, Malik S, Avenal C, Bulau P, Rüger P, Reusch D (2015) In vitro glycoengineering of IgG1 and its effect on Fc receptor binding and ADCC activity. PLoS One 10(8):e0134949. https://doi.org/10.1371/journal.pone.0134949
160. Yu X, Vasiljevic S, Mitchell DA, Crispin M, Scanlan CN (2013) Dissecting the molecular mechanism of IVIg therapy: the interaction between serum IgG and DC-SIGN is independent of antibody glycoform or Fc domain. J Mol Biol 425:1253–1258
161. Bournazos S, Ravetch JV (2015) Fcγ receptor pathways during active and passive immunization. Immunol Rev 268(1):88–103
162. Zhou Q, Shankara S, Roy A, Qiu H, Estes S, McVie-Wylie A, Culm-Merdek K, Park A, Pan C, Edmunds T (2008) Development of a simple and rapid method for producing non-fucosylated oligomannose containing antibodies with increased effector function. Biotechnol Bioeng 99:652–665
163. Zhang P, Chan KF, Haryadi R, Bardor M, Song Z (2013) CHO glycosylation mutants as potential host cells to produce therapeutic proteins with enhanced efficacy. Adv Biochem Eng Biotechnol 131:63–87
164. Pacis E, Yu M, Autsen J, Bayer R, Li F (2011) Effects of cell culture conditions on antibody N-linked glycosylation-what affects high mannose 5 glycofor. Biotechnol Bioeng 108 (10):2348–2358
165. Zhong X, Cooley C, Seth N, Juo ZS, Presman E, Resendes N, Kumar R, Allen M, Mosyak L, Stahl M, Somers W, Kriz R (2012) Engineering novel Lec1 glycosylation mutants in CHO–DUKX cells: molecular insights and effector modulation of *N*-acetylglucosaminyltransferase I. Biotechnol Bioeng 109:1723–1734
166. Coss KP, Vasiljevic S, Pritchard LK, Krumm SA, Glaze M, Madzorera S, Moore PL, Crispin M, Doores KJ (2016) HIV-1 Glycan density drives the persistence of the mannose patch within an infected individual. J Virol 90(24):11132–11144
167. Van Patten SM, Hughes H, Huff MR, Piepenhagen PA, Waire J, Qiu H, Ganesa C, Reczek D, Ward PV, Kutzko JP, Edmunds T (2007) Effect of mannose chain length on targeting of glucocerebrosidase for enzyme replacement therapy of Gaucher disease. Glycobiology 17:467–478

168. Lingg N, Zhang P, Song Z, Bardor M (2012) The sweet tooth of biopharmaceuticals: importance of recombinant protein glycosylation analysis. Biotechnol J 7:1462–1472
169. Jaumouillé V, Grinstein S (2016) Molecular mechanisms of phagosome formation. Microbiol Spectr 4(3). https://doi.org/10.1128/microbiolspec.MCHD-0013-2015
170. Brändén CI, Deisenhofer J (1997) Proteins. Curr Opin Struct Biol 7(6):819–820
171. Padlan EA (1990) In: Metzger H (ed) Fc receptors and the action of antibodies. American Society for Microbiology, Washington, pp 12–30
172. Sauer-Eriksson AE, Kleywegt GJ, Uhlén M, Jones TA (1995) Crystal structure of the C2 fragment of streptococcal protein G in complex with the Fc domain of human IgG. Structure 3 (3):265–278
173. Corper AL, Sohi MK, Bonagura VR, Steinitz M, Jefferis R, Feinstein A, Beale D, Taussig MJ, Sutton BJ (1997) Structure of human IgM rheumatoid factor Fab bound to its autoantigen IgG Fc reveals a novel topology of antibody-antigen interaction. Nat Struct Biol 4(5):374–381
174. Ramsland PA, Farrugia W, Bradford TM, Sardjono CT, Esparon S, Trist HM, Powell MS, Tan PS, Cendron AC, Wines BD, Scott AM, Hogarth PM (2011) Structural basis for Fc gammaRIIa recognition of human IgG and formation of inflammatory signaling complexes. J Immunol 187(6):3208–3217
175. Sondermann P, Huber R, Oosthuizen V et al (2000) The 3.2-A crystal structure of the human IgG1 Fc-FcγRIIIb complex. Nature 406:267–273
176. Radaev S, Motyka S, Fridman WH et al (2001) The structure of human type FcγIII receptor in complex with Fc. J Biol Chem 276:16469–16477
177. Radaev S, Sun P (2002) Recognition of immunoglobulins by Fcgamma receptors. Mol Immunol 38(14):1073–1083
178. Acuner Ozbabacan SE, Engin HB, Keskin O (2011) Transient protein-protein interactions. Protein Eng Des Sel 24(9):635–648
179. Mimoto F, Kadono S, Katada H, Igawa T, Kamikawa T, Hattori K (2014) Crystal structure of a novel asymmetrically engineered Fc variant with improved affinity for FcγRs. Mol Immunol 58(1):132–138
180. Davis DL, Schiel J, Borisov O (eds) (2016) Current state of the art and emerging technologies for the characterisation of monoclonal antibodies Volume 1. Monoclonal antibody therapeutics: structure, function, and regulatory space. ACS symposium series. American Chemical Society, Washington. ISBN: 9780841230262
181. Schiel JE, Davis DL, Borisov OV (eds) (2016) State-of-the-art and emerging technologies for therapeutic monoclonal antibody characterization Volume 2. Biopharmaceutical characterization the NISTmAb case study. ACS symposium series. American Chemical Society, Washington. ISBN: 9780841230293
182. Schiel JE, Davis DL, Borisov OV (eds) (2016) State-of-the-art and emerging technologies for therapeutic monoclonal antibody characterization Volume 3. Defining the next generation of analytical and biophysical techniques. ACS symposium series. American Chemical Society, Washington. ISBN: 9780841230316
183. Coloma MJ, Trinh RK, Martinez AR, Morrison SL (1999) Position effects of variable region carbohydrate on the affinity and in vivo behavior of an anti-(1→6) dextran antibody. J Immunol 162:2162–2170
184. Jacquemin M (2010) Variable region heavy chain glycosylation determines the anticoagulant activity of a factor VIII antibody. Haemophilia 16(102):16–19
185. Wu SJ, Luo J, O'Neil KT, Kang J, Lacy ER, Canziani G, Baker A, Huang M, Tang QM, Raju TS, Jacobs SA, Teplyakov A, Gilliland GL, Feng Y (2010) Structure-based engineering of a monoclonal antibody for improved solubility. Protein Eng Des Sel 23:643–651
186. Stork R, Zettlitz KA, Müller D, Rether M, Hanisch FG, Kontermann RE (2008) N-glycosylation as novel strategy to improve pharmacokinetic properties of bispecific single-chain diabodies. J Biol Chem 283(12):7804–7812
187. Qian J, Liu T, Yang L, Daus A, Crowley R, Zhou Q (2007) Structural characterization of N-linked oligosaccharides on monoclonal antibody cetuximab by the combination of orthogonal matrix-assisted laser desorption/ionization hybrid quadrupole-quadrupole time-of-flight tandem mass spectrometry and sequential enzymatic digestion. Anal Biochem 364:8–18

188. Wiegandt A, Meyer B (2014) Unambiguous characterization of N-glycans of monoclonal antibody cetuximab by integration of LC-MS/MS and $^1$H NMR spectroscopy. Anal Chem 86 (10):4807–4814
189. Chung CH, Mirakhur B, Chan E, Le QT, Berlin J, Morse M, Murphy BA, Satinover SM, Hosen J, Mauro D, Slebos RJ, Zhou Q, Gold D, Hatley T, Hicklin DJ, Platts-Mills TA (2008) Cetuximab-induced anaphylaxis and IgE specific for galactose-alpha-1,3-galactose. N Engl J Med 358(11):1109–1117
190. Lammerts van Bueren JJ, Rispens T, Verploegen S, van der Palen-Merkus T, Stapel S, Workman LJ, James H, van Berkel PH, van de Winkel JG, Platts-Mills TA, Parren PW (2011) Anti-galactose-α-1,3-galactose IgE from allergic patients does not bind α-galactosylated glycans on intact therapeutic antibody Fc domains. Nat Biotechnol 29 (7):574–576
191. Daguet A, Watier H (2011) 2nd Charles Richet et Jules Héricourt workshop: therapeutic antibodies and anaphylaxis; May 31–June 1, 2011, Tours, France. MAbs 3(5):417–421
192. Pointreau Y, Commins SP, Calais G, Watier H, Platts-Mills TA (2012) Fatal infusion reactions to cetuximab: role of immunoglobulin e-mediated anaphylaxis. J Clin Oncol 30 (3):334
193. Mullins RJ, James H, Platts-Mills TA, Commins S (2012) Relationship between red meat allergy and sensitization to gelatin and galactose-α-1,3-galactose. J Allergy Clin Immunol 129(5):1334–1342
194. Berg EA, Platts-Mills TA, Commins SP (2014) Drug allergens and food–the cetuximab and galactose-α-1,3-galactose story. Ann Allergy Asthma Immunol 112(2):97–101
195. Huang L, Biolsi S, Bales KR, Kuchibhotla U (2006) Impact of variable domain glycosylation on antibody clearance: an LC/MS characterization. Anal Biochem 349(2):197–207
196. Lim A, Reed-Bogan A, Harmon BJ (2008) Glycosylation profiling of a therapeutic recombinant monoclonal antibody with two N-linked glycosylation sites using liquid chromatography coupled to a hybrid quadrupole time-of-flight mass spectrometer. Anal Biochem 375 (2):163–172
197. Millward TA, Heitzmann M, Bill K, Längle U, Schumacher P, Forrer K (2008) Effect of constant and variable domain glycosylation on pharmacokinetics of therapeutic antibodies in mice. Biologicals 36(1):41–47

# Impact of Protein Glycosylation on the Design of Viral Vaccines

**Kathleen Schön, Bernd Lepenies, and Guillaume Goyette-Desjardins**

## Contents



K. Schön
Immunology Unit and Research Center for Emerging Infections and Zoonoses, University of Veterinary Medicine Hannover, Hanover, Germany

Institute for Parasitology, Centre for Infection Medicine, University of Veterinary Medicine Hannover, Hanover, Germany

B. Lepenies (✉) and G. Goyette-Desjardins (✉)
Immunology Unit and Research Center for Emerging Infections and Zoonoses, University of Veterinary Medicine Hannover, Hanover, Germany
e-mail: bernd.lepenies@tiho-hannover.de; guillaume.goyette-desjardins@tiho-hannover.de

**Abstract** Glycans play crucial roles in various biological processes such as cell proliferation, cell-cell interactions, and immune responses. Since viruses co-opt cellular biosynthetic pathways, viral glycosylation mainly depends on the host cell glycosylation machinery. Consequently, several viruses exploit the cellular glycosylation pathway to their advantage. It was shown that viral glycosylation is strongly dependent on the host system selected for virus propagation and/or protein expression. Therefore, the use of different expression systems results in various glycoforms of viral glycoproteins that may differ in functional properties. These differences clearly illustrate that the choice of the expression system can be important, as the resulting glycosylation may influence immunological properties. In this review, we will first detail protein *N*- and *O*-glycosylation pathways and the resulting glycosylation patterns; we will then discuss different aspects of viral glycosylation in pathogenesis and in vaccine development; and finally, we will elaborate on how to harness viral glycosylation in order to optimize the design of viral vaccines. To this end, we will highlight specific examples to demonstrate how glycoengineering approaches and exploitation of different expression systems could pave the way towards better self-adjuvanted glycan-based viral vaccines.

**Graphical Abstract**

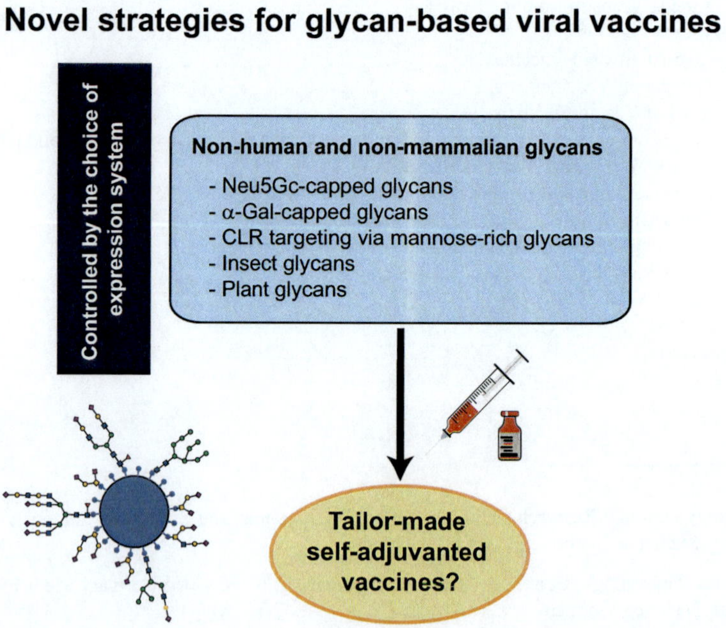


**Keywords** Glycoengineering, Immunity, Lectins, *N*-Glycosylation, *O*-Glycosylation, Vaccine, Virus

## Abbreviations

| | |
|---|---|
| APC | Antigen-presenting cell |
| Asn or N | Asparagine |
| CHO | Chinese hamster ovary |
| CLR | C-type lectin receptor |
| DC-SIGN | Dendritic cell-specific intercellular adhesion molecule-3-grabbing non-integrin |
| EBOV | Ebola virus |
| ER | Endosplasmic reticulum |
| FcR | Fc receptor |
| FDL | Fused lobes |
| Fuc | L-Fucose |
| Gal | D-Galactose |
| GalNAc | *N*-Acetyl-D-galactosamine |
| Glc | D-Glucose |
| GlcNAc | *N*-Acetyl-D-glucosamine |
| GP | Glycoprotein |
| HA | Hemagglutinin |
| HBV | Hepatitis B virus |
| HCV | Hepatitis C virus |
| HIV-1 | Human immunodeficiency virus type 1 |
| HSV-1 | Herpes simplex virus type 1 |
| HSV-2 | Herpes simplex virus type 2 |
| JEV | Japanese encephalitis virus |
| LacNAc | *N*-Acetyllactosamine (β-D-galactopyranosyl-(1→4)-2-acetamido-2-deoxy-β-D-glucopyranose) |
| Man | D-Mannose |
| MDCK | Madin-Darby canine kidney |
| MDL-1 | Myeloid DAP12-associating lectin 1 |
| MMR | Macrophage mannose receptor |
| MPL | 3-*O*-Desacyl-4′-monophosphoryl lipid |
| NA | Neuraminidase |
| nAb | Neutralizing antibody |
| Neu5Ac | *N*-Acetylneuraminic acid |
| Neu5Gc | *N*-Glycolylneuraminic acid |
| NIPV | Nipah virus |
| PRR | Pattern recognition receptor |
| RVFV | Rift Valley fever phlebovirus |
| Ser or S | Serine |
| sGP | Secreted glycoprotein |
| Sia | Sialic acid |
| SIV | Simian immunodeficiency virus |
| SNFG | Symbol Nomenclature for Glycans |

| | |
|---|---|
| Thr or T | Threonine |
| TLR | Toll-like receptor |
| VLP | Virus-like particle |
| WNV | West Nile virus |
| Xyl | D-Xylose |

# 1 Introduction

"Glycan" is a general term encompassing most carbohydrate polymers encountered in biology, whether alone (polysaccharide) or as part of a glycoconjugate, namely, an oligo-/polysaccharide linked to a lipid (glycolipid) or to an amino acid (glycopeptide/glycoprotein). Glycans are known to play essential roles in various biological processes, such as cell proliferation and differentiation, organism development, cell communication, cellular migration, and immunity [1, 2]. However, and in clear contrast to nucleic acids and peptides, glycan structures are more complex as they may be branched (instead of strictly being linear) and their monomeric units may be linked to one another by two different anomeric forms and multiple branching points [3]. While this great complexity has hampered the study of glycobiology in the past, considerable technological and methodological advancements in recent decades have provided a tremendous throve of novel knowledge and opened new avenues for research into glycan-based diagnostics, prophylactic, and therapeutic applications.

Viruses are considered to be obligatory intracellular pathogens: for a successful infection, they need to inject their genetic material into the host cell, highjack its machinery to replicate themselves, assemble new virions, and then release those to infect more cells and/or organisms [4, 5]. As such, viral glycoproteins are produced through the secretory pathway (like eukaryotic glycoproteins) and will become glycosylated in the same fashion as host glycoproteins. Therefore, viruses are dependent on the glycosylation machinery present in the infected cell. Furthermore, any alteration made to the cell's glycan synthesis will also be reflected onto the viral particles, just as different cell types and species possess different and/or unique glycosylation patterns. While certain aspects of viral glycosylation are well appreciated, such as the role of host glycans in cell recognition and entry, the function of glycan shielding in escaping immunity, and the contribution of differential glycosylation patterns to infectivity, still little is known about the impact of differential glycosylation on viral immunogenicity and its implications in vaccine design [4–7].

The effects of glycoforms (defined as differently glycosylated versions of a given glycoprotein [8]) are well-known for biotherapeutics production, where tremendous biotechnological and glycoengineering advancements are being developed to ensure increased tolerability, safety, and efficacy by creating human-like glycans [9–12]. Consistently, by exploiting non-human or non-mammalian-like glycans, self-adjuvanted glycoprotein-based vaccines may be obtained. Exploiting the impact of glycans on the immunogenicity of vaccines may offer the potential of lowering the dose itself or the number of doses required to achieve protective immunity and/or eliminate the need to use other adjuvants. Thus, exploiting immunogenic glycans in glycoprotein-based vaccines would result in better safety profiles and potentially lower production costs. On the other hand, the clinical use of non-human glycans requires careful consideration against potential side effects caused by immunologically active glycans.

In this review, we are going to focus on viral protein glycosylation and more specifically on how it could be harnessed in order to better the design of viral vaccines. We will also highlight specific examples to demonstrate how glycoengineering approaches and exploitation of different expression systems could pave the way towards self-adjuvanted glycan-based viral vaccines.

## 2 Protein *N*- and *O*-Glycosylation

Generally, glycoproteins are known to show considerable heterogeneity in their glycosylation. For a given glycoprotein, macroheterogeneity refers to the different positions that are found to be glycosylated or not (occupancy of glycosylation sites), while microheterogeneity refers to the various glycan structures to be found at a given site [4].

This variability in glycans is the combined result of many factors that can be characterized as being cell-dependent (according to species, genome, cell type, metabolic and physiological state of the cell affecting localization and activity of glycosylation enzymes and transporters, transport rate of the protein through the lumen of the endoplasmic reticulum [ER] and Golgi) or conformation-dependent (sequon localization and conformational issues affecting glycan availability to subsequent enzymatic modification) [2, 8]. Also, protein glycosylation is considered to be non-template driven, unlike the biosynthesis of nucleic acids and peptides/proteins. As will be explained below in more details, the localization of glycosyltransferases within the different sub-compartments of the Golgi can determine which enzymes encounter acceptors. Significantly, most glycosyltransferases and glycosidases require the prior actions of other enzymes before they can act, and also some are competing for the same acceptor.

Of all the types of protein glycosylation that exist [13], in this review focus will be given to *N*-glycans and *O*-glycans as they are the most studied in viral research [4]. Both types of glycosylation mainly occur in the secretory pathway (ER and Golgi), although other glycosylation reactions (such as *O*-GlcNAc) can also occur in the cytoplasm, nucleus, and mitochondria [14, 15].

## 2.1 Biosynthesis of N-Glycans in Mammals

*N*-Glycans are formed by the attachment through a *N*-glycosidic bond to a protein at asparagine (Asn, N) residues. The recognized amino acid sequence is Asn-X-Ser/Thr in which "X" is any amino acid except proline. In eukaryotes, the first phase of the process is highly conserved and consists of the assembly of the precursor oligosaccharide $Glc_3Man_9GlcNAc_2$ at the ER membrane on the lipid carrier dolichol-phosphate. During the passage of newly synthesized polypeptides through the translocon into the ER, the oligosaccharyltransferase (OST) catalyzes the transfer of the 14-sugar glycan in a β-linkage onto the amine group of a receptive Asn (Fig. 1) [4, 8, 13, 14, 16].

Early processing of *N*-glycans begins in the ER with the sequential removal of the Glc residues by α-glucosidases I and II (MOGS and GANAB) to yield $Man_9GlcNAc_2$-N (Fig. 1). Before exiting the ER, ER α-mannosidase I (MAN1B1) trims the terminal α1,2-Man residue on the central arm to yield $Man_8GlcNAc_2$-N. Further trimming occurs in the *cis*-Golgi with the actions of

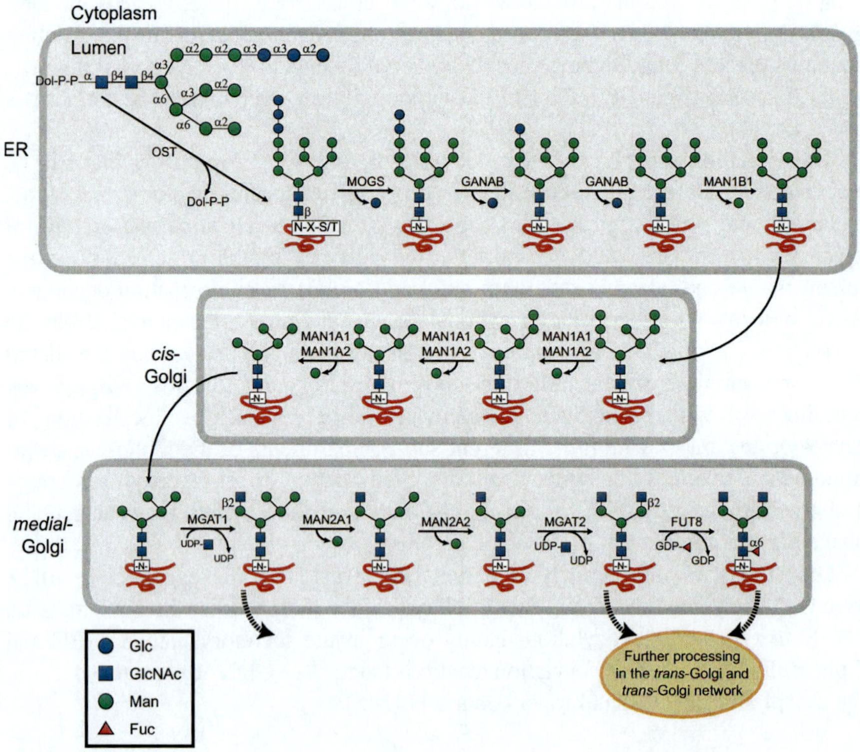


**Fig. 1** Biosynthesis of mammalian *N*-glycans, from oligomannose to hybrid and complex glycans. The red line represents the polypeptidic chain being glycosylated. Monosaccharide symbols follow the Symbol Nomenclature for Glycans (SNFG) system [222]

α-mannosidases IA and IB (MAN1A1, MAN1A2) on the three remaining α1,2-Man residues to generate $Man_5GlcNAc_2$-N, a key intermediate in the pathway to hybrid and complex *N*-glycans. While the majority of *N*-glycans are completely processed during these early steps, the incapacity of the mannosidases or of MGAT1 (see below) to act upon their substrates will prevent further processing and results in glycoproteins bearing **oligomannose *N*-glycans** (also known as high-mannose; $Man_{5-9}GlcNAc_2$-N). One should note that most mature glycoproteins possess some oligomannose *N*-glycans [4, 8, 13, 14, 16].

Late processing of *N*-glycans is initiated in the *medial*-Golgi by the addition of a GlcNAc residue to the C-2 of the α1,3-Man in $Man_5GlcNAc_2$-N by the *N*-acetylglucosaminyltransferase I (MGAT1) (Fig. 1). This step is essential for the synthesis of hybrid and complex *N*-glycans. In turn, the majority of $GlcNAcMan_5GlcNAc_2$-N are to be further trimmed down by α-mannosidase II enzymes MAN2A1 and MAN2A2 to yield $GlcNAcMan_3GlcNAc_2$-N. **Hybrid *N*-glycans** result when α-mannosidase II fails to act on $GlcNAcMan_5GlcNAc_2$-N; the terminal GlcNAc residue will be further processed like complex *N*-glycans (Fig. 2a). Once both Man residues are removed, *N*-acetylglucosaminyltransferase II (MGAT2) can act by adding GlcNAc to the C-2 of the α1,6-Man (Fig. 1). As such, $GlcNAc_2Man_3GlcNAc_2$-N, the precursor for all **biantennary complex *N*-glycans**, is synthesized and where each one of the terminal GlcNAc residues constitute the beginning of an antenna or branch. Further processing of *N*-glycans will occur in the *trans*-Golgi and its network. *N*-Acetylglucosaminyltransferases IV (MGAT4A, MGAT4B) and V (MGAT5, MGAT5B) can also initiate additional branches by transferring β-GlcNAc residues at C-4 of the core α1,3-Man and at C-6 of the core α1,6-Man, respectively, yielding **tri- and tetra-antennary complex *N*-glycans** (Fig. 2c) [4, 8, 13, 14, 16]. A fifth branch, initiated by *N*-acetylglucosaminyltransferases VI (MGAT6) at C-4 of the core α1,6-Man, is usually found in birds and fish, but can also be found in cancerous cells [8, 17–19].

Complex and hybrid *N*-glycan cores can also be modified by *N*-acetylglucosaminyltransferase III (MGAT3) to yield a bisecting GlcNAc on the core β1,4-Man residue (Fig. 2b) to form **bisected *N*-glycans** [8]. While the bisecting GlcNAc does not prevent further branching, it cannot be elongated itself, and it induces conformational changes that may suppress the elongation and termination of other branches of the *N*-glycan, thereby altering the tertiary structure and the function of the glycoprotein [20]. Additionally, the most important core modification in vertebrate *N*-glycans is core fucosylation, where the fucosyltransferase FUT8 catalyzes the addition of **core α1,6-Fuc** to the Asn-linked GlcNAc (Fig. 1). Core Fuc has been shown to play important roles in organism development and in the functional activities of immunoglobulins [21].

Finally, the majority of initiating β-GlcNAc in complex and hybrid *N*-glycans will be elongated by the addition of Gal to form the Gal-β1,4-GlcNAc (*N*-acetyllactosamine; LacNAc) building block (Fig. 3). Sequential additions of GlcNAc (in β1,3-) and Gal (in β1,4-) will yield tandem repeats termed poly-LacNAc. Branch elongation is terminated by capping reactions that involve the addition, usually with an α conformation, of sialic acids, Fuc, Gal, GlcNAc, and sulfate [4, 8, 13, 14, 16].

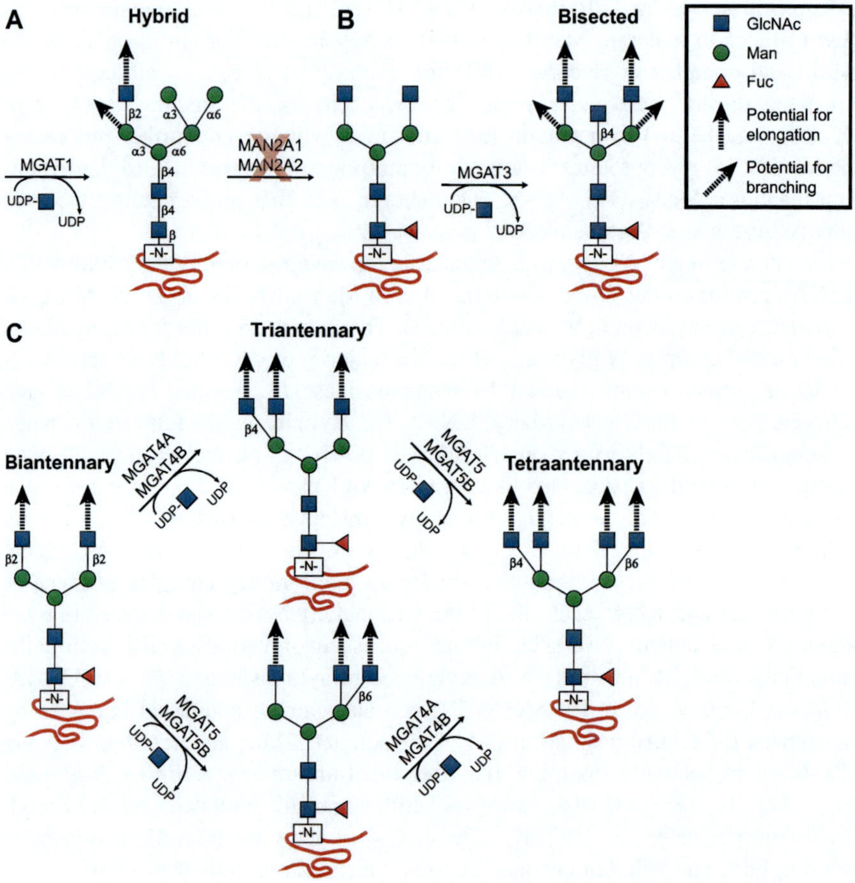


**Fig. 2** Branching and core modifications of complex *N*-glycans. Biosynthesis and common core structures of hybrid (**a**) and bisected (b) *N*-glycans. Biosynthesis of multiantennary of complex *N*-glycans (**c**). The red line represents the polypeptidic chain being glycosylated. Monosaccharide symbols follow the Symbol Nomenclature for Glycans (SNFG) system [222]

## 2.2 *Biosynthesis of N-Glycans in Insects*

As insects can be important vectors for viral infections (like arthropod-transmitted arboviruses) and also with the popular use of baculovirus-insect cell expression systems for producing *N*-glycosylated recombinant proteins, it is crucial to address the differences between the mammalian and insect pathways and the resulting *N*-glycans. In insects, the *N*-glycosylation pathway proceeds just like in mammalian cells from the transfer of the precursor oligosaccharide through the early and late processing steps until reaching $GlcNAcMan_3GlcNAc_2$-N, the intermediary for hybrid *N*-glycans (Fig. 1). The core of this intermediary can usually become α1,6-fucosylated, and only for some species can a second core Fuc be

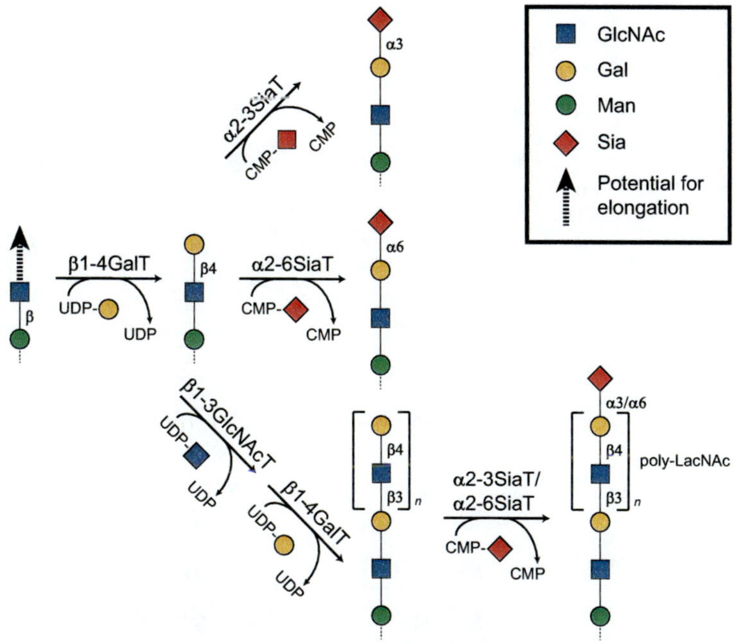


**Fig. 3** Branch elongation and capping reactions of complex *N*-glycans. One should note that most of these reactions also occur in *O*-glycans. Monosaccharide symbols follow the Symbol Nomenclature for Glycans (SNFG) system [222]. Here, sialic acids (Sia) are depicted as the family and, according to the context, could either be *N*-acetylneuraminic acid (Neu5Ac) or *N*-glycolylneuraminic acid (Neu5Gc)

transferred in α1,3- (**core α1,3-Fuc**), thanks to the two fucosyltransferases FucT6 and FucTA (Fig. 4). It should be noted that core α1,3-Fuc is also present in plant *N*-glycans [22]. From there, and independently of core fucosylation status, the terminal GlcNAc residue is removed by Fused lobes (FDL), a *N*-acetylglucosaminidase, to yield $Man_3GlcNAc_2$-N **paucimannose *N*-glycans** (Fig. 4). It is accepted that the activity of FDL is responsible for yielding only oligomannose and paucimannose *N*-glycans, which account for >90% of total *N*-glycans in insect and insect-produced glycoproteins. It also explains the inability of classical baculovirus-insect cell expression systems to produce complex and sialylated *N*-glycans [23, 24].

## 2.3 Biosynthesis of O-Glycans

*O*-Glycans are formed by the attachment through an *O*-glycosidic bond to a protein at serine (Ser, S) and threonine (Thr, T) residues. Unlike *N*-glycans, no conserved protein sequence motifs have been identified for *O*-glycosylation, making it difficult

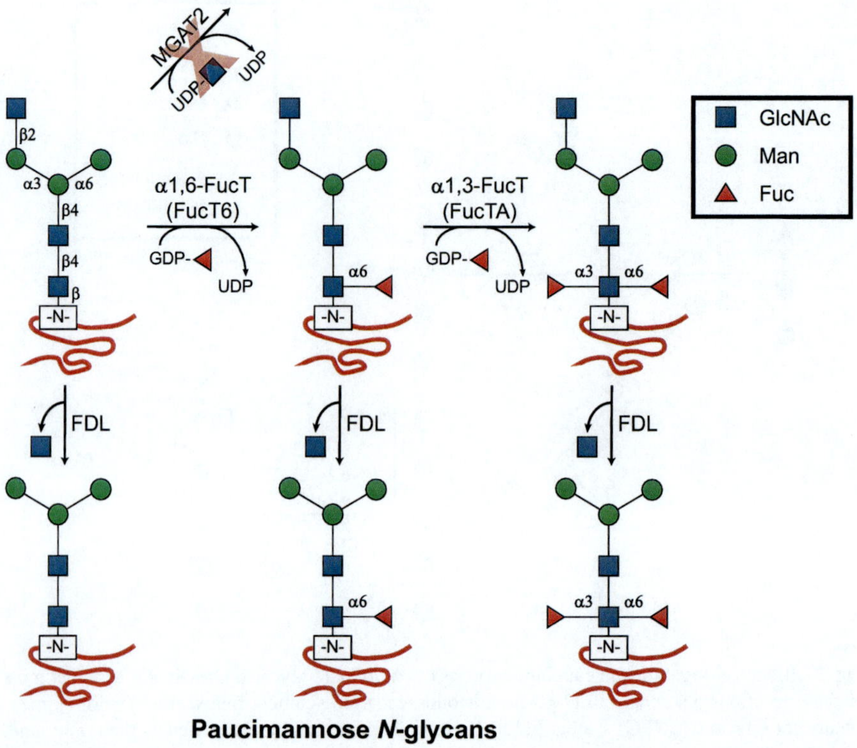


**Fig. 4** Biosynthesis of paucimannose *N*-glycans in insects. The red line represents the polypeptidic chain being glycosylated. Monosaccharide symbols follow the Symbol Nomenclature for Glycans (SNFG) system [222]

to predict this modification without experimental evidence [4]. Yet, prediction algorithms exist and are constantly improving in order to predict *O*-glycosylation sites in both mucin-like and non-mucin-like domains [5]. Moreover, many types of *O*-glycosylation exist and are classified based on the initiating monosaccharide, including, but not limited to, GalNAc, GlcNAc, Man, Fuc, and Xyl [13, 25–27]. In the context of this review, the terms *O*-glycosylation and *O*-glycans will refer to *O*-GalNAc glycosylation (also known as mucin-type), which is the most common form of protein *O*-glycosylation.

*O*-Glycosylation is initiated in the Golgi apparatus by the transfer of GalNAc in an α-linkage to the hydroxyl group of a receptive Ser or Thr protein residue and is catalyzed by a polypeptide *N*-acetylgalactosaminetransferase (GALNT) (Fig. 5). GALNTs constitute a family of conserved isoforms (up to 20 in humans) that are sequentially and functionally conserved throughout the animal kingdom. They are differentially expressed, and each possesses its own substrate specificity (in terms of amino acid sequons), which explains the high diversity in density and site occupancy of *O*-glycans [4, 14, 27–29].

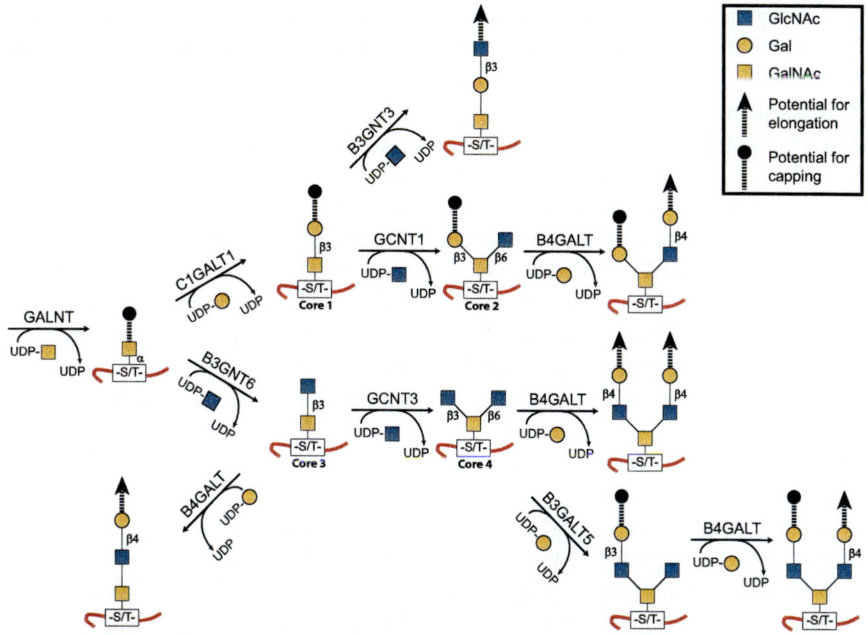


**Fig. 5** Biosynthesis of cores 1 to 4 *O*-GalNAc glycans. The red line represents the polypeptidic chain being glycosylated. Monosaccharide symbols follow the Symbol Nomenclature for Glycans (SNFG) system [222]

Following initiation, the other glycosyltransferases involved may act sequentially to build the *O*-glycan, which may consist of a single GalNAc residue up to 20 sugars, with a linear or biantennary branched structure. The addition of one or two neutral sugar residues to the *O*-GalNAc will lead to the formation of the eight cores of *O*-glycans. Since they are the most abundant, we will focus in this review on cores 1–4. For more information regarding cores 5–8, please refer to some excellent chapters and reviews [5, 14, 28, 30]. **Core 1** is ubiquitously found and formed by core 1 β1,3-galactosyltransferase I (C1GALT1) which adds a Gal to the *O*-GalNAc (Fig. 5). **Core 2** *O*-glycans, which are more cell-type specific and whose expression is tightly regulated, are formed by the core 2 β1,6-*N*-acetylglucosaminyltransferase (GCNT1). The expression of cores 3 and 4 *O*-glycans is mainly restricted to mucus epithelia from the gastrointestinal and respiratory tracts and to salivary glands. **Core 3** β1,3-*N*-acetylglucosaminyltransferase (B3GNT6) acts on the *O*-GalNAc, resulting in GlcNAc-β1,3-GalNAc-α-S/T, the core 3 *O*-glycan. Subsequent action by core 2/4 β1,6-*N*-acetylglucosaminyltransferase (GCNT3) results in the **core 4** *O*-glycan formation [4, 14, 27–29].

Once synthesized, these cores can be elongated to form complex *O*-glycans. One should note that most of these reactions (elongation, branching and capping) also occur in *N*-glycans. The elongation is catalyzed by families of β1,3-*N*-acetylglucosaminyltransferases, β1,3-galactosyltransferases, and β1,4-galactosyltransferases. Notably, the β1,3-Gal residue of cores 1 and 2 can be elongated by the β1,3-*N*-acetylglucosaminyltransferase B3GNT3, which allows for building further LacNAc unit(s). Unsubstituted β-GlcNAc residues on cores 2–4 can become galactosylated by the action of β1,4-galactosyltransferases (B4GALT1-B4GALT6) and may further lead to poly-LacNAc synthesis (Fig. 3) [4, 14, 27–29]. The Gal residues in those linear poly-LacNAc can also be modified by β1,6-*N*-acetylglucosaminyltransferases (like GCNT2), and these new GlcNAc residues may also be transformed into LacNAc/poly-LacNAc units by the same processes [31]. Also, the β1,3-GlcNAc residue of cores 3 and 4 can become β1,3-galactosylated by the galactosyltransferase B3GALT5. However, note that β1,3-Gal residue can only be capped afterwards as it cannot be used as a substrate for poly-LacNAc synthesis [29].

Possible capping reactions for complex *O*-glycans may involve sialylation, sulfation and fucosylation, which will stop branch elongation. Different families of those enzymes possess varying specificities: while some prefer *O*-glycan substrates, many have overlapping activities with other glycans. For example, some α2,6-sialyltransferases (ST6GALNACI to ST6GALNAC4) are involved in the formation of the sialyl-Tn epitope (Neu5Ac-α2,6-GalNAc-α-S/T) and of sialylated core 1 *O*-glycans. Some α2,3-sialyltransferases are strictly specific to *O*-glycans, including ST3GAL1 which is mainly responsible for the sialylation of the β1,3-Gal residue of cores 1 and 2 *O*-glycans [29]. Additionally, the blood group H determinant of *O*-glycans is formed by the α1,2-fucosyltransferases FUT1 and FUT2, which can be converted to blood group A or group B by further enzymatic modification [21].

## 3 Viral Glycosylation

Viruses require and therefore hijack the host cell machinery for replication. Usually after a virus enters a living cell, its genome gets transcribed and translated, and the proteins are formed by the host cell protein biosynthesis machinery. At the end, newly formed virus particles, also known as virions, are assembled then released to infect other cells. Viral glycoproteins are formed during passage through the host cell secretory pathway. As described previously, *N*-glycosylation of viral proteins begins with the addition of the oligomannose precursor at the ER-membrane. Then, those *N*-glycans are trimmed, branched, elongated, and capped as they progress to the Golgi and move through its sub-compartments. Additionally, *O*-glycosylation occurs, while the protein is moving through the Golgi sub-compartments. In this section, we will discuss the different roles of glycans in virus entry, protein formation, and viral release as well as glycan-mediated viral immune evasion to gain a better insight into viral glycosylation.

## *3.1 Viral Glycans in the Replication of Enveloped Viruses*

Enveloped viruses surround themselves with the former host cell membrane in which its viral surface proteins are embedded. Many of these proteins are *N*- and *O*-glycosylated, such as the heavily glycosylated gp120, a much studied glycoprotein from the human immunodeficiency virus-1 (HIV-1) envelope [32]. It is well-known that protein glycosylation influences virus-cell interactions, virus replication, and recognition of viral epitopes by the host immune system [4–6, 33]. Thus, interest in viral glycosylation has increased in the last decades, and its implications have only yet started to be considered in vaccine design.

### 3.1.1 Virus Binding and Cell Entry

Many studies have shown the impact of glycosylation on the entry of enveloped viruses into host cells. For many different virus families like the *Retroviridae* (HIV-1 and simian immunodeficiency virus [SIV]), *Phenuiviridae* (Rift Valley fever phlebovirus [RVFV] and Uukuniemi phlebovirus), *Flaviviridae* (hepatitis-C virus [HCV] and West Nile virus [WNV]), and *Filoviridae* (Ebola virus [EBOV]), binding and/or cell entry via host cell-expressed lectin receptors has been demonstrated [34–38]. One example is human dendritic cell-specific intercellular adhesion molecule-3-grabbing non-integrin (DC-SIGN), a C-type lectin receptor (CLR) mainly expressed by dendritic cells that recognizes oligomannose and fucose-containing glycans [38–40]. Many examples support the important role of this CLR in the recognition, binding, and internalization of enveloped viruses. For instance, DC-SIGN is required for the entry of phleboviruses into host cells [37], binds to the viral envelope glycoproteins of HIV-1 and SIV [34, 41], and enhances the infectivity of EBOV [42, 43]. In the case of HIV-1, recognition of the oligomannose *N*-glycans on the gp120 by DC-SIGN is necessary for T-cell migration [44, 45]. Another CLR involved in HIV-1 recognition is the macrophage mannose receptor (MMR) which can also bind gp120 via its oligomannose *N*-glycans [46, 47]. MMR was also shown to recognize Dengue virus surface glycoprotein in a *N*-glycan-dependent manner [48].

Not only are *N*-linked glycans involved in virus-cell interactions, but so are viral *O*-glycans, as they are known to interact with cell surface proteins as well. For instance, removal of *O*-glycan structures in herpes simplex virus type 1 (HSV-1) attachment factor gC resulted in lower cell binding [49] and removal of the *O*-glycosylation sites of the paramyxoviruses Nipah virus (NIPV) and Hendra virus led to altered virus entry and cell-to-cell spread [50].

### 3.1.2 Virus Assembly and Release

In addition to binding and entry, virus assembly and release from the cell may also be regulated by glycosylation. Luo et al. [51] showed that specific *N*-glycosylation sites in the herpes simplex virus type 2 (HSV-2) glycoprotein B are important for cell-cell fusion and HSV-2 entry, while another *N*-glycosylation site (N133) affected viral protein formation and virus release, as removing this site markedly inhibited viral protein transport through ER and Golgi. In hepatitis B virus (HBV), elimination of *N*-linked glycosylation sites in the small envelope protein also inhibited virus secretion. Interestingly, the addition of a new glycosylation site into the same protein restored viral release [52]. Another example is the EBOV GP, which is essential for virus-cell entry and infection. This glycoprotein is encoded as a precursor protein and later matured to a fully *N*- and *O*-glycosylated surface protein [53]. It was demonstrated that this maturation is regulated by two *N*-glycosylation sites in the transmembrane subunit of the protein. Removal of both sites resulted in misfolded proteins and yielded noninfectious virus particles [54]. The abovementioned examples highlight that glycosylation is essential in the viral life cycle. The presence of *N*- and *O*-glycosylation sites may either enhance or inhibit interactions of the virus with its host cell.

## *3.2 Viral Glycans in Immune Evasion*

Another crucial factor in virus replication is host innate and adaptive immune responses against viral infection. One of the many ways viruses have found to circumvent immunity is by exploiting viral glycans for antigen shielding and mimicry.

### 3.2.1 Glycan Shielding

Important antigen epitopes at the viral surface can be masked with glycans to prevent them from detection by components of the host immune system, such as neutralizing antibodies (nAbs) and complement. Figure 6 illustrates how heavily some viral surface proteins can be glycosylated. This immune evasion strategy is employed by many virus families [5, 6], like Epstein-Barr virus [55], Lassa virus [56], HCV [57], and EBOV [58]. Here, we want to highlight just a few prominent examples. Probably the best-characterized glycan shield of all is the one from the HIV-1 envelope protein, for which numerous studies have investigated the functional role of the glycans in neutralizing antibody responses against HIV-1 [59]. It is a trimer of non-covalently associated gp120-gp41 heterodimers [60] and covered with many *N*-glycans as well as *O*-glycans [61]. The heterodimer contains more than 90 potential *N*-glycosylation sites,

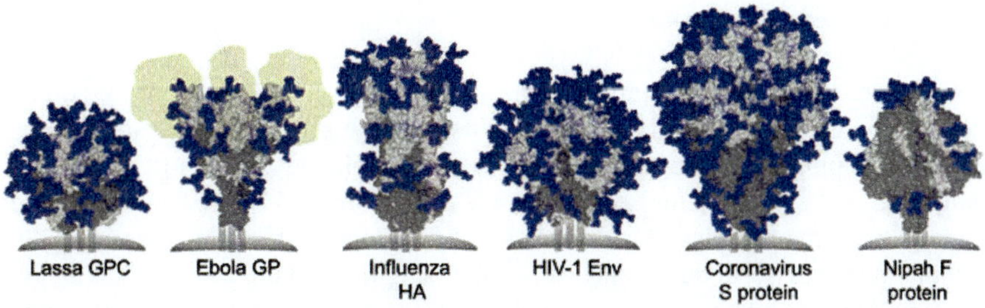


**Fig. 6** Glycan Shielding of Viral Class I Fusion Proteins. Left to right: Glycan shield models of Lassa virus glycoprotein complex (PDB ID: 5VK2) [223, 224], Ebola glycoprotein (PDB ID: 5JQ3) [225], A/H3N2/361/Victoria/2011 H3N2 Influenza virus hemagglutinin (PDB ID: 4O5N) [226, 227], BG505 SOSIP.664 HIV-1 Env (PDB ID: 4ZMJ) [228, 229], human coronavirus-NL63 (HCoV-NL63) S protein (PDB ID: 5SZS) [230], Nipah F protein (PDB ID: 5EVM) [231]. Glycans and proteins are shown in blue and gray, respectively. The fusion protein subunit is shown in dark gray. The positions of mucin-like domains of Ebola glycoprotein are shown in yellow. Most predominant sugar compositions were modelled onto each *N*-linked glycan site, using pre-existing GlcNAc residues, if possible, with $Man_5GlcNAc_2$ modelled on if compositional information was lacking. This figure, by Watanabe et al. [5], is licensed under the CC BY 4.0 license (http://creativecommons.org/licenses/BY/4.0/) and can be accessed at https://doi.org/10.1016/j.bbagen.2019.05.012

and *N*-glycans make up approximately half of the trimer mass [62]. These glycans are suggested to cover large surface areas at the envelope protein by physical shielding [63]. Glycan sites of the HIV-1 envelope protein are added and deleted frequently, resulting in constantly renewed epitopes for nAbs [64–67]. This dynamic is one of the reasons why HIV-1 can easily evade the nAb response. Wagh et al. [68] showed that the resistance to autologous nAbs was increased in vivo when potential *N*-glycosylation sites were added to fill holes in the HIV-1 glycan shield. In vitro, removal of potential *N*-glycosylation sites led to increased virus neutralization [69–72]. It is suggested that there is generally an advantage for viral fitness if the virus harbors less potential *N*-glycosylation sites; however, protection of key viral epitopes from the nAb response needs to be maintained by glycan shielding [59, 73]. Similar results were observed for other viruses like NIPV [74], HBV [75], and EBOV [76]. It was shown that, on one hand, over-glycosylation of surface structures masked antigenic epitopes from recognition by the host immune system but, on the other hand, that this resulted in higher costs in viral fitness due to decreased binding affinity and virus production. Not only *N*- but also *O*-linked glycans can shield important epitopes from recognition by B-cells; thus, removal of glycosylation sites may elicit altered host immune responses [57, 77–80]. Besides B-cell recognition, also cytotoxic T-cell activity may be affected by the alteration of viral glycosylation [81].

### 3.2.2 Secreted Glycoproteins

Beside glycan shielding and mimicry, viruses can evade the host immune system by secreting or shedding viral glycoproteins. Firstly, these glycoproteins can misdirect the humoral immune response by favoring non-neutralizing epitopes [82, 83]. Secondly, secreted glycoproteins neutralize nAbs (by adsorption) before they could bind the targeted virus particles [84]. In the HIV-1 envelope protein, monomeric gp120 subunits that may form due to incorrect processing or shedding assists the virus in escaping from the host immune system by exposing epitopes that are not present on the functional spike [85]. Thus, non-neutralizing antibodies are produced against this monomer during HIV-1 infection [82, 83]. EBOV expresses three different glycoproteins from the same genetic region caused by frame shift: the full-length membrane-bound fusion GP; a small soluble GP, whose function is yet unknown; and a dimeric secreted glycoprotein (sGP) [84, 86]. The sGP is secreted by EBOV-infected cells and acts as an antibody decoy [84]. Mohan et al. [84] proposed a model where sGP adsorbs anti-GP antibodies by sharing the same epitopes with GP. In a murine model, sGP thus eliminated protection in animals that had previously been immunized with GP. By investigating the specificity of antibodies isolated from human survivors of Ebola disease and experimentally infected macaques, researchers showed that most of them were directed against sGP, compared to the lower expressed surface protein GP, and that those antibodies directed against sGP are non-neutralizing [87, 88]. As such, the abundantly expressed EBOV sGP is a potent secreted antigen that effectively neutralizes nAbs by adsorbing them before they can bind to GP on the viral surface [5, 89]. This example demonstrates that viral glycoproteins are able to misdirect the humoral immune response, thus highlighting viral glycoprotein secretion as a potent viral mechanism to evade immunity. Figure 7 resumes the different aspects of viral glycosylation in the viral life cycle, from their replication to evading the host immune response, that were discussed here.

## 4 Viral Vaccine Design

The development of vaccines is one of the greatest triumphs in medicine. Worldwide, vaccination against pathogens has prevented an estimated two to three millions of deaths annually [90]. Particularly for viral diseases, one of the biggest achievements in human medicine is the worldwide eradication of smallpox in 1980 [91]. Furthermore, the reduced incidence of major diseases such as measles and poliomyelitis represents a huge success of human vaccine programs [92, 93]. In veterinary medicine, the viral disease rinderpest, also known as cattle plague, was declared eradicated in May 2011 [94]. With high mortality rates up to 100% in buffalo and cattle herds, this disease alone was responsible for heavy economic losses over centuries [95].

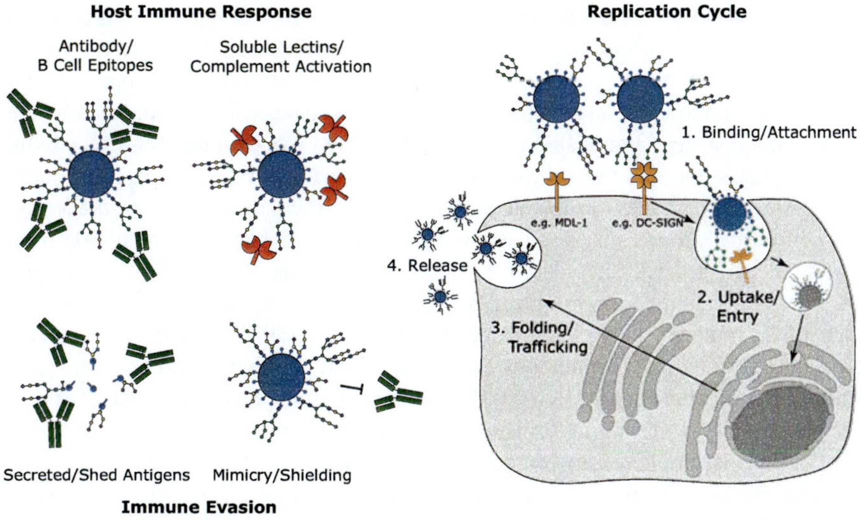


**Fig. 7** The different roles of viral glycosylation in enveloped virus biology. Glycans at the viral surface are involved in viral infection and immune evasion and also affect the host immune response. They serve to bind to surface lectin receptors (1) which in turn results in virus entry into the host cell (2). During protein processing in the ER and Golgi, glycosylation sites on viral proteins become glycosylated (3). Those glycans become important for correct protein trafficking and, finally, virus release (4). By secreting or shedding antigens, along with antigen mimicry and shielding (bottom left), the virus particle is able to counter/evade the host immune response. Furthermore, viral glycans can be targeted by neutralizing antibodies as well as serve as ligands for receptors of the innate immune system (top left). Abbreviations: myeloid DAP12-associating lectin 1 (MDL-1) and dendritic cell-specific intercellular adhesion molecule-3-grabbing non-integrin (DC-SIGN)

A good vaccine must initiate an innate immune response in the vaccinated human/animal that will lead to a protective antigen-specific adaptive immune response and also induce an immunological memory. Once generated, those memory cells can rapidly reactivate upon encountering their antigen and help to eliminate the infection in its early stages. Traditional vaccination methods that are employed to immunize and protect against viral diseases include live-attenuated, inactivated, and subunits vaccines.

## 4.1 *Live-Attenuated Vaccines*

Live-attenuated viral vaccines contain weakened forms of the natural virus, thus reducing their infectious potential [96]. Importantly, their immunogenicity is similar to the wild-type virus so that attenuated strains induce a strong and mostly long-lasting immune response. Prominent examples for attenuated viral vaccines include measles, rubella, varicella, influenza, and mumps [93, 96–98]. The most common method to generate an attenuated virus strain is by passaging the natural strain several times in cell culture to accumulate weakening mutations. Additionally, chemicals or other mutagenic means can be used to accelerate this process. For immunization purposes, an attenuated virus strain needs to still be recognized by the host immune system and unable to cause disease [98]. An example for a successful live-attenuated virus is the yellow fever vaccine strain [96]. By undergoing more than 200 serial passages of the yellow fever virus in monkeys as well as mouse and chicken tissue culture, an attenuated virus strain was obtained in the 1930s from which all licensed vaccines derive to this day [99]. The RVFV MP12 strain is a vaccine candidate that was produced by passaging in the presence of 5-fluorouracil, a chemical mutagen [100]. Compared to the wild-type RVFV ZH548 strain, this procedure resulted in 23 mutations across the whole genome including nine stable amino acid substitutions [101, 102]. Those substitutions ensure that MP12 is at a very low risk of reverting to virulence by a single reversion mutation [102]. While this vaccine is employed to protect against virulent RVFV in sheep, cattle, and macaques [103–105], the live-attenuated vaccine strain MP12 is still not licensed for human use [106, 107]. This example highlights the efficacy of live-attenuated viral vaccines but also that their production can be labor-intensive and time-consuming, thus rendering their development costly [96]. The main advantage of live-attenuated vaccines is their ability to generate long-lasting immunity as they carry the same or similar epitopes on their surface as the wild-type virus [96]. A potential drawback of live-attenuated vaccines is that immunocompromised individuals are at a potential risk of developing diseases caused by revertants [93, 98].

## 4.2 Inactivated Vaccines

Another traditional vaccine strategy is immunization with inactivated/killed whole virus particles. The first successful inactivated viral vaccine was generated against the influenza in 1936 [108]. Later, inactivated vaccines were developed against polio [109], hepatitis A [110], tick-borne encephalitis virus [111], and Japanese encephalitis virus (JEV) [112]. By treating viruses with heat, chemicals (such as formaldehyde, detergents, etc.), or radiation (such as ultraviolet light or $\gamma$-irradiation), they will become inactivated. Thus, compared to live-attenuated vaccine strains, inactivated vaccines are safer since they pose no risk of generating revertants or of causing disease. However, immunization with inactivated virus particles usually does not provide such a long-lasting protection as live-attenuated vaccines do, thus often requiring boost immunizations to attain the desired protection efficacy [93]. Examples for inactivated whole virus vaccines include poliovirus, rabies, and hepatitis A [92, 93]. One benefit is that, like live-attenuated vaccines, inactivated vaccines induce a broad immune response due to the presence of multiple antigens. However, key epitopes may be denatured by the inactivation procedure [113–115]. Immunization of rats with a live-attenuated or a formaldehyde-inactivated respiratory syncytial virus (RSV) vaccine resulted in antibody responses with different specificities [116]. Sera from rats immunized with the live-attenuated vaccine strain reacted with three different antigenic epitopes, whereas only one epitope was recognized by the formaldehyde-inactivated RSV vaccinated animals [116]. Thus, in this case, formaldehyde treatment enabled antibody responses to only one immunogenic epitope and/or downregulated the response to other epitopes [116]. Additionally, Ibsen [115] showed that formaldehyde inactivation of JEV altered the epitope binding pattern for murine monoclonal antibodies. In this particular case, hydrogen peroxide ($H_2O_2$) seemed to be preferable to other inactivation methods since JEV retained its antigenicity after treatment [115]. It becomes apparent that not all inactivation methods preserve natural epitopes and potential consequences must be considered during vaccine design.

## 4.3 Subunit Vaccines

Viral subunit vaccines, like proteins or virus-like particles (VLP), may serve as alternatives to attenuated or inactivated virus particles. Protein subunits can be produced recombinantly using genetic approaches [117, 118] or purified from whole virus preparations [119]. The optimal expression system and purification methods have to be considered in terms of preservation of antigenic epitopes and glycosylation patterns as the purification procedure often eliminates important immunogenic structures [120]. One must take into account that a virus passaged in cell culture does not necessarily display the same glycosylation pattern as a virus replicated in the infected host. Even cells in the same organism can exhibit different glycosylation patterns due to a differential expression of glycoenzymes, such as different cell types or cells from different tissues/organs. Hendra virus glycoprotein G, for example, when expressed in HeLa and HEK293 cells (both of human origin), yields two different glycoforms [121]. One must also consider that cancerous and immortalized cell lines may express different glycosylation patterns when compared with their normal/healthy counterpart in vivo [122]. As such, glycosylation may differ when viruses are propagated or viral glycoproteins are produced in different host species or cell lines. Many viral diseases are zoonotic and therefore circulate between animal and human hosts. Insects also represent important disease vectors, and, as described above, their *N*- as well as *O*-glycosylation machinery markedly differs from mammals [123–126].

Purified viral proteins are often used for vaccination since they are well-tolerated, also by immune-compromised patients [93]. However, subunit vaccines are often less immunogenic compared to live-attenuated vaccines [93]. Virus-like particles are genome-free particles that form spontaneously by the assembly of viral proteins following recombinant production [127, 128]. Compared to using purified proteins, one of the main advantages for employing VLPs for vaccination is their ability to present viral epitopes in a natural and multivalent fashion, mimicking the surface of a native virus, which in turn yields stronger and longer-lasting immune responses [127, 128]. The first VLP-vaccine was licensed in 1981 for hepatitis B [129]. This was followed later by vaccines against human papillomavirus with several other candidate vaccines currently in clinical trials [128].

To circumvent the lower immunogenicity of subunit vaccines, in many cases, adjuvants as well as boost immunizations are required [130]. Adjuvants are substances that enhance the antigen immunogenicity, but are not antigenic themselves, and thereby help to induce a desired immune response and immunological memory [130]. Aluminum salts (also commonly referred as "alum") are a well-known family that has been employed in human vaccines for over 90 years [131]. They are able to induce strong antibody responses, but are limited in provoking cellular immune responses [130]. In some cases, adjuvants are formulated by combining different immunostimulating compounds to effectively shape the immune response [93]. AS04, for example, is a combination of alum and 3-*O*-desacyl-4′-monophosphoryl lipid (MPL, a Toll-like receptor (TLR) 4 agonist) designed to enhance adaptive immune responses [132], whereas AS03, used in influenza and HBV vaccines, is formulated with α-tocopherol, a metabolizable oil, to produce an oil-in-water emulsion [93, 133, 134].

Nowadays, the knowledge of how adjuvants and vaccine formulations can be tailored to optimize vaccination efficacy is increasing, yet most of that research is still empirical in nature [130]. Potential undesired side effects caused by the selected adjuvants always need to be carefully considered [135], along with the vaccine production system and processes, to ensure optimal antigenicity and immunogenicity.

## 4.4 Influenza Virus: A Representative Example of the Importance of Glycosylation

Influenza viruses are zoonotic pathogens and belong to the family *Orthomyxoviridae*. The single-stranded, negative-sense RNA genome consists of eight segments, each encoding for 1–3 structural and non-structural proteins. The surface glycoproteins hemagglutinin (HA) and neuraminidase (NA), comprised of 18 (H1–18) and 11 (N1–11) subtypes [136–138], are necessary for virus-cell interaction and virus replication. HA, the major surface glycoprotein, binds sialylated host cell glycans to facilitate fusion, and it is an important neutralization target for the humoral immune response [139–141], whereas NA is involved in sialic acid cleavage to enable virus particles the release from the host cell membrane [142].

Like all other viruses, influenza viruses utilize the host cell machinery for biosynthesis and are under pressure to evade the host immune response. Due to antigenic shift, many potential combinations of NA and HA subtypes are possible, and almost all have been reported in birds [138, 143]. Over the past 100 years, new influenza virus strains are emerging constantly, some of which were able to cause pandemics, as in 1918 (H1N1), 1957 (H2N2), 1968 (H3N2), and 2009 (H1N1), for example, with millions of deaths each [144–146]. Additionally, the seasonal flu period causes around 3 to 5 million human cases of severe illness worldwide resulting in around 290,000–650,000 respiratory deaths, annually [147]. Today, licensed influenza vaccines are typically live-attenuated viruses or subunits composed by surface antigens [145]. Most commonly, the vaccine strains are propagated in embryonated chicken eggs [145]. Annual influenza vaccines are generally prepared as tri- and quadrivalent formulations designed to protect against different subtypes of influenza A and influenza B viruses [145]. Which antigen combination to recommend is estimated annually upon surveillance of worldwide seasonal circulating viruses, as well as on clinical and laboratory observations [145, 148]. On the one hand, influenza vaccination saves human lives every year, but the vaccine effectivity from 2009 to 2019 varied just around 40 to 60% [145]. The need for a better and, of course, universal vaccine is highly desirable. However, due to antigenic shift, a multitude of influenza subtypes are circulating and rapidly evolving. Furthermore, as this evolution yields frequently mutations that affect vaccine epitopes, we have no choice but to redevelop vaccines to include new strains every year.

One of today's challenges in vaccine production is the selection of the best-suited cell system for virus propagation. It is known that influenza viruses are able to adapt to the cell type in which they are cultivated [149], as illustrated by the discovery of mutations near the receptor-binding site of HA in egg-cultivated strains [150, 151]. De Vries et al. showed that changes in glycan formation of recombinant HA due to different expression systems have a direct impact on its immunogenicity [152]. Furthermore, the reference antigens used for the 2014–2015 influenza vaccine showed diverse glycan profiles when produced in different cell lines, which in turn may influence the resulting immune response [153]. The egg-cultivated H3N2 vaccine strain, for example, exhibited a substitution in H3 followed by a conformation change that resulted in a very low antibody response against the naturally circulating virus [154]. Another example for the importance of the choice of cell line for vaccine production is that the HA antigen expressed in a new porcine suspension cell line displayed clear differences in glycosylation compared to other common mammalian cell lines [155]. Evolutionary, each 5–7 years glycans are added to H1 and H3 until they reach a limit and then are swapped, but two times more slowly [156]. Those mutations can have a huge role in antibody evasion and therefore vaccine effectiveness. In order to develop a universal vaccine [144], one has to account for all these changes in glycosylation as well as all the subtype-specific differences.

## 5 Novel Glycan-Exploiting Vaccine Strategies

In recent years, some authors have discussed glycan-based viral vaccine approaches, but mainly focused on generating anti-glycan immunity [4, 5, 7, 12, 157, 158]. While the generation of anti-glycan antibodies would help counter virus glycan shielding and broaden protective immunity, major drawbacks are that glycans are poorly immunogenic and that, in the context of viral infections, those glycans are also part of the "self," meaning that the host may exhibit tolerance towards those antigens.

It is well-known that viral glycoproteins expressed in human cell lines are characterized by complex, oligomannose, and hybrid-type *N*-glycans. In contrast, glycoproteins derived from insect cell lines carry oligomannose *N*-glycans as well as truncated paucimannose *N*-glycans. While mammalian cell lines (such as Chinese hamster ovary (CHO) cells) are generally able to produce glycoproteins with complex human-like *N*-glycans, they also exhibit the immunogenic α-Gal epitope as well as the non-human sialic acid *N*-glycolylneuraminic acid [2]. These differences clearly illustrate that the choice of the appropriate glycoprotein expression system can be important, as the resulting *N*-glycosylation may influence recognition by antigen-presenting cells and immunological properties [2]. To date, however, this effect has remained largely unexplored. In one study, two glycovariants of the influenza A virus were propagated in either Vero or Madin-Darby canine kidney (MDCK) cells [159]. Analysis of *N*-glycosylation profiles of the hemagglutinin showed that the MDCK-glycovariant was mainly composed by complex *N*-glycans, while the Vero-glycovariant was mainly composed of oligomannose *N*-glycans. In vitro and in vivo immunological assays showed that the Vero-glycovariant may favor cellular immune responses, while the MDCK-glycovariant led to higher antibody production. Another study has explored the role of different glycoforms of influenza hemagglutinins on dendritic cell activation in vitro [160]. Other studies comparing differentially glycosylated subunit vaccines for HCV show contradictory results regarding the advantage conferred by insect over mammalian glycans [157, 161]. Taken together, those results illustrate that glycosylation of viral glycoproteins plays a major role in immunogenicity and that further research is warranted to understand this phenomenon and to exploit it in order to design better vaccines.

Nowadays, such studies can be easily conducted with the great diversity of glycoprotein-expression systems and genetic engineering tools available to explore the differential role of glycosylation on viral glycoprotein immunogenicity [9, 162–167]. By identifying non-human and/or non-mammalian glycans and by glycoengineering expression systems (by knocking-out, knocking-in, or overexpressing target enzymes and/or by adjusting cell culture conditions, nutrients, and supplements), self-adjuvanted glycoprotein-based vaccines may be obtained. Moreover, glycoengineering can also contribute to improve the efficacy of those vaccines by reducing the heterogeneity in displayed glycans [168]. In the following subsections, we will briefly highlight a non-exhaustive list of non-human glycans that can be exploited to design glycan-based self-adjuvanted vaccines and to enhance vaccine responses and immunity.

## 5.1 *Sialylated Glycans with N-Glycolylneuraminic Acid*

Most complex glycans encountered are found to be sialylated. The sialic acid family comprises about 43 derivatives of the acidic 9-carbon monosaccharide neuraminic acid [169]. The amino group is usually found to be acetylated in nature, leading to *N*-acetylneuraminic acid (Neu5Ac), the most widespread form of sialic acid. If a hydroxyl group is substituted onto the methyl group of the *N*-acetyl moiety, *N*-glycolylneuraminic (Neu5Gc) acid is obtained. Furthermore, the hydroxyl groups of sialic acids can be further modified by acetylation, methylation, sulfonation, and phosphorylation. All those structural variations are known to influence the biological properties of sialic acids, most notably their immunogenicity [170].

Neu5Gc is frequently referred to as "non-human sialic acid," the reason being that while this sugar is widely expressed across mammalian species, an inactivating mutation got fixed in the human lineage more than two million years ago which caused the inability of humans to biosynthesize this sugar from Neu5Ac [171, 172]. Therefore, all human adults possess varying levels of anti-Neu5Gc antibodies that were shown to promote chronic inflammation [171]. Those antibodies constitute a well-known problem for biotherapeutics and have prompted the development of many strategies to ensure the bioproduction of Neu5Gc-free glycoproteins [10, 173].

In the context of vaccination, viral glycoproteins sialylated with Neu5Gc could however exploit those anti-Neu5Gc antibodies: through an effective internalization of those immune complexes by antigen-presenting cells (APCs) and through increased transport to lymph nodes and processing by APCs, one could obtain an effective activation of vaccine antigen-specific lymphocytes and a strong cellular and humoral immune response [174, 175]. It has also been shown that cross talk between Fc receptors (FcRs) and other pattern recognition receptors (PRRs), such as TLRs, leads to improved and fine-tuned adaptive immune responses [176]. As such, vaccine antigen delivery strategies involving opsonization by host natural antibodies represent a promising avenue, which warrants considerable interest.

## 5.2 *The α-Gal Epitope*

The α-Gal epitope (Gal-α1,3-Gal-β1,4-GlcNAc-R) is ubiquitously expressed on the glycoproteins and glycolipids of non-primate mammals, marsupials, and New World monkeys, while remaining absent in apes, Old World monkeys and humans [177, 178]. As such, it is not surprising that anti-Gal antibodies constitute the most abundant natural antibody in humans with approximately 1% of all serum IgGs [179]. Previous studies have already demonstrated the potential of the α-Gal epitope in enhancing vaccine responses against influenza viruses and HIV [179–183]. Importantly, the mechanism described above for the antibody-mediated enhancement of vaccinal responses was initially described in the context of anti-Gal antibodies [174, 175]. Despite their great potential for vaccination, careful evaluation of vaccine safety is warranted since anti-Gal antibodies have notably been linked with allergic reactions in humans [184–186].

## 5.3 Targeting CLRs: Mannose-Rich and Other Fungal Glycans

One promising avenue for antigen-delivery technologies and vaccination purposes is the directed targeting of APCs through CLRs [2, 187–190]. While the CLR superfamily is diversified in terms of ligand specificity, cellular distribution, signalling pathways, and effector functions [191, 192], some receptors have emerged as promising targets for antigen delivery and immune modulation [193, 194]. For example, DC-SIGN, Dectin-1, Dectin-2, Mincle, and MMR are well-known sensors of fungal pathogens and can mediate antigen uptake and initiate adaptative immune responses. Among their known ligands, they bind to fungal glycans such as β-glucans, α-mannans, oligomannose, and hypermannosylated *N*-glycans [191, 192, 195, 196]. This explains in part why yeast-produced glycoproteins are considered highly immunogenic and has sparked increased interest in the use of yeast-based systems as vaccine antigens and adjuvants [197, 198]. Therefore, strategies aiming at exploiting mannose-rich glycans or other fungal glycans, such as using yeast expression systems or glycoengineering cell lines to strictly produce oligomannose *N*-glycans (e.g., as a result of the deletion of MGAT1 [199, 200]) may hold great potential for glycan-based vaccine design.

## 5.4 Insect-Produced Glycans

Examples of insect cell-based subunit viral vaccines that are approved or under investigation include FluBlok® (targeting the HA of influenza viruses) and vaccines based on the E2 protein of HCV [157, 161, 201]. As it was previously explained, insect *N*-glycans are quite different from their mammalian counterparts as they are in majority paucimannosidic with or without core α1,6-Fuc and/or core α1,3-Fuc. The popularity of the baculovirus-insect cell system has prompted considerable efforts to glycoengineer those insect cells to produce human-like glycans in order to obtain safer and more efficient biotherapeutics [23]. One drawback from those insect *N*-glycans is that core α1,3-fucosylation is considered to be an immunogenic epitope and to be involved in allergy development [202, 203]. These safety concerns will be addressed below for plant-produced glycans.

Additionally, improvement in analytical methods led to the observation that insect cells and other invertebrates can carry previously unrecognized *N*- and *O*-glycans displaying unusual (non-mammalian) capping modifications [24, 204–207]. Glycoengineering approaches aimed at producing homogeneously some of those novel glycans would allow to conduct immunological studies to evaluate their use as potential self-adjuvanted vaccines.

### 5.5 *Plant-Produced Glycans*

In recent years, plant expression systems have been established to produce biopharmaceuticals such as enzymes, growth factors, peptides, or antibodies, but also recombinant subunit vaccines and peptides as antigens [208]. While there are still no plant-based vaccines approved for humans, many candidates are undergoing clinical trials, notably against influenza viruses, HBV, and rabies viruses [208, 209]. A vaccine to protect against Newcastle disease (caused by the Newcastle disease virus in poultry) has been approved for veterinary use [208]. Another promising plant-based vaccine strategy is the production of hemagglutinin-based VLP vaccines against influenza viruses [210–213].

One key feature of plant glycans is that only two major glycoforms account for more than 90% of their *N*-glycans: the GnGnXF and MMXF structures [22]. The GnGnXF structures consist of complex biantennary *N*-glycans where the core ($GlcNAc_2Man_3GlcNAc_2$) is both α1,3-fucosylated (on the Asn-linked β-GlcNAc) and β1,2-xylosylated (on the β1,4-Man) and naturally lacks mammalian β1,4-Gal and sialic acid residues [214]. The MMXF structures consist of paucimannose structures that also display core α1,3-Fuc and β1,2-Xyl. While early studies demonstrated that those glycans were immunogenic in humans and sparked a debate over the safety of plant glyco-epitopes, recent large-scale clinical trials conducted in humans have shown those concerns to be mostly unwarranted [22, 203, 215–217].

In summary, plant-based and insect-based glycans are markedly different from their mammalian/human counterparts. To help decide whether those differences might be worth exploiting for glycan-based self-adjuvanted vaccination approaches, more in-depth immunological and mechanistic studies are warranted.

## 6 Perspectives

Without doubt, viral glycosylation represents a promising target to interfere with viral infectivity and to modulate the host immune response that has largely been underexplored yet. On the one hand, viruses utilize glycans to escape immune surveillance, whereas on the other hand vaccinologists may exploit viral glycosylation to design “next-generation vaccines”. Cutting-edge methods such as the glycan array technology or glyco-nanotechnology offer new tools to elucidate the role of viral glycans in host cell attachment and entry or to selectively interfere with virus-cell interactions [218, 219], respectively. The design of tailor-made self-adjuvanted vaccines, in which the glycan part serves as an intrinsic adjuvant, holds great potential for antigen cell-specific vaccine delivery and/or the stimulation of desired immune responses. The choice of appropriate virus production systems and expression systems for viral glycoproteins as well as the targeted glycoengineering of cell lines constitute promising strategies to access immunogenic glycosylation patterns. Additionally, glycoengineering would allow further studies into the role of specific glycan structures into virus biology. In the present review, we have provided an overview of viral glycosylation and have highlighted examples of how glycan-exploiting strategies can be harnessed for vaccine design. These strategies may also prove useful for current global challenges, such as the development of a protective vaccine against the SARS-CoV-2 that newly emerged in late 2019 [220, 221]. Future studies and joint efforts of virologists, biochemists, (glyco) biologists, and immunologists, among others, are critical to pave the way towards glyco-optimized next-generation vaccines.

**Acknowledgments** G. Goyette-Desjardins is a recipient of a postdoctoral research fellowship from the "Fonds de recherche du Québec - Nature et technologies" (FRQNT, Canada). K. Schön is funded by the "Deutsche Forschungsgemeinschaft" (DFG, Germany; #398066876/GRK 2485/1).

**Conflict of Interest Statement** The authors declare that they have no conflict of interest.

## References

1. Varki A, Gagneux P (2015) Chapter 7 – Biological functions of glycans. In: Varki A, Cummings RD, Esko JD et al (eds) Essentials of glycobiology, 3rd edn. Cold Spring Harbor Laboratory Press, Cold Spring Harbor, pp 77–88
2. Johannssen T, Lepenies B (2017) Glycan-based cell targeting to modulate immune responses. Trends Biotechnol 35(4):334–346
3. Dwek RA (1996) Glycobiology: toward understanding the function of sugars. Chem Rev 96 (2):683–720
4. Bagdonaite I, Wandall HH (2018) Global aspects of viral glycosylation. Glycobiology 28 (7):443–467
5. Watanabe Y, Bowden TA, Wilson IA et al (2019) Exploitation of glycosylation in enveloped virus pathobiology. Biochim Biophys Acta 1863(10):1480–1497
6. Bagdonaite I, Vakhrushev SY, Joshi HJ et al (2018) Viral glycoproteomes: technologies for characterization and outlook for vaccine design. FEBS Lett 592(23):3898–3920
7. Crispin M, Doores KJ (2015) Targeting host-derived glycans on enveloped viruses for antibody-based vaccine design. Curr Opin Virol 11:63–69
8. Stanley P, Taniguchi N, Aebi M (2015) Chapter 9 – *N*-glycans. In: Varki A, Cummings RD, Esko JD et al (eds) Essentials of glycobiology, 3rd edn. Cold Spring Harbor Laboratory Press, Cold Spring Harbor, pp 99–111
9. Dicker M, Strasser R (2015) Using glyco-engineering to produce therapeutic proteins. Expert Opin Biol Ther 15(10):1501–1516
10. Gupta SK, Shukla P (2018) Glycosylation control technologies for recombinant therapeutic proteins. Appl Microbiol Biotechnol 102(24):10457–10468
11. Wang Q, Chung CY, Chough S et al (2018) Antibody glycoengineering strategies in mammalian cells. Biotechnol Bioeng 115(6):1378–1393

12. Buettner MJ, Shah SR, Saeui CT et al (2018) Improving immunotherapy through glycodesign. Front Immunol 9:2485
13. Spiro RG (2002) Protein glycosylation: nature, distribution, enzymatic formation, and disease implications of glycopeptide bonds. Glycobiology 12(4):43R–56R
14. Corfield A (2017) Eukaryotic protein glycosylation: a primer for histochemists and cell biologists. Histochem Cell Biol 147(2):119–147
15. Zachara N, Akimoto Y, Hart GW (2015) Chapter 19 – the *O*-GlcNAc modification. In: Varki A, Cummings RD, Esko JD et al (eds) Essentials of glycobiology, 3rd edn. Cold Spring Harbor Laboratory Press, Cold Spring Harbor, pp 239–251
16. Kornfeld R, Kornfeld S (1985) Assembly of asparagine-linked oligosaccharides. Annu Rev Biochem 54(1):631–664
17. Brockhausen I, Hull E, Hindsgaul O et al (1989) Control of glycoprotein synthesis. Detection and characterization of a novel branching enzyme from hen oviduct, UDP-*N*-acetylglucosamine:GlcNAc beta 1-6 (GlcNAc beta 1-2)Man alpha-R (GlcNAc to Man) beta-4-*N*-acetylglucosaminyltransferase VI. J Biol Chem 264(19):11211–11221
18. Taguchi T, Ogawa T, Inoue S et al (2000) Purification and characterization of UDP-GlcNAc: GlcNAcbeta 1-6(GlcNAcbeta 1-2)Manalpha 1-R [GlcNAc to Man]-beta 1, 4-*N*-acetylglucosaminyltransferase VI from hen oviduct. J Biol Chem 275(42):32598–32602
19. Watanabe T, Ihara H, Miyoshi E et al (2006) A specific detection of GlcNAcbeta1-6Manalpha1 branches in *N*-linked glycoproteins based on the specificity of *N*-acetylglucosaminyltransferase VI. Glycobiology 16(5):431–439
20. Nakano M, Mishra SK, Tokoro Y et al (2019) Bisecting GlcNAc is a general suppressor of terminal modification of *N*-glycan. Mol Cell Proteomics 18(10):2044–2057
21. Schneider M, Al-Shareffi E, Haltiwanger RS (2017) Biological functions of fucose in mammals. Glycobiology 27(7):601–618
22. Montero-Morales L, Steinkellner H (2018) Advanced plant-based glycan engineering. Front Bioeng Biotechnol 6(81):81
23. Shi X, Jarvis DL (2007) Protein *N*-glycosylation in the baculovirus-insect cell system. Curr Drug Targets 8(10):1116–1125
24. Tiemeyer M, Nakato H, Esko JD (2015) Chapter 26 – Arthropoda. In: Varki A, Cummings RD, Esko JD et al (eds) Essentials of glycobiology, 3rd edn. Cold Spring Harbor Laboratory Press, Cold Spring Harbor, pp 335–349
25. Haltiwanger RS, Wells L, Freeze HH et al (2015) Chapter 13 – Other classes of eukaryotic glycans. In: Varki A, Cummings RD et al (eds) Essentials of glycobiology, 3rd edn. Cold Spring Harbor Laboratory Press, Cold Spring Harbor, pp 151–160
26. Varki A, Kornfeld S (2015) Chapter 1 – Historical background and overview. In: Varki A, Cummings RD, Esko JD et al (eds) Essentials of glycobiology, 3rd edn. Cold Spring Harbor Laboratory Press, Cold Spring Harbor, pp 1–18
27. Jensen PH, Kolarich D, Packer NH (2010) Mucin-type *O*-glycosylation - putting the pieces together. FEBS J 277(1):81–94
28. Marth JD (1999) Chapter 8 – *O*-glycans. In: Varki A, Cummings R, Esko J, Freeze H, Hart G, Marth J (eds) Essentials of glycobiology, 1st edn. Cold Spring Harbor Laboratory Press, Cold Spring Harbor
29. Brockhausen I, Stanley P (2015) Chapter 10 – *O*-GalNAc glycans. In: Varki A, Cummings RD, Esko JD et al (eds) Essentials of glycobiology, 3rd edn. Cold Spring Harbor Laboratory Press, Cold Spring Harbor, pp 113–123
30. Corfield AP, Berry M (2015) Glycan variation and evolution in the eukaryotes. Trends Biochem Sci 40(7):351–359
31. Stanley P, Cummings RD (2015) Chapter 14 – Structures common to different glycans. In: Varki A, Cummings RD, Esko JD et al (eds) Essentials of glycobiology, 3rd edn. Cold Spring Harbor Laboratory Press, Cold Spring Harbor, pp 161–178
32. Fenouillet E, Gluckman JC, Bahraoui E (1990) Role of *N*-linked glycans of envelope glycoproteins in infectivity of human immunodeficiency virus type 1. J Virol 64(6):2841–2848

33. Vigerust DJ, Shepherd VL (2007) Virus glycosylation: role in virulence and immune interactions. Trends Microbiol 15(5):211–218
34. Lin G, Simmons G, Pohlmann S et al (2003) Differential *N*-linked glycosylation of human immunodeficiency virus and Ebola virus envelope glycoproteins modulates interactions with DC-SIGN and DC-SIGNR. J Virol 77(2):1337–1346
35. Lozach P-Y, Amara A, Bartosch B et al (2004) C-type lectins L-SIGN and DC-SIGN capture and transmit infectious hepatitis C virus pseudotype particles. J Biol Chem 279 (31):32035–32045
36. Leger P, Tetard M, Youness B et al (2016) Differential use of the C-type lectins L-SIGN and DC-SIGN for phlebovirus endocytosis. Traffic 17(6):639–656
37. Lozach PY, Kuhbacher A, Meier R et al (2011) DC-SIGN as a receptor for phleboviruses. Cell Host Microbe 10(1):75–88
38. Monteiro J, Lepenies B (2017) Myeloid C-type lectin receptors in viral recognition and antiviral immunity. Viruses 9(3):59
39. van Liempt E, Bank CM, Mehta P et al (2006) Specificity of DC-SIGN for mannose- and fucose-containing glycans. FEBS Lett 580(26):6123–6131
40. Mitchell DA, Fadden AJ, Drickamer K (2001) A novel mechanism of carbohydrate recognition by the C-type lectins DC-SIGN and DC-SIGNR. J Biol Chem 276(31):28939–28945
41. Curtis BM, Scharnowske S, Watson AJ (1992) Sequence and expression of a membrane-associated C-type lectin that exhibits CD4-independent binding of human immunodeficiency virus envelope glycoprotein gp120. Proc Natl Acad Sci U S A 89(17):8356–8360
42. Alvarez CP, Lasala F, Carrillo J et al (2002) C-type lectins DC-SIGN and L-SIGN mediate cellular entry by Ebola virus in *cis* and in *trans*. J Virol 76(13):6841–6844
43. Simmons G, Reeves JD, Grogan CC et al (2003) DC-SIGN and DC-SIGNR bind Ebola glycoproteins and enhance infection of macrophages and endothelial cells. Virology 305 (1):115–123
44. Geijtenbeek TBH, Torensma R, Van Vliet SJ et al (2000) Identification of DC-SIGN, a novel dendritic cell–specific ICAM-3 receptor that supports primary immune responses. Cell 100 (5):575–585
45. Hong PW, Flummerfelt KB, de Parseval A et al (2002) Human immunodeficiency virus envelope (gp120) binding to DC-SIGN and primary dendritic cells is carbohydrate dependent but does not involve 2G12 or cyanovirin binding sites: implications for structural analyses of gp120-DC-SIGN binding. J Virol 76(24):12855–12865
46. Nguyen DG, Hildreth JEK (2003) Involvement of macrophage mannose receptor in the binding and transmission of HIV by macrophages. Eur J Immunol 33(2):483–493
47. Lai J, Bernhard OK, Turville SG et al (2009) Oligomerization of the macrophage mannose receptor enhances gp120-mediated binding of HIV-1. J Biol Chem 284(17):11027–11038
48. Miller JL, Dewet BJM, Martinez-Pomares L et al (2008) The mannose receptor mediates dengue virus infection of macrophages. PLoS Pathog 4(2):e17
49. Altgärde N, Eriksson C, Peerboom N et al (2015) Mucin-like region of herpes simplex virus type 1 attachment protein glycoprotein C (gC) modulates the virus-glycosaminoglycan interaction. J Biol Chem 290(35):21473–21485
50. Stone JA, Nicola AV, Baum LG et al (2016) Multiple novel functions of henipavirus *O*-glycans: the first *O*-glycan functions identified in the paramyxovirus family. PLoS Pathog 12 (2):e1005445
51. Luo S, Hu K, He S et al (2015) Contribution of *N*-linked glycans on HSV-2 gB to cell–cell fusion and viral entry. Virology 483:72–82
52. Ito K, Qin Y, Guarnieri M et al (2010) Impairment of hepatitis B virus virion secretion by single-amino-acid substitutions in the small envelope protein and rescue by a novel glycosylation site. J Virol 84(24):12850–12861
53. Volchkov VE, Feldmann H, Volchkova VA et al (1998) Processing of the Ebola virus glycoprotein by the proprotein convertase furin. Proc Natl Acad Sci U S A 95(10):5762–5767

54. Wang B, Wang Y, Frabutt DA et al (2017) Mechanistic understanding of *N*-glycosylation in Ebola virus glycoprotein maturation and function. J Biol Chem 292(14):5860–5870
55. Szakonyi G, Klein MG, Hannan JP et al (2006) Structure of the Epstein-Barr virus major envelope glycoprotein. Nat Struct Mol Biol 13(11):996–1001
56. Sommerstein R, Flatz L, Remy MM et al (2015) Arenavirus glycan shield promotes neutralizing antibody evasion and protracted infection. PLoS Pathog 11(11):e1005276
57. Falkowska E, Kajumo F, Garcia E et al (2007) Hepatitis C virus envelope glycoprotein E2 glycans modulate entry, CD81 binding, and neutralization. J Virol 81(15):8072–8079
58. Beniac DR, Booth TF (2017) Structure of the Ebola virus glycoprotein spike within the virion envelope at 11 Å resolution. Sci Rep 7:46374
59. Seabright GE, Doores KJ, Burton DR et al (2019) Protein and glycan mimicry in HIV vaccine design. J Mol Biol 431(12):2223–2247
60. Hallenberger S, Bosch V, Angliker H et al (1992) Inhibition of furin-mediated cleavage activation of HIV-1 glycoprotein gp160. Nature 360(6402):358–361
61. Silver ZA, Antonopoulos A, Haslam SM et al (2020) Discovery of *O*-linked carbohydrate on HIV-1 envelope and its role in shielding against one category of broadly neutralizing antibodies. Cell Rep 30(6):1862–1869.e1864
62. Lasky LA, Groopman JE, Fennie CW et al (1986) Neutralization of the AIDS retrovirus by antibodies to a recombinant envelope glycoprotein. Science 233(4760):209–212
63. Lee JH, Ozorowski G, Ward AB (2016) Cryo-EM structure of a native, fully glycosylated, cleaved HIV-1 envelope trimer. Science 351(6277):1043–1048
64. Stewart-Jones GB, Soto C, Lemmin T et al (2016) Trimeric HIV-1-Env structures define glycan shields from clades A, B, and G. Cell 165(4):813–826
65. Wei X, Decker JM, Wang S et al (2003) Antibody neutralization and escape by HIV-1. Nature 422(6929):307–312
66. Moore PL, Gray ES, Wibmer CK et al (2012) Evolution of an HIV glycan-dependent broadly neutralizing antibody epitope through immune escape. Nat Med 18(11):1688–1692
67. Dacheux L, Moreau A, Ataman-Onal Y et al (2004) Evolutionary dynamics of the glycan shield of the human immunodeficiency virus envelope during natural infection and implications for exposure of the 2G12 epitope. J Virol 78(22):12625–12637
68. Wagh K, Kreider EF, Li Y et al (2018) Completeness of HIV-1 envelope glycan shield at transmission determines neutralization breadth. Cell Rep 25(4):893–908.e897
69. McCaffrey RA, Saunders C, Hensel M et al (2004) *N*-linked glycosylation of the V3 loop and the immunologically silent face of gp120 protects human immunodeficiency virus type 1 SF162 from neutralization by anti-gp120 and anti-gp41 antibodies. J Virol 78(7):3279–3295
70. Koch M, Pancera M, Kwong PD et al (2003) Structure-based, targeted deglycosylation of HIV-1 gp120 and effects on neutralization sensitivity and antibody recognition. Virology 313 (2):387–400
71. Li Y, Cleveland B, Klots I et al (2008) Removal of a single *N*-linked glycan in human immunodeficiency virus type 1 gp120 results in an enhanced ability to induce neutralizing antibody responses. J Virol 82(2):638–651
72. Back NK, Smit L, De Jong JJ et al (1994) An *N*-glycan within the human immunodeficiency virus type 1 gp120 V3 loop affects virus neutralization. Virology 199(2):431–438
73. Lynch RM, Wong P, Tran L et al (2015) HIV-1 fitness cost associated with escape from the VRC01 class of CD4 binding site neutralizing antibodies. J Virol 89(8):4201–4213
74. Aguilar HC, Matreyek KA, Filone CM et al (2006) *N*-Glycans on Nipah virus fusion protein protect against neutralization but reduce membrane fusion and viral entry. J Virol 80 (10):4878–4889
75. Julithe R, Abou-Jaoude G, Sureau C (2014) Modification of the hepatitis B virus envelope protein glycosylation pattern interferes with secretion of viral particles, infectivity, and susceptibility to neutralizing antibodies. J Virol 88(16):9049–9059
76. Lennemann NJ, Rhein BA, Ndungo E et al (2014) Comprehensive functional analysis of *N*-linked glycans on Ebola virus GP1. mBio 5(1):e00862–e00813

77. Sodora DL, Cohen GH, Eisenberg RJ (1989) Influence of asparagine-linked oligosaccharides on antigenicity, processing, and cell surface expression of herpes simplex virus type 1 glycoprotein D. J Virol 63(12):5184–5193
78. Hobman TC, Qiu ZY, Chaye H et al (1991) Analysis of rubella virus E1 glycosylation mutants expressed in COS cells. Virology 181(2):768–772
79. Fournillier A, Wychowski C, Boucreux D et al (2001) Induction of hepatitis C virus E1 envelope protein-specific immune response can be enhanced by mutation of *N*-glycosylation sites. J Virol 75(24):12088–12097
80. Helle F, Vieyres G, Elkrief L et al (2010) Role of *N*-linked glycans in the functions of hepatitis C virus envelope proteins incorporated into infectious virions. J Virol 84(22):11905–11915
81. Liu M, Chen H, Luo F et al (2007) Deletion of *N*-glycosylation sites of hepatitis C virus envelope protein E1 enhances specific cellular and humoral immune responses. Vaccine 25 (36):6572–6580
82. Sattentau QJ, Moore JP (1995) Human immunodeficiency virus type 1 neutralization is determined by epitope exposure on the gp120 oligomer. J Exp Med 182(1):185–196
83. Sanders RW, Derking R, Cupo A et al (2013) A next-generation cleaved, soluble HIV-1 Env trimer, BG505 SOSIP.664 gp140, expresses multiple epitopes for broadly neutralizing but not non-neutralizing antibodies. PLoS Pathog 9(9):e1003618
84. Mohan GS, Li W, Ye L et al (2012) Antigenic subversion: a novel mechanism of host immune evasion by Ebola virus. PLoS Pathog 8(12):e1003065
85. Moore PL, Crooks ET, Porter L et al (2006) Nature of nonfunctional envelope proteins on the surface of human immunodeficiency virus type 1. J Virol 80(5):2515–2528
86. Trefry JC, Wollen SE, Nasar F et al (2015) Ebola virus infections in nonhuman primates are temporally influenced by glycoprotein poly-U editing site populations in the exposure material. Viruses 7(12):6739–6754
87. Maruyama T, Parren PW, Sanchez A et al (1999) Recombinant human monoclonal antibodies to Ebola virus. J Infect Dis 179(s1):S235–S239
88. Druar C, Saini SS, Cossitt MA et al (2005) Analysis of the expressed heavy chain variable-region genes of *Macaca fascicularis* and isolation of monoclonal antibodies specific for the Ebola virus' soluble glycoprotein. Immunogenetics 57(10):730–738
89. Cook JD, Lee JE (2013) The secret life of viral entry glycoproteins: moonlighting in immune evasion. PLoS Pathog 9(5):e1003258
90. WHO (2019) Ten threats to global health in 2019. https://www.who.int/news-room/feature-stories/ten-threats-to-global-health-in-2019. Accessed 20 Apr 2020
91. Tognotti E (2010) The eradication of smallpox, a success story for modern medicine and public health: what lessons for the future? J Infect Dev Ctries 4(5):264–266
92. Delany I, Rappuoli R, De Gregorio E (2014) Vaccines for the 21st century. EMBO Mol Med 6 (6):708–720
93. Vetter V, Denizer G, Friedland LR et al (2018) Understanding modern-day vaccines: what you need to know. Ann Med 50(2):110–120
94. Morens DM, Holmes EC, Davis AS et al (2011) Global rinderpest eradication: lessons learned and why humans should celebrate too. J Infect Dis 204(4):502–505
95. Hamilton K, Baron MD, Matsuo K et al (2017) Rinderpest eradication: challenges for remaining disease free and implications for future eradication efforts. Rev Sci Tech 36 (2):579–588
96. Minor PD (2015) Live attenuated vaccines: historical successes and current challenges. Virology 479-480:379–392
97. Plotkin S (2014) History of vaccination. Proc Natl Acad Sci U S A 111(34):12283–12287
98. Hajj Hussein I, Chams N, Chams S et al (2015) Vaccines through centuries: major cornerstones of global health. Front Public Health 3:269
99. Barrett ADT (2017) Yellow fever live attenuated vaccine: a very successful live attenuated vaccine but still we have problems controlling the disease. Vaccine 35(44):5951–5955

100. Caplen H, Peters CJ, Bishop DH (1985) Mutagen-directed attenuation of Rift Valley fever virus as a method for vaccine development. J Gen Virol 66(10):2271–2277
101. Lokugamage N, Freiberg AN, Morrill JC et al (2012) Genetic subpopulations of Rift Valley fever virus strains ZH548 and MP-12 and recombinant MP-12 strains. J Virol 86 (24):13566–13575
102. Ikegami T, Hill TE, Smith JK et al (2015) Rift Valley fever virus MP-12 vaccine is fully attenuated by a combination of partial attenuations in the S, M, and L segments. J Virol 89 (14):7262–7276
103. Morrill JC, Jennings GB, Caplen H et al (1987) Pathogenicity and immunogenicity of a mutagen-attenuated Rift Valley fever virus immunogen in pregnant ewes. Am J Vet Res 48 (7):1042–1047
104. Morrill JC, Peters CJ (2003) Pathogenicity and neurovirulence of a mutagen-attenuated Rift Valley fever vaccine in rhesus monkeys. Vaccine 21(21–22):2994–3002
105. Morrill JC, Mebus CA, Peters CJ (1997) Safety and efficacy of a mutagen-attenuated Rift Valley fever virus vaccine in cattle. Am J Vet Res 58(10):1104–1109
106. Lang Y, Li Y, Jasperson D et al (2019) Identification and evaluation of antivirals for Rift Valley fever virus. Vet Microbiol 230:110–116
107. Ikegami T (2019) Candidate vaccines for human Rift Valley fever. Expert Opin Biol Ther 19 (12):1333–1342
108. Thomas Jr F, Magill T (1936) Vaccination of human subjects with virus of human influenza. Proc Soc Exp Biol Med 33(4):604–606
109. Salk JE, Krech U, Youngner JS et al (1954) Formaldehyde treatment and safety testing of experimental poliomyelitis vaccines. Am J Public Health Nations Health 44(5):563–570
110. Provost PJ, Hughes JV, Miller WJ et al (1986) An inactivated hepatitis A viral vaccine of cell culture origin. J Med Virol 19(1):23–31
111. Kunz C (1962) Aktiv und passive Immunoprophylaxe der Fruhsommer-Meningoencephalitis (FSME). Arzneimittelforschung 28:1806
112. Yamashita T, Ishikawa N, Hojo F et al (1970) Japanese encephalitis purified vaccine. II. Purity of the mouse brain vaccine purified by ultracentrifugation. Biken J 13(1):25–38
113. Fan YC, Chiu HC, Chen LK et al (2015) Formalin inactivation of Japanese encephalitis virus vaccine alters the antigenicity and immunogenicity of a neutralization epitope in envelope protein domain III. PLoS Negl Trop Dis 9(10):e0004167
114. di Tommaso A, de Magistris MT, Bugnoli M et al (1994) Formaldehyde treatment of proteins can constrain presentation to T cells by limiting antigen processing. Infect Immun 62 (5):1830–1834
115. Ibsen PH (1996) The effect of formaldehyde, hydrogen peroxide and genetic detoxification of pertussis toxin on epitope recognition by murine monoclonal antibodies. Vaccine 14 (5):359–368
116. Widjaja I, Wicht O, Luytjes W et al (2016) Characterization of epitope-specific anti-respiratory syncytial virus (anti-RSV) antibody responses after natural infection and after vaccination with formalin-inactivated RSV. J Virol 90(13):5965–5977
117. Clark TG, Cassidy-Hanley D (2005) Recombinant subunit vaccines: potentials and constraints. Dev Biol (Basel) 121:153–163
118. Michel ML, Tiollais P (2010) Hepatitis B vaccines: protective efficacy and therapeutic potential. Pathol Biol (Paris) 58(4):288–295
119. Soema PC, Kompier R, Amorij J-P et al (2015) Current and next generation influenza vaccines: formulation and production strategies. Eur J Pharm Biopharm 94:251–263
120. Rappuoli R, Pizza M, Del Giudice G et al (2014) Vaccines, new opportunities for a new society. Proc Natl Acad Sci U S A 111(34):12288–12293
121. Colgrave ML, Snelling HJ, Shiell BJ et al (2012) Site occupancy and glycan compositional analysis of two soluble recombinant forms of the attachment glycoprotein of Hendra virus. Glycobiology 22(4):572–584

122. Orntoft TF, Vestergaard EM (1999) Clinical aspects of altered glycosylation of glycoproteins in cancer. Electrophoresis 20(2):362–371
123. Rendić D, Wilson IB, Paschinger K (2008) The glycosylation capacity of insect cells. Croat Chem Acta 81(1):7–21
124. Vandenborre G, Smagghe G, Ghesquiere B et al (2011) Diversity in protein glycosylation among insect species. PLoS One 6(2):e16682
125. Walski T, De Schutter K, Van Damme EJM et al (2017) Diversity and functions of protein glycosylation in insects. Insect Biochem Mol Biol 83:21–34
126. Joshi HJ, Narimatsu Y, Schjoldager KT et al (2018) SnapShot: *O*-glycosylation pathways across kingdoms. Cell 172(3):632–632.e632
127. Hill BD, Zak A, Khera E et al (2018) Engineering virus-like particles for antigen and drug delivery. Curr Protein Pept Sci 19(1):112–127
128. Roldão A, Mellado MCM, Castilho LR et al (2010) Virus-like particles in vaccine development. Expert Rev Vaccines 9(10):1149–1176
129. Krugman S (1982) The newly licensed hepatitis B vaccine. Characteristics and indications for use. JAMA 247(14):2012–2015
130. Pasquale A, Preiss S, Silva F et al (2015) Vaccine adjuvants: from 1920 to 2015 and beyond. Vaccine 3(2):320–343
131. Kool M, Fierens K, Lambrecht BN (2012) Alum adjuvant: some of the tricks of the oldest adjuvant. J Med Microbiol 61(7):927–934
132. Didierlaurent AM, Morel S, Lockman L et al (2009) AS04, an aluminum salt- and TLR4 agonist-based adjuvant system, induces a transient localized innate immune response leading to enhanced adaptive immunity. J Immunol 183(10):6186–6197
133. Morel S, Didierlaurent A, Bourguignon P et al (2011) Adjuvant system AS03 containing α-tocopherol modulates innate immune response and leads to improved adaptive immunity. Vaccine 29(13):2461–2473
134. Cohet C, van der Most R, Bauchau V et al (2019) Safety of AS03-adjuvanted influenza vaccines: a review of the evidence. Vaccine 37(23):3006–3021
135. Batista-Duharte A, Martínez DT, Carlos IZ (2018) Efficacy and safety of immunological adjuvants. Where is the cut-off? Biomed Pharmacother 105:616–624
136. Air GM (1981) Sequence relationships among the hemagglutinin genes of 12 subtypes of influenza A virus. Proc Natl Acad Sci U S A 78(12):7639–7643
137. Wu Y, Wu Y, Tefsen B et al (2014) Bat-derived influenza-like viruses H17N10 and H18N11. Trends Microbiol 22(4):183–191
138. Sautto GA, Kirchenbaum GA, Ross TM (2018) Towards a universal influenza vaccine: different approaches for one goal. Virol J 15(1):17
139. Air GM (2014) Influenza virus–glycan interactions. Curr Opin Virol 7:128–133
140. Epstein SL, Misplon JA, Lawson CM et al (1993) Beta 2-microglobulin-deficient mice can be protected against influenza A infection by vaccination with vaccinia-influenza recombinants expressing hemagglutinin and neuraminidase. J Immunol 150(12):5484–5493
141. Angeletti D, Gibbs JS, Angel M et al (2017) Defining B cell immunodominance to viruses. Nat Immunol 18(4):456–463
142. Doyle TM, Jaentschke B, Van Domselaar G et al (2013) The universal epitope of influenza A viral neuraminidase fundamentally contributes to enzyme activity and viral replication. J Biol Chem 288(25):18283–18289
143. Webster RG, Govorkova EA (2014) Continuing challenges in influenza. Ann N Y Acad Sci 1323(1):115–139
144. Erbelding EJ, Post DJ, Stemmy EJ et al (2018) A universal influenza vaccine: the strategic plan for the National Institute of allergy and infectious diseases. J Infect Dis 218(3):347–354
145. Wei CJ, Crank MC, Shiver J et al (2020) Next-generation influenza vaccines: opportunities and challenges. Nat Rev Drug Discov 19(4):239–252
146. Wendel I, Matrosovich M, Klenk HD (2015) SnapShot: evolution of human influenza A viruses. Cell Host Microbe 17(3):416–416.e411

147. WHO (2020) Influenza (seasonal) fact sheet. https://www.who.int/news-room/fact-sheets/detail/influenza-(seasonal). Accessed 18 Apr 2020
148. Chang D, Zaia J (2019) Why glycosylation matters in building a better flu vaccine. Mol Cell Proteomics 18(12):2348–2358
149. Schwarzer J, Rapp E, Hennig R et al (2009) Glycan analysis in cell culture-based influenza vaccine production: influence of host cell line and virus strain on the glycosylation pattern of viral hemagglutinin. Vaccine 27(32):4325–4336
150. Schild GC, Oxford JS, de Jong JC et al (1983) Evidence for host-cell selection of influenza virus antigenic variants. Nature 303(5919):706–709
151. Robertson JS, Bootman JS, Newman R et al (1987) Structural changes in the haemagglutinin which accompany egg adaptation of an influenza A (H1N1) virus. Virology 160(1):31–37
152. de Vries RP, Smit CH, de Bruin E et al (2012) Glycan-dependent immunogenicity of recombinant soluble trimeric hemagglutinin. J Virol 86(21):11735–11744
153. An Y, Parsons LM, Jankowska E et al (2019) *N*-glycosylation of seasonal influenza vaccine hemagglutinins: implication for potency testing and immune processing. J Virol 93(2): e01693–e01618
154. Wu NC, Zost SJ, Thompson AJ et al (2017) A structural explanation for the low effectiveness of the seasonal influenza H3N2 vaccine. PLoS Pathog 13(10):e1006682
155. Granicher G, Coronel J, Pralow A et al (2019) Efficient influenza A virus production in high cell density using the novel porcine suspension cell line PBG.PK2.1. Vaccine 37 (47):7019–7028
156. Altman MO, Angel M, Kosik I et al (2019) Human influenza A virus hemagglutinin glycan evolution follows a temporal pattern to a glycan limit. mBio 10(2):e00204–e00219
157. Li D, von Schaewen M, Wang X et al (2016) Altered glycosylation patterns increase immunogenicity of a subunit hepatitis C virus vaccine, inducing neutralizing antibodies which confer protection in mice. J Virol 90(23):10486–10498
158. Go EP, Ding H, Zhang S et al (2017) Glycosylation benchmark profile for HIV-1 envelope glycoprotein production based on eleven Env trimers. J Virol 91(9):e02428–e02416
159. Hutter J, Rodig JV, Hoper D et al (2013) Toward animal cell culture-based influenza vaccine design: viral hemagglutinin *N*-glycosylation markedly impacts immunogenicity. J Immunol 190(1):220–230
160. Liu WC, Lin YL, Spearman M et al (2016) Influenza virus hemagglutinin glycoproteins with different *N*-glycan patterns activate dendritic cells *in vitro*. J Virol 90(13):6085–6096
161. Urbanowicz RA, Wang R, Schiel JE et al (2019) Antigenicity and immunogenicity of differentially glycosylated hepatitis C virus E2 envelope proteins expressed in mammalian and insect cells. J Virol 93(7):e01403–e01418
162. Ronda C, Pedersen LE, Hansen HG et al (2014) Accelerating genome editing in CHO cells using CRISPR Cas9 and CRISPy, a web-based target finding tool. Biotechnol Bioeng 111 (8):1604–1616
163. Toth AM, Kuo C-W, Khoo K-H et al (2014) A new insect cell glycoengineering approach provides baculovirus-inducible glycogene expression and increases human-type glycosylation efficiency. J Biotechnol 182-183:19–29
164. Heffner KM, Wang Q, Hizal DB et al (2018) Glycoengineering of mammalian expression systems on a cellular level. In: Advances in biochemical engineering/biotechnology. Springer, Berlin. https://doi.org/10.1007/1010_2017_1057
165. Mabashi-Asazuma H, Jarvis DL (2017) CRISPR-Cas9 vectors for genome editing and host engineering in the baculovirus-insect cell system. Proc Natl Acad Sci U S A 114 (34):9068–9073
166. Narimatsu Y, Joshi HJ, Nason R et al (2019) An atlas of human glycosylation pathways enables display of the human glycome by gene engineered cells. Mol Cell 75(2):394–407.e395
167. Yang Z, Wang S, Halim A et al (2015) Engineered CHO cells for production of diverse, homogeneous glycoproteins. Nat Biotechnol 33:842
168. Lepenies B, Seeberger PH (2014) Simply better glycoproteins. Nat Biotechnol 32(5):443–445

169. Traving C, Schauer R (1998) Structure, function and metabolism of sialic acids. Cell Mol Life Sci 54(12):1330–1349
170. Varki A, Gagneux P (2012) Multifarious roles of sialic acids in immunity. Ann N Y Acad Sci 1253(1):16–36
171. Dhar C, Sasmal A, Varki A (2019) From "serum sickness" to "xenosialitis": past, present, and future significance of the non-human sialic acid Neu5Gc. Front Immunol 10:807
172. Altman MO, Gagneux P (2019) Absence of Neu5Gc and presence of anti-Neu5Gc antibodies in humans-an evolutionary perspective. Front Immunol 10:789
173. Ghaderi D, Taylor RE, Padler-Karavani V et al (2010) Implications of the presence of *N*-glycolylneuraminic acid in recombinant therapeutic glycoproteins. Nat Biotechnol 28 (8):863–867
174. Benatuil L, Kaye J, Rich RF et al (2005) The influence of natural antibody specificity on antigen immunogenicity. Eur J Immunol 35(9):2638–2647
175. Abdel-Motal UM, Wigglesworth K, Galili U (2009) Mechanism for increased immunogenicity of vaccines that form *in vivo* immune complexes with the natural anti-gal antibody. Vaccine 27(23):3072–3082
176. Bakema JE, Tuk CW, van Vliet SJ et al (2015) Antibody-opsonized bacteria evoke an inflammatory dendritic cell phenotype and polyfunctional Th cells by cross-talk between TLRs and FcRs. J Immunol 194(4):1856–1866
177. Huai G, Qi P, Yang H et al (2016) Characteristics of α-gal epitope, anti-gal antibody, α1,3 galactosyltransferase and its clinical exploitation (review). Int J Mol Med 37(1):11–20
178. Macher BA, Galili U (2008) The Galα1,3Galβ1,4GlcNAc-R (α-gal) epitope: a carbohydrate of unique evolution and clinical relevance. Biochim Biophys Acta 1780(2):75–88
179. Abdel-Motal UM, Guay HM, Wigglesworth K et al (2007) Immunogenicity of influenza virus vaccine is increased by anti-gal-mediated targeting to antigen-presenting cells. J Virol 81 (17):9131–9141
180. Abdel-Motal U, Wang S, Lu S et al (2006) Increased immunogenicity of human immunodeficiency virus gp120 engineered to express Galalpha1-3Galbeta1-4GlcNAc-R epitopes. J Virol 80(14):6943–6951
181. Abdel-Motal UM, Wang S, Awad A et al (2010) Increased immunogenicity of HIV-1 p24 and gp120 following immunization with gp120/p24 fusion protein vaccine expressing alpha-gal epitopes. Vaccine 28(7):1758–1765
182. Henion TR, Gerhard W, Anaraki F et al (1997) Synthesis of alpha-gal epitopes on influenza virus vaccines, by recombinant alpha-1,3-galactosyltransferase, enables the formation of immune complexes with the natural anti-gal antibody. Vaccine 15(11):1174–1182
183. Galili U, Repik PM, Anaraki F et al (1996) Enhancement of antigen presentation of influenza virus hemagglutinin by the natural human anti-gal antibody. Vaccine 14(4):321–328
184. Steinke JW, Platts-Mills TA, Commins SP (2015) The alpha-gal story: lessons learned from connecting the dots. J Allergy Clin Immunol 135(3):589–596
185. Chinuki Y, Morita E (2019) Alpha-gal-containing biologics and anaphylaxis. Allergol Int 68 (3):296–300
186. Román-Carrasco P, Lieder B, Somoza V et al (2019) Only α-gal bound to lipids, but not to proteins, is transported across enterocytes as an IgE-reactive molecule that can induce effector cell activation. Allergy 74(10):1956–1968
187. Lepenies B, Lee J, Sonkaria S (2013) Targeting C-type lectin receptors with multivalent carbohydrate ligands. Adv Drug Del Rev 65(9):1271–1281
188. Brzezicka K, Vogel U, Serna S et al (2016) Influence of core beta-1,2-xylosylation on glycoprotein recognition by murine C-type lectin receptors and its impact on dendritic cell targeting. ACS Chem Biol 11(8):2347–2356
189. Johannssen T, Lepenies B (2015) Identification and characterization of carbohydrate-based adjuvants. Methods Mol Biol 1331:173–187

190. Maglinao M, Eriksson M, Schlegel MK et al (2014) A platform to screen for C-type lectin receptor-binding carbohydrates and their potential for cell-specific targeting and immune modulation. J Control Release 175:36–42
191. Mayer S, Raulf M-K, Lepenies B (2017) C-type lectins: their network and roles in pathogen recognition and immunity. Histochem Cell Biol 147(2):223–237
192. Goyal S, Castrillon-Betancur JC, Klaile E et al (2018) The interaction of human pathogenic fungi with C-type lectin receptors. Front Immunol 9:1261
193. van Kooyk Y, Unger WWJ, Fehres CM et al (2013) Glycan-based DC-SIGN targeting vaccines to enhance antigen cross-presentation. Mol Immunol 55(2):143–145
194. Hu J, Wei P, Seeberger PH et al (2018) Mannose-functionalized nanoscaffolds for targeted delivery in biomedical applications. Chem Asian J 13(22):3448–3459
195. Gemmill TR, Trimble RB (1999) Overview of *N*- and *O*-linked oligosaccharide structures found in various yeast species. Biochim Biophys Acta 1426(2):227–237
196. Kottom TJ, Hebrink DM, Monteiro JT et al (2019) Myeloid C-type lectin receptors that recognize fungal mannans interact with *Pneumocystis* organisms and major surface glycoprotein. J Med Microbiol 68(11):1649–1654
197. Angrand G, Quillévéré A, Loaëc N et al (2019) Sneaking out for happy hour: yeast-based approaches to explore and modulate immune response and immune evasion. Genes 10(9):667
198. Vetvicka V, Vannucci L, Sima P (2020) Beta-glucan as a new tool in vaccine development. Scand J Immunol 91(2):e12833
199. Stanley P, Chen W (2003) Five Lec1 CHO cell mutants have distinct Mgat1 gene mutations that encode truncated *N*-acetylglucosaminyltransferase I. Glycobiology 13(1):43–50
200. Byrne G, O'Rourke SM, Alexander DL et al (2018) CRISPR/Cas9 gene editing for the creation of an MGAT1-deficient CHO cell line to control HIV-1 vaccine glycosylation. PLoS Biol 16(8):e2005817
201. Cox MMJ, Hollister JR (2009) FluBlok, a next generation influenza vaccine manufactured in insect cells. Biologicals 37(3):182–189
202. Wilson IBH (2002) Glycosylation of proteins in plants and invertebrates. Curr Opin Struct Biol 12(5):569–577
203. Altmann F (2007) The role of protein glycosylation in allergy. Int Arch Allergy Immunol 142 (2):99–115
204. Gaunitz S, Jin C, Nilsson A et al (2013) Mucin-type proteins produced in the *Trichoplusia ni* and *Spodoptera frugiperda* insect cell lines carry novel *O*-glycans with phosphocholine and sulfate substitutions. Glycobiology 23(7):778–796
205. Kurz S, Aoki K, Jin C et al (2015) Targeted release and fractionation reveal glucuronylated and sulphated *N*- and *O*-glycans in larvae of dipteran insects. J Proteome 126:172–188
206. Wilson IBH, Cummings RD, Aebi M (2015) Chapter 25 – Nematoda. In: Varki A, Cummings RD, Esko JD et al (eds) Essentials of glycobiology, 3rd edn. Cold Spring Harbor Laboratory Press, Cold Spring Harbor, pp 321–333
207. Martini F, Eckmair B, Stefanic S et al (2019) Highly modified and immunoactive *N*-glycans of the canine heartworm. Nat Commun 10(1):75
208. Shim BS, Hong KJ, Maharjan PM et al (2019) Plant factory: new resource for the productivity and diversity of human and veterinary vaccines. Clin Exp Vaccine Res 8(2):136–139
209. Takeyama N, Kiyono H, Yuki Y (2015) Plant-based vaccines for animals and humans: recent advances in technology and clinical trials. Ther Adv Vaccine 3(5–6):139–154
210. D'Aoust M-A, Couture MMJ, Charland N et al (2010) The production of hemagglutinin-based virus-like particles in plants: a rapid, efficient and safe response to pandemic influenza. Plant Biotechnol J 8(5):607–619
211. Landry N, Ward BJ, Trépanier S et al (2010) Preclinical and clinical development of plant-made virus-like particle vaccine against avian H5N1 influenza. PLoS One 5(12):e15559
212. Landry N, Pillet S, Favre D et al (2014) Influenza virus-like particle vaccines made in *Nicotiana benthamiana* elicit durable, poly-functional and cross-reactive T cell responses to influenza HA antigens. Clin Immunol 154(2):164–177

213. Le Mauff F, Mercier G, Chan P et al (2015) Biochemical composition of haemagglutinin-based influenza virus-like particle vaccine produced by transient expression in tobacco plants. Plant Biotechnol J 13(5):717–725
214. Margolin E, Chapman R, Williamson A-L et al (2018) Production of complex viral glycoproteins in plants as vaccine immunogens. Plant Biotechnol J 16(9):1531–1545
215. Ward BJ, Landry N, Trépanier S et al (2014) Human antibody response to *N*-glycans present on plant-made influenza virus-like particle (VLP) vaccines. Vaccine 32(46):6098–6106
216. Shaaltiel Y, Tekoah Y (2016) Plant specific *N*-glycans do not have proven adverse effects in humans. Nat Biotechnol 34(7):706–708
217. Rup B, Alon S, Amit-Cohen B-C et al (2017) Immunogenicity of glycans on biotherapeutic drugs produced in plant expression systems—the taliglucerase alfa story. PLoS One 12(10): e0186211
218. Smith DF, Cummings RD, Song X (2019) History and future of shotgun glycomics. Biochem Soc Trans 47(1):1–11
219. Tamburrini A, Colombo C, Bernardi A (2019) Design and synthesis of glycomimetics: recent advances. Med Res Rev 40(2):495–531
220. Guan WJ, Ni ZY, Hu Y et al (2020) Clinical characteristics of coronavirus disease 2019 in China. N Engl J Med. https://doi.org/10.1056/NEJMoa2002032
221. Wu F, Zhao S, Yu B et al (2020) A new coronavirus associated with human respiratory disease in China. Nature 579(7798):265–269
222. Varki A, Cummings RD, Aebi M et al (2015) Symbol nomenclature for graphical representations of glycans. Glycobiology 25(12):1323–1324
223. Hastie KM, Zandonatti MA, Kleinfelter LM et al (2017) Structural basis for antibody-mediated neutralization of Lassa virus. Science 356(6341):923–928
224. Watanabe Y, Raghwani J, Allen JD et al (2018) Structure of the Lassa virus glycan shield provides a model for immunological resistance. Proc Natl Acad Sci U S A 115(28):7320–7325
225. Zhao Y, Ren J, Harlos K et al (2016) Toremifene interacts with and destabilizes the Ebola virus glycoprotein. Nature 535(7610):169–172
226. Lee PS, Ohshima N, Stanfield RL et al (2014) Receptor mimicry by antibody F045-092 facilitates universal binding to the H3 subtype of influenza virus. Nat Commun 5(1):3614
227. An Y, McCullers JA, Alymova I et al (2015) Glycosylation analysis of engineered H3N2 influenza A virus hemagglutinins with sequentially added historically relevant glycosylation sites. J Proteome Res 14(9):3957–3969
228. Struwe WB, Chertova E, Allen JD et al (2018) Site-specific glycosylation of virion-derived HIV-1 Env is mimicked by a soluble trimeric immunogen. Cell Rep 24(8):1958–1966.e1955
229. Kwon YD, Pancera M, Acharya P et al (2015) Crystal structure, conformational fixation and entry-related interactions of mature ligand-free HIV-1 Env. Nat Struct Mol Biol 22 (7):522–531
230. Walls AC, Tortorici MA, Frenz B et al (2016) Glycan shield and epitope masking of a coronavirus spike protein observed by cryo-electron microscopy. Nat Struct Mol Biol 23 (10):899–905
231. Xu K, Chan YP, Bradel-Tretheway B et al (2015) Crystal structure of the pre-fusion Nipah virus fusion glycoprotein reveals a novel hexamer-of-trimers assembly. PLoS Pathog 11(12): e1005322

Adv Biochem Eng Biotechnol (2021) 175: 355–378
DOI: 10.1007/10_2018_71

Published online: 25 August 2018

# Interplay of Carbohydrate and Carrier in Antibacterial Glycoconjugate Vaccines

**Tyler D. Moeller, Kevin B. Weyant, and Matthew P. DeLisa**

## Contents



**Abstract** Bacterial infections are a serious health concern and are responsible for millions of illnesses and deaths each year in communities around the world. Vaccination is an important public health measure for reducing and eliminating this burden, and regions with comprehensive vaccination programs have achieved significant reductions in infection and mortality. This is often accomplished by immunization with bacteria-derived carbohydrates, typically in conjunction with other biomolecules, which induce immunological memory and durable protection against bacterial human pathogens. For many species, however, vaccines are currently unavailable or have suboptimal efficacy characterized by short-lived memory and incomplete protection, especially among at-risk populations. To address this challenge, new tools and techniques have emerged for engineering carbohydrates and

T. D. Moeller, K. B. Weyant, and M. P. DeLisa (✉)
Robert Frederick Smith School of Chemical and Biomolecular Engineering, Cornell University, Ithaca, NY, USA
e-mail: md255@cornell.edu

conjugating them to carrier molecules in a tractable and scalable manner. Collectively, these approaches are yielding carbohydrate-based vaccine designs with increased immunogenicity and protective efficacy, thereby opening up new opportunities for this important class of antigens. In this chapter we detail the current understanding of how carbohydrates interact with the immune system to provide immunity; how glycoengineering, especially in the context of glycoconjugate vaccines, can be used to modify and enhance immune responses; and current trends and strategies being pursued for the rational design of next-generation glycosylated antibacterial vaccines.

**Graphical Abstract**

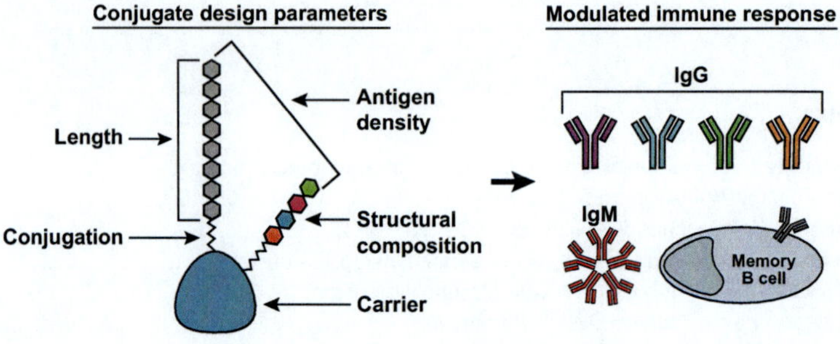


**Keywords** Adaptive immune response, Bacterial vaccines, Carbohydrate antigens, Glycoconjugate vaccines, Glycoimmunology, Immunity, Oligosaccharides, Polysaccharides

## Abbreviations

| | |
|---|---|
| APC | Antigen-presenting cell |
| BCR | B cell receptor |
| CPS | Capsular polysaccharide |
| DT | Diphtheria toxoid |
| GBS | Group B *Streptococcus* |
| glycOMV | Glycosylated outer membrane vesicle |
| Hib | *Haemophilus influenzae* type b |
| IgG | Immunoglobulin G |
| IgM | Immunoglobulin M |
| iNKT | Invariant natural killer T cell |
| LOS | Lipooligosaccharide |
| LPS | Lipopolysaccharide |
| MAMP | Microbe-associated molecular pattern |
| MenB | *Neisseria meningitidis* serogroup B |
| MHC | Major histocompatibility complex |

| | |
|---|---|
| MW | Molecular weight |
| NKT | Natural killer T cell |
| O-PS | *O*-antigen polysaccharide |
| OST | Oligosaccharyltransferase |
| PolySia | Polysialic acid |
| PRP | Polyribosylribitol phosphate |
| PRR | Pattern recognition receptor |
| RU | Repeating unit |
| TD | T cell-dependent |
| TI | T cell-independent |
| TT | Tetanus toxoid |
| ZPS | Zwitterionic polysaccharide |
| αGalCer | α-Galactosylceramide |

# 1 Combating Bacterial Disease with Advances in Glycobiotechnology

Bacteria are the causative agents of numerous diseases, including bubonic plague, cholera, meningitis, tuberculosis, and many others, which have imposed an enormous burden on human health throughout history and continue to affect various regions of the world today. Pneumonia remains a leading cause of childhood death worldwide, with over half of the fatalities resulting from infection by *Streptococcus pneumoniae* or *Haemophilus influenzae* type b (Hib) [1]. However, reduction and eventual elimination of many bacterial infectious diseases is an achievable possibility thanks to significant progress in many countries toward the implementation of effective public health and sanitation measures, access to effective medical treatments such as antibiotics, and prophylactic intervention through widespread adoption of vaccines. The deployment of vaccines, which are used to elicit a protective immune response, has played a critical role in preventing or ameliorating bacterial infection with impressive results. The 2017 recommended immunization schedule published by the United States Centers for Disease Control and Prevention includes vaccines for Hib, *S. pneumoniae*, and *Neisseria meningitidis*, with some of the most effective formulations protecting susceptible populations of both young and old [2]. These vaccines have led to dramatic reductions in infectious disease, and nearly all of them have contributed to this feat by eliciting protective antibodies against carbohydrates on the bacterial cell surface [3].

Various glycan structures coat the surfaces of both gram-negative and gram-positive bacteria (Fig. 1). In gram-negative bacteria, glycolipids such as lipopolysaccharide (LPS) or lipooligosaccharide (LOS) are ubiquitous, and capsular polysaccharide (CPS) composes the capsule that envelopes many species. Likewise, gram-positive bacteria are surrounded by a thick layer of peptidoglycan with teichoic acid embedded, and the cell may also be enveloped in CPS. A unique class of gram-

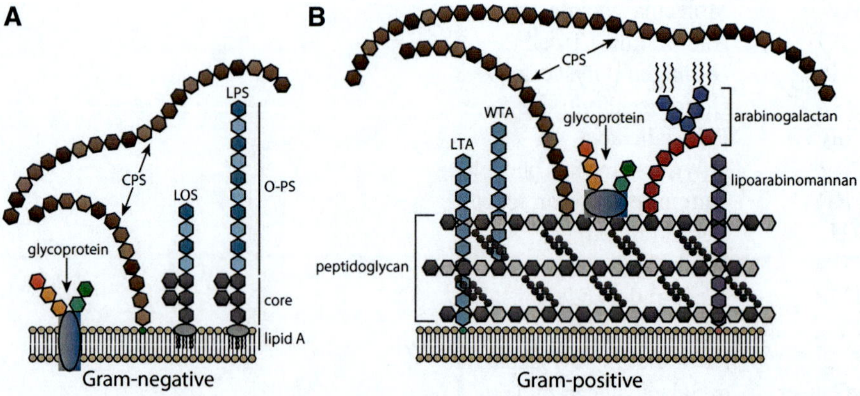


**Fig. 1** Glycan expression on the bacterial cell surface. Carbohydrate motifs on the surface of bacteria accessible to cells of the immune system are useful targets for raising a protective immune response. Some, including CPS and glycoproteins, are ubiquitous and present on many bacterial species. Other glycan structures are found primarily in (**a**) gram-negative bacteria, such as LOS and LPS, or (**b**) gram-positive bacteria, such as wall teichoic acid (WTA) and lipoteichoic acid (LTA). Some glycans, including lipoarabinomannan and arabinogalactan, are unique to mycobacteria, a subclass of gram-positive bacteria. Peptidoglycan is present on the surface of gram-positive but not gram-negative bacteria, where it is instead found in the periplasmic space between inner and outer lipid membranes

positive bacteria called mycobacteria contains additional glycans such as lipoarabinomannan and arabinogalactan [4, 5]. Moreover, bacteria can also contain surface-exposed glycoproteins [6, 7]. To elicit a protective immune response against such glycan-coated bacteria, it is common practice to isolate or synthesize these glycans and then use them to formulate subunit vaccines. CPS- and LPS-derived glycans are most commonly targeted in vaccines because of their high cell surface density and the fact that these structures are often unique to a particular species or strain.

Several factors influence the development and efficacy of carbohydrate-based bacterial vaccines, with significant variation even among those currently on the market. In addition to the antigenic target, the presence or absence of particular immunopotentiators such as proteins, and how they associate with the carbohydrate antigen, can dictate which populations are protected and the strength of immunological memory. Practical considerations, such as the ease of manufacture, safety, and ease of transport, also need to be considered. In this chapter we describe the current understanding of the complex interplay between glycans and the immune system and some of the key design factors that are critical for creating an effective vaccine. We also highlight recent technologies that facilitate rational vaccine design. On the basis of a deeper understanding of these and other related issues, we anticipate improvements to existing formulations and the release of approved vaccines against deadly pathogens such as *Francisella tularensis*, group B *Streptococcus* (GBS), *Shigella* sp., and *Staphylococcus aureus*, for which no licensed vaccines currently exist [8–11].

## 2 Unconjugated Polysaccharide Vaccines

Carbohydrate involvement in immune responses was established in the 1920s after the discovery that *S. pneumoniae* infection yielded immune responses to CPS that were serotype-specific [12]. Purified CPS injection of patients infected with the corresponding pneumococcal serotype elicited an immune reaction [13], and protection against several distinct serotypes could be achieved with a single immunization of different CPS [14]. Accordingly, a 14-valent CPS-based vaccine against *S. pneumoniae* was approved in 1977 [15], and the effectiveness of this vaccine prompted its widespread adoption among the general population [16]. A 23-valent pneumococcal polysaccharide vaccine is available today for adults, and it provides coverage of 90% of the disease-causing *S. pneumoniae* serotypes in the United States, reducing infection by 65% [17].

The advent of CPS vaccines marked the establishment of carbohydrates as compelling, viable targets of a directed immune response. CPS vaccines are now well-established with more than four decades of clinical usage. In addition, multivalent vaccine formulation is relatively straightforward from a technical perspective, and the polysaccharide composition can be increased or changed should the most pathologically relevant serotypes vary across geographical region or over time. The main disadvantage of an unconjugated polysaccharide is the lack of protective and memory responses, particularly in populations most susceptible to bacterial infection, including the elderly, immune-compromised, and children less than 2 years of age [18].

Many bacterial carbohydrates, including CPS, LPS, peptidoglycan, and other glycans found on the bacterial surface, are potent stimulators of the fast-responding innate immune system. They contain microbe-associated molecular patterns (MAMPs) recognized by pattern recognition receptors (PRRs) including toll-like receptors, nod-like receptors, and C-type lectin receptors found on the surface of macrophages, dendritic cells, and other innate immune cells. Activation of PRRs by MAMP binding results in cytokine production that promotes inflammation and recruits effector cells. However, although effective activators of the innate immune system, most polysaccharides are T cell-independent (TI) antigens that develop adaptive immune responses characterized by a lack of glycan-specific high-affinity antibodies as well as limited memory responses. The long, repetitive sugar motifs present in these polysaccharides crosslink B cell receptors (BCR) on the surface of glycan-specific B cells to elicit the production of predominately low-affinity and short-lived immunoglobulin M (IgM). In contrast, T cell-dependent (TD) antigens elicit antibody class-switching and affinity-maturation processes that result in more high-affinity and long-lasting immunoglobulin G (IgG) antibodies (Fig. 2).

Although polysaccharides are generally classified as TI antigens, there are notable cases where T cell activation and subsequent high-affinity class-switched antibody production and memory cell development can occur. Zwitterionic polysaccharides (ZPS), which contain both positively and negatively charged residues, can be processed by antigen-presenting cells (APCs) onto a cell-surface major

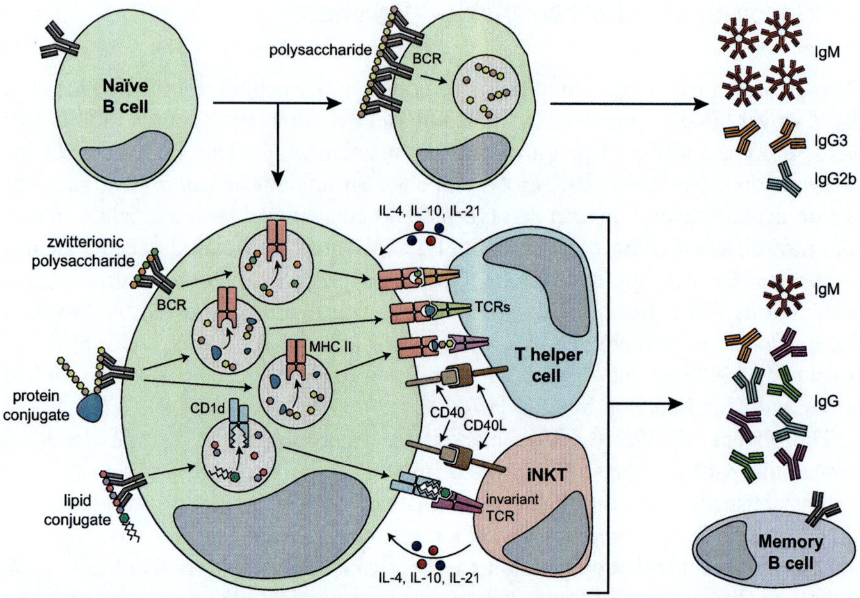


**Fig. 2** Humoral immune response to carbohydrates. Antigen recognition and processing by carbohydrate-specific B cells is important in determining the nature of the corresponding immune response. Unconjugated polysaccharides can bind to multiple B cell receptors to elicit a TI response characterized by production of IgM with some class-switched IgGs (IgG2b and IgG3 in mice) as well as limited memory responses. Glycans conjugated to other biomolecules or that meet specific structural requirements may be processed intracellularly and bind to surface proteins on B cells for display and recognition by cognate T cells. Zwitterionic polysaccharides, peptides, and glycopeptides can be loaded onto MHC II for recognition by epitope-specific T cell receptors on T helper cells. Lipids or glycolipids, including analogues of α-galactosylceramide, are loaded onto CD1d and recognized by semi-invariant TCRs on invariant natural killer T cells. Co-stimulation between surface-displayed molecules including CD40 and CD40L on B cells and T cells, respectively, facilitates the release of cytokines that activate the B cell. TD responses are characterized by high-affinity, class-switched antibodies and memory cell production

histocompatibility complex (MHC) and recognized by cognate T cells, leading to cell activation similar to that of a classical protein antigen. Interestingly, teichoic acid, a carbohydrate-containing cell wall component of gram-positive bacteria, was originally an uncharacterized impurity known as C polysaccharide that elicited high antibody titers in early CPS studies by Heidelberger and colleagues. Teichoic acid has however been shown to have a zwitterionic state and activate T cells in an MHC-dependent manner [19]. Polysaccharides from some serotypes of *S. aureus*, *S. pneumoniae*, and *Bacteriodes fragilis* exhibit ZPS properties [20, 21], and *B. fragilis* ZPS PSA1 has been modified to serve as a carrier molecule for small, non-immunogenic carbohydrates associated with cancer [22].

## 3 Conjugate Polysaccharide Vaccines

Bacterial glycans, traditionally a heterogeneous mixture of polysaccharides from a targeted bacterial serotype, can be harvested and purified before undergoing chemical activation and conjugation to a carrier protein such as tetanus toxoid (TT) from *Clostridium tetani*, diphtheria toxoid (DT) from *Corynebacterium diphtheriae*, and $CRM_{197}$, a DT mutant with a single amino acid change that ameliorates toxicity [23]. These glycoconjugates have been demonstrated to elicit more desirable immune responses, including long-lasting immunological memory and protective IgG antibodies that are often absent in polysaccharide-only vaccines. For much of their history, glycoconjugate vaccines were presumed to elicit a TD immune response through availability of T cell epitopes derived from the protein carrier. These short peptides, derived from intracellular processing of the carrier and binding to MHC molecules on the APC surface, allow for the activation of corresponding T cells. B cells specific for the glycan antigen associate with these activated T cells and form an immune synapse before undergoing downstream processes such as antibody class-switching that underpins high-affinity antibody production. However, it was recently demonstrated that conjugate-derived glycopeptide epitopes can also be presented to T cells. At least in some cases, these epitopes can elicit more potent immune responses than peptide epitopes alone [24].

Glycoconjugate design requires careful manipulation of several key design parameters (highlighted in Fig. 3). Each of these characteristics is capable of modulating the immune response and can be analyzed through an iterative process whereby a design specification is rationally made followed by in vitro and in vivo studies to analyze antigen binding, antibody titers, and protection. In the following sections, polysaccharide length, structural composition, and density, as well as conjugation method and carrier molecule, are examined for their impact on vaccine function. This is followed by a discussion of several strategies that improve upon

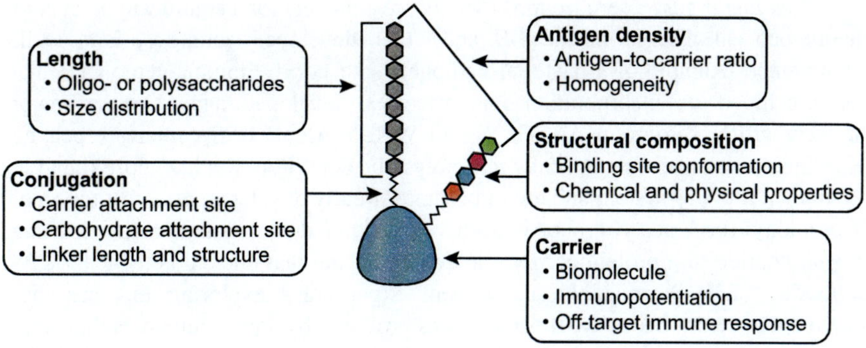


**Fig. 3** Design parameters in glycoconjugate development. Several key characteristics of a glycoconjugate vaccine can affect its immunogenicity and efficacy. Manipulation of one variable can limit or otherwise affect another aspect of the glycoconjugate design. These parameters must be carefully identified, optimized, and validated for a new vaccine candidate

traditional glycoconjugate design, including new synthesis and screening methods for the rational design of a 'minimal' glycan antigen and the use of novel lipid-based carrier molecules.

## 4 Glycan Structural Composition as a Vaccine Design Parameter

The structure and exposed residues of a glycan immunogen, determined by the monosaccharide subunits that comprise it, play a significant role in shaping the subsequent immune response. Chemical modification of sugar residues has been found to enhance the immune response against carbohydrates of varying size, structure and source, including tumor-associated carbohydrates and viral shield glycans, which normally exhibit little or insufficient immunogenicity [25–27]. Changing the underlying glycan composition can be used to alter existing B cell receptor-binding epitopes. Although large structures such as polysaccharides may have numerous epitopes or features that can bind the variable region of immunoglobulin chains, one or several sites are often most relevant for immunization and drive the elicitation of antibodies with desirable characteristics such as high affinity, specificity, and protection against the target pathogen. Modulation of immune response can therefore be realized from modification of existing epitopes or addition of new sites. Even manipulation of a single chemical moiety is sufficient to alter the response in many cases. For example, the absence of pyruvate ketal and phosphate groups was shown to reduce or abolish protection by conjugated *S. pneumoniae* serogroup 4 and *Clostridium difficile* polysaccharides, respectively, [28–30]. Similarly, polysaccharides that were not *O*-acetylated significantly reduced titers to *N. meningitidis* serogroup A and prevented cross-protection between *S. pneumoniae* serotypes 15B and 15C [31, 32].

The *N. meningitidis* serogroup B (MenB), responsible for a significant percent of meningococcal disease in the US and other developed countries, is a well-documented example of a bacterial pathogen with poorly immunogenic CPS that has precluded its incorporation into vaccines developed against other major *N. meningitidis* serogroups, A, C, Y, and W-135. MenB CPS contains repeating α2,8-linked sialic acid chains known as polysialic acid (PolySia) like those found in the human nervous system and are important for early development. Substitution of *N*-propionyl for *N*-acetyl groups in MenB PolySia, and conjugation of the resulting polysaccharide to a protein carrier, yielded a vaccine that elicited anti-MenB CPS antibodies [33]. Surprisingly, subsequent experiments exploring this response showed that serum bactericidal activity was provided by some antibodies that were not cross-reactive with human PolySia [34, 35]. Later work revealed that this is the result of antibodies generated against de-*N*-acetylated or otherwise modified sialic acid that avoid auto-reactivity towards host PolySia [36, 37].

Although glycan composition can be manipulated to modify adaptive immune recognition, other properties relevant in vaccine production and development can also be improved. The Vi antigen of *Salmonella enterica* serovar Typhi Vi is an α-(1,4)-*N*-acetylgalactosaminuronate polymer with high levels of carbon 3 - *O*-acetylation. Plant-derived pectin is abundant and composed of polygalacturonic acid that can undergo *O*-acetylation at carbons 2 and 3. This modified pectin (OAcPec) is similar to Vi antigen with the exception of *O*-acetylation in lieu of *N*-acetylation at carbon 2. However, both elicited similar levels of antibodies when immunized as a TT conjugate. OAcPec is more soluble than Vi antigen, which aids in production, and an OAcPec-based Typhi vaccine represents a safer and cheaper alternative to using natively sourced polysaccharide [38, 39].

Other opportunities may exist for modification of glycans. Behavior such as flexibility and conformation of glycoconjugates in solution is dependent on the attached carbohydrates [40–42]. Additionally, CPS *O*-acetylation is preferentially bound by lectins that initiate complement pathways [43], raising the possibility that carefully considered changes to carbohydrates could be a strategy for harnessing glyco-antigen interactions with the innate immune system and glyco-antigen processing in the body.

## 5 Polysaccharide Length as a Vaccine Design Parameter

The carbohydrate component of many glycoconjugates is derived from surface-exposed polysaccharides on bacterial pathogens. Extraction typically results in a polysaccharide mixture of varying lengths because of the heterogeneity of naturally occurring CPS and the *O*-antigen polysaccharide (O-PS) component of LPS antigens. Chain length of the polysaccharide following conjugation to a carrier molecule is dependent on the exact extraction method and subsequent preparation and activation steps, and is usually reported as an average molecular weight (MW) or degree of polymerization of repeating units (RU) of oligosaccharide. The immunological consequences of this size distribution, and whether an optimal length exists for optimal vaccine efficacy, are important considerations that have been actively investigated for several decades. Polysaccharide length was first established as an important glycoconjugate design variable in studies where small chains of dextran conjugated to TT were found to provoke higher carbohydrate-specific class-switched antibody titers compared to larger dextran-TT conjugates [44]. In more recent studies, higher MW O-PS (70–95 RU) from *S. enterica* serovar Typhimurium conjugated to $CRM_{197}$ was found to be significantly less immunogenic than lower MW (25–35 RU) conjugates [45].

The inverse relationship between polysaccharide length and protection is not a universal phenomenon, with many glycoconjugate vaccines showing different behavior. For example, *Vibrio cholerae* O-PS length was not found to affect the immune response [46] whereas 150,000 MW *S. pneumonia*-type 14 polysaccharide conjugated to TT was found to induce higher antibody titers than TT conjugates bearing a range of smaller (1,337–70,000 MW) polysaccharides [47, 48]. Mixed results have also been reported for conjugation of the well-studied Hib polysaccharide polyribosylribitol phosphate (PRP) unit. Studies in both mice [49] and humans [50] failed to identify significant correlation between PRP length and antibody titers in most cases. Interestingly, Anderson and coworkers noticed that polysaccharide length affected the anamnestic response for adults and 1-year-old infant vaccine recipients differently. The greatest antibody titers for infants were obtained with $CRM_{197}$ protein conjugated to 7 RU, compared to 20 RU for adults. This is reminiscent of polysaccharide vaccines that protect adults but are poor inducers of immunity in children, suggesting that increasing the size of polysaccharide conjugated to a carrier molecule might, in some cases, promote a more TI-like response possibly as a result of increased crosslinking of BCR on the surface of antigen-specific B cells or interference in antigen processing.

A minimal polysaccharide length is required for proper epitope recognition and binding by polysaccharide-specific BCR. This suggests that optimal polysaccharide size may be constrained by a Goldilocks principle in which chains that are too large or too short provide suboptimal immune responses. Indeed, it was shown that TT conjugates bearing intermediate sized oligosaccharides (14 RU) performed better in generating antigen-specific protective antibodies compared to shorter (7 RU) or longer (27 RU) oligosaccharides against GBS III [51].

In practice, determination of what is 'just right' for optimal polysaccharide length is highly dependent on the polysaccharide serotype and likely requires empirical testing and validation for each new vaccine candidate. Efforts to improve protection through exhaustive examination of various polysaccharide lengths are hindered by the different experimental conditions, molecular structures, and control of other design parameters that make generalization based on results in the published literature difficult. Comprehensive screens of glycoconjugate variants where experimental conditions are kept identical and only one parameter, for example polysaccharide length, is manipulated at a time can be used to make recommendations more conclusively regarding the effect of chain size. This is exemplified in a recent study that tested the antibody response of a vaccine containing Vi antigen from the CPS of *S. enterica* serovar Typhi across several key design parameters, including full-length PS (165,000 MW) versus a smaller fragment (43,000 MW). Different response kinetics were observed between these sizes and only lower sized polysaccharide fragments conjugated to $CRM_{197}$ and DT showed significantly increased secondary antibody responses [52].

## 6 Conjugation as a Vaccine Design Parameter

A key concept of glycoconjugate vaccines is the attachment of sugar to protein. Different reaction mechanisms can accomplish this task, and the chemistry used plays an important role in determining what parameters, such as antigen density, can be achieved. In general, conjugation is performed by reacting derivatized or activated glycans with functional groups on the carrier molecule, often utilizing a linker moiety to facilitate attachment. The most common means of conjugation is by reductive amination, whereby aldehyde groups from oxidized glycans or linkers are reacted to bind covalently the amine side group of lysine residues on the protein [53, 54]. This attachment process is often considered random, although conjugation by reductive amination of $CRM_{197}$ with different linkers [55] or carbohydrates [56] has been found to favor certain lysine residues. Crotti and coworkers observed that preferential conjugation to specific lysine residues could be achieved by limiting the amount of linker present in the conjugation reaction. With several decades of notable successes, reductive amination is the current standard conjugation method.

The emergence of novel site-selective chemistries has fueled interest in developing glycoconjugates that are more homogenous and consistent between preparations, simultaneously decreasing the likelihood of interfering with relevant epitopes on the carrier [57]. Copper-mediated and copper-free azide-alkyne cycloaddition reactions have been used to add glycans selectively onto tyrosine residues of the carrier protein. Immunization with identical doses by protein content of glycoconjugates synthesized by tyrosine ligation or random conjugation elicited similar anti-carbohydrate IgG titers and potency as measured by opsonophagocytic killing assay [58]. Thiol-reacting maleimide moieties have also been used as conjugation linkers to help induce IgG antibodies [59]. Significant advances in the identification and production of the enzymes involved in natural glycan conjugation in bacteria have provided new avenues for glycoconjugate vaccine production [60, 61]. For example, the oligosaccharyltransferase (OST) enzyme PglB from *Campylobacter jejuni* has been used to attach bacterial polysaccharides to carrier proteins through both in vitro chemoenzymatic synthesis [62] and in vivo conjugation [63]. OSTs offer a tractable platform that directly attaches sugars to a defined amino acid motif in a highly specific and controllable manner without the need for linkers. The carrier can be designed with the desired attachment site(s) and configured through established protein engineering techniques. Both *N*-linked [64] and *O*-linked [65] glycosylation mechanisms have been used to attach bacterial polysaccharides to asparagine or serine/threonine residues, respectively. Moreover, the discovery and engineering of OSTs to expand enzymatic conjugation capabilities is ongoing [66, 67]. Other enzymes have also been utilized for selective conjugation strategies, even working in concert with click chemistries. For example, microbial transglutaminase (MTG) obtained from *Streptoverticillium mobaraense* has been leveraged to attach functional linkers enzymatically to lysine residues that can subsequently be used in azide-alkyne cycloaddition chemistry for adding polysaccharide antigens [68].

To date, most conjugation strategies focus on direct covalent attachment of glycans to carrier proteins, a process that has historically been considered a key requirement for successful TD immune response. However, the need for covalent conjugation has been challenged by recent studies, including work by Malley and coworkers that utilized a recombinantly produced biotin-binding protein, rhizavidin, to capture biotinylated polysaccharide antigens, resulting in an affinity-based approach for linking glycans to carrier molecules. The *S. pneumoniae*-derived protein and CPS immunogens that were non-covalently assembled in this manner elicited a robust protective response comparable to traditional whole cell vaccines [69]. Along similar lines, it was recently demonstrated that polysaccharides embedded in a protein matrix made class-switched anti-carbohydrate antibodies upon immunization. The memory-inducing response of this antigen elicited antibody titers similar to the commercially available glycoconjugate but was abolished when the same polysaccharide was co-administered but not entrapped with the protein matrix [70]. This suggests that the close proximity, but not necessarily covalent linkage, of glycans to T cell epitopes such as those found in proteins is the crucial requirement for efficacious delivery of the glyco-antigen to the immune system. Hence, although covalent conjugation may be a convenient means for ensuring this association following delivery in the host, novel glycoconjugate designs that do not rely on linkers or conjugation steps are emerging as viable alternatives.

## 7 Carrier Molecule as a Vaccine Design Parameter

Commonly used carrier molecules in commercially available bacterial vaccines that promote a TD immune response are TT, diphtheria toxoid DT, and $CRM_{197}$ [23]. Because of the prominent role they play in current vaccine formulations and schedules around the world, the immunological properties of these proteins have been extensively investigated [71, 72]. Continued use of these proteins is promoted by their history as safe, effective carriers and how, in some cases, a 'carrier priming' benefit can boost vaccine efficacy. Infants are routinely immunized with detoxified TT/DT as part of the diphtheria and tetanus vaccine and are likely to be exposed to these antigens before or during glycoconjugate administration. Immunization of mice with DT/TT/$CRM_{197}$ prior to $CRM_{197}$-conjugated N. meningitidis serogroup A and C polysaccharides has been found to improve anti-polysaccharide IgG titers significantly. Additional experiments showed heightened T helper cell responses and carbohydrate-specific plasmablast numbers [73]. The activation of carrier-specific T helper cells from priming could result in more effective activation of glycan-specific B cells with carrier-derived fragments presented on their surface.

However, in other situations, prior or simultaneous exposure to a protein can result in vaccine interference that actually decreases glycoconjugate efficacy. Several mechanisms have been proposed for this, including the steric hindrance of glycan-specific B cell binding by carrier-specific antibodies, competition for carrier-specific T helper cells by the expanded carrier-specific B cells population, and/or alteration of the immune environment by regulatory T cells, cytokines, and other factors [74, 75]. In practice, the immunogenicity impact of glycoconjugate carrier proteins can be difficult to predict and rationalize, particularly when multiple vaccines are being administered in a similar time window. Following separate immunization with DT/TT/$CRM_{197}$-containing vaccines, English infants exhibited a decrease in anti-MenC antibody titers in a two-dose administration of MenC-$CRM_{197}$ and then MenC-TT, but not when this order was reversed [76]. Several factors could be responsible for this result, including interference between carrier proteins and differences in carrier preparation. TT and DT undergo a detoxification process that results in protein crosslinking and a heterogeneous mixture of connected toxoid with potentially altered conformation and binding epitopes [77].

Efforts are underway to expand beyond these toxoid carriers for use as immunogenic glycan attachment scaffolds. In addition to mitigating unwanted vaccine interference, new protein carriers can themselves be antigens that, in tandem with the conjugated glycan component, provide or enhance protection. A survey of the pathogen-derived protein carriers currently being developed for glycoconjugate vaccines against several bacterial species has recently been described [78]. Computational prediction tools can be used with sequenced genomes of pathogens to identify protein candidates for generating immunity [79]. A GBS pilus protein previously selected by this reverse immunology approach and shown to confer protection was conjugated to GBS polysaccharide II. Antibodies were successfully raised against both the polysaccharide and protein components and conferred protection in mice [58]. The flagellin protein of *Burkholderia pseudomallei* was conjugated to its O-PS and induced glycan-specific IgGs and increased survival [80], with a similar result obtained for *B. pseudomallei* glycoconjugates containing different proteins identified from genome analysis [81]. Despite the fact that each new carrier protein needs to undergo testing for safety and efficacy, and requires optimization for conjugation, glycan density, and other parameters, their development as scaffolds for next-generation glycoconjugates appears promising.

## 8 Glycan Density as a Vaccine Design Parameter

The amount of carbohydrate antigen attached to each carrier molecule, referred to here as glycan density, is another important consideration that can affect glycoconjugate performance. In cases with well-defined, single attachment site carbohydrates conjugated to protein, glycan density can simply be described as a mass/mass ratio of carbohydrate to protein. However, for many formulations, including glycoconjugates with lattice-type structures or multiple attachment site carbohydrates, mass/mass ratio alone is not an adequate characterization. Glycan density is also heavily impacted by other design parameters such as conjugation method or protein carrier, both of which can exhibit batch-to-batch variation in glycan attachment. Care should be taken when comparing glycan density across studies to account for these differences.

Indeed, the literature suggests that the effect of glycan density is case-specific and dependent on the production, immunization, and analytical methods used for each glycoconjugate, leading to a range of conclusions. Nonetheless, some studies have determined that it is an important factor for improving immunogenicity as measured by post-immunization serum titers [82, 83]. In one such case, immunization with an intermediate density of ~9–10 saccharides per protein elicited the highest titers of LPS-specific antibodies relative to conjugates with 4 or 19–23 saccharides per protein [83]. In contrast, other studies concluded that varying the amount of glycan did not play a significant role in enhancing immune responses [44]. For example, when a bivalent linker was used to double the amount of polysaccharide attached to a protein carrier, no change in antibody titers was observed [84]. Although no general trend has emerged to define the relationship between glycan density and vaccine efficacy, the potentially significant effect that glycan density has been shown to play in some work necessitates the determination of an optimal density on a case-by-case basis.

## 9 Minimal Epitope Synthetic Vaccines

Improving glycoconjugate vaccine efficacy has traditionally relied on varying one or several of the aforementioned parameters through established techniques in protein engineering, conjugation chemistry, and glycan extraction. Glycan extraction in particular has been a major development impediment because of the heterogeneous mixture of glycans and contaminants it can produce, making characterization difficult and limiting scientists' ability to link important molecular-level design parameters with immunological performance. Moreover, practical constraints concerning the ability to culture pathogenic bacteria and isolate glycans at sufficient yield have prevented the wide application of glycoconjugates. Synthesizing glycans in a bottom-up approach via chemical or chemo-enzymatic means offers opportunities for cheaper and safer production as well as increased vaccine homogeneity and more precise control over the immune response. Essential to these efforts is the identification of the minimal carbohydrate structures and motifs required for binding and eliciting protection-conferring antibodies, small enough for synthesis yet providing immunity comparable to full-length polysaccharides.

Identifying minimal epitopes relies on a laborious screening process against pre-existing antibodies from animals that were immunized with full-length polysaccharides. Competitive binding assays of post-immunization serum against whole polysaccharides and polysaccharide-derived fragments are used to determine the fragments most critical for antibody recognition. In this way, it was found that the branched rhamnopyranoside moiety of the tetrasaccharide RU of *S. pneumoniae* serotype 23F was necessary for recognition and opsonization by reactive polyclonal sera [85] as well as antibody fragments derived from immunized adults [86]. However, until relatively recently, challenges in glycan synthesis generally limited the scope of these studies to identify the one or two most important residues for binding.

Improvements in synthetic glycan production and higher-throughput glycan array technologies allow for more accurate epitope identification. Two serotypes of *S. pneumoniae* (7F and 23F) were investigated by Seeberger and coworkers by screening multiple CPS-derived oligosaccharides. The RU of *S. pneumoniae* serotype 7F CPS is a heptasaccharide containing two branched side chains, and a glycan microarray-based screen showed that both chains were required for binding by anti-7F CPS antibodies, confirming the importance of retaining these structures in new synthetic 7F glycoconjugate vaccines [87]. However, binding to CPS-specific antibodies does not necessarily indicate that the epitope elicits protective antibodies; rather, it must be empirically validated through vaccination. For example, depyruvated tetrasaccharide and trisaccharide derivatives of the *S. pneumoniae* serotype 4 CPS tetrasaccharide RU were found to elicit glycan-specific IgG antibodies, but notably these did not bind full-length CPS [30].

Oligosaccharide synthesis and minimal epitope discovery techniques have been applied to other bacterial pathogens including *N. meningitidis* serotype W135 [88] and *C. difficile* [89]. The RU of *C. difficile* PS-I is a glucose- and rhamnose-containing pentasaccharide. A screen of RU-derived oligosaccharides with patient sera demonstrated binding toward rhamnose-(1→3)-glucose, presumably in its capacity as the minimal antigen, and immunization with $CRM_{197}$ conjugated to this disaccharide elicited IgG antibodies that recognized PS-I. Attachment of five disaccharides to a synthetic scaffold induced a glycan-specific IgG response comparable to the disaccharide glycoconjugate [82].

Minimal epitope identification is greatly aided by detailed structural analysis and modeling of glycan-antibody binding. Solved crystal structures have helped determine epitope binding of *V. cholerae* O1 [90] and *Shigella flexneri* serotype 2a [91]. Initial studies of GBS type III polysaccharide-specific antibodies suggested that a 4 RU polysaccharide fragment was necessary to form a helical structure thought to be essential for antibody binding [92–94]. However, more recent work suggests that other binding motifs are possible. A competitive binding assay of GBS III polysaccharide structures using both synthesized oligosaccharides and fragments isolated from full-length polysaccharide showed 2 RU to be sufficient for antibody binding. NMR analysis of this 2 RU oligosaccharide confirmed that a terminal sialic acid side chain residue had a significant role in antibody binding in concert with three other side chain and backbone residues. Moreover, X-ray crystallography of a polysaccharide-specific antibody fragment bound to the 2 RU oligosaccharide revealed a binding motif consisting of the side chain and backbone of one RU and one backbone sugar of the second RU. With validation from NMR and X-ray crystallography, a novel minimal epitope has been proposed that is both simpler and smaller than previously proposed minimal epitopes [95]. Other studies of GBS III polysaccharide determined that a different epitope was recognized by monoclonal IgM [96]. It is possible, and perhaps likely, that multiple antibody-binding epitopes exist for a single glycan. Identifying and synthesizing epitope(s) that contribute most to immunological protection and eliminate less important sites may allow for biasing of the immune response by rational glycan design in new glycoconjugate vaccines.

## 10 Lipid-Linked Carbohydrate Vaccines

The human immune system regularly encounters carbohydrates displayed on the bacterial cell surface. Although the exact topography encountered by APCs can vary greatly between species, it typically contains a high density of glycan-containing structures that are attached to the membrane via an embedded lipid anchor (e.g., LPS in gram-negative bacteria). Accordingly, some lipid-linked glycans are able to activate the immune system through recognition by CD1, a family of proteins on APCs that are homologous to MHC class I (Fig. 2). CD1 possesses a hydrophobic antigen-binding pocket that can hold related lipid structures and an opening that allows for glycan head groups of varying size. Uptake and processing of the antigen is analogous to that for proteins being loaded into MHC. However, canonical T helper cells do not recognize CD1. Instead, CD1-bound glycolipids are recognized by invariant T cell receptor chains present on other T cell populations, including natural killer T cells (NKTs) and γδ T cells. Upon recognition of a glycolipid epitope, these cells release cytokines that can enhance conventional T cell help to B cells as well as activate other components of the immune system [97].

Recently, two research groups independently identified a mechanism of direct B cell help that involves a subset of invariant NKTs (iNKTs) and CD1d, a CD1 protein that binds to analogues of the marine sponge-derived glycolipid α-galactosylceramide (αGalCer). These studies demonstrated that the immune synapse between iNKTs and B cells induces a strong primary IgG response along with some affinity maturation. However, few memory B cells and no long-lived plasma cells were produced, two cell populations critical for vaccine efficacy [98, 99]. Nonetheless, researchers have exploited this unique mechanism of B cell help to enhance immune responses against carbohydrate antigens without the presence of a protein carrier. For example, glycoconjugates with *S. pneumoniae* serotype 4 CPS attached to αGalCer as well as liposomes containing *S. pneumonia* serotype 14 CPS RU along with an αGalCer analogue both elicited carbohydrate-specific TD IgGs [100, 101]. Interestingly, both formulations also elicited a long-lasting memory response, in contrast to earlier studies on B cell help from iNKTs, and challenge after immunization with the *S. pneumonia* serotype 4 conjugate resulted in significantly greater protection than unconjugated CPS in mice. By incorporating many of the design parameters established for protein glycoconjugates including polysaccharide length and linker chemistry to enhance immune responses further, future αGalCer-based carbohydrate vaccines may provide protection equal to or in some cases greater than their protein conjugated counterparts.

Several pathogen-derived glycolipids have also been shown to bind CD1 molecules, including lipoarabinomannan and diacylsulfoglycolipid in mycobacteria and α-glucosyldiacylglycerol in *S. pneumonia* [102]. However, although many of these glycolipids have been shown to enhance immune responses through cytokine release, the elicitation of direct CD1-dependent B cell help similar to analogues of αGalCer has not been firmly established. In addition, other common bacterial lipids such as lipid A, a vital component of LOS and LPS in gram-negative bacteria, are not believed to bind CD1. Indeed, conjugation of the TI cancer antigen GM2 to monophosphoryl lipid A, a detoxified lipid A molecule commonly used as an adjuvant, only induced elevated levels of TI antibodies [103, 104].

Interestingly, administration of whole-bacteria vaccines can elicit strong TD responses against LPS in at least some species, including *F. tularensis* [105]. Likewise, immunization of mice with pathogen-derived outer membrane vesicles (OMVs), which are 20–250 nm proteoliposomes derived from the bacterial periplasm and outer membrane, resulted in TD O-PS-specific responses [106]. Activation of O-PS-specific B cells in these vaccines is presumably provided by other associated lipids or proteins on the membrane, a phenomenon recently exploited in laboratory engineered glycosylated OMV vaccines (glycOMVs) [107–109]. GlycOMVs are produced by genetic engineering of tractable host organisms including non-pathogenic strains of *Escherichia coli* to display heterologous O-PS antigens on their exteriors. These cell surface-expressed O-PS molecules become constituents of released OMVs that are constitutively shed from the outer membrane of the host bacteria. In one notable example, *E. coli* K12-derived glycOMVs displaying structural mimics of O-PS from highly virulent *F. tularensis* Schu S4 provoked strong O-PS-specific IgG responses and conferred protection against any challenge with multiple strains of pathogenic *F. tularensis* [107]. Although much work remains to be done with respect to clinical translation of these vaccine candidates, glycOMVs are emerging as attractive alternatives to conventional glycoconjugates [110, 111].

## 11 Concluding Remarks

Utilization of carbohydrates as targets for immune recognition has led to the development of increasingly elegant and effective vaccines for protection against and reduction of bacterial disease. Protein-based glycoconjugate vaccines have played a prominent role in these efforts thanks to the strong TD response and modifiable platform they provide. The current need to optimize glycoconjugates for each glycan/pathogen target using general guidelines but few definitive design principles creates a number of challenges but also provides an opening for the development and application of new approaches and technologies. Further refinement of protein-based glycoconjugate vaccines that elicit precisely tailored anti-carbohydrate immune responses should be made possible by a deeper understanding of glycoimmunology mechanisms and more rapid production and characterization of vaccine candidates. At the same time, new rationally designed synthetic and lipid-based approaches potentially offer an exciting future of safer, cheaper, more diverse, and more effective carbohydrate-based bacterial vaccines.

## References


1. Walker CLF, Rudan I, Liu L, Nair H, Theodoratou E, Bhutta ZA, O'Brien KL, Campbell H, Black RE (2013) Global burden of childhood pneumonia and diarrhoea. Lancet 381 (9875):1405–1416
2. Robinson CL, Romero JR, Kempe A, Pellegrini C (2017) Advisory committee on immunization practices recommended immunization schedule for children and adolescents aged 18 years or younger — United States, 2017. MMWR Morb Mortal Wkly Rep 66(5):134–135
3. Thigpen MC, Whitney CG, Messonnier NE, Zell ER, Lynfield R, Hadler JL, Harrison LH, Farley MM, Reingold A, Bennett NM, Craig AS, Schaffner W, Thomas A, Lewis MM, Scallan E, Schuchat A (2011) Bacterial meningitis in the United States, 1998–2007. N Engl J Med 364(21):2016–2025
4. Alderwick LJ, Harrison J, Lloyd GS, Birch HL (2015) The mycobacterial cell wall--peptidoglycan and arabinogalactan. Cold Spring Harb Perspect Med 5(8):a021113
5. Fukuda T, Matsumura T, Ato M, Hamasaki M, Nishiuchi Y, Murakami Y, Maeda Y, Yoshimori T, Matsumoto S, Kobayashi K, Kinoshita T, Morita YS (2013) Critical roles for lipomannan and lipoarabinomannan in cell wall integrity of mycobacteria and pathogenesis of tuberculosis. mBio 4(1):e00472–e00412
6. Fulton KM, Smith JC, Twine SM (2016) Clinical applications of bacterial glycoproteins. Expert Rev Proteomics 13(4):345–353
7. Lindenthal C, Elsinghorst EA (1999) Identification of a glycoprotein produced by enterotoxigenic Escherichia coli. Infect Immun 67(8):4084–4091
8. Giersing BK, Dastgheyb SS, Modjarrad K, Moorthy V (2016) Status of vaccine research and development of vaccines for Staphylococcus aureus. Vaccine 34(26):2962–2966
9. Kobayashi M, Vekemans J, Baker CJ, Ratner AJ, Le Doare K, Schrag SJ (2016) Group B Streptococcus vaccine development: present status and future considerations, with emphasis on perspectives for low and middle income countries. F1000Res 5:2355
10. Mani S, Wierzba T, Walker RI (2016) Status of vaccine research and development for Shigella. Vaccine 34(26):2887–2894
11. Sunagar R, Kumar S, Franz BJ, Gosselin EJ (2016) Tularemia vaccine development: paralysis or progress? Vaccine (Auckl) 6:9–23
12. Heidelberger M, Avery OT (1924) The soluble specific substance of pneumococcus: second paper. J Exp Med 40(3):301–317
13. Tillett WS, Francis T (1929) Cutaneous reactions to the polysaccharides and proteins of pneumococcus in lobar pneumonia. J Exp Med 50(5):687–701
14. MacLeod CM, Hodges RG, Heidelberger M, Bernhard WG (1945) Prevention of pneumococcal pneumonia by immunization with specific capsular polysaccharides. J Exp Med 82 (6):445–465
15. Smit P, Oberholzer D, Hayden-Smith S, Koornhof HJ, Hilleman MR (1977) Protective efficacy of pneumococcal polysaccharide vaccines. JAMA 238(24):2613–2616
16. Shapiro ED, Berg AT, Austrian R, Schroeder D, Parcells V, Margolis A, Adair RK, Clemens JD (1991) The protective efficacy of polyvalent pneumococcal polysaccharide vaccine. N Engl J Med 325(21):1453–1460
17. Daniels CC, Rogers PD, Shelton CM (2016) A review of pneumococcal vaccines: current polysaccharide vaccine recommendations and future protein antigens. J Pediatr Pharmacol Ther 21(1):27–35
18. Laferriere C (2011) The immunogenicity of pneumococcal polysaccharides in infants and children: a meta-regression. Vaccine 29(40):6838–6847
19. Weidenmaier C, McLoughlin RM, Lee JC (2010) The zwitterionic cell wall teichoic acid of Staphylococcus aureus provokes skin abscesses in mice by a novel CD4+ T-cell-dependent mechanism. PLoS One 5(10):e13227
20. Johnson JL, Jones MB, Cobb BA (2015) Polysaccharide A from the capsule of Bacteroides fragilis induces clonal CD4+ T cell expansion. J Biol Chem 290(8):5007–5014

21. Kalka-Moll WM, Tzianabos AO, Bryant PW, Niemeyer M, Ploegh HL, Kasper DL (2002) Zwitterionic polysaccharides stimulate T cells by MHC class II-dependent interactions. J Immunol 169(11):6149–6153
22. De Silva RA, Wang Q, Chidley T, Appulage DK, Andreana PR (2009) Immunological response from an entirely carbohydrate antigen: design of synthetic vaccines based on Tn-PS A1 conjugates. J Am Chem Soc 131(28):9622–9623
23. Giannini G, Rappuoli R, Ratti G (1984) The amino-acid sequence of two non-toxic mutants of diphtheria toxin: CRM45 and CRM197. Nucleic Acids Res 12(10):4063–4069
24. Avci FY, Li X, Tsuji M, Kasper DL (2011) A mechanism for glycoconjugate vaccine activation of the adaptive immune system and its implications for vaccine design. Nat Med 17(12):1602–1609
25. Doores KJ, Fulton Z, Hong V, Patel MK, Scanlan CN, Wormald MR, Finn MG, Burton DR, Wilson IA, Davis BG (2010) A nonself sugar mimic of the HIV glycan shield shows enhanced antigenicity. Proc Natl Acad Sci U S A 107(40):17107–17112
26. Yang F, Zheng XJ, Huo CX, Wang Y, Zhang Y, Ye XS (2011) Enhancement of the immunogenicity of synthetic carbohydrate vaccines by chemical modifications of STn antigen. ACS Chem Biol 6(3):252–259
27. Zheng XJ, Yang F, Zheng M, Huo CX, Zhang Y, Ye XS (2015) Improvement of the immune efficacy of carbohydrate vaccines by chemical modification on the GM3 antigen. Org Biomol Chem 13(22):6399–6406
28. Adamo R, Romano MR, Berti F, Leuzzi R, Tontini M, Danieli E, Cappelletti E, Cakici OS, Swennen E, Pinto V, Brogioni B, Proietti D, Galeotti CL, Lay L, Monteiro MA, Scarselli M, Costantino P (2012) Phosphorylation of the synthetic hexasaccharide repeating unit is essential for the induction of antibodies to Clostridium difficile PSII cell wall polysaccharide. ACS Chem Biol 7(8):1420–1428
29. Danieli E, Lay L, Proietti D, Berti F, Costantino P, Adamo R (2011) First synthesis of C. difficile PS-II cell wall polysaccharide repeating unit. Org Lett 13(3):378–381
30. Geissner A, Pereira CL, Leddermann M, Anish C, Seeberger PH (2016) Deciphering antigenic determinants of Streptococcus pneumoniae serotype 4 capsular polysaccharide using synthetic oligosaccharides. ACS Chem Biol 11(2):335–344
31. Berry DS, Lynn F, Lee CH, Frasch CE, Bash MC (2002) Effect of O acetylation of Neisseria meningitidis serogroup A capsular polysaccharide on development of functional immune responses. Infect Immun 70(7):3707–3713
32. Rajam G, Carlone GM, Romero-Steiner S (2007) Functional antibodies to the O-acetylated pneumococcal serotype 15B capsular polysaccharide have low cross-reactivities with serotype 15C. Clin Vaccine Immunol 14(9):1223–1227
33. Jennings HJ, Roy R, Gamian A (1986) Induction of meningococcal group B polysaccharide-specific IgG antibodies in mice by using an N-propionylated B polysaccharide-tetanus toxoid conjugate vaccine. J Immunol 137(5):1708–1713
34. Jennings HJ, Gamian A, Ashton FE (1987) N-propionylated group B meningococcal polysaccharide mimics a unique epitope on group B Neisseria meningitidis. J Exp Med 165 (4):1207–1211
35. Jennings HJ, Gamian A, Michon F, Ashton FE (1989) Unique intermolecular bactericidal epitope involving the homosialopolysaccharide capsule on the cell surface of group B Neisseria meningitidis and Escherichia coli K1. J Immunol 142(10):3585–3591
36. Granoff DM, Bartoloni A, Ricci S, Gallo E, Rosa D, Ravenscroft N, Guarnieri V, Seid RC, Shan A, Usinger WR, Tan S, McHugh YE, Moe GR (1998) Bactericidal monoclonal antibodies that define unique meningococcal B polysaccharide epitopes that do not cross-react with human polysialic acid. J Immunol 160(10):5028–5036
37. Moe GR, Dave A, Granoff DM (2005) Epitopes recognized by a nonautoreactive murine anti-N-propionyl meningococcal group B polysaccharide monoclonal antibody. Infect Immun 73 (4):2123–2128

38. Szewczyk B, Taylor A (1980) Immunochemical properties of Vi antigen from Salmonella typhi Ty2: presence of two antigenic determinants. Infect Immun 29(2):539–544
39. Szu SC, Bystricky S, Hinojosa-Ahumada M, Egan W, Robbins JB (1994) Synthesis and some immunologic properties of an O-acetyl pectin [poly(1-->4)-alpha-D-GalpA]-protein conjugate as a vaccine for typhoid fever. Infect Immun 62(12):5545–5549
40. Abdelhameed AS, Adams GG, Morris GA, Almutairi FM, Duvivier P, Conrath K, Harding SE (2016) A glycoconjugate of Haemophilus influenzae Type b capsular polysaccharide with tetanus toxoid protein: hydrodynamic properties mainly influenced by the carbohydrate. Sci Rep 6:22208
41. Abdelhameed AS, Morris GA, Almutairi F, Adams GG, Duvivier P, Conrath K, Harding SE (2016) Solution conformation and flexibility of capsular polysaccharides from Neisseria meningitidis and glycoconjugates with the tetanus toxoid protein. Sci Rep 6:35588
42. Harding SE, Abdelhameed AS, Gillis RB, Morris GA, Adams GG (2015) Characterization of capsular polysaccharides and their glycoconjugates by hydrodynamic methods. Methods Mol Biol 1331:211–227
43. Brady AM, Calix JJ, Yu J, Geno KA, Cutter GR, Nahm MH (2014) Low invasiveness of pneumococcal serotype 11A is linked to ficolin-2 recognition of O-acetylated capsule epitopes and lectin complement pathway activation. J Infect Dis 210(7):1155–1165
44. Seppala I, Makela O (1989) Antigenicity of dextran-protein conjugates in mice. Effect of molecular weight of the carbohydrate and comparison of two modes of coupling. J Immunol 143(4):1259–1264
45. Rondini S, Micoli F, Lanzilao L, Gavini M, Alfini R, Brandt C, Clare S, Mastroeni P, Saul A, MacLennan CA (2015) Design of glycoconjugate vaccines against invasive African Salmonella enterica serovar Typhimurium. Infect Immun 83(3):996–1007
46. Ftacek P, Nelson V, Szu SC (2013) Immunochemical characterization of synthetic hexa-, octa- and decasaccharide conjugate vaccines for Vibrio cholerae O:1 serotype Ogawa with emphasis on antigenic density and chain length. Glycoconj J 30(9):871–880
47. Daum RS, Hogerman D, Rennels MB, Bewley K, Malinoski F, Rothstein E, Reisinger K, Block S, Keyserling H, Steinhoff M (1997) Infant immunization with pneumococcal CRM197 vaccines: effect of saccharide size on immunogenicity and interactions with simultaneously administered vaccines. J Infect Dis 176(2):445–455
48. Laferriere CA, Sood RK, de Muys JM, Michon F, Jennings HJ (1998) Streptococcus pneumoniae type 14 polysaccharide-conjugate vaccines: length stabilization of opsonophagocytic conformational polysaccharide epitopes. Infect Immun 66(6):2441–2446
49. Rana R, Dalal J, Singh D, Kumar N, Hanif S, Joshi N, Chhikara MK (2015) Development and characterization of Haemophilus influenzae type b conjugate vaccine prepared using different polysaccharide chain lengths. Vaccine 33(23):2646–2654
50. Anderson PW, Pichichero ME, Stein EC, Porcelli S, Betts RF, Connuck DM, Korones D, Insel RA, Zahradnik JM, Eby R (1989) Effect of oligosaccharide chain length, exposed terminal group, and hapten loading on the antibody response of human adults and infants to vaccines consisting of Haemophilus influenzae type b capsular antigen unterminally coupled to the diphtheria protein CRM197. J Immunol 142(7):2464–2468
51. Paoletti LC, Kasper DL, Michon F, DiFabio J, Jennings HJ, Tosteson TD, Wessels MR (1992) Effects of chain length on the immunogenicity in rabbits of group B Streptococcus type III oligosaccharide-tetanus toxoid conjugates. J Clin Investig 89(1):203–209
52. Arcuri M, Di Benedetto R, Cunningham AF, Saul A, MacLennan CA, Micoli F (2017) The influence of conjugation variables on the design and immunogenicity of a glycoconjugate vaccine against Salmonella Typhi. PLoS One 12(12):e0189100
53. Gray GR (1974) The direct coupling of oligosaccharides to proteins and derivatized gels. Arch Biochem Biophys 163(1):426–428
54. Wessels MR, Paoletti LC, Kasper DL, DiFabio JL, Michon F, Holme K, Jennings HJ (1990) Immunogenicity in animals of a polysaccharide-protein conjugate vaccine against type III group B Streptococcus. J Clin Investig 86(5):1428–1433

55. Crotti S, Zhai H, Zhou J, Allan M, Proietti D, Pansegrau W, Hu QY, Berti F, Adamo R (2014) Defined conjugation of glycans to the lysines of CRM197 guided by their reactivity mapping. ChemBioChem 15(6):836–843
56. Moginger U, Resemann A, Martin CE, Parameswarappa S, Govindan S, Wamhoff EC, Broecker F, Suckau D, Pereira CL, Anish C, Seeberger PH, Kolarich D (2016) Cross Reactive Material 197 glycoconjugate vaccines contain privileged conjugation sites. Sci Rep 6:20488
57. Hu QY, Berti F, Adamo R (2016) Towards the next generation of biomedicines by site-selective conjugation. Chem Soc Rev 45(6):1691–1719
58. Nilo A, Morelli L, Passalacqua I, Brogioni B, Allan M, Carboni F, Pezzicoli A, Zerbini F, Maione D, Fabbrini M, Romano MR, Hu QY, Margarit I, Berti F, Adamo R (2015) Anti-group B Streptococcus glycan-conjugate vaccines using pilus protein GBS80 as carrier and antigen: comparing lysine and tyrosine-directed conjugation. ACS Chem Biol 10(7):1737–1746
59. van der Put RM, Kim TH, Guerreiro C, Thouron F, Hoogerhout P, Sansonetti PJ, Westdijk J, Stork M, Phalipon A, Mulard LA (2016) A synthetic carbohydrate conjugate vaccine candidate against shigellosis: improved bioconjugation and impact of alum on immunogenicity. Bioconjug Chem 27(4):883–892
60. Cuccui J, Wren B (2015) Hijacking bacterial glycosylation for the production of glycoconjugates, from vaccines to humanised glycoproteins. J Pharm Pharmacol 67 (3):338–350
61. Terra VS, Mills DC, Yates LE, Abouelhadid S, Cuccui J, Wren BW (2012) Recent developments in bacterial protein glycan coupling technology and glycoconjugate vaccine design. J Med Microbiol 61(7):919–926
62. Glover KJ, Weerapana E, Numao S, Imperiali B (2005) Chemoenzymatic synthesis of glycopeptides with PglB, a bacterial oligosaccharyl transferase from Campylobacter jejuni. Chem Biol 12(12):1311–1315
63. Feldman MF, Wacker M, Hernandez M, Hitchen PG, Marolda CL, Kowarik M, Morris HR, Dell A, Valvano MA, Aebi M (2005) Engineering N-linked protein glycosylation with diverse O antigen lipopolysaccharide structures in Escherichia coli. Proc Natl Acad Sci U S A 102 (8):3016–3021
64. Kampf MM, Braun M, Sirena D, Ihssen J, Thony-Meyer L, Ren Q (2015) In vivo production of a novel glycoconjugate vaccine against Shigella flexneri 2a in recombinant Escherichia coli: identification of stimulating factors for in vivo glycosylation. Microb Cell Factories 14:12
65. Pan C, Sun P, Liu B, Liang H, Peng Z, Dong Y, Wang D, Liu X, Wang B, Zeng M, Wu J, Zhu L, Wang H (2016) Biosynthesis of conjugate vaccines using an O-linked glycosylation system. mBio 7(2):e00443–e00416
66. Ihssen J, Haas J, Kowarik M, Wiesli L, Wacker M, Schwede T, Thony-Meyer L (2015) Increased efficiency of Campylobacter jejuni N-oligosaccharyltransferase PglB by structure-guided engineering. Open Biol 5(4):140227
67. Ollis AA, Chai Y, Natarajan A, Perregaux E, Jaroentomeechai T, Guarino C, Smith J, Zhang S, DeLisa MP (2015) Substitute sweeteners: diverse bacterial oligosaccharyltransferases with unique N-glycosylation site preferences. Sci Rep 5:15237
68. Nilo A, Passalacqua I, Fabbrini M, Allan M, Usera A, Carboni F, Brogioni B, Pezzicoli A, Cobb J, Romano MR, Margarit I, Hu QY, Berti F, Adamo R (2015) Exploring the effect of conjugation site and chemistry on the immunogenicity of an anti-group B Streptococcus glycoconjugate vaccine based on GBS67 pilus protein and type V polysaccharide. Bioconjug Chem 26(8):1839–1849
69. Zhang F, Lu YJ, Malley R (2013) Multiple antigen-presenting system (MAPS) to induce comprehensive B- and T-cell immunity. Proc Natl Acad Sci U S A 110(33):13564–13569
70. Thanawastien A, Cartee RT, Griffin TJ, Killeen KP, Mekalanos JJ (2015) Conjugate-like immunogens produced as protein capsular matrix vaccines. Proc Natl Acad Sci U S A 112(10): E1143–E1151
71. Broker M, Costantino P, DeTora L, McIntosh ED, Rappuoli R (2011) Biochemical and biological characteristics of cross-reacting material 197 CRM197, a non-toxic mutant of

diphtheria toxin: use as a conjugation protein in vaccines and other potential clinical applications. Biologicals 39(4):195–204
72. Pecetta S, Lo Surdo P, Tontini M, Proietti D, Zambonelli C, Bottomley MJ, Biagini M, Berti F, Costantino P, Romano MR (2015) Carrier priming with CRM 197 or diphtheria toxoid has a different impact on the immunogenicity of the respective glycoconjugates: biophysical and immunochemical interpretation. Vaccine 33(2):314–320
73. Pecetta S, Tontini M, Faenzi E, Cioncada R, Proietti D, Seubert A, Nuti S, Berti F, Romano MR (2016) Carrier priming effect of CRM197 is related to an enhanced B and T cell activation in meningococcal serogroup A conjugate vaccination. Immunological comparison between CRM197 and diphtheria toxoid. Vaccine 34(20):2334–2341
74. Dagan R, Poolman J, Siegrist CA (2010) Glycoconjugate vaccines and immune interference: a review. Vaccine 28(34):5513–5523
75. Findlow H, Borrow R (2016) Interactions of conjugate vaccines and co-administered vaccines. Hum Vaccin Immmunother 12(1):226–230
76. Ladhani SN, Andrews NJ, Waight P, Hallis B, Matheson M, England A, Findlow H, Bai X, Borrow R, Burbidge P, Pearce E, Goldblatt D, Miller E (2015) Interchangeability of meningococcal group C conjugate vaccines with different carrier proteins in the United Kingdom infant immunisation schedule. Vaccine 33(5):648–655
77. Dagan R, Poolman JT, Zepp F (2008) Combination vaccines containing DTPa-Hib: impact of IPV and coadministration of CRM197 conjugates. Expert Rev Vaccines 7(1):97–115
78. Broker M, Berti F, Schneider J, Vojtek I (2017) Polysaccharide conjugate vaccine protein carriers as a "neglected valency" - potential and limitations. Vaccine 35(25):3286–3294
79. Rappuoli R (2001) Reverse vaccinology, a genome-based approach to vaccine development. Vaccine 19(17–19):2688–2691
80. Brett PJ, Woods DE (1996) Structural and immunological characterization of Burkholderia pseudomallei O-polysaccharide-flagellin protein conjugates. Infect Immun 64(7):2824–2828
81. Muruato LA, Tapia D, Hatcher CL, Kalita M, Brett PJ, Gregory AE, Samuel JE, Titball RW, Torres AG (2017) The use of reverse vaccinology in the design and construction of nanoglycoconjugate vaccines against Burkholderia pseudomallei. Clin Vaccine Immunol 24(11): e00206–e00217
82. Broecker F, Hanske J, Martin CE, Baek JY, Wahlbrink A, Wojcik F, Hartmann L, Rademacher C, Anish C, Seeberger PH (2016) Multivalent display of minimal Clostridium difficile glycan epitopes mimics antigenic properties of larger glycans. Nat Commun 7:11224
83. Pozsgay V, Chu C, Pannell L, Wolfe J, Robbins JB, Schneerson R (1999) Protein conjugates of synthetic saccharides elicit higher levels of serum IgG lipopolysaccharide antibodies in mice than do those of the O-specific polysaccharide from Shigella dysenteriae type 1. Proc Natl Acad Sci U S A 96(9):5194–5197
84. Adamo R, Hu Q-Y, Torosantucci A, Crotti S, Brogioni G, Allan M, Chiani P, Bromuro C, Quinn D, Tontini M, Berti F (2014) Deciphering the structure-immunogenicity relationship of anti-Candida glycoconjugate vaccines. Chem Sci 5(11):4302–4311
85. Alonso de Velasco E, Verheul AF, van Steijn AM, Dekker HA, Feldman RG, Fernandez IM, Kamerling JP, Vliegenthart JF, Verhoef J, Snippe H (1994) Epitope specificity of rabbit immunoglobulin G (IgG) elicited by pneumococcal type 23F synthetic oligosaccharide- and native polysaccharide-protein conjugate vaccines: comparison with human anti-polysaccharide 23F IgG. Infect Immun 62(3):799–808
86. Reason DC, Zhou J (2004) Correlation of antigenic epitope and antibody gene usage in the human immune response to Streptococcus pneumoniae type 23F capsular polysaccharide. Clin Immunol 111(1):132–136
87. Menova P, Sella M, Sellrie K, Pereira C, Seeberger PH (2018) Identification of the minimal glycotope of Streptococcus pneumoniae 7F capsular polysaccharide using synthetic oligosaccharides. Chem Eur J 24(16):4181–4187
88. Wang CH, Li ST, Lin TL, Cheng YY, Sun TH, Wang JT, Cheng TJ, Mong KK, Wong CH, Wu CY (2013) Synthesis of Neisseria meningitidis serogroup W135 capsular oligosaccharides

for immunogenicity comparison and vaccine development. Angew Chem Int Ed Engl 52 (35):9157–9161
89. Martin CE, Broecker F, Oberli MA, Komor J, Mattner J, Anish C, Seeberger PH (2013) Immunological evaluation of a synthetic Clostridium difficile oligosaccharide conjugate vaccine candidate and identification of a minimal epitope. J Am Chem Soc 135 (26):9713–9722
90. Villeneuve S, Souchon H, Riottot MM, Mazie JC, Lei P, Glaudemans CP, Kovac P, Fournier JM, Alzari PM (2000) Crystal structure of an anti-carbohydrate antibody directed against Vibrio cholerae O1 in complex with antigen: molecular basis for serotype specificity. Proc Natl Acad Sci 97(15):8433–8438
91. Vulliez-Le Normand B, Saul FA, Phalipon A, Belot F, Guerreiro C, Mulard LA, Bentley GA (2008) Structures of synthetic O-antigen fragments from serotype 2a Shigella flexneri in complex with a protective monoclonal antibody. Proc Natl Acad Sci U S A 105 (29):9976–9981
92. Brisson JR, Uhrinova S, Woods RJ, van der Zwan M, Jarrell HC, Paoletti LC, Kasper DL, Jennings HJ (1997) NMR and molecular dynamics studies of the conformational epitope of the type III group B Streptococcus capsular polysaccharide and derivatives. Biochemistry 36 (11):3278–3292
93. Gonzalez-Outeirino J, Kadirvelraj R, Woods RJ (2005) Structural elucidation of type III group B Streptococcus capsular polysaccharide using molecular dynamics simulations: the role of sialic acid. Carbohydr Res 340(5):1007–1018
94. Zou W, Mackenzie R, Therien L, Hirama T, Yang Q, Gidney MA, Jennings HJ (1999) Conformational epitope of the type III group B Streptococcus capsular polysaccharide. J Immunol 163(2):820–825
95. Carboni F, Adamo R, Fabbrini M, De Ricco R, Cattaneo V, Brogioni B, Veggi D, Pinto V, Passalacqua I, Oldrini D, Rappuoli R, Malito E, Margarit IYR, Berti F (2017) Structure of a protective epitope of group B Streptococcus type III capsular polysaccharide. Proc Natl Acad Sci U S A 114(19):5017–5022
96. Johnson MA, Jaseja M, Zou W, Jennings HJ, Copie V, Pinto BM, Pincus SH (2003) NMR studies of carbohydrates and carbohydrate-mimetic peptides recognized by an anti-group B Streptococcus antibody. J Biol Chem 278(27):24740–24752
97. Brigl M, Brenner MB (2004) CD1: antigen presentation and T cell function. Annu Rev Immunol 22:817–890
98. Chang PP, Barral P, Fitch J, Pratama A, Ma CS, Kallies A, Hogan JJ, Cerundolo V, Tangye SG, Bittman R, Nutt SL, Brink R, Godfrey DI, Batista FD, Vinuesa CG (2011) Identification of Bcl-6-dependent follicular helper NKT cells that provide cognate help for B cell responses. Nat Immunol 13(1):35–43
99. King IL, Fortier A, Tighe M, Dibble J, Watts GF, Veerapen N, Haberman AM, Besra GS, Mohrs M, Brenner MB, Leadbetter EA (2011) Invariant natural killer T cells direct B cell responses to cognate lipid antigen in an IL-21-dependent manner. Nat Immunol 13(1):44–50
100. Bai L, Deng S, Reboulet R, Mathew R, Teyton L, Savage PB, Bendelac A (2013) Natural killer T (NKT)-B-cell interactions promote prolonged antibody responses and long-term memory to pneumococcal capsular polysaccharides. Proc Natl Acad Sci U S A 110 (40):16097–16102
101. Cavallari M, Stallforth P, Kalinichenko A, Rathwell DC, Gronewold TM, Adibekian A, Mori L, Landmann R, Seeberger PH, De Libero G (2014) A semisynthetic carbohydrate-lipid vaccine that protects against S. pneumoniae in mice. Nat Chem Biol 10(11):950–956
102. Mori L, Lepore M, De Libero G (2016) The immunology of CD1- and MR1-restricted T cells. Annu Rev Immunol 34:479–510
103. Collins AM (2016) IgG subclass co-expression brings harmony to the quartet model of murine IgG function. Immunol Cell Biol 94(10):949–954

104. Zhou Z, Mandal SS, Liao G, Guo J, Guo Z (2017) Synthesis and evaluation of GM2-monophosphoryl lipid A conjugate as a fully synthetic self-adjuvant cancer vaccine. Sci Rep 7(1):11403
105. Dreisbach VC, Cowley S, Elkins KL (2000) Purified lipopolysaccharide from Francisella tularensis live vaccine strain (LVS) induces protective immunity against LVS infection that requires B cells and gamma interferon. Infect Immun 68(4):1988–1996
106. Wang Z, Lazinski DW, Camilli A (2017) Immunity provided by an outer membrane vesicle cholera vaccine is due to O-antigen-specific antibodies inhibiting bacterial motility. Infect Immun 85(1):e00626–e00616
107. Chen L, Valentine JL, Huang CJ, Endicott CE, Moeller TD, Rasmussen JA, Fletcher JR, Boll JM, Rosenthal JA, Dobruchowska J, Wang Z, Heiss C, Azadi P, Putnam D, Trent MS, Jones BD, DeLisa MP (2016) Outer membrane vesicles displaying engineered glycotopes elicit protective antibodies. Proc Natl Acad Sci U S A 113(26):E3609–E3618
108. Price NL, Goyette-Desjardins G, Nothaft H, Valguarnera E, Szymanski CM, Segura M, Feldman MF (2016) Glycoengineered outer membrane vesicles: a novel platform for bacterial vaccines. Sci Rep 6:24931
109. Valentine JL, Chen L, Perregaux EC, Weyant KB, Rosenthal JA, Heiss C, Azadi P, Fisher AC, Putnam D, Moe GR, Merritt JH, DeLisa MP (2016) Immunization with outer membrane vesicles displaying designer glycotopes yields class-switched, glycan-specific antibodies. Cell Chem Biol 23(6):655–665
110. Valguarnera E, Feldman MF (2017) Glycoengineered outer membrane vesicles as a platform for vaccine development. Methods Enzymol 597:285–310
111. Weyant KB, Mills DC, DeLisa MP (2018) Engineering a new generation of carbohydrate-based vaccines. Curr Opin Chem Eng 19:77–85

# State-of-the-Art Glycomics Technologies in Glycobiotechnology

**Alexander Pralow, Samanta Cajic, Kathirvel Alagesan, Daniel Kolarich, and Erdmann Rapp**

## Contents



A. Pralow and S. Cajic
Max Planck Institute for Dynamics of Complex Technical Systems, Magdeburg, Germany

K. Alagesan
Institute for Glycomics, Griffith University, Gold Coast, QLD, Australia

D. Kolarich
Institute for Glycomics, Griffith University, Gold Coast, QLD, Australia

ARC Centre of Excellence in Nanoscale Biophotonics, Griffith University, Gold Coast, QLD, Australia

E. Rapp (✉)
Max Planck Institute for Dynamics of Complex Technical Systems, Magdeburg, Germany

glyXera GmbH, Magdeburg, Germany
e-mail: rapp@mpi-magdeburg.mpg.de; e.rapp@glyxera.com

**Abstract** Glycosylation affects the properties of biologics; thus regulatory bodies classified it as critical quality attribute and force biopharma industry to capture and control it throughout all phases, from R&D till end of product lifetime. The shift from originators to biosimilars further increases importance and extent of glycoanalysis, which thus increases the need for technology platforms enabling reliable high-throughput and in-depth glycan analysis. In this chapter, we will first summarize on established glycoanalytical methods based on liquid chromatography focusing on hydrophilic interaction chromatography, capillary electrophoresis focusing on multiplexed capillary gel electrophoresis, and mass spectrometry focusing on matrix-assisted laser desorption; we will then highlight two emerging technologies based on porous graphitized carbon liquid chromatography and on ion-mobility mass spectrometry as both are highly promising tools to deliver an additional level of information for in-depth glycan analysis; additionally we elaborate on the advantages and challenges of different glycoanalytical technologies and their complementarity; finally, we briefly review applications thereof to biopharmaceutical products. This chapter provides an overview of current state-of-the-art analytical approaches for glycan characterization of biopharmaceuticals that can be employed to capture glycoprotein heterogeneity in a biopharmaceutical context.

## Graphical Abstract

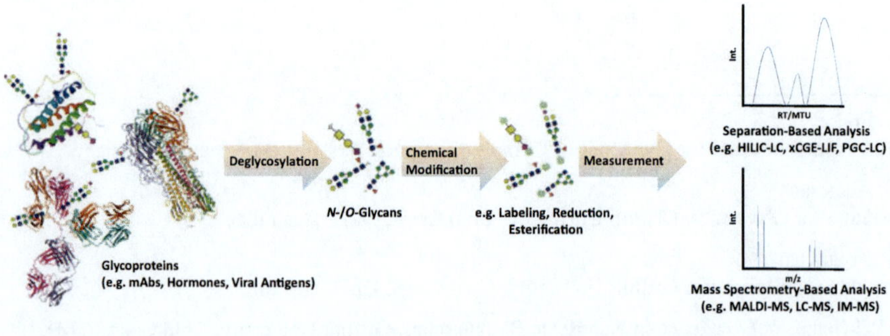


**Keywords** Biopharmaceuticals, Biosimilars, Capillary electrophoresis, Glycomics, Glycoprofile, Glycosylation, Liquid Chromatography, Mass spectrometry, Vaccines

## Abbreviations

| | |
|---|---|
| 2-AA | 2-anthranilic acid |
| 2-AB | 2-aminobenzamide |
| APTS | 8-aminopyrene-1,3,6-trisulfonic acid |
| AQC | 6-aminoquinoyl-*N*-hydroxysuccinimidyl carbamate |
| BPC | Base peak chromatogram |
| CCS | Collisional cross sections |
| CDG | Congenital disorders of glycosylation |
| CE | Capillary electrophoresis |
| CGE | Capillary gel electrophoresis |
| CQA | Critical quality attribute |
| CZE | Capillary zone electrophoresis |
| DHB | 2,5-dihydroxybenzoic acid |
| EIC | Extracted ion chromatogram |
| EOF | Electroosmotic flow |
| EPO | Erythropoietin |
| ESI | Electrospray ionization |
| ETD | Electron-transfer dissociation |
| FDA | Food and Drug Administration |
| FLR | Fluorescence detection |
| Fuc | Fucose |
| Gal | Galactose |
| GalNAc | *N*-acetylgalactosamine |
| GlcNAc | *N*-acetylglucosamine |
| HILIC | Hydrophilic interaction chromatography |
| HPAEC | High-performance anion-exchange chromatography |
| HT | High throughput |
| IgG | Immunoglobulin |
| IM | Ion mobility |
| LC | Liquid chromatography |
| mAbs | Monoclonal antibodies |
| MALDI | Matrix-assisted laser desorption/ionization mass spectrometry |
| Man | Mannose |
| MS | Mass spectrometry |
| MS/MS | Tandem mass spectrometry |
| N | Asparagine |
| NeuAc | *N*-acetylneuraminic acid |
| NIST | National Institute of Standards and Technology |
| NMR | Nuclear magnetic resonance |
| NP | Normal phase |
| P | Proline |
| PA | 2-amino-pyridine |
| PAD | Pulsed amperometric detection |

| | |
|---|---|
| PCMF | Post-column make-up flow |
| PGC | Porous graphitized carbon |
| PNGase F | Peptide-*N*-glycosidase F |
| Procainamide | 4-amino-*N*-(2-diethylaminoethyl) benzamide |
| PTM | Post-translational modification |
| QC | Quality control |
| RP | Reversed phase |
| S | Serine |
| Sia | Sialic acid |
| sIgA | Secretory immunoglobulin A |
| SNFG | Symbol nomenclature for glycans |
| T | Threonine |
| UHPLC | Ultrahigh-performance liquid chromatography |
| WAX | Weak anion-exchange |
| xCGE-LIF | Multiplexed capillary-gel electrophoresis with laser-induced fluorescence detection |

## 1 Introduction

Glycosylation is one of the most important critical quality attributes (CQAs) for optimal efficacy and safety of a biopharmaceutical [1]. Regulatory bodies such as the Food and Drug Administration (FDA) explicitly require that the glycoprofile of a biotherapeutic remains stable and consistent from the trial phase until the final product and beyond, and deviations of the approved glycoprofile can result in revoking the license [2–4]. In consequence, the ability to determine, monitor, and control the glycosylation of biotherapeutic products such as monoclonal antibodies (mAbs) or other hormones, fusion proteins, growth factors, cytokines, and therapeutic enzymes is critical for product safety and proper function [2–4]. Interestingly, glycosylation is currently not classified as a CQA in the production of vaccines such as the influenza vaccine [5], although both major antigens, the hemagglutinin (HA) as well as the neuraminidase (NA), are both well known to be heavily glycosylated [6]. For influenza vaccine production based on recombinant major antigens (e.g., Flublok, recombinant hemagglutinin), however, glycosylation needs to be monitored as it is the case for any other recombinant glycoprotein product. The main focus of this book is on glycoproteins in the biopharmaceutical industry; nevertheless it is important to note that glycoanalyses are also highly relevant in other biotechnology fields such as functional foods (e.g., milk oligosaccharides) or pharmaceutical polysaccharides (e.g., glycosaminoglycans like hyaluron and heparin) [7–10].

The shift of approved products from originators to biosimilars also pushes the analytical monitoring of a product to be the most cost-intensive factor (Fig. 1) [11, 12]. This is highly relevant in the context of glycosylation, as a similar glycoprofile needs to be demonstrated for a product to be classified as a biosimilar,

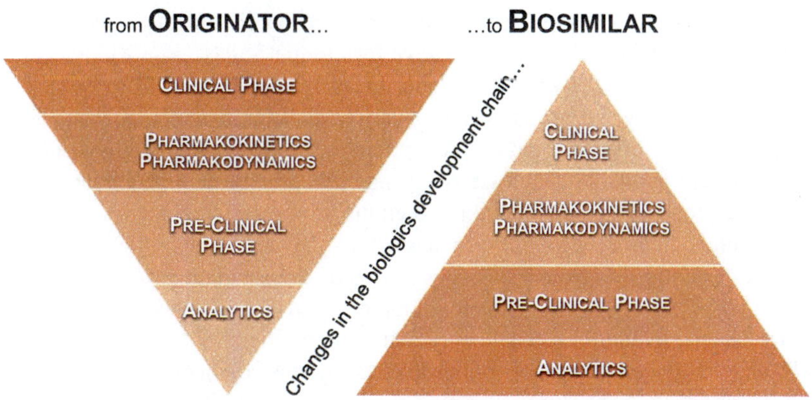


**Fig. 1** Originator vs. biosimilar. The development of a biosimilar reverses the common distribution of the necessary entities to bring a biopharmaceutical product through the regulatory agencies and on the market

and deviations thereof can jeopardize the biosimilar status. Therefore, reliable and suitable analytical workflows to determine and monitor glycosylation have been gaining tremendous interest.

Protein glycosylation is a co-/post-translational modification (PTM) involved in several biological key functions [13]. The "right" type and form of glycosylation is crucial for the functionality of proteins and cells, and in consequence changes have frequently been associated with major diseases such as cancer [14], inflammatory as well as infectious diseases [15]. Next to these "dynamic" changes, a large number of rare genetic and metabolic disorders, so-called congenital disorders of glycosylation (CDG), are disrupting proper glycosylation pathways due to mutation defects in one or more genes responsible to translate crucial enzymes of the glycosylation pathways [16].

Two main forms of glycosylation are frequently observed in therapeutic glycoprotein products: *N*- and *O*-type glycosylation, indicating the amino acid side chain atom; these glycans are attached to their proteins. *N*-glycans are linked to the side chain amino group of asparagine (N) if occurring within a specific consensus sequence N-X-S/T/C; X $\neq$ P [serine (S), threonine (T), proline (P)]. *N*-glycans are also characterized by a common core structure $GlcNAc_2Man_3$ [*N*-acetylglucosamine (GlcNAc), mannose (Man)], which builds the basis for four major types of *N*-glycan classes: oligomannose, paucimannose, hybrid-type, and complex-type *N*-glycans [17]. Next to these *N*-glycans, *O*-glycans are forming the second, major type of glycan PTM found on therapeutic glycoproteins. There, an *N*-acetylgalactosamine (GalNAc) is attached to the hydroxyl group of S or T residues before that is being extended with the monosaccharides galactose (Gal), GlcNAc, GalNAc, fucose (Fuc), and sialic acid (Sia) into more or less complex variations of mucin-type *O*-glycans. Generally, no single consensus sequence is known for the attachment of *O*-glycans compared to *N*-glycans, but site-specific *O*-glycosylation is highly cell-type and protein-specific. These factors impede in silico prediction of *O*-glycosylation sites and make detailed analyses imperative for product safety [17, 18].

Traditional analytical methods include releasing the glycans from the protein backbone before they can be analyzed by a variety of different methods. While for *N*-glycans mild and effective enzymatic options such as Peptide-*N*-Glycosidase F (PNGase F) are available to release *N*-glycans from the protein backbone, to date chemical approaches such as reductive β-elimination or hydrazinolysis remain the only reliable quantitative options to release *O*-glycans from proteins [19]. The market is currently dominated by *N*-glycosylated, recombinant glycoproteins such as mAbs [12]. Therefore, this chapter will mainly focus on the relevant aspects of *N*-glycan analysis, including its challenges and opportunities.

*N*- and *O*-glycans are not only defined by the composition of different monosaccharides. Each glycosidic bond can occur in different forms regarding linkage-type, branching, and confirmation [18]. Realizing this enormous diversity of different glycan structures states the question for suitable instrumentational techniques and bioanalytical methods to gain solid structural, qualitative, as well as quantitative analytical data. This chapter wants to break down the current state-of-the-art analytical methods and technologies in use for the glycan characterization of biopharmaceutical products and extend this to highlight the most recent developments of novel approaches [20–24].

## 2 Established Technologies for the Analysis of Glycans in the Biopharmaceutical Industry

The following section describes established technologies for the analysis of glycans, like liquid chromatography (LC), matrix-assisted laser desorption/ionization mass spectrometry (MALDI-MS), and capillary electrophoresis (CE) in the biopharmaceutical industry, providing general as well as advanced method information together with commercialized glycan analysis kits.

### *2.1 Liquid Chromatography-Based Glycan Analysis*

Different LC-based separation modes have been employed over the past decades to capture and analyze glycans: reversed phase (RP) [25], normal phase (NP) [26], and high-performance anion-exchange chromatography (HPAEC) [27, 28] have become standard approaches to separate and analyze released glycans [29]. Hydrophilic interaction chromatography (HILIC), in particular with fluorescence detection (HILIC-FLR), however, is possibly the most widespread used "gold standard" method within the biopharmaceutical industry to monitor and document glycoprofiles of recombinant glycoproteins [21].

Since the pioneering contribution of Dwek, Rudd, Hase, and others [30–33], HILIC-based LC systems have been most widely used in pharmaceutical glycoconjugate analysis [21, 34–37]. HILIC refers to the basic separation principle, where compounds are largely separated based on the strength of their hydrophilic interactions with a stationary phase within a hydrophobic environment (mobile phase) [37, 38]. Separation is usually achieved by developing a gradient that increases hydrophilic conditions in the mobile phase to disrupt the hydrophilic interactions between the molecules and the stationary phase [29]. Depending on the stationary phase and the individual gradient, certain isobaric glycans (= same mass, but different structure) can be separated using HILIC. HILIC-based LC systems are in principle also suitable for coupling with MS for compound detection [21, 36, 39, 40]. A wide variety of different stationary phase chemistries are available for HILIC separation and can contain one or more amine, amide, diol, or sulfobetaine zwitterionic functionalities [38, 41], offering a wide range of opportunities that can be tailored for different applications. In consequence, most HILIC-based separation strategies reflect a combination of electrostatic interaction, adsorption, and liquid portioning effects that are considered the predominant factors influencing HILIC separation [42–44]. With the GlycanPac AXH-1, Thermo Fisher recently released specific mixed-mode columns for glycan analyses that combine weak anion-exchange (WAX) and HILIC retention mechanisms within a single stationary phase [45]. Another type of mixed-mode column is the GlycanPac AXR-1, which combines RP with WAX properties in a single stationary phase [46]. Both columns are available with a particle size of 1.9 μm for UHPLC and 3 μm for HPLC applications and have shown great promise to increase the analytical depth of LC-based glycan analyses. Readers interested in more details on the molecular basis of HILIC separation are referred to some excellent reviews on that topic [38, 43, 44].

HILIC generally provides highly reproducible data with respect to peak width, symmetry, resolution, and retention time stability [21, 47]. These factors were welcomed by the biopharmaceutical industry and clearly contributed to the current widespread distribution of this technique. The example of a typical HILIC-UHPLC-FLR [ultrahigh-performance liquid chromatography (UHPLC)] analysis of 2-AB-labeled *N*-glycans derived from human serum exemplifies the separation capacity of this approach (Fig. 2). The overall time consumption regarding sample preparation, measurement time, and data interpretation makes such HILIC-LC-based approaches also feasible for high-throughput (HT) applications [21]. With the exception of HPAEC in combination with pulsed amperometric detection (PAD) or if detection is achieved by mass spectrometry (MS) [29], all LC-based approaches share the requirement that glycans need to be chemically labeled with a fluorescent dye for sensitive detection due to the lack of reasonable chromogenic molecules in native *N*- or *O*-glycans [55]. Hence, efficient glycan labeling with chromogenic dyes forms an essential part of most LC-based glycan analysis protocols, and thus the most important aspects of glycan labeling, their advantages, drawbacks, and challenges are discussed in more detail.

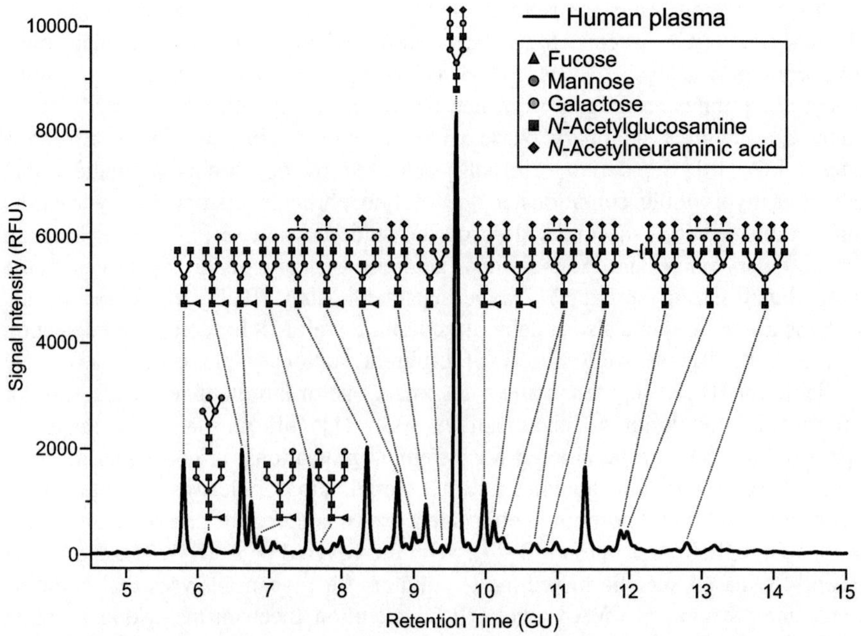


**Fig. 2** HILIC-UHPLC-FLR chromatogram of 2-AB-labeled *N*-glycans released from human plasma glycoproteins. Larger glycans with charged residues such as Sia usually show stronger retention compared to smaller, neutral ones. Peak quantitation is achieved as in any other HPLC-based analysis using the area under the curve. De novo structure determination usually requires additional experiments that include digestion with specific exoglycosidases [48–52]. Symbolic representation of *N*-glycans was drawn with GlycoWorkbench Version 1.1, following the symbol nomenclature for glycans (SNFG) [53, 54]

### 2.1.1 Glycan Labeling

Since the beginning of LC-based glycan analysis, extensive research has been performed how to modify glycans for more sensitive and selective detection. This has delivered a variety of different labels that are now available to suite the downstream separation and detection technology of choice. Over the years, 2-aminobenzamide (2-AB) [56, 57], 2-anthranilic acid (2-AA) [58, 59], 2-amino-pyridine (PA) [60], and procainamide (4-amino-*N*-(2-diethylaminoethyl) benzamide) [61, 62] have evolved to be the most widely used labels, but new ones are still being introduced, in particular with respect to dual detection by FLR and MS [63], as most labels are not well suited for highly sensitive detection by MS due to weak ionization efficiency.

While a plethora of different labeling protocols are found in research labs, most are unsuitable for use in an industrial environment with strict quality control (QC) compliance requirements. A number of different companies have addressed that challenge and have developed commercially available, validated glycan labeling

**Table 1** Commercialized glycan labeling kits for LC-FLR/(MS) analysis. AQC – 6-aminoquinoyl-*N*-hydroxysuccinimidyl carbamate

| Company | Trade name | Fluorophore | MS compatibility | Reference |
|---|---|---|---|---|
| Ludger Ltd | LudgerTag | 2-AB | No | [64] |
| | | 2-AA | Yes, negative mode | |
| | | Procainamide | Yes | [64, 65] |
| Sigma Aldrich | GlycoProfile | 2-AB | No | [66] |
| | | 2-AA | Yes, negative mode | [67] |
| Waters | GlycoWorks | 2-AB | No | [68] |
| | | RapiFluor-MS | Yes | [69] |
| ProZyme | GlycoPrep | 2-AB | No | [70] |
| | | InstantAB | No | [71] |
| | | InstantPC | Yes | [72] |
| Synchem | – | AQC | Yes | [22, 57, 73] |

kits, which clearly facilitated introduction of these into the biopharmaceutical industry (Table 1).

The labels listed in Table 1 are probably the most widespread used labels in the biopharmaceutical industry, but by far not the only ones available. For a more comprehensive list of possible fluorophores for glycan labeling, the inclined reader is referred to the following references [55, 74, 75].

### 2.1.2 Automation of Sample Preparation and Data Analysis

A number of systems are on the market that enable automatization of sample preparation and, partially, also analysis. Agilent is offering an automated liquid handling system called AssayMAP Bravo [76] that can be used for automated sample preparation including enzymatic glycan release, derivatization with 2-AB, and purification. Purified samples are subsequently directed to a HILIC system for analysis. Ludger developed another automated analytical workflow using the Hamilton STARlet liquid handling robot. Their automated workflow combines glycan release, 2-AB labeling, post-labeling clean-up for HILIC-FLR-based *N*-glycan analysis [77]. Stockmann et al. demonstrated an automated workflow performed on the Hamilton Star liquid handling robot combining release, 2-AB labeling, and solid-phase extraction (SPE), for HILIC-FLR-based separation and quantification of IgG *N*-glycans [78].

Pushing HILIC applications at least to medium throughput (>50 analyses/day) requires assistance for the analysis of the increasing amount of data which is produced. GlycoBase (originally developed from the NIBRT, now in cooperation with Waters) is a relational database which contains HPLC and UPLC elution positions of a huge set of different 2-AB-labeled *N*-glycans together with the predicted positions of potential exoglycosidase digestion products. AutoGU acts as a tool to use such database to semi-automatically assign HPLC peaks for glycan identification [79]. Water is providing a comprehensive workflow: comprising RapiFluor-MS Kit for sample preparation, LC-FLR and/or LC-MS for separation and detection, as well as GlycoBase 3+ database and software applications for data analysis [80].

## 2.2 *Matrix-Assisted Laser Desorption/Ionization Mass Spectrometry-Based Glycan Analysis*

Next to HILIC, MALDI-MS is possibly the second most commonly used analytical technique for the analysis of different types of glycans in biopharma industry. For a long time, MS-based approaches have been considered limited, in particular with respect to quantitation. Improvements in terms of stability automation and throughput that were achieved over the past years in MALDI-MS make this technique to an increasingly interesting tool for the screening of glycans [81]. In consequence, this technique has gained traction in the biopharmaceutical industry for the screening of glycans and specific glycoforms, from initial screening up to final product validation.

The principle of a MALDI-MS experiment is fairly simple: one μL or less of analyte is spotted onto a target plate and allowed to dry. This dry spot is then covered with an appropriate matrix [e.g., 2,5-dihydroxybenzoic acid (DHB)] and allowed to dry. Alternatively, sample and matrix can also be pre-mixed prior loading onto the target plate. These dry, matrix/sample spots are then subjected to a pulsed-laser irradiation that vaporizes the sample-matrix mixture. The matrix fulfills the job of gently transferring energy to the analyte and in consequence generates ions for analysis by MS [37]. For more details on the principles of MALDI-MS analyses and tips and tricks for specific type of molecules, we recommend the excellent reviews and work by Harvey et al., which are possibly the most comprehensive pieces of work with respect to MALDI-MS analysis of glycans [81–85]. Nevertheless, one of the biggest challenges in MALDI-MS-based analyses is the appropriate quantitative representation of all compounds present in a sample. Molecules with different chemical properties (e.g., charged, neutral, hydrophilic/hydrophobic) can result in different ionization efficiencies, and effectively, the detected signal might not be an appropriate representation of a molecules' actual concentration within a sample [86]. To overcome these issues, a number of different strategies have been developed.

### 2.2.1 Derivatization Strategies to Improve Detection and Deliver Accurate Relative Quantitation

A number of different derivatization approaches can be employed to improve the ionization properties of glycans and remove negative ionization effects introduced by Sia. While certain reducing end tags such as 2-AA or PA can also improve ionization efficiency [41, 81], they are seldom used for this purpose but more so in the context of making glycans fluorescent for detection after LC separation (see above). The two most widely applied glycan derivatization approaches for MS detection purposes are permethylation [87] and linkage-specific Sia esterification [88].

During the permethylation reaction, all free hydroxyl groups of a glycan are being methylated, which effectively also neutralizes any negative charges present on Sia and thus results in similar ionization efficiencies for all glycans present in a given sample. Nevertheless, like any chemical derivatization procedure, possible disadvantages manifest themselves in incomplete derivatization and the use of highly toxic chemicals and water-free conditions that need to be maintained during permethylation [89]. As a consequence of methylating hydroxyl groups, the overall hydrophobicity increases, which allows an fractionation via C18 reversed-phase LC [90, 91] and under certain elevated temperature conditions, also by porous graphitized carbon (PGC)-LC [92]. In addition, permethylation also improves the generation of prevalent cross-ring fragments in tandem mass spectrometry (MS/MS) experiments facilitating structure assignment [93]. In principle, permethylation can also be implemented in an automated workflow and thus can also be employed in HT applications [20].

Sia-specific esterification originally invented by Harvey's group [94] and later improved by Wuhrer's group [95, 96] represents an alternative approach to overcome the ionization issues of Sia containing glycans. In addition, Wuhrer's improved approach provides an easy opportunity to differentiate the Sia linkage (2–3 or 2–6) from the detected mass alone as α2–3-linked *N*-acetylneuraminic acid (NeuAc) residues undergo a lactonization in the presence of carboxylic acid activators in ethanol, while α2–6-linked NeuAc will experience an ethyl esterification [95, 96]. In any case, the negative impact of the Sia on the ionization is neutralized and as an additional benefit information on the Sia linkage is obtained within a single MS experiment, as demonstrated in the example of human serum *N*-glycans (Fig. 3). This protocol has also been adapted for HT performance and even extended to the analysis of intact glycopeptides using a dimethyl amidation instead of the use of alcohols [97, 98]. Within such an automated sample preparation, both protocols (permethylation and Sia esterification) require similar sample preparation times between 5 h for 96 samples (permethylation) and 5.5 h for 384 samples (esterification) [20, 97].

Several commercial kits are available on the market, ensuring reliable quality and efficacy of the chemicals for qualified usage in the biopharmaceutical industry. For MALDI-MS analysis, the permethylation kit from Ludger (LudgerTag Permethylation of Glycans) is already combined with the automation workflow [20, 99], for an easy implementation within a biopharmaceutical industry setting to achieve a comprehensive, HT-ready, reproducible, and "easy-to-use" analytical strategy employing MALDI-MS for glycan analyses. The improved patented linkage-specific Sia esterification approach of Wuhrer was exclusively licensed and commercialized by glyXera [100].

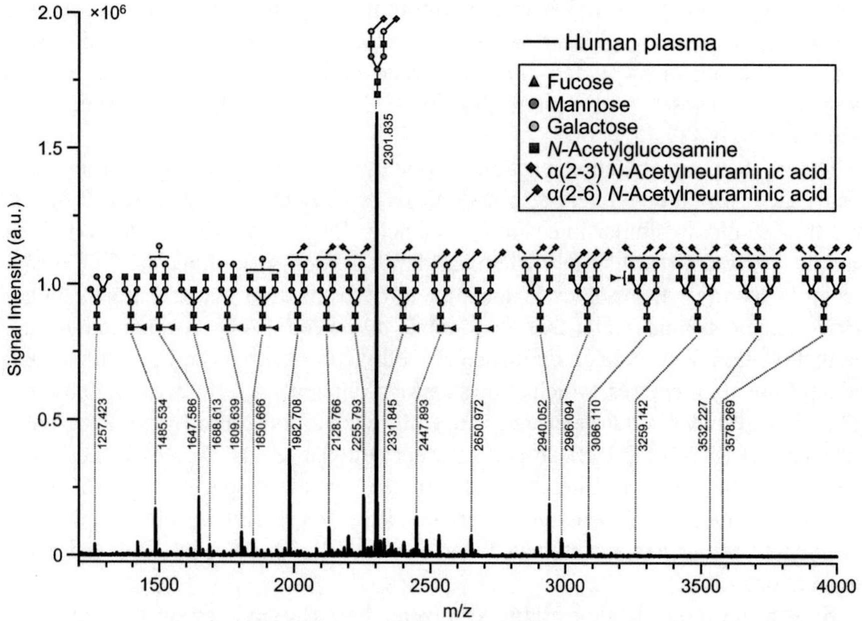


**Fig. 3** MALDI-MS spectrum of esterified *N*-glycans released from human serum glycoproteins. The illustration of the *N*-glycans results from the detected molecular masses. Measurement was performed in positive ionization mode using DHB matrix supplemented with sodium, resulting in sodium adducts, respectively. Structural differentiation between Siaα2–3 and Siaα2–6 is based on specific mass shifts resulting from the esterification of the *N*-glycans

While a detailed discussion of automated data analyses goes beyond the framework of this chapter, an elegant and freely available tool that can automate MALDI-MS data analysis shall quickly be mentioned: MassyTools enables HT data processing via automated annotation and quantification of MALDI-MS data and is capable of calibrating spectra, extracting data, and performing QC calculations based on a user-defined list of molecule masses. Compared to the commercial software FlexAnalysis from Bruker, MassyTools has been demonstrated to yield better calibration with a comparable relative standard deviation in terms of relative quantitation [101], closing one crucial bottleneck in any HT data analysis.

## 2.3 Capillary Electrophoresis-Based Glycan Analysis

CE is the umbrella term for the different capillary electrokinetic separation technologies such as capillary zone electrophoresis (CZE) or capillary gel electrophoresis (CGE). Electrophoresis was born more than 200 years ago [102], even long before the concept of chromatography was described. However, it took about 150 more years before the use of capillaries was introduced [103] and an additional decade to establish CE as it is most widely known today and to perform separations that seemed unattainable at that time [104, 105]. Despite its superior performance regarding separation power, sensitivity, and its potential for real high throughput by multiplexing the separation capillaries, the use of CE for glycan analysis lagged behind other commonly used analytical techniques, as stability, reliability, and repeatability issues could technically not be solved. The first attempts to use CE for glycan analysis started in the 1990s were rather humble and not broadly embraced by the glycoscientific community, but – with the availability of robust and reliable genetic analyzers based on multiplexed CGE with laser-induced fluorescence detection (xCGE-LIF) – by the genetics community, where they were crucial in the early stages, as they built the vital backbone of HT genome sequencing.

After their use in genomics, these xCGE-LIF-based genetic analyzers found their application in glycomics, which turned out to be an even larger complex challenge. The gel buffer increases the viscosity of the electrophoresis medium and consequently suppresses the electroosmotic flow (EOF) inside the capillaries, decreases the electrophoretic mobility of charged analytes, and thereby improves their separation [106–108]. As glycans lack any endogenous fluorescence necessary for LIF-detection [109] and in majority also charges, 8-aminopyrene-1,3,6-trisulfonic-acid (APTS) fluorescent dye [110–112] is almost universally employed in CE-based glycan analysis today, as it meets these requirements. APTS-labeled glycans will, therefore, be separated based on charge and molecular size/shape with high resolution [113–115]. Recent work has shown that xCGE-LIF can separate even challenging positional and linkage isomers in a single analysis run. For example, the method is capable of distinguishing the position and linkage of Fuc (α1–6 core Fuc from α1–3/1–4 Fuc on antenna GlcNAc or α1–2 Fuc on Gal) [116–120], Sia type (*N*-acetylneuraminic acid from *N*-glycolylneuraminic acid) [121], Sia linkage (α2–3 from α2–6) [116, 117, 120, 122, 123], Gal linkage (β1–3 from β1–4) [117, 119, 120], and even position isomers of glycan structures (e.g., Gal on α1–3 from Gal on α1–6 arm of the core structure) [116, 119, 124–127]. This feature becomes especially advantageous when thinking about the importance of determining immunogenic α-Gal and *N*-glycolylneuraminic acid on glycoprotein therapeutics [128–130]. Additionally, the fact that only α2–6, and not α2–3 Sia, affects the anti-inflammatory activity of an IgG antibody [131] makes it crucial to have a method capable of their differentiation. Example of high-resolution separation of *N*-glycans released from a complex sample utilizing xCGE-LIF technology is given in Fig. 4.

Besides this extraordinary separation power, this method is attractive due to the impressive sensitivity (low attomole range) [109, 132] of LIF detection. In addition, xCGE-LIF-based DNA sequencers employ a multicapillary format incorporating up to 96 capillaries so that hundreds to thousands of samples can be measured by CE per day [133], which is an amazing "real" high-throughput capability. Another attractive option provided by those instruments allows recording of different fluorescent dyes at different wavelengths at the same time (within one run). This special feature has been exploited for internal normalization of migration times, giving long-time stable migration times [134–136]. Beside migration time normalization and instant

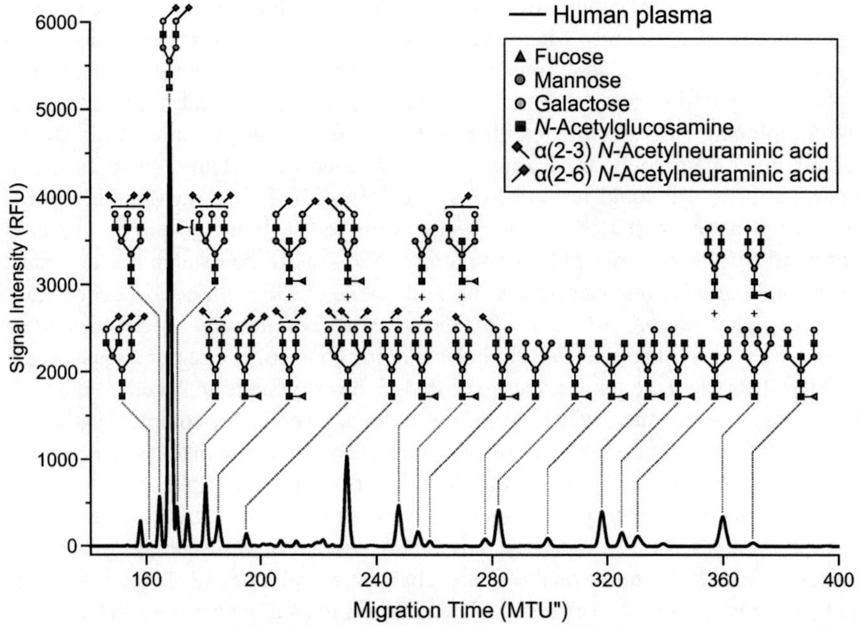


**Fig. 4** xCGE-LIF electropherogram of APTS-labeled *N*-glycans released from human plasma glycoproteins. *N*-glycans were released from human plasma, labeled, purified, and subsequently analyzed by xCGE-LIF as previously published[28]. *N*-glycan structures are annotated via migration time matching with a database (glyXbase from glyXera) and confirmed by various exoglycosidase digests

structural assignment search in a built-in database (more than 300 *N*-glycan entries), glyXtool software (glyXera [100]) allows automated background subtraction, raw data smoothing, peak picking, integration, relative quantification, and sample comparison [116, 137]. Meanwhile, also ready-to-use methods and kits are on the market [64, 138–141], enabling fast and robust comparison of glycoprofiles, which makes the technology more and more attractive for the industry.

As for other separation-based glycoanalytical techniques, like HILIC-FLR, one bottleneck of this powerful analytical technique is that it does not provide any direct structural information. Thus, like for LC-based glycan analyses, for explorative glycoanalytical investigations, additional techniques, such as exoglycosidase digests [117, 120, 121, 133, 135, 142] or MS [143–147], need to be used together with CE.

Clearly, CE has matured to the point that can stay side by side to other more traditionally used techniques for glycan analysis. Because of its unbeatable speed, resolution, sensitivity, and simplicity, the method is gaining more and more attention. Its unprecedented separation power, together with HT capability by multiplexing, and femto- to nanoliter injection make CE to more than a proper competitor to other techniques for glycan analysis. Commercially available analysis kits and software solutions including glycan databases make method appealable to the biopharmaceutical industry. Thus, today, almost 40 years since the potential of

CE was recognized, advantages of the method for analysis of glycans are becoming increasingly obvious and appreciated. The recent and upcoming developments in miniaturization and analysis toolboxes show that there is exciting future in glycan analysis for this recently revived technology.

# 3 Emerging Technologies for the Analysis of Glycans in the Biopharmaceutical Industry

Next to HILIC-FLR, the "gold-standard" approach for glycan analysis in biopharma, and the more newly established high-throughput methods like MALDI-MS and xCGE-LIF introduced above, we want to spot on two exciting technology developments that offer unique opportunities for structure characterization of *N*- and *O*-glycans.

## *3.1 Porous Graphitized Carbon Liquid Chromatography Online Coupled to MS for N- and O-Glycomics*

PGC is a stationary phase chemistry with unique properties for glycan separation and analysis. The specific interactions involved in PGC-LC are only vaguely understood but are believed to be a combination of hydrophobic, ionic, polar, and molecular features that together influence analyte retention [148–155]. PGC has shown a unique ability to resolve native, non-labeled, glycans in a LC setting (Fig. 5). Since PGC separation works best when non-labeled, reduced glycans are analyzed, MS has been the preferred detection approach due to its sensitivity and versatility. Here, glycan reduction provides the additional advantage as it introduces a specific mass tag on the reducing end that facilitates fragment peak assignment in tandem MS spectra [29].

The PGC-LC-MS approach has successfully been employed by several laboratories around the world to investigate either protein-specific or tissue/body fluid-specific, global glycosylation patterns [151, 157, 158] but also for biopharma mAb products [159]. In a recent National Institute of Standards and Technology (NIST)-organized multi-laboratory effort to evaluate comparability of mAb glycan analyses across biopharma and research labs, PGC-LC-MS performed equally well if not a bit better compared to most standard methods employed [160]. Since no glycan labeling is required, it is equally suitable to separate any glycan type.

Thus, PGC-LC-MS has been applied to analyze *N*- and *O*-glycans but also to characterize the glycan portion of glycolipids and glycosaminoglycan fragments, as also reviewed recently by Stavenhagen et al. in the context of clinical glycomics [150].

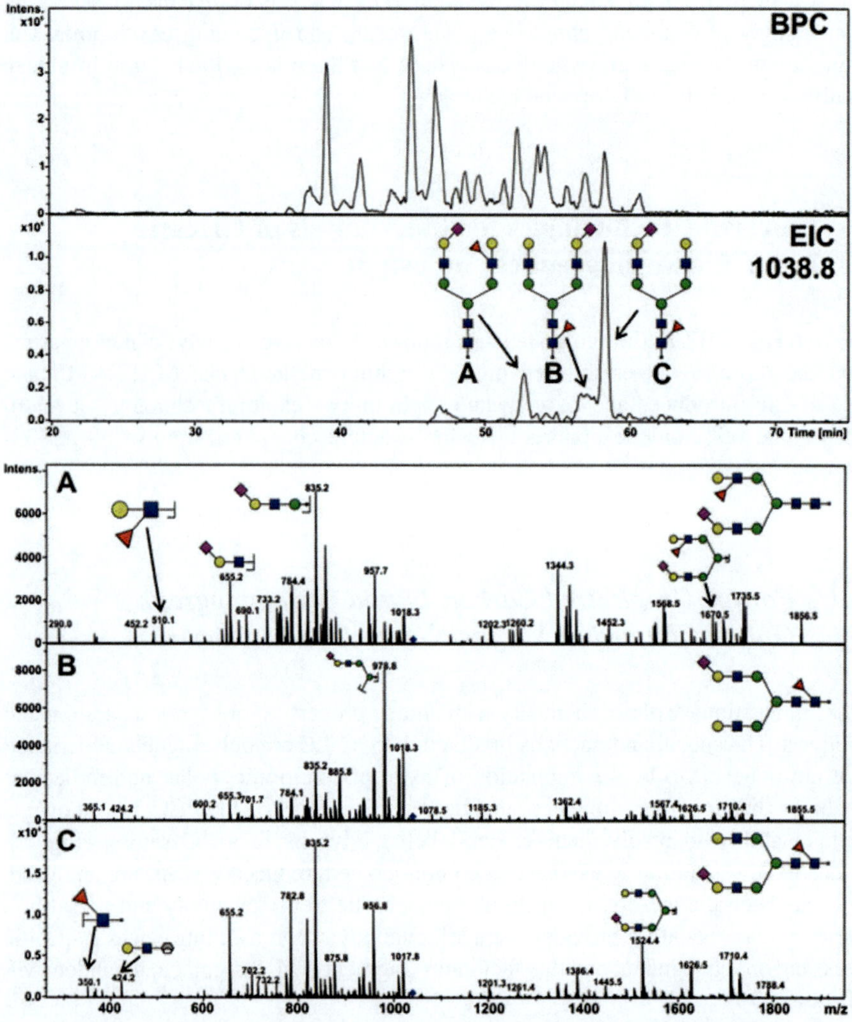


**Fig. 5** *N*-glycans released from human secretory immunoglobulin A (sIgA) and analyzed by PGC nanoLC-ESI MS/MS [figure taken from [156] with permission from publisher]. Top panel: The base peak chromatogram (BPC) provides a global overview of the present structures. An extracted ion chromatogram (EIC), here presented for *m/z* 1038.8, shows three distinct isobaric *N*-glycan alditols with individual LC elution properties. Due to this chromatographic separation, individual fingerprint MS/MS spectra can be acquired that subsequently allow differentiation and structural characterization (e.g., Fuc linkage) of these isobaric oligosaccharide alditols

The PGC-LC-MS glycomics approach delivers three independent sets of information within a single analysis that are used to qualitatively and semi-quantitatively describe glycan structures: glycan composition and type (by MS), glycan structure (by retention time and MS/MS), and relative quantity of glycans (by signal intensity). Hence, PGC-LC-MS has clear advantages if just very limited amounts of material are available as this is often the case when working with clinical histopathology sections [161, 162]. But it also can easily be combined to study protein-specific glycosylation after 1D or 2D electrophoretic separation, providing unique opportunities to obtain protein-specific glycosylation information without the requirement of complex protein purification steps [163]. This capacity allows establishing in-depth glycoprofiles from less than 500 ng of glycoprotein [161, 162]. PGC-LC-MS has successfully been employed to study the components of secretory IgA from human breast milk [164], human plasma proteins such as butyrylcholinesterase [165] or corticosteroid-binding globulin [166, 167], human as well as recombinant immunoglobulins [159], human cancer-associated glycoproteins such as E-cadherin [168], bacterial flagellins [169], or influenza virus antigens (Fig. 6), to name a few.

Recently, the analytical performance of the conventional PGC-LC-MS setup was systematically evaluated and compared to an improved setup with a tailored post-column make-up flow (PCMF). The PCMF-based setup showed an increased electrospray stability, glycomics sensitivity (30–100×), coverage and quantitative accuracy not least for the difficult-to-detect early-eluting and low-abundance glycans detached from *N*- and *O*-glycoproteins [170, 171].

Although mostly used for the analysis of non-labeled, non-derivatized glycans, recently the separation of permethylated glycans by PGC has also been reported [93, 172]. The packing of PGC into nanoscale chromatography chips for nanoLC-MS based analysis of permethylated glycans [173] has shown a considerable sensitivity promise in the rapid analysis of glycans, albeit sacrificing isomer separation due to the significantly increased hydrophobicity of the permethylated glycan molecules.

### 3.2 *Ion-Mobility Mass Spectrometry-Based Glycan Analysis*

Over the past decade, MS has become the most powerful tool for in-depth carbohydrate sequencing. New instrument developments in fragmentation approaches and gas-phase ion handling such as electron-transfer dissociation (ETD) or ion mobility (IM), respectively, have become available in commercial instruments, further pushing the limits in glycoanalytics (also reviewed recently by Everest-Dass and co-workers [174, 175]).

Ion-mobility mass spectrometry (IM-MS) is one of the most recent additions introduced in commercial MS-instruments that has the potential to revolutionize glycan structure characterization by MS. In IM-MS, charged ions are separated based on their charge and gas-phase structure/size before the ions are detected by traditional MS analyzers. This provides an additional, entirely orthogonal layer of separation that is able to differentiate isobaric ions that just differ in their structure but not composition [176–179]. These structural differences are expressed in so-called collisional cross sections (CCS), which are independent of instrument

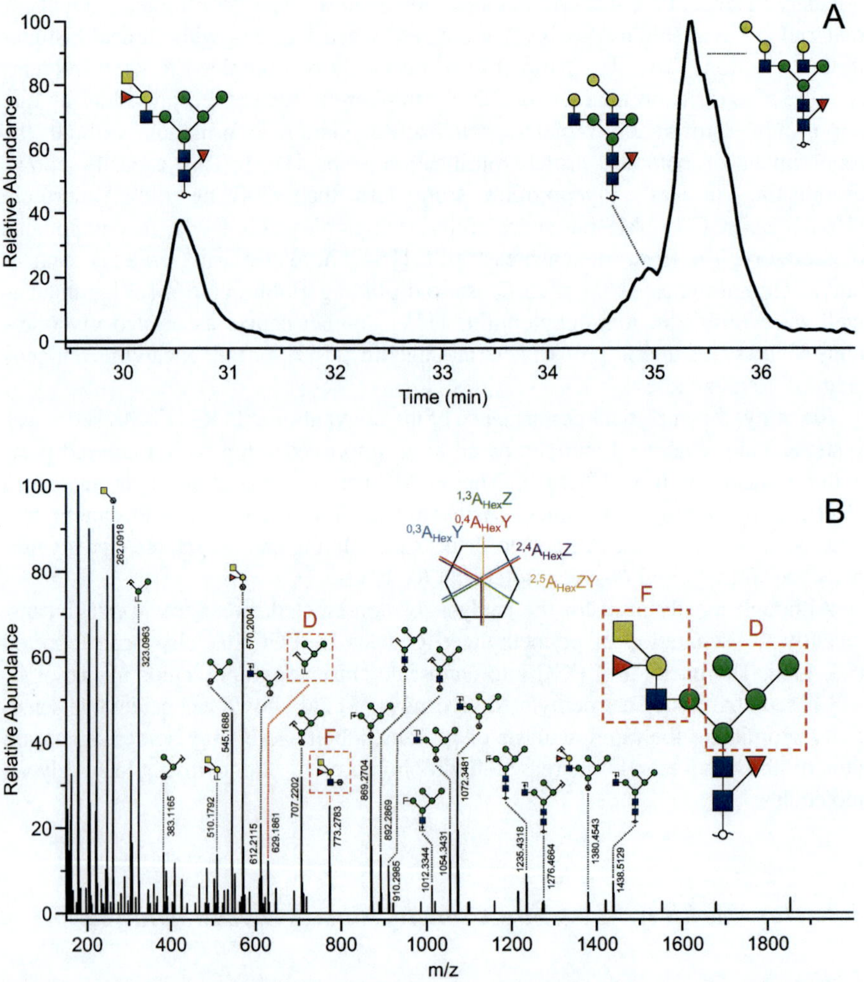


**Fig. 6** Separation and identification of isobaric *N*-glycan isomers released from influenza virus antigens. (**a**) nanoPGC-LC-MS extracted ion-chromatogram showing peaks representing three *N*-glycan isomers that are isobaric but differ in their structure. They were identified via specific cross-ring fragment ions due to negative mode ionization in combination with higher-energy collisional energy dissociation (HCD) as (**b**) exemplarily demonstrated

parameters under controlled conditions and correlate to the shape of an ion [180]. To receive appropriate IM-MS data, the choice of the drift gas and a suitable calibration is essential to record CCS data that can be used for glycan structure characterization. Nitrogen has so far delivered the best results for the analysis of complex carbohydrates, while dextran ladders have shown promising results to calibrate IM-MS instruments for accurate CCS determination [181].

Since the initial work of Gabryelski and Froese in 2003 on disaccharide structures [182], a number of publications have demonstrated how IM-MS can be used to dissect structure details otherwise impossible to differentiate by MS analyses on complex type [183] or oligomannose type *N*-glycans [184]. IM-MS has also been used to differentiate Sia linkage directly on glycopeptides, opening novel opportunities to simultaneously collect site-specific *N*-glycan or *O*-glycan structure information beyond composition alone [178] (see also chapter 'Glycoproteomics Technologies in Glycobiotechnology'). For more details the interested reader is referred to the excellent reviews from Hofmann and Pagel [180] and Gray et al. [185].

In a first step toward a semi-automated integration of CCS values for glycan structure determination, the online tool GlycoMob was recently published, which stores the CCS values of different glycan structures from IM-MS experiments [186]. This reduces the time for data interpretation and facilitates de novo sequencing of the corresponding glycans for structural evaluation. This is currently still a time-consuming approach using parameters such as precursor mass, CCS values, and diagnostic cross-ring fragment ions obtained after CID fragmentation of positively or negatively charged precursor ions for glycan structure evaluation [187].

While these promising technologies are still requiring a considerable level of user expertise and input compared to established technologies such as HILIC, MALDI, as well as CE, IM-MS and PGC-LC-MS are highly promising tools to deliver an additional level of information for in-depth glycan analysis. Factors such as high-throughput have been less of a focus for these methods, but ongoing work to develop decent databases and standardization procedures has shown promising progress that will increase their future capabilities also for biopharmaceutical industry applications [186, 188]. The choice of method depends on the analytical question that needs to be solved, and the most sophisticated technology is not always the most suitable one to answer a specific question. The fact that scientists can now choose from a selection of different methodologies with different strengths and limitations means that the optimal analytical solution can be employed for a specific problem.

# 4 Applications in the Biopharmaceutical Industry

## *4.1 Glycomics of Monoclonal Antibodies*

MAbs are still the major biopharmaceutical product class with the most approvals during the last years [12]. As of May 2017, 74 antibody-based molecules had been approved by authorities together with 70 molecules in Phase III and 575 in Phase I/II clinical trials [189]. Glycosylation is of essential interest in antibody production regarding safety, efficacy, immunogenicity, toxicity, and affinity [190, 191]. Especially in terms of safety, one prominent example is the cetuximab-induced anaphylaxis, caused by alpha-1,3-Gal binding IgE in some individuals [128]. In contrast to other mammals, humans do not have alpha-1,3-galactosyltransferase. Therefore, alpha-1,3-Gal on mAbs and in general recombinant proteins can cause devastating immunological reactions in humans [128]. Glycoengeneering of mAbs is an increasing field in the scientific community and the industry [1, 192–194]. Most recombinant antibody molecules harbor a single *N*-glycosylation site at the Fc-domain. However, it has been established that about 30% of polyclonal human IgG molecules bear *N*-linked oligosaccharides within the variable regions of the kappa (Vκ), lambda ($V_\lambda$), or heavy ($V_H$) chains, and sometimes both [190, 195]. Dealing with such complexity requires site-specific, as well as structural glycan and glycopeptide, analysis to cover macro- and microheterogeneity of the antibody glycoprofile. This brings new emerging LC-MS-based analytical methods for comprehensive site-specific and structural analysis more into the focus of interest. As stated by the authorities, antibody glycosylation is a CQA and needs to be monitored [3, 4]. To fulfill the requirements given by the regulatory authority guidelines, the analytical standard includes HILIC-UHPLC-FLR separation/detection of AB- or AA-labeled *N*-glycans coupled with exoglycosidase digestion experiments for structural identification. An exemplary workflow is nicely given for the production of mAbs in CHO cells by Carillo et al. (2017) [196]. MALDI-MS, CE-LIF, and PGC-MS-based application for the analysis of *N*-glycans derived from mAb's are also available and well discussed in the method comparison study from Reusch et al. [197, 198], as well in the recent NIST-organized study [160]. However, HILIC-UHPLC still is the standard measurement method in the industry in terms of mAbs glycosylation analysis.

## 4.2 Glycomics of Pharmaceutically Relevant Glycoproteins Beyond Monoclonal Antibodies

Beside mAbs as the leading biopharmaceutical recombinant glycoproteins, other glycoprotein products like hormones, fusion proteins, growth factors, cytokines, and therapeutic enzymes give new challenges to the glycan analysis. These recombinant proteins tend to be more complex in terms of protein structure, containing multiple *N*-glycosylation sites and other oligosaccharide motives like *O*-linked glycans [12]. To fulfill the requirements to monitor all potential PTMs influencing the efficacy of a therapeutic recombinant protein, site-specific glycoprotein and glycopeptide analysis is crucial beside glycan analysis. Detailed strategies for site-specific glycan analysis are given in this book, as well. So far, glycan analysis of recombinant proteins is performed similarly to mAbs. Exemplarily, Ludger is offering a comprehensive workflow for the analysis of the glycosylation of biopharmaceutical follicle-stimulating hormone (FSH) as CQAs, using a procainamide labeling system with separation/detection via HILIC-UHPLC-FLR online coupled to ESI-MS/MS for structural glycan analysis and reversed-phase (RP)-LC-ESI-MS for glycopeptide mapping [199]. Another application is the analysis of released *N*-glycans of recombinant human erythropoietin (EPO) that was achieved using HILIC-UHPLC-FLR of

RFMS-labeled *N*-glycans on a GlycanPac AXH-1 column [200]. Another group used MALDI-MS to detect *N*-glycans released from EPO via linkage-specific Sia esterification [201]. CE-MS, CE-LIF, and PGC-LC-MS are also suitable for the *N*-glycan analysis of more complex recombinant proteins [157, 202], but PGC-LC-MS is the only suitable approach for the structural analysis of *O*-glycans released by reductive β-elimination as non-reductive approaches suffer from significant peeling artifacts [203].

### *4.3 Glycomics of Vaccine Antigens*

Vaccines represent a further big sector of biopharmaceutical products. In contrast to recombinantly produced therapeutic proteins, vaccines are natural products. Vaccines are traditionally produced in embryonated chicken eggs. Ongoing efforts in establishing mammalian cell-based cultivation systems led already to approved vaccine products as well [204, 205]. Viruses can harbor glycoproteins as antigens. In case of influenza virus antigen glycosylation (hemagglutinin (HA) and neuraminidase (NA)), especially differences of HA glycosylation in terms of macro- and microheterogeneity are demonstrated to have a great impact on vaccine efficacy regarding immunogenicity and virulence [1, 206–209]. As an example, contrary to mAbs glycosylation, in vitro alpha-1,3-galactosylation of influenza virus antigens propagated in embryonated chicken eggs was demonstrated to harbor an increased vaccine efficiency for the treatment of alpha-1,3-galactosyltranferase negative mice [210]. Interestingly, glycan analysis of viral antigens is quite underrepresented. This could be due to the fact that the glycosylation of viral antigens from original viruses is not stated as a CQA by the regulatory agencies so far [5]. However, it can be expected to see changes regarding CQAs of biopharmaceutical vaccine products in terms of antigen glycosylation in the future. Especially CE was used in the past for the in-depth analysis of vaccine antigen *N*-glycosylation [136, 211]. Hennig et al. describe the *N*-glycan analysis of influenza virus antigens using xCGE-LIF very detailed [24]. Briefly, after SDS-PAGE separation of viral antigens HA and NA, an in-gel *N*-glycan release is performed using PNGase F digestion. Afterward, the glycans are labeled with APTS and enriched using HILIC solid-phase extraction (SPE). Structural separation and detection of labeled *N*-glycans are achieved using xCGE-LIF. The structural assignment of specific peaks is performed using sequential exoglycosidase digestion strategies and database-assisted comparison of specific migration times due to double internal normalization. For more details on released *N*-glycan and glycopeptide analyses from viral glycoproteins, we recommend the comprehensive reviews by Harvey et al. (2018) [212] and others [213–215].

Recombinant viral glycoproteins are an area of growing interest. Ensuring the correct glycosylation of recombinantly generated vaccines is equally mandatory in the biopharmaceutical industry (e.g., Flublok). We expect that the number of novel methods and publications specifically dealing with the aspects of recombinant vaccine glycosylation profiling will increase due to the growing requirements.

## 5 Conclusion

The methods discussed have been demonstrated to allow medium to HT glycomics and deliver good to high sensitivity, performance, and reproducibility. However, they provide different levels of composition/structure assignment certainty. The huge variety of different analytical techniques result in different types of data which require technology-specific, individual approaches for data analysis and mining. In consequence, this data-output heterogeneity still forms a major bottleneck compared to other -omics techniques. This is partially due to the intrinsic nature of glycans that are seldom linear constructed molecules and do occur in various linkages despite having the same "sequence" but also due to the fact that hardly a single technology provides all necessary information for unambiguous structure determination within a single analysis. In combination these factors clearly put challenges for any automated, software-assisted data analysis tools. Nevertheless, many of these challenges can be overcome when focusing on specific aspects or specific sample types. In the biopharmaceutical context, IgG represents the most widely analyzed glycoprotein, and its glycosylation is comparably simple in most expression systems used.

Several reviews compare these different state-of-the-art analytical techniques for the analysis of released glycans to more or less extent [160, 216, 217] and consistently report that the methods delivered reliable qualitative and comparable quantitative data. Thus, in summary, a wide selection of analytical approaches is available for glycan characterization of biopharmaceuticals. The latest NIST study showed that glycoanalytical approaches based on fluorescently labeled *N*-glycans with non-laser induced fluorescence detection are still the preferred choice for biopharmaceutical applications, maybe also due the fact that these have been already well-established for years. However, the NIST study also showed that these approaches usually deliver a lower identification rate compared to the methods applied in research laboratories that are more likely to apply methods based on highly sensitive LIF-detection or on MS-based detection of non-fluorescently-labeled glycans or glycopeptides [160].

The choice which analytical methodology is chosen to determine the glycoprofile of a biopharmaceutical depends on a variety of factors such as the type of protein, the expected complexity, and the preference for HT or in-depth structure elucidation, to name a few. With continuous development of novel methods and technologies, new advances in glycoanalytical technologies can be expected to overcome current limitations. The ongoing enhancement and improvement of the glycoanalytical toolbox will help the biopharmaceutical industry to face regulations from the authorities regarding glycosylation of recombinant proteins, beyond the relatively simple one of mAbs. Furthermore, it is expected that glycosylation of vaccines will become a CQA as well. With respect to more complex molecules and other cultivation systems (prokaryotes, plant cells, etc.), the importance and necessity of in-depth glycan analysis, facing comprehensive structural, site-specific, as well as quantitative information, will increase as this next generation of biologics pushes toward regulatory approval. The community is challenged to provide tools to fulfill the increasing requirements in the future, including production relevant parameters like high throughput, high performance, and high resolution. Next to the more reliably established *N*-glycan analysis, there will be an increased requirement to capture *O*-glycans on biopharmaceutical products.

## References


1. Butler M, Spearman M (2014) The choice of mammalian cell host and possibilities for glycosylation engineering. Curr Opin Biotechnol 30:107–112
2. Fournier J (2015) A review of glycan analysis requirements. Biopharm Int 28(10):32
3. EMA – Guideline on development, production, characterisation and specifications for monoclonal antibodies and related products – draft (2009) 2013 updated 2013/07/21. http://www.ema.europa.eu/docs/en_GB/document_library/Scientific_guideline/2009/09/WC500003074.pdf
4. Quality Guidelines: ICH (2014) 2018. updated 2018/05/17. http://www.ich.org/products/guidelines/quality/article/quality-guidelines.html
5. EMA – Guideline on Influenza Vaccines – Quality module (2017) 2017. updated 2017/07/28. http://www.ema.europa.eu/docs/en_GB/document_library/Scientific_guideline/2017/07/WC500232414.pdf
6. York IA, Stevens J, Alymova IV (2019) Influenza virus N-linked glycosylation and innate immunity. Biosci Rep 39(1):BSR20171505
7. Newburg DS, Grave G (2014) Recent advances in human milk glycobiology. Pediatr Res 75 (5):675–679
8. Yu ZT, Chen C, Kling DE, Liu B, McCoy JM, Merighi M et al (2013) The principal fucosylated oligosaccharides of human milk exhibit prebiotic properties on cultured infant microbiota. Glycobiology 23(2):169–177
9. Liu B, Newburg DS (2013) Human milk glycoproteins protect infants against human pathogens. Breastfeed Med 8(4):354–362
10. Kowitsch A, Zhou G, Groth T (2017) Medical application of glycosaminoglycans: a review. J Tissue Eng Regen Med
11. Hajba L, Szekrényes Á, Borza B, Guttman A (2018) On the glycosylation aspects of biosimilarity. Drug Discov Today
12. Walsh G (2014) Biopharmaceutical benchmarks 2014. Nat Biotechnol 32(10):992–1000
13. Varki A, Gagneux P (2015) Biological functions of glycans. In: Varki A, Cummings RD, Esko JD, Stanley P, Hart GW, Aebi M et al (eds) Essentials of glycobiology3rd edn. Cold Spring Harbor, New York
14. Christiansen MN, Chik J, Lee L, Anugraham M, Abrahams JL, Packer NH (2014) Cell surface protein glycosylation in cancer. Proteomics 14(4–5):525–546
15. Rhodes J, Campbell BJ, Yu L-G (2010) Glycosylation and disease. Encyclopedia of life sciences (ELS). Wiley, Chichester
16. Jaeken J (2013) Congenital disorders of glycosylation. Handb Clin Neurol 113:1737–1743
17. Moremen KW, Tiemeyer M, Nairn AV (2012) Vertebrate protein glycosylation: diversity, synthesis and function. Nat Rev Mol Cell Biol 13(7):448–462
18. Varki A, Hart GW (2017) Essentials of glycobiology. Cold Spring Harbor Laboratory Press, New York
19. Kolarich D, Packer NH (2012) Mass spectrometry for glycomics analysis of N- and O-linked glycoproteins. In: Yuriev E, Ramsland PA (eds) Structural glycobiology. CRC Press, Boca Raton, pp 141–161

20. Shubhakar A, Kozak RP, Reiding KR, Royle L, Spencer DI, Fernandes DL et al (2016) Automated high-throughput permethylation for glycosylation analysis of biologics using MALDI-TOF-MS. Anal Chem 88(17):8562–8569
21. Aich U, Lakbub J, Liu A (2016) State-of-the-art technologies for rapid and high-throughput sample preparation and analysis of N-glycans from antibodies. Electrophoresis 37(11):1468–1488
22. Stockmann H, Duke RM, Millan Martin S, Rudd PM (2015) Ultrahigh throughput, ultrafiltration-based N-glycomics platform for ultraperformance liquid chromatography (ULTRA (3)). Anal Chem 87(16):8316–8322
23. Hennig R (2009) Development of high-throughput method for the analysis of total plasma N-glycans by CGE-LIF. Otto von Guericke Universität, Magdeburg, Magdeburg
24. Hennig R, Rapp E, Kottler R, Cajic S, Borowiak M, Reichl U. N-glycosylation fingerprinting of viral glycoproteins by xCGE-LIF. 2015
25. Vreeker GC, Wuhrer M (2017) Reversed-phase separation methods for glycan analysis. Anal Bioanal Chem 409(2):359–378
26. Wuhrer M, Koeleman CA, Deelder AM, Hokke CH (2004) Normal-phase nanoscale liquid chromatography– mass spectrometry of underivatized oligosaccharides at low-femtomole sensitivity. Anal Chem 76(3):833–838
27. Rohrer JS, Basumallick L, Hurum D (2013) High-performance anion-exchange chromatography with pulsed amperometric detection for carbohydrate analysis of glycoproteins. Biochem Moscow 78(7):697–709
28. Rohrer JS, Basumallick L, Hurum DC (2016) Profiling N-linked oligosaccharides from IgG by high-performance anion-exchange chromatography with pulsed amperometric detection. Glycobiology 26(6):582–591
29. Alagesan K, Everest-Dass A, Kolarich D (2018) Isomeric separation and characterisation of glycoconjugates. Adv Exp Med Biol 1104:77–99
30. Rudd PM, Guile GR, Kuster B, Harvey DJ, Opdenakker G, Dwek RA (1997) Oligosaccharide sequencing technology. Nature 388(6638):205–207
31. Alpert AJ, Shukla M, Shukla AK, Zieske LR, Yuen SW, Ferguson MA et al (1994) Hydrophilic-interaction chromatography of complex carbohydrates. J Chromatogr A 676(1):191–122
32. Guile GR, Rudd PM, Wing DR, Prime SB, Dwek RA (1996) A rapid high-resolution high-performance liquid chromatographic method for separating glycan mixtures and analyzing oligosaccharide profiles. Anal Biochem 240(2):210–226
33. Natsuka S, Hase S (1998) Analysis of N- and O-glycans by pyridylamination. Methods Mol Biol 76:101–113
34. Tharmalingam T, Adamczyk B, Doherty MA, Royle L, Rudd PM (2013) Strategies for the profiling, characterisation and detailed structural analysis of N-linked oligosaccharides. Glycoconj J 30(2):137–146
35. Zauner G, Deelder AM, Wuhrer M (2011) Recent advances in hydrophilic interaction liquid chromatography (HILIC) for structural glycomics. Electrophoresis 32(24):3456–3466
36. Melmer M, Stangler T, Premstaller A, Lindner W (2011) Comparison of hydrophilic-interaction, reversed-phase and porous graphitic carbon chromatography for glycan analysis. J Chromatogr A 1218(1):118–123
37. Briggs JB (2017) Glycan characterization. Anal Char Biother:257–332
38. Jandera P, Janas P (2017) Recent advances in stationary phases and understanding of retention in hydrophilic interaction chromatography. A review. Anal Chim Acta 967:12–32
39. Huang Y, Nie Y, Boyes B, Orlando R (2016) Resolving isomeric glycopeptide glycoforms with hydrophilic interaction chromatography (HILIC). J Biomol Tech 27(3):98–104
40. Fu Q, Liang T, Li ZY, Xu XY, Ke YX, Jin Y et al (2013) Separation of carbohydrates using hydrophilic interaction liquid chromatography. Carbohydr Res 379:13–17
41. Ruhaak LR, Zauner G, Huhn C, Bruggink C, Deelder AM, Wuhrer M (2010) Glycan labeling strategies and their use in identification and quantification. Anal Bioanal Chem 397(8):3457–3481

42. Buszewski B, Noga S (2012) Hydrophilic interaction liquid chromatography (HILIC)–a powerful separation technique. Anal Bioanal Chem 402(1):231–247
43. Nagy G, Peng T, Pohl NLB (2017) Recent liquid chromatographic approaches and developments for the separation and purification of carbohydrates. Anal Methods 9(24):3579–3593
44. Guo Y (2015) Recent progress in the fundamental understanding of hydrophilic interaction chromatography (HILIC). Analyst 140(19):6452–6466
45. GlycanPac™ AXH-1 LC-Säulen (2018) Updated 2018/05/17. http://www.thermofisher.com/order/catalog/product/082468
46. GlycanPac™ AXR-1 LC-Säulen (2018) Updated 2018/05/17. http://www.thermofisher.com/order/catalog/product/088136
47. Rodriguez-Sanchez S, Garcia-Sarrio MJ, Quintanilla-Lopez JE, Soria AC, Sanz ML (2015) Analysis of amino sugars and other low molecular weight carbohydrates in Aglaonema sp extracts by hydrophilic interaction liquid chromatography coupled to mass spectrometry. J Chromatogr A 1423:104–110
48. Jacob GS, Scudder P (1994) Glycosidases in structural analysis. Methods Enzymol 230:280–299
49. Rudd PM, Dwek RA (1997) Rapid, sensitive sequencing of oligosaccharides from glycoproteins. Curr Opin Biotechnol 8(4):488–497
50. Kobata A, Takasaki S (1993) Fukuda M, Kobata A (eds) Structural characterization of oligosaccharides from glycoproteins. Glycobiology: a practical approach. Oxford University Press, New York, pp 165–185
51. Kobata A (2013) Exo-and endoglycosidases revisited. Proc Jpn Acad Ser B 89(3):97–117
52. Prime S, Merry T (1998) Exoglycosidase sequencing of N-linked glycans by the reagent array analysis method (RAAM). Glycoanalysis protocols. Springer, Berlin, pp 53–69
53. Varki A, Cummings RD, Aebi M, Packer NH, Seeberger PH, Esko JD et al (2015) Symbol nomenclature for graphical representations of glycans. Glycobiology 25(12):1323–1324
54. Neelamegham S, Aoki-Kinoshita K, Bolton E, Frank M, Lisacek F, Lütteke T et al (2019) Updates to the symbol nomenclature for glycans guidelines. Glycobiology 29(9):620–624
55. Pabst M, Kolarich D, Poltl G, Dalik T, Lubec G, Hofinger A et al (2009) Comparison of fluorescent labels for oligosaccharides and introduction of a new postlabeling purification method. Anal Biochem 384(2):263–273
56. Melmer M, Stangler T, Schiefermeier M, Brunner W, Toll H, Rupprechter A et al (2010) HILIC analysis of fluorescence-labeled N-glycans from recombinant biopharmaceuticals. Anal Bioanal Chem 398(2):905–914
57. Stockmann H, O'Flaherty R, Adamczyk B, Saldova R, Rudd PM (2015) Automated, high-throughput serum glycoprofiling platform. Integr Biol 7(9):1026–1032
58. Anumula KR, Dhume ST (1998) High resolution and high sensitivity methods for oligosaccharide mapping and characterization by normal phase high performance liquid chromatography following derivatization with highly fluorescent anthranilic acid. Glycobiology 8 (7):685–694
59. Ruhaak LR, Huhn C, Waterreus WJ, de Boer AR, Neususs C, Hokke CH et al (2008) Hydrophilic interaction chromatography-based high-throughput sample preparation method for N-glycan analysis from total human plasma glycoproteins. Anal Chem 80(15):6119–6126
60. Deguchi K, Keira T, Yamada K, Ito H, Takegawa Y, Nakagawa H et al (2008) Two-dimensional hydrophilic interaction chromatography coupling anion-exchange and hydrophilic interaction columns for separation of 2-pyridylamino derivatives of neutral and sialylated N-glycans. J Chromatogr A 1189(1–2):169–174
61. Klapoetke S, Zhang J, Becht S, Gu X, Ding X (2010) The evaluation of a novel approach for the profiling and identification of N-linked glycan with a procainamide tag by HPLC with fluorescent and mass spectrometric detection. J Pharm Biomed Anal 53(3):315–324
62. Thomson RI, Gardner RA, Strohfeldt K, Fernandes DL, Stafford GP, Spencer DIR et al (2017) Analysis of three epoetin alpha products by LC and LC-MS indicates differences in glycosylation critical quality attributes including sialic acid content. Anal Chem 89(12):6455–6462

63. Lauber MA, Yu YQ, Brousmiche DW, Hua Z, Koza SM, Magnelli P et al (2015) Rapid preparation of released N-glycans for HILIC analysis using a labeling reagent that facilitates sensitive fluorescence and ESI-MS detection. Anal Chem 87(10):5401–5409
64. Glycan Labeling Kits – Ludger Products (2018) Updated 2018/05/17. https://www.ludger.com/products/glycan_labeling_kits.php
65. Kozak RP, Tortosa CB, Fernandes DL, Spencer DI (2015) Comparison of procainamide and 2-aminobenzamide labeling for profiling and identification of glycans by liquid chromatography with fluorescence detection coupled to electrospray ionization-mass spectrometry. Anal Biochem 486:38–40
66. GlycoProfile™ 2-AB Labeling Kit PP0520 (2018) Updated 2018/05/17. https://www.sigmaaldrich.com/catalog/product/sigma/pp0520?lang=de&region=DE
67. GlycoProfile™ 2-AA Labeling Kit PP0530 (2018) Updated 2018/05/17. https://www.sigmaaldrich.com/catalog/product/sigma/pp0530?lang=de&region=DE
68. Waters. GlycoWorks 2-AB N-Glycan Kits: Waters (2018) Updated 2018/05/17. http://www.waters.com/waters/en_US/GlycoWorks-2-AB-N-Glycan-Kits/nav.htm?cid=134914425&locale=en_US
69. Waters. GlycoWorks RapiFluor-MS N-Glycan Kit: Waters (2018) Updated 2018/05/17. http://www.waters.com/waters/en_CA/GlycoWorks-RapiFluor-MS-N-Glycan-Kit/nav.htm?locale=en_CA&cid=134828150
70. ProZyme. GlykoPrep® 2-AB (2018) Updated 2018/05/17. https://prozyme.com/collections/glykoprep-2-ab
71. ProZyme. GlykoPrep® InstantAB™ (2018) Updated 2018/05/17. https://prozyme.com/collections/glykoprep-instantab
72. ProZyme. GlykoPrep InstantPC (2018) Updated 2018/05/17. https://prozyme.com/collections/glykoprep-instantpc
73. AQC (6-Aminoquinolyl-N-hydroxysuccinimidyl carbamate) – Synchem (2018) Updated 2018/05/17. https://www.synchem.de/product/aqc-6-aminoquinolyl-n-hydroxysuccinimidyl-carbamate
74. Zhou S, Veillon L, Dong X, Huang Y, Mechref Y (2017) Direct comparison of derivatization strategies for LC-MS/MS analysis of N-glycans. Analyst 142(23):4446–4455
75. Keser T, Pavić T, Lauc G, Gornik O (2018) Comparison of 2-aminobenzamide, procainamide and RapiFluor-MS as derivatizing agents for high-throughput HILIC-UPLC-FLR-MS N-glycan analysis. Front Chem 6:324
76. AssayMAP Bravo Platform. Agilent (2018) Updated 2018/05/17. https://www.agilent.com/en/products/automation-solutions/protein-sample-preparation/assaymap-bravo-platform
77. Ludger (2020) Ludger high throughput N-glycan sample preparation. Updated 07/30/2020. https://www.ludger.com/products/high-throughput-sample-preparation.php
78. Stockmann H, Adamczyk B, Hayes J, Rudd PM (2013) Automated, high-throughput IgG-antibody glycoprofiling platform. Anal Chem 85(18):8841–8849
79. Campbell MP, Royle L, Rudd PM (2015) GlycoBase and autoGU: resources for interpreting HPLC-glycan data. Methods Mol Biol 1273:17–28
80. Waters. Released glycan analysis: waters (2018) Updated 2018/05/17. http://www.waters.com/waters/de_DE/Released-Glycan-Analysis/nav.htm?cid=10116552&locale=127
81. Harvey DJ (2017) Analysis of carbohydrates and glycoconjugates by matrix-assisted laser desorption/ionization mass spectrometry: an update for 2011-2012. Mass Spectrom Rev 36 (3):255–422
82. Harvey DJ (2015) Analysis of carbohydrates and glycoconjugates by matrix-assisted laser desorption/ionization mass spectrometry: an update for 2009-2010. Mass Spectrom Rev 34 (3):268–422
83. Harvey DJ (2012) Analysis of carbohydrates and glycoconjugates by matrix-assisted laser desorption/ionization mass spectrometry: an update for 2007-2008. Mass Spectrom Rev 31 (2):183–311
84. Harvey DJ (2009) Analysis of carbohydrates and glycoconjugates by matrix-assisted laser desorption/ionization mass spectrometry: an update for 2003-2004. Mass Spectrom Rev 28 (2):273–361

85. Harvey DJ (2008) Analysis of carbohydrates and glycoconjugates by matrix-assisted laser desorption/ionization mass spectrometry: an update covering the period 2001-2002. Mass Spectrom Rev 27(2):125–201
86. Stavenhagen K, Hinneburg H, Thaysen-Andersen M, Hartmann L, Varon Silva D, Fuchser J et al (2013) Quantitative mapping of glycoprotein micro-heterogeneity and macro-heterogeneity: an evaluation of mass spectrometry signal strengths using synthetic peptides and glycopeptides. J Mass Spectrom 48(6):627–639
87. Zhou S, Dong X, Veillon L, Huang Y, Mechref Y (2017) LC-MS/MS analysis of permethylated N-glycans facilitating isomeric characterization. Anal Bioanal Chem 409 (2):453–466
88. Jiang K, Zhu H, Li L, Guo Y, Gashash E, Ma C et al (2017) Sialic acid linkage-specific permethylation for improved profiling of protein glycosylation by MALDI-TOF MS. Anal Chim Acta 981:53–61
89. Alley Jr WR, Madera M, Mechref Y, Novotny MV (2010) Chip-based reversed-phase liquid chromatography-mass spectrometry of permethylated N-linked glycans: a potential methodology for cancer-biomarker discovery. Anal Chem 82(12):5095–5106
90. Ritamo I, Räbinä J, Natunen S, Valmu L (2013) Nanoscale reversed-phase liquid chromatography–mass spectrometry of permethylated N-glycans. Anal Bioanal Chem 405 (8):2469–2480
91. Zhou S, Hu Y, Mechref Y (2016) High-temperature LC-MS/MS of permethylated glycans derived from glycoproteins. Electrophoresis 37(11):1506–1513
92. Zhou S, Huang Y, Dong X, Peng W, Veillon L, Kitagawa DAS et al (2017) Isomeric separation of permethylated glycans by porous graphitic carbon (PGC)-LC-MS/MS at high temperatures. Anal Chem 89(12):6590–6597
93. Costello CE, Contado-Miller JM, Cipollo JF (2007) A glycomics platform for the analysis of permethylated oligosaccharide alditols. J Am Soc Mass Spectrom 18(10):1799–1812
94. Wheeler SF, Domann P, Harvey DJ (2009) Derivatization of sialic acids for stabilization in matrix-assisted laser desorption/ionization mass spectrometry and concomitant differentiation of alpha(2 –> 3)- and alpha(2 –> 6)-isomers. Rapid Commun Mass Spectrometry 23(2):303–312
95. Reiding KR, Blank D, Kuijper DM, Deelder AM, Wuhrer M (2014) High-throughput profiling of protein N-glycosylation by MALDI-TOF-MS employing linkage-specific sialic acid esterification. Anal Chem 86(12):5784–5793
96. de Haan N, Reiding KR, Haberger M, Reusch D, Falck D, Wuhrer M (2015) Linkage-specific sialic acid derivatization for MALDI-TOF-MS profiling of IgG glycopeptides. Anal Chem 87 (16):8284–8291
97. Bladergroen MR, Reiding KR, Hipgrave Ederveen AL, Vreeker GC, Clerc F, Holst S et al (2015) Automation of high-throughput mass spectrometry-based plasma N-glycome analysis with linkage-specific sialic acid esterification. J Proteome Res 14(9):4080–4086
98. Reiding KR, Lonardi E, Hipgrave Ederveen AL, Wuhrer M (2016) Ethyl esterification for MALDI-MS analysis of protein glycosylation. Methods Mol Biol 1394:151–162
99. Ludger Glycan Permethylation (2018) Updated 2018/05/17. https://www.ludger.com/permethylation
100. glyXera. Updated 08/18/2020. https://www.glyxera.com/
101. Jansen BC, Reiding KR, Bondt A, Hipgrave Ederveen AL, Palmblad M, Falck D et al (2015) MassyTools: a high-throughput targeted data processing tool for relative quantitation and quality control developed for glycomic and glycoproteomic MALDI-MS. J Proteome Res 14 (12):5088–5098
102. Reuss FF (1809) Sur un nouvel effet de l'électricité galvanique. Mem Soc Imp Nat Moscou 2:327–337
103. Hjertén S (1967) Free zone electrophoresis. Chromatogr Rev 9(2):122–219
104. Jorgenson JW, Lukacs KD (1981) Zone electrophoresis in open-tubular glass capillaries. Anal Chem 53(8):1298–1302

105. Jorgenson JW, Lukacs K (1981) Free-zone electrophoresis in glass capillaries. Clin Chem 27 (9):1551–1553
106. Watson DG (2012) Pharmaceutical analysis: control of the quality of analytical methods; Chapter 2 Physical and chemical properties of drug molecules; Chapter 3 Titrimetic and chemical analysis methods; Chapter 4 Ultraviolet and visible spectroscopy; Chapter 5 Infrared spectrophotometry; Chapter 6 Atomic spectrophotometry; Chapter 7 Molecular emission spectroscopy; Chapter 8 Nuclear magnetic resonance spectroscopy; Chapter 9 Mass spectrometry; Chapter 10 Chromatographic theory; Chapter 11 Gas chromatography; Chapter 12 High pressure liquid chromatography; Chapter 13 Thin layer chromatography; Chapter 14 High performance capillary electrophoresis; Chapter 15 Extraction methods in pharmaceutical analysis. Elsevier Churchill Livingstone
107. Guttman A, Cooke N, Starr CM (1994) Capillary electrophoresis separation of oligosaccharides: I. effect of operational variables. Electrophoresis 15(1):1518–1522
108. Luo R, Archer-Hartmann SA, Holland LA (2010) Transformable capillary electrophoresis for oligosaccharide separations using phospholipid additives. Anal Chem 82(4):1228–1233
109. Guttman A, Chen F-TA, Evangelista RA, Cooke N (1996) High-resolution capillary gel electrophoresis of reducing oligosaccharides labeled with 1-aminopyrene-3, 6, 8-trisulfonate. Anal Biochem 233(2):234–242
110. Briggs JB, Keck RG, Ma S, Lau W, Jones AJ (2009) An analytical system for the characterization of highly heterogeneous mixtures of N-linked oligosaccharides. Anal Biochem 389 (1):40–51
111. Evangelista RA, Liu M-S, Chen F-TA (1995) Characterization of 9-aminopyrene-1, 4, 6-trisulfonate derivatized sugars by capillary electrophoresis with laser-induced fluorescence detection. Anal Chem 67(13):2239–2245
112. Ruhaak LR, Hennig R, Huhn C, Borowiak M, Dolhain RJ, Deelder AM et al (2010) Optimized workflow for preparation of APTS-labeled N-glycans allowing high-throughput analysis of human plasma glycomes using 48-channel multiplexed CGE-LIF. J Proteome Res 9 (12):6655–6664
113. Guttman A, Chen FTA, Evangelista RA (1996) Separation of 1-aminopyrene-3, 6, 8-trisulfonate-labeled asparagine-linked fetuin glycans by capillary gel electrophoresis. Electrophoresis 17(2):412–417
114. Guttman A, Herrick S (1996) Effect of the quantity and linkage position of mannose (α1, 2) residues in capillary gel electrophoresis of high-mannose-type oligosaccharides. Anal Biochem 235(2):236–239
115. Kottler R, Mank M, Hennig R, Muller-Werner B, Stahl B, Reichl U et al (2013) Development of a high-throughput glycoanalysis method for the characterization of oligosaccharides in human milk utilizing multiplexed capillary gel electrophoresis with laser-induced fluorescence detection. Electrophoresis 34(16):2323–2336
116. Hennig R, Cajic S, Borowiak M, Hoffmann M, Kottler R, Reichl U et al (2016) Towards personalized diagnostics via longitudinal study of the human plasma N-glycome. Biochim Biophys Acta 1860(8):1728–1738
117. Thiesler CT, Cajic S, Hoffmann D, Thiel C, van Diepen L, Hennig R et al (2016) Glycomic characterization of induced pluripotent stem cells derived from a patient suffering from phosphomannomutase 2 congenital disorder of glycosylation (PMM2-CDG). Mol Cell Proteomics 15(4):1435–1452
118. Weiz S, Wieczorek M, Schwedler C, Kaup M, Braicu EI, Sehouli J et al (2016) Acute-phase glycoprotein N-glycome of ovarian cancer patients analyzed by CE-LIF. Electrophoresis 37 (11):1461–1467
119. Schwedler C, Kaup M, Weiz S, Hoppe M, Braicu EI, Sehouli J et al (2014) Identification of 34 N-glycan isomers in human serum by capillary electrophoresis coupled with laser-induced fluorescence allows improving glycan biomarker discovery. Anal Bioanal Chem 406 (28):7185–7193
120. Konze SA, Cajic S, Oberbeck A, Hennig R, Pich A, Rapp E et al (2017) Quantitative assessment of sialo-glycoproteins and N-glycans during cardiomyogenic differentiation of human induced pluripotent stem cells. Chembiochem 18(13):1317–1331

121. Abeln M, Borst KM, Cajic S, Thiesler H, Kats E, Albers I et al (2017) Sialylation is dispensable for early murine embryonic development in vitro. Chembiochem 18(13):1305
122. Donczo B, Szarka M, Tovari J, Ostoros G, Csanky E, Guttman A (2017) Molecular glycopathology by capillary electrophoresis: analysis of the N-glycome of formalin-fixed paraffin-embedded mouse tissue samples. Electrophoresis 38(12):1602–1608
123. Meininger M, Stepath M, Hennig R, Cajic S, Rapp E, Rotering H et al (2016) Sialic acid-specific affinity chromatography for the separation of erythropoietin glycoforms using serotonin as a ligand. J Chromatogr B 1012:193–203
124. Schwedler C, Kaup M, Petzold D, Hoppe B, Braicu EI, Sehouli J et al (2014) Sialic acid methylation refines capillary electrophoresis laser-induced fluorescence analyses of immunoglobulin GN-glycans of ovarian cancer patients. Electrophoresis 35(7):1025–1031
125. Chen J, Fang M, Chen X, Yi C, Ji J, Cheng C et al (2017) N-glycosylation of serum proteins for the assessment of patients with IgD multiple myeloma. BMC Cancer 17(1):881
126. Huang C, Liu Y, Wu H, Sun D, Li Y (2017) Characterization of IgG glycosylation in rheumatoid arthritis patients by MALDI-TOF-MS n and capillary electrophoresis. Anal Bioanal Chem 409(15):3731–3739
127. Guttman M, Váradi C, Lee KK, Guttman A (2015) Comparative glycoprofiling of HIV gp120 immunogens by capillary electrophoresis and MALDI mass spectrometry. Electrophoresis 36 (11–12):1305–1313
128. Chung CH, Mirakhur B, Chan E, Le Q-T, Berlin J, Morse M et al (2008) Cetuximab-induced anaphylaxis and IgE specific for galactose-α-1, 3-galactose. N Engl J Med 358(11):1109–1117
129. Teranishi K, Manez R, Awwad M, Cooper DK (2002) Anti-Galα1-3Gal IgM and IgG antibody levels in sera of humans and old world non-human primates. Xenotransplantation 9(2):148–154
130. Brinkman-Van der Linden EC, Sjoberg ER, Juneja LR, Crocker PR, Varki N, Varki A (2000) Loss of N-glycolylneuraminic acid in human evolution implications for sialic acid recognition by siglecs. J Biol Chem 275(12):8633–8640
131. Anthony RM, Nimmerjahn F, Ashline DJ, Reinhold VN, Paulson JC, Ravetch JV (2008) Recapitulation of IVIG anti-inflammatory activity with a recombinant IgG Fc. Science 320 (5874):373–376
132. Hennig R, Borowiak M, Ruhaak L, Wuhrer M, Rapp E (2011) High-throughput CGE-LIF based analysis of APTS-labeled N-glycans, utilizing a multiplex capillary DNA sequencer
133. Mittermayr S, Bones J, Guttman AS (2013) Unraveling the glyco-puzzle: glycan structure identification by capillary electrophoresis. Anal Chem 85(9):4228–4238
134. Callewaert N, Geysens S, Molemans P, Contreras R (2001) Ultrasensitive profiling and sequencing of N-linked oligosaccharides using standard DNA-sequencing equipment. Glycobiology 11(4):275–281
135. Laroy W, Contreras R, Callewaert N (2006) Glycome mapping on DNA sequencing equipment. Nat Protoc 1(1):397
136. Schwarzer J, Rapp E, Reichl U (2008) N-glycan analysis by CGE-LIF: profiling influenza A virus hemagglutinin N-glycosylation during vaccine production. Electrophoresis 29 (20):4203–4214
137. Hennig R, Reichl U, Rapp E (2011) A software tool for automated high-throughput processing of CGE-LIF based glycoanalysis data, generated by a multiplexing capillary DNA sequencer. 5th Glycan Forum
138. glyxera. glyxera kits (2018) Updated 2018/05/16. http://www.glyxera.com/?pageid=PRODUCTS
139. Prozyme. GlykoPrep® Rapid N-Glycan Preparation with APTS (24-ct) [GP24NG-APTS] (2018) Updated 2018/5/16. https://prozyme.com/products/gp24ng-apts
140. Scientific TF (2018) GlycanAssure APTS kit. Updated 2018/5/16. http://www.thermofisher.com/order/catalog/product/A28676

141. SCIEX (2018) Fast glycan analysis and labeling for the PA 800 plus. Updated 2018/5/16. https://sciex.com/products/consumables-and-standards/fast-glycan-analysis-and-labeling-for-the-pa-800-plus
142. Szigeti M, Guttman A (2017) Automated N-glycosylation sequencing of biopharmaceuticals by capillary electrophoresis. Sci Rep 7(1):1–7
143. Bunz S-C, Rapp E, Neusüss C (2013) Capillary electrophoresis/mass spectrometry of APTS-labeled glycans for the identification of unknown glycan species in capillary electrophoresis/ laser-induced fluorescence systems. Anal Chem 85(21):10218–10224
144. Dotz V, Haselberg R, Shubhakar A, Kozak RP, Falck D, Rombouts Y et al (2015) Mass spectrometry for glycosylation analysis of biopharmaceuticals. TrAC Trends Anal Chem 73:1–9
145. Snyder CM, Zhou X, Karty JA, Fonslow BR, Novotny MV, Jacobson SC (2017) Capillary electrophoresis–mass spectrometry for direct structural identification of serum N-glycans. J Chromatogr A 1523:127–139
146. Khatri K, Klein JA, Haserick JR, Leon DR, Costello CE, McComb ME et al (2017) Microfluidic capillary electrophoresis–mass spectrometry for analysis of monosaccharides, oligosaccharides, and glycopeptides. Anal Chem 89(12):6645–6655
147. Lageveen-Kammeijer GS, de Haan N, Mohaupt P, Wagt S, Filius M, Nouta J et al (2019) Highly sensitive CE-ESI-MS analysis of N-glycans from complex biological samples. Nat Commun 10(1):1–8
148. Everest-Dass AV, Abrahams JL, Kolarich D, Packer NH, Campbell MP (2013) Structural feature ions for distinguishing N- and O-linked glycan isomers by LC-ESI-IT MS/MS. J Am Soc Mass Spectrom 24(6):895–906
149. Campbell MP, Nguyen-Khuong T, Hayes CA, Flowers SA, Alagesan K, Kolarich D et al (2014) Validation of the curation pipeline of UniCarb-DB: building a global glycan reference MS/MS repository. Biochim Biophys Acta 1844(1 Pt A):108–116
150. Stavenhagen K, Kolarich D, Wuhrer M (2015) Clinical glycomics employing graphitized carbon liquid chromatography-mass spectrometry. Chromatographia 78(5–6):307–320
151. Kolarich D, Windwarder M, Alagesan K, Altmann F (2015) Isomer-specific analysis of released N-glycans by LC-ESI MS/MS with porous graphitized carbon. Methods Mol Biol 1321:427–435
152. Pabst M, Altmann F (2008) Influence of electrosorption, solvent, temperature, and ion polarity on the performance of LC-ESI-MS using graphitic carbon for acidic oligosaccharides. Anal Chem 80(19):7534–7542
153. Pabst M, Grass J, Toegel S, Liebminger E, Strasser R, Altmann F (2012) Isomeric analysis of oligomannosidic N-glycans and their dolichol-linked precursors. Glycobiology 22(3):389–399
154. Miller RL, Guimond SE, Prescott M, Turnbull JE, Karlsson N (2017) Versatile separation and analysis of heparan sulfate oligosaccharides using graphitized carbon liquid chromatography and electrospray mass spectrometry. Anal Chem 89(17):8942–8950
155. Adamczyk B, Jin C, Polom K, Munoz P, Rojas-Macias MA, Zeeberg D et al (2018) Sample handling of gastric tissue and O-glycan alterations in paired gastric cancer and non-tumorigenic tissues. Sci Rep 8(1):242
156. Kolarich D, Packer NH (2013) Mass spectrometry for glycomics analysis of N-and O-linked glycoproteins
157. Jensen PH, Karlsson NG, Kolarich D, Packer NH (2012) Structural analysis of N- and O-glycans released from glycoproteins. Nat Protoc 7(7):1299–1310
158. Wongtrakul-Kish K, Kolarich D, Pascovici D, Joss JL, Deane E, Packer NH (2013) Characterization of N- and O-linked glycosylation changes in milk of the tammar wallaby (*Macropus eugenii*) over lactation. Glycoconj J 30(5):523–536
159. Stadlmann J, Pabst M, Kolarich D, Kunert R, Altmann F (2008) Analysis of immunoglobulin glycosylation by LC-ESI-MS of glycopeptides and oligosaccharides. Proteomics 8(14):2858–2871

160. De Leoz MLA, Duewer DL, Fung A, Liu L, Yau HK, Potter O et al (2020) NIST interlaboratory study on glycosylation analysis of monoclonal antibodies: comparison of results from diverse analytical methods. Mol Cell Proteomics 19(1):11–30
161. Hinneburg H, Korac P, Schirmeister F, Gasparov S, Seeberger PH, Zoldos V et al (2017) Unlocking cancer glycomes from histopathological formalin-fixed and paraffin-embedded (FFPE) tissue microdissections. Mol Cell Proteomics 16(4):524–536
162. Moginger U, Grunewald S, Hennig R, Kuo CW, Schirmeister F, Voth H et al (2018) Alterations of the human skin N- and O-glycome in basal cell carcinoma and squamous cell carcinoma. Front Oncol 8:70
163. Wilson NL, Schulz BL, Karlsson NG, Packer NH (2002) Sequential analysis of N- and O-linked glycosylation of 2D-PAGE separated glycoproteins. J Proteome Res 1(6):521–529
164. Deshpande N, Jensen PH, Packer NH, Kolarich D (2010) GlycoSpectrumScan: fishing glycopeptides from MS spectra of protease digests of human colostrum sIgA. J Proteome Res 9(2):1063–1075
165. Kolarich D, Weber A, Pabst M, Stadlmann J, Teschner W, Ehrlich H et al (2008) Glycoproteomic characterization of butyrylcholinesterase from human plasma. Proteomics 8 (2):254–263
166. Sumer-Bayraktar Z, Kolarich D, Campbell MP, Ali S, Packer NH, Thaysen-Andersen M (2011) N-glycans modulate the function of human corticosteroid-binding globulin. Mol Cell Proteomics 10(8):M111 009100
167. Sumer-Bayraktar Z, Nguyen-Khuong T, Jayo R, Chen DD, Ali S, Packer NH et al (2012) Micro- and macroheterogeneity of N-glycosylation yields size and charge isoforms of human sex hormone binding globulin circulating in serum. Proteomics 12(22):3315–3327
168. Carvalho S, Catarino TA, Dias AM, Kato M, Almeida A, Hessling B et al (2016) Preventing E-cadherin aberrant N-glycosylation at Asn-554 improves its critical function in gastric cancer. Oncogene 35(13):1619–1631
169. Rath CB, Schirmeister F, Figl R, Seeberger PH, Schaffer C, Kolarich D (2018) Flagellin glycoproteomics of the periodontitis associated pathogen Selenomonas sputigena reveals previously not described O-glycans and rhamnose fragment rearrangement occurring on the glycopeptides. Mol Cell Proteomics 17(4):721–736
170. Nguyen-Khuong T, Pralow A, Reichl U, Rapp E (2018) Improvement of electrospray stability in negative ion mode for nano-PGC-LC-MS glycoanalysis via post-column make-up flow. Glycoconj J 35(6):499–509
171. Hinneburg H, Chatterjee S, Schirmeister F, Nguyen-Khuong T, Packer NH, Rapp E et al (2019) Post-column make-up flow (PCMF) enhances the performance of capillary-flow PGC-LC-MS/MS-based glycomics. Anal Chem 91(7):4559–4567
172. Huang Y, Zhou S, Zhu J, Lubman DM, Mechref Y (2017) LC-MS/MS isomeric profiling of permethylated N-glycans derived from serum haptoglobin of hepatocellular carcinoma (HCC) and cirrhotic patients. Electrophoresis 38(17):2160–2167
173. Hua S, An HJ, Ozcan S, Ro GS, Soares S, DeVere-White R et al (2011) Comprehensive native glycan profiling with isomer separation and quantitation for the discovery of cancer biomarkers. Analyst 136(18):3663–3671
174. Everest-Dass AV, Moh ESX, Ashwood C, Shathili AMM, Packer NH (2018) Human disease glycomics: technology advances enabling protein glycosylation analysis – part 2. Expert Rev Proteomics 15(4):341–352
175. Everest-Dass AV, Moh ESX, Ashwood C, Shathili AMM, Packer NH (2018) Human disease glycomics: technology advances enabling protein glycosylation analysis – part 1. Expert Rev Proteomics 15(2):165–182
176. Pagel K, Harvey DJ (2013) Ion mobility-mass spectrometry of complex carbohydrates: collision cross sections of sodiated N-linked glycans. Anal Chem 85(10):5138–5145
177. Hofmann J, Hahm HS, Seeberger PH, Pagel K (2015) Identification of carbohydrate anomers using ion mobility-mass spectrometry. Nature 526(7572):241–244

178. Hinneburg H, Hofmann J, Struwe WB, Thader A, Altmann F, Varon Silva D et al (2016) Distinguishing N-acetylneuraminic acid linkage isomers on glycopeptides by ion mobility-mass spectrometry. Chem Commun 52(23):4381–4384
179. Manz C, Pagel K (2017) Glycan analysis by ion mobility-mass spectrometry and gas-phase spectroscopy. Curr Opin Chem Biol 42:16–24
180. Hofmann J, Pagel K (2017) Glycan analysis by ion mobility-mass spectrometry. Angew Chem Int Ed Engl 56(29):8342–8349
181. Hofmann J, Struwe WB, Scarff CA, Scrivens JH, Harvey DJ, Pagel K (2014) Estimating collision cross sections of negatively charged N-glycans using traveling wave ion mobility-mass spectrometry. Anal Chem 86(21):10789–10795
182. Gabryelski W, Froese KL (2003) Rapid and sensitive differentiation of anomers, linkage, and position isomers of disaccharides using high-field asymmetric waveform ion mobility spectrometry (FAIMS). J Am Soc Mass Spectrom 14(3):265–277
183. Harvey DJ, Scarff CA, Edgeworth M, Pagel K, Thalassinos K, Struwe WB et al (2016) Travelling-wave ion mobility mass spectrometry and negative ion fragmentation of hybrid and complex N-glycans. J Mass Spectrom 51(11):1064–1079
184. Harvey DJ, Scarff CA, Edgeworth M, Struwe WB, Pagel K, Thalassinos K et al (2016) Travelling-wave ion mobility and negative ion fragmentation of high-mannose N-glycans. J Mass Spectrom 51(3):219–235
185. Gray CJ, Thomas B, Upton R, Migas LG, Eyers CE, Barran PE et al (2016) Applications of ion mobility mass spectrometry for high throughput, high resolution glycan analysis. Biochim Biophys Acta 1860(8):1688–1709
186. Struwe WB, Pagel K, Benesch JL, Harvey DJ, Campbell MP (2016) GlycoMob: an ion mobility-mass spectrometry collision cross section database for glycomics. Glycoconj J 33 (3):399–404
187. Harvey DJ, Abrahams JL (2016) Fragmentation and ion mobility properties of negative ions from N-linked carbohydrates: part 7. Reduced glycans. Rapid Commun Mass Spectrom 30 (5):627–634
188. Pabst M, Bondili JS, Stadlmann J, Mach L, Altmann F (2007) Mass + retention time = structure: a strategy for the analysis of N-glycans by carbon LC-ESI-MS and its application to fibrin N-glycans. Anal Chem 79(13):5051–5057
189. Strohl WR (2018) Current progress in innovative engineered antibodies. Protein Cell 9(1):86–120
190. Jefferis R. Recombinant proteins and monoclonal antibodies. 2017
191. Jennewein MF, Alter G (2017) The immunoregulatory roles of antibody glycosylation. Trends Immunol 38(5):358–372
192. Liu CP, Tsai TI, Cheng T, Shivatare VS, Wu CY, Wu CY et al (2018) Glycoengineering of antibody (Herceptin) through yeast expression and in vitro enzymatic glycosylation. Proc Natl Acad Sci U S A 115(4):720–725
193. Wang Q, Stuczynski M, Gao Y, Betenbaugh MJ (2015) Strategies for engineering protein N-glycosylation pathways in mammalian cells. Glyco-Eng Methods Protocols 1321:287–305
194. Keys TG, Aebi M (2017) Engineering protein glycosylation in prokaryotes. Curr Opin Syst Biol 5:23–31
195. van de Bovenkamp FS, Hafkenscheid L, Rispens T, Rombouts Y (2016) The emerging importance of IgG fab glycosylation in immunity. J Immunol 196(4):1435–1441
196. Carillo S, Mittermayr S, Farrell A, Albrecht S, Bones J (1603) Glycosylation analysis of therapeutic glycoproteins produced in CHO cells. Methods Mol Biol 2017:227–241
197. Reusch D, Haberger M, Falck D, Peter B, Maier B, Gassner J et al (2015) Comparison of methods for the analysis of therapeutic immunoglobulin G Fc-glycosylation profiles-part 2: mass spectrometric methods. MAbs 7(4):732–742
198. Reusch D, Haberger M, Maier B, Maier M, Kloseck R, Zimmermann B et al (2015) Comparison of methods for the analysis of therapeutic immunoglobulin G Fc-glycosylation profiles-part 1: separation-based methods. MAbs 7(1):167–179

199. Biopharmaceutical follicle-stimulating hormone (FSH) characterisation – ludger-wcbp-2018-fsh-characterisation.pdf (2018) Updated 2018/01/30. https://www.ludger.com/docs/posters/ludger-wcbp-2018-fsh-characterisation.pdf
200. Cowper B, Li X, Yu L, Zhou Y, Fan W, Rao C (2018) Comprehensive glycan analysis of twelve recombinant human erythropoietin preparations from manufacturers in China and Japan. J Pharm Biomed Anal 153:214–220
201. Falck D, Haberger M, Plomp R, Hook M, Bulau P, Wuhrer M et al (2017) Affinity purification of erythropoietin from cell culture supernatant combined with MALDI-TOF-MS analysis of erythropoietin N-glycosylation. Sci Rep 7(1):5324
202. Grass J, Pabst M, Chang M, Wozny M, Altmann F (2011) Analysis of recombinant human follicle-stimulating hormone (FSH) by mass spectrometric approaches. Anal Bioanal Chem 400(8):2427–2438
203. Wada Y, Dell A, Haslam SM, Tissot B, Canis K, Azadi P et al (2010) Comparison of methods for profiling O-glycosylation: human proteome organisation human disease glycomics/proteome initiative multi-institutional study of IgA1. Mol Cell Proteom 9(4):719–727
204. Lohr V, Genzel Y, Behrendt I, Scharfenberg K, Reichl U (2010) A new MDCK suspension line cultivated in a fully defined medium in stirred-tank and wave bioreactor. Vaccine 28 (38):6256–6264
205. Milian E, Kamen AA (2015) Current and emerging cell culture manufacturing technologies for influenza vaccines. Biomed Res Int 2015:504831
206. Zhang X, Chen S, Jiang Y, Huang K, Huang J, Yang D et al (2015) Hemagglutinin glycosylation modulates the pathogenicity and antigenicity of the H5N1 avian influenza virus. Vet Microbiol 175(2–4):244–256
207. Suptawiwat O, Boonarkart C, Chakritbudsabong W, Uiprasertkul M, Puthavathana P, Wiriyarat W et al (2015) The N-linked glycosylation site at position 158 on the head of hemagglutinin and the virulence of H5N1 avian influenza virus in mice. Arch Virol 160 (2):409–415
208. Sun X, Jayaraman A, Maniprasad P, Raman R, Houser KV, Pappas C et al (2013) N-linked glycosylation of the hemagglutinin protein influences virulence and antigenicity of the 1918 pandemic and seasonal H1N1 influenza A viruses. J Virol 87(15):8756–8766
209. Hutter J, Rodig JV, Hoper D, Seeberger PH, Reichl U, Rapp E et al (2013) Toward animal cell culture-based influenza vaccine design: viral hemagglutinin N-glycosylation markedly impacts immunogenicity. J Immunol 190(1):220–230
210. Abdel-Motal UM, Guay HM, Wigglesworth K, Welsh RM, Galili U (2007) Immunogenicity of influenza virus vaccine is increased by anti-gal-mediated targeting to antigen-presenting cells. J Virol 81(17):9131–9141
211. Schwarzer J, Rapp E, Hennig R, Genzel Y, Jordan I, Sandig V et al (2009) Glycan analysis in cell culture-based influenza vaccine production: influence of host cell line and virus strain on the glycosylation pattern of viral hemagglutinin. Vaccine 27(32):4325–4336
212. Harvey DJ (2018) Mass spectrometric analysis of glycosylated viral proteins. Expert Rev Proteomics 15(5):391–412
213. Hargett AA, Renfrow MB (2019) Glycosylation of viral surface proteins probed by mass spectrometry. Curr Opin Virol 36:56–66
214. Bagdonaite I, Wandall HH (2018) Global aspects of viral glycosylation. Glycobiology 28 (7):443–467
215. Bagdonaite I, Vakhrushev SY, Joshi HJ, Wandall HH (2018) Viral glycoproteomes: technologies for characterization and outlook for vaccine design. FEBS Lett 592(23):3898–3920
216. Szekrenyes A, Park SS, Santos M, Lew C, Jones A, Haxo T et al (2016) Multi-site N-glycan mapping study 1: capillary electrophoresis – laser induced fluorescence. MAbs 8(1):56–64
217. Reiding KR, Bondt A, Hennig R, Gardner RA, O'Flaherty R, Trbojevic-Akmacic I et al (2019) High-throughput serum N-glycomics: method comparison and application to study rheumatoid arthritis and pregnancy-associated changes. Mol Cell Proteom 18(1):3–15

# Glycoproteomics Technologies in Glycobiotechnology

**Kathirvel Alagesan, Marcus Hoffmann, Erdmann Rapp, and Daniel Kolarich**

## Contents



K. Alagesan
Institute for Glycomics, Griffith University, Gold Coast, QLD, Australia

Max Planck Unit for the Science of Pathogens, Berlin, Germany
e-mail: k.alagesan@griffith.edu.au

M. Hoffmann
Max Planck Institute for Dynamics of Complex Technical Systems, Magdeburg, Germany
e-mail: mhoffmann@mpi-magdeburg.mpg.de

E. Rapp
Max Planck Institute for Dynamics of Complex Technical Systems, Magdeburg, Germany

glyXera GmbH, Magdeburg, Germany
e-mail: rapp@mpi-magdeburg.mpg.de; e.rapp@glyxera.com

D. Kolarich (✉)
Institute for Glycomics, Griffith University, Gold Coast, QLD, Australia

ARC Centre of Excellence in Nanoscale Biophotonics, Griffith University, Gold Coast, QLD, Australia
e-mail: d.kolarich@griffith.edu.au

**Abstract** Glycosylation is a key factor determining the pharmacological properties of biotherapeutics, including their stability, solubility, bioavailability, pharmacokinetics, and immunogenicity. As such, comprehensive information about glycosylation of biotherapeutics is critical to demonstrate similarity. Regulatory agencies also require extensive documentation of the comprehensive analyses of glycosylation-related critical quality attributes (CQAs) during the development, manufacturing, and release of biosimilars. Mass spectrometry has catalysed tremendous advancements in the characterisation of glycosylation CQAs of biotherapeutics. Here we provide a perspective overview on the MS-based technologies relevant for biotherapeutic product characterisation with an emphasis on the recent developments that allow determination of glycosylation features such as site of glycosylation, sialic acid linkage, glycan structure, and content.

**Graphical Abstract**

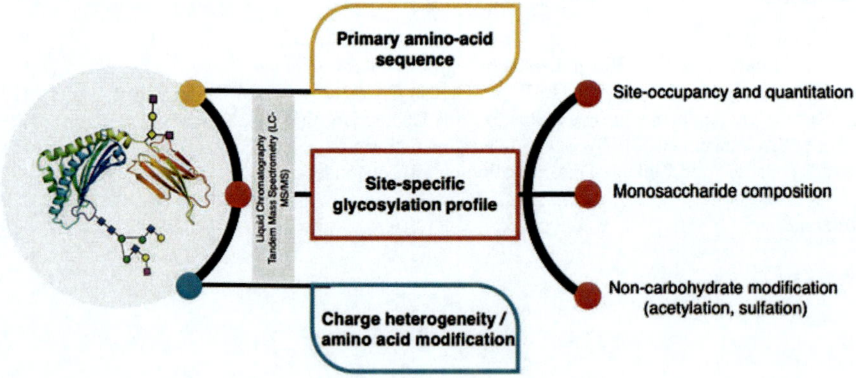


**Keywords** Analytics, Biopharmaceuticals, Glycoproteomics, Mass spectrometry

## Abbreviations

| | |
|---|---|
| ADCC | Antibody-dependent cell-mediated cytotoxicity |
| AI-ETD | Activated ion electron transfer dissociation |
| CE-MS | Capillary electrophoresis–mass spectrometry |
| CID | Collision-induced dissociation |
| CQA | Critical quality attribute |
| ECD | Electron-capture dissociation |
| EIC | Extracted-ion chromatogram |
| EIE | Extracted ion electropherogram |
| EMA | European Medicines Agency |
| EPO | Erythropoietin |
| ESI | Electrospray ionisation |
| ETD | Electron-transfer dissociation |

| | |
|---|---|
| FcγR | Fc-γ receptor |
| FDA | Food and Drug Administration |
| FSH | Follicle-stimulating hormone |
| FSHR | Follicle-stimulating hormone receptor |
| FT-ICR MS | Fourier transform–ion cyclotron resonance mass spectrometry |
| HCD | Higher-energy collision-induced dissociation |
| IgGs | Immunoglobulin G |
| IM-MS | Ion-mobility mass spectrometry |
| LC | Liquid chromatography |
| *m/z* | Mass-to-charge ratio |
| MALDI | Matrix-assisted laser desorption/ionisation |
| MAM | Multi-attribute monitoring |
| MS | Mass spectrometry |
| PNGase F | Peptide:*N*-glycosidase F |
| PSA | Prostate-specific antigen |
| PTM | Post-translational modification |
| QbD | Quality by design |
| RP | Reversed phase |
| TOF | Time of flight |

## 1 Introduction

In contrast to defined small generic molecules and reproducible structures, protein pharmaceuticals are large heterogeneous molecules prone to numerous enzymatic and chemical post-translational modifications (PTMs) during production, formulation, and storage [1, 2]. As such, an inherent amount of heterogeneity is always present in biotherapeutica that could affect their immunogenicity, structure, function, and pharmaceutical properties and requires an arsenal of orthogonal analytical techniques for reliable and accurate product characterisation during development and quality control. Modern high-resolution and sensitive mass spectrometry (MS)-based multi-attribute methods offer unprecedented opportunities to evaluate a combination of quality attributes to assess batch-to-batch stability, similarity, and safety of these products from early development onwards through to lot release [3, 4].

The FDA guidelines state that evaluation of the comparability of biosimilar products is a stepwise approach; thus, the comparability of the quality profiles of biosimilar and originator products should be considered the first priority. Normally, biosimilar evaluation will include (but not limited to) the following analyses in comparison to the originator reference [5]:

1. Primary structure, amino acid sequence
2. Higher-order structures, including secondary, tertiary, and quaternary structure (if appropriate)

3. Intended post-translational modifications such as glycosylation as described for the originator
4. Unintended post-translational modifications occurring during production and downstream processing, such as protein deamidation and oxidation

## *1.1 Glycosylation of Biosimilars and Regulatory Requirements*

Although glycosylation is mentioned only as one subset among various other criteria, it is probably one of the most challenging primary structure attributes for which biosimilarity of glycoprotein therapeutics needs to be demonstrated. In compliance with the US Food and Drug Administration (FDA), European Medicines Agency (EMA), and the International Conference on Harmonization guidelines, evaluation of glycosylation is a critical quality attribute (CQA) and constitutes a quality by design (QbD) parameter [6]. QbD is a scientific, risk-based, proactive, and systematic approach to drug development, which promotes understanding of the product and manufacturing process in such a way that quality is built-in (from product development and through manufacturing process), rather than just tested after development.

Owing to the fact that most biopharmaceuticals, which include monoclonal antibodies (mAbs) and other recombinant protein products such as fusion proteins, growth factors, cytokines, therapeutic enzymes, and hormones, are glycoproteins, it is imperative to elucidate their glycosylation pattern and confirm consistency across batches. The FDA and EMA mandate in-depth characterisation of glycosylation features of biopharmaceuticals and require information on carbohydrate composition, structure, and site of attachment including site occupancy as all these factors are relevant for product efficacy, safety, stability, and functionality [7–13].

The pharmaceutical or functional relevance of glycosylation is strongly protein-dependent. In the case of erythropoietin (EPO), the serum half-life depends on the presence of sialic acid on the *N*-glycans [14]; the site-specific glycosylation at asparagine (Asn) 52 of the α-subunit of follicle-stimulating hormone (FSH) plays a pivotal role in FSH receptor (FSHR) activation/signalling [15]; antibody-dependent cell-mediated cytotoxicity (ADCC) of immunoglobulins (IgGs) after binding of Fc-γ receptors (FcγR) is influenced by *N*-glycosylation in the IgG CH2 domain. Also, presence of glycan residues, such as α1–3 galactose, β1–2 xylose, α1–3 fucose, and *N*-glycolylneuraminic acid have negative impact on the safety and/or immunogenicity of biotherapeutics (Table 1). Therefore, next to confirming the "right" glycosylation profile of a biosimilar product, glycosylation analyses are also relevant to identify such immunogenic glycan species already from the early stages of originator and biosimilar drug development through to final lot release to ensure product safety. They also can result in the development of glycosylation-improved biosimilars, so-called biobetters [16, 17]. The methods of choice for glycoprotein characterisation rely heavily on modern mass spectrometers in

**Table 1** Overview on analytical techniques used for glycan characterisation of therapeutic proteins

| Level of analysis | Method | Obtained information | Site-specific information | Advantages | Disadvantages |
|---|---|---|---|---|---|
| Intact and subunit protein level | RPLC-MS | Glycan heterogeneity | Yes | Fast and robust technique for routine analysis at subunit level | Limited resolving power for glycoforms and introduction of artefacts by MS analysis |
| | CE/cIEF | Sialylation heterogeneity | Yes | High resolution due to the electro-driven separation | Strong protein adsorption to capillary wall |
| Glycopeptides | LC-ESI-MS | Glycoform determination | Yes | Information on site specific glycosylation, primary sequence information and other PTMs | Limited interaction with small polar glycopeptides (RP-LC) and introduction of artefacts by MS analysis (e.g. fucose migration) |
| | MALDI-MS | Glycoform determination | Yes | Highly automated procedure with limited hands-on time and linkage specific information on terminal sialic acid | Limited information on site-specific glycan composition information is available |
| | CE-MS | Glycoform determination | Yes | Complementary to RPLC to achieve complete sequence coverage | Introduction of experimental artefacts by MS analysis |

combination with different ionisation methods and fragmentation techniques as they often enable a multi-attribute monitoring (MAM) simultaneous monitoring of several CQAs within a single analytical approach.

## *1.2 Analytical Strategies for In-Depth Glycan Characterisation of Biosimilars*

Protein glycosylation is dynamic and influenced by the cell line type, cell cultivation conditions, as well as downstream purification and product formulation conditions. Unlike DNA, RNA, or proteins, glycan biosynthesis does not follow any template, but is accomplished by the concerted action of over 250 different enzymes such as glycosyltransferases and glycosidases that attach and assemble activated monosaccharides into *N*- and *O*-glycan oligosaccharides. From a biopharma perspective, the situation is further complicated as the very same protein construct can exhibit a very different glycosylation profile when expressed in different cell types or under different process conditions. In consequence, protein glycosylation cannot be predicted based on the gene sequence or from protein chemistry alone but needs to be determined analytically. These glycosylation analyses of therapeutic proteins can be performed either at intact glycoprotein, at glycopeptide, or at released glycan level. All these approaches have their merits and shortcomings. It depends on the type of glycoprotein, the depth of characterisation required, and the complexity of protein-specific glycosylation if one single approach is sufficient or several orthogonal characterisation techniques need to be employed. In a previous chapter, we have already discussed the opportunities and challenges associated with released *N*- and *O*-glycan analyses (Chap. 13). In this chapter we focus on recent advances in the characterisation of biosimilar glycosylation on intact glycopeptide level to meet the requirements of regulatory guidelines. This is in particular relevant for any glycoproteins exhibiting more than a single site of glycosylation.

The analysis of intact glycopeptides and glycoproteins has experienced an unprecedented boost with the development of the soft ionisation techniques such as electrospray ionisation (ESI) or matrix-assisted laser desorption/ionisation (MALDI). Although it is possible to directly infuse the analyte solution to an MS, biotherapeutic analyses often require separation strategies that facilitate the removal of buffer components (e.g. salts) or fractionate different species in a mixture to increase the overall dynamic range, which especially supports detection of low-abundance analytes. Analytical approaches based on hyphenated separation techniques such as liquid chromatography (LC) or capillary electrophoresis (CE) can easily be coupled with online mass spectrometric detection. These have been proven to be indispensable tools to monitor not only glycosylation but also an extensive array of product quality attributes such as oxidation or glycation starting from early drug development through to final lot release.

In combination with different ionisation methods and fragmentation techniques, LC-MS-based multi-attribute methods allow in-depth characterisation of peptide sequence, identification of the glycosylation sites, and glycan composition/structures within a single analysis. Over the past decades, reversed phase (RP) -LC-ESI-MS has been the most widely used technique for glycoprotein and glycopeptide characterisation [18–20]. In our overview, we roughly divide glycopeptide characterisation into three categories based on the MS techniques used: (1) bottom-up peptide-mapping strategies that analyse a biosimilar after an extensive proteolytic digestion; (2) middle-down strategies that characterise large fragments produced by disulphide reduction, chemical cleavage and/or limited digestion with enzymes such as Lys-C, and IdeS protease (FabRICATOR); (3) top-down strategies analyse the entire, intact protein by MS, often performed under native conditions [21] (Fig. 1 and Table 1).

Bottom-up (glyco-)proteomics relies on enzymatic proteolytic digestion before online liquid chromatography-coupled tandem mass spectrometry analysis. The (glyco-)peptides obtained after protease digestion are the primary unit of

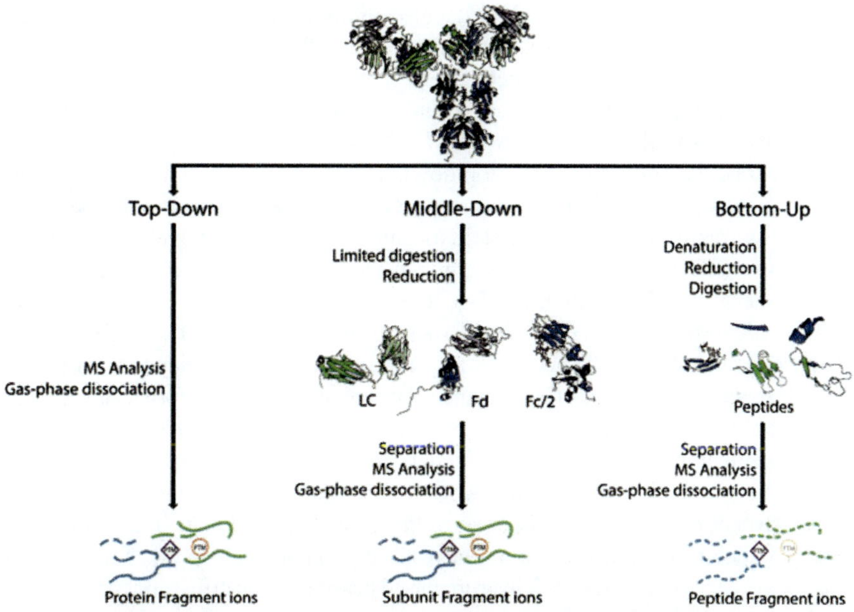


**Fig. 1** A schematic diagram comparing top-down (left) and middle-down (middle) MS workflows with bottom-up MS protocols (right) for mAb sequencing. For bottom-up MS approaches, (glyco-) proteins are digested into small (glyco-)peptides for LC separation and MS analysis, where (glyco-) peptides are selected and sequenced. Some labile PTMs may be lost during bottom-up workflows. In top-down MS, all proteoforms are directly sequenced in the gas-phase using advanced MS/MS strategies. For middle-down workflows, MS/MS analysis is performed on large fragments or mAb subunits after limited proteolysis in order to maximise both sequence coverage and PTM retention. (*LC* light chain, *Fd* heavy chain fragment generated from reduction of the antigen binding fragment, *Fc/2* heavy chain fragment obtained after reducing the Fc fragment) [22]

measurement in bottom-up (glyco-)proteomics, but their relatively small size (typically ~8–25 residues long) leads to problems such as sample complexity, difficulties in assigning their amino acid sequences to specific gene products rather than protein groups, and loss of single and combinatorial PTM information. Often, glycopeptides are less abundant than their peptide counterpart due to their microheterogeneity thus requiring highly selective enrichment techniques prior LC-MS analysis [23]. This enrichment approach separates the glycopeptides from the non-glycosylated peptides and increases their ionisation and detection properties [24]. Montacir and co-workers employed a multi-protease bottom-up workflow using Lys-C, trypsin, and Asp-N to demonstrate a very high *N*- and *O*-glycosylation similarity between etanercept originator and biosimilar [25]. Etanercept is a soluble fusion protein of the tumour necrosis factor receptor extracellular domain, linked to an Fc part of IgG1. Using their approach, they were able to map the two *N*-glycosylation sites present in the tumour necrosis factor receptor region carrying highly sialylated glycans, while the Fc-part of etanercept bears Fc-specific glycosylation features such as Man5, G0, G0F, and G1F. Recently, Pralow et al. redefined the cleavage specificity of flavastacin. They demonstrated for the first time the unique cleavage specificity that flavastacin exhibits towards the C-terminus of *N*-glycosylated asparagine residues. They developed a "*N*-glyco-specific" proteolytic strategy specific for *N*-glycosylated asparagine at the C-terminus by sequential digestion with trypsin and flavastacin to increase the confidence of MS-based site-specific analyses [26].

In contrast to bottom-up MS, top-down MS analyses intact (glyco-)proteins without proteolytic digestion. If necessary, (glyco-)protein sequence information is obtained by fragmentation of the protein ions in the gas phase using electron capture dissociation (ECD) in Fourier transform ion cyclotron resonance (FT-ICR) mass spectrometers or ETD in hybrid FT-ICR MS, Orbitrap FTMS, or time-of-flight (qTOF) MS instruments [27]. A distinct advantage of top-down over (glyco-)peptide-based approaches is the access to the entire (glyco-)protein sequence including amino acid sequence variants (i.e. mutants, alternatively spliced isoforms, amino acid polymorphisms) and PTMs [28]. However, top-down ESI-MS analysis of glycoproteins is challenging due to the less efficient desolvation resulting from the glycan heterogeneity and adduct formation. Also, the degree of glycosylation limits detection, as glycans cover large surface areas of the amino acid backbone of glycoproteins thereby reducing efficient ionisation. The resulting decreased charge state reduces the range of *m/z* analysed by ESI-MS instruments. These technical difficulties are to some extent overcome by the application of nano-electrospray as shown by Wilm and Mann in the analysis of ovalbumin glycoforms [29]. The coupling of nano-ESI with high-resolution mass analysers such as time of flight (TOF) analysers has also produced well-resolved glycoforms of bovine α1-acid glycoprotein [30] and cellulases purified from *Trichoderma reesei* [31]. Similarly, Nagy et al. showed the high resolution α1-acid glycoprotein glycoforms by ESI-FT-ICR MS [32]. Heck and co-workers demonstrated the glycosylation analysis of native human erythropoietin using high-resolution native MS. The work elegantly demonstrated the characterisation of site-specific glycans with minimal sample preparation and analysis time required to quantify glycan composition without ionisation bias [33].

Last but not least the middle-down approach is an emerging high-throughput strategy to define PTM co-existence frequency. It is a variant of the top-down approach yet involves analysis of truncated (glyco-)peptides (instead of "intact (glyco-)proteins") obtained by limited proteolysis or chemical degradation steps (which is characteristic of bottom-up approach). The main benefit of the middle-down analysis is the reduction in sample preparation time/steps, in comparison with the bottom-up approach, while still providing sufficient resolving power for glycan characterisation [34]. Use of specific enzymes such as papain or IgG-degrading enzyme of *Streptococcus pyogenes* (IdeS) [35] can facilitate the characterisation of large fragments of therapeutic mAbs. Middle-down analysis of Fab and F(ab′)2 fragments of mAb offer advantages over intact analysis by reducing nonspecific antigen binding to Fc region. On the other hand, the Fc region can be used for in-depth glycosylation analysis [36, 37]. Möginger et al. [38] established an integrated middle-down LC-MALDI-TOF-MS assay utilising cyanogen bromide for the in-depth characterisation of glycoconjugate vaccines. Production of glycoconjugate vaccines involves the chemical conjugation of glycans to an immunogenic carrier protein such as Cross-Reactive-Material-197 (CRM197). While the glycan is structurally defined, the attachment sites on the protein are not. Their approach of chemical cleavage of glycoconjugates followed by a middle-down LC-MALDI-ISD detection strategy provided several advantages towards any protease-based assays for comprehensive and in-depth semi-quantitative evaluation of region-specific conjugation efficiency providing virtually complete sequence coverage [38].

Capillary electrophoresis-mass spectrometry (CE-MS) provides the highest success for resolving the glycoforms of highly glycosylated proteins. Several studies have shown the application of CE-MS to almost completely resolving the various glycoforms of biologically relevant glycoproteins such as human plasma antithrombin [39] and recombinant erythropoietin [40, 41]. The high resolving power of CE in analysing glycoforms has immense potential for high throughput screening of recombinant glycoproteins although little is known about the structural features of the attached glycans by this type of analysis. However, the integration of CE-MS with other orthogonal methodologies can mitigate this issue. Takur et al. [42] demonstrated the characterisation of 60 glycoforms of recombinant human chorionic gonadotrophin using CE coupled to a high-resolution FT-ICR MS; subsequent analysis of the tryptic glycopeptides enabled site-specific glycan variant identification.

Native MS in combination with IM-MS can provide information on the shape/size of biosimilars [43]. Native IM-MS is an emerging method to characterise intact antibodies and can be used for routine batch-to-batch characterisation of therapeutics, mAbs glycosylation analysis, and higher order structures [44, 45]. Also, they can be used to gain structural insights on conformational changes induced upon antigen binding to the individual mAB [46–48] (Fig. 2).

# 2 Recent Developments in Intact Glycopeptide Characterisation

## *2.1 Advancements in Glycopeptide Fragmentation*

A variety of different glycopeptide fragmentation techniques are available to obtain structural data on glycopeptides (Fig. 3). Collision-induced dissociation (CID) preferentially yields glycan product ion fragments by cleaving the glycosidic bonds between carbohydrate units (B- and Y-ions). This information can be used to confirm the glycan composition present on a specific glycopeptide but rarely provides sufficient cleavage of the peptide backbone to deliver peptide b- and y-type product ions (cleavage of peptide bond) that would allow unambiguous peptide identification [49, 50]. This can be achieved using higher-energy CID (HCD) that mainly results in b/y-type peptide product ions next to glycan oxonium ions and fewer Y-type ions. This approach is widely applied to obtain peptide sequence data [51], even from complex samples. Under optimal collision energy settings, HCD fragmentation of glycopeptides results in distinct Y1 ions (peptide + GlcNAc in the case of *N*-glycans), which allows determination of the putative peptide mass of the *N*-glycopeptide.

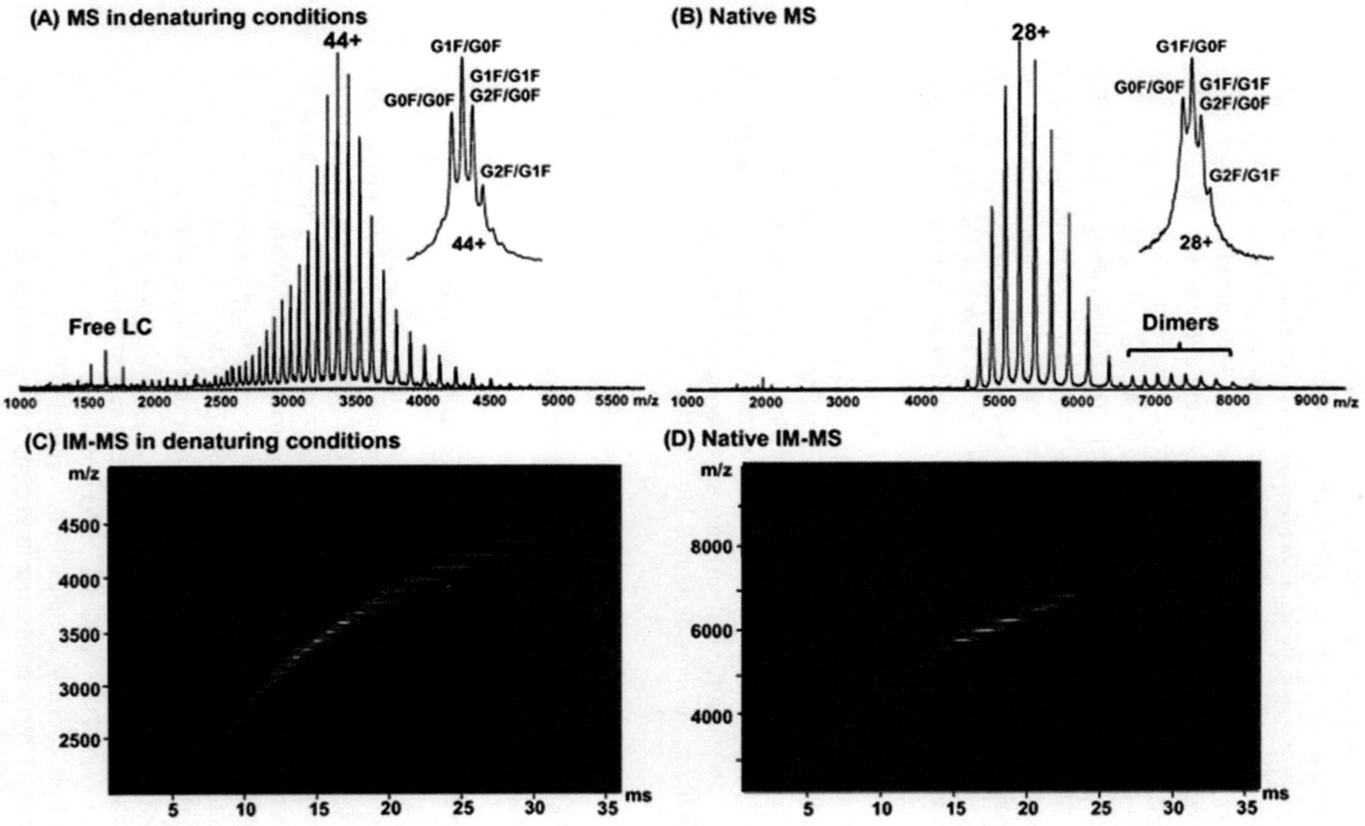


**Fig. 2** Native MS and IM-MS analysis of rituximab. ESI-MS mass spectra of rituximab obtained in classical denaturing (**a**) or in native (**b**) conditions. In denaturing conditions (**a**), two species are detected: the most intense one corresponds to the mass of intact rituximab and its main glycoforms; free non-covalent bound light chain is also detected as a minor specie. In native conditions (**b**), rituximab is mostly detected as a monomer and a minor ion series corresponding to rituximab dimers is detected. IM-MS analysis of rituximab in classical denaturing (**c**) or native (**d**) conditions. Native IM-MS analysis allows the detection of minor amounts of dimeric rituximab [46]

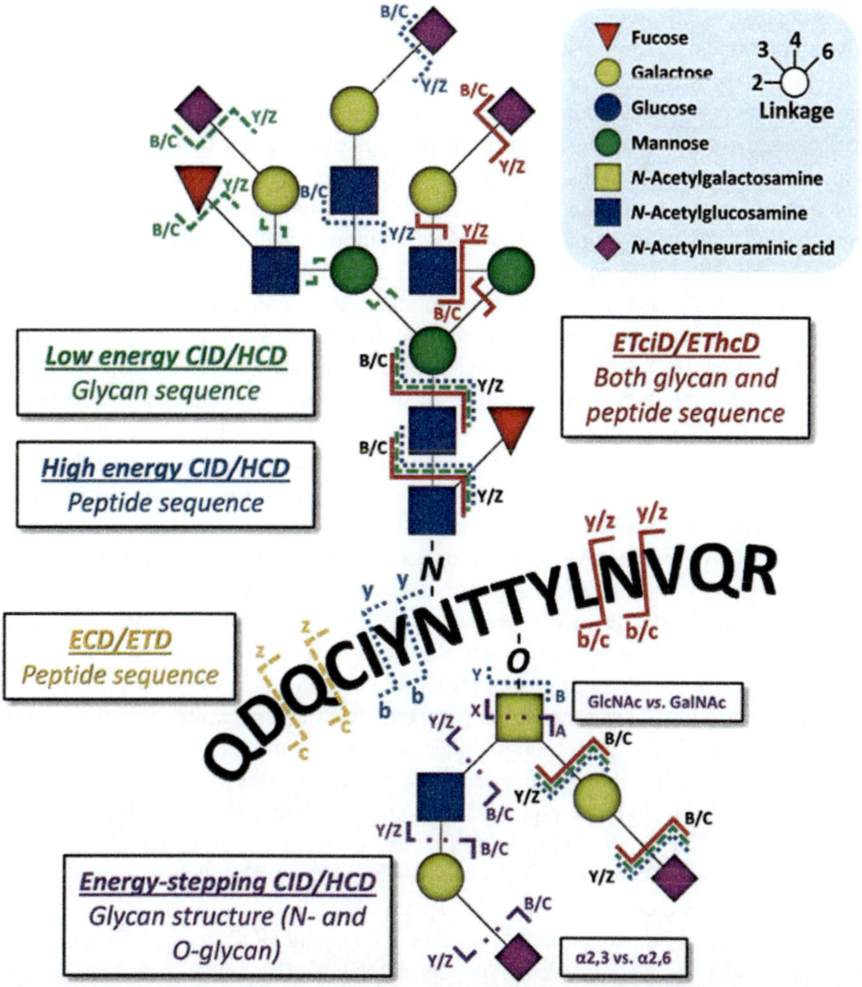


**Fig. 3** Overview of the most widely applied glycopeptide fragmentation methods and their preferred sites of action. The displayed peptide sequence matches that of a tryptic *N*-glycopeptide from alpha-1-acid glycoprotein, whereas the *O*-glycan has been added for illustrative purposes. Locations of fragmentation are exemplary, and dissociation of glycosidic linkages and peptide bonds can also occur elsewhere on the molecule. The actual observed fragments will depend highly on both the glycan and the peptide in question and the particular energy deposited in the precursor ions. Figure reproduced with permission from [71]

The standard high-energy conditions, however, impede acquisition of comprehensive glycan backbone data. Hinneburg et al. developed a workflow that allowed simultaneous acquisition of MS spectra at lower and higher collision energies (collision energy stepping CID) [52]. Fragment ions covering both the glycan and the peptide moieties of glycopeptides can be generated by this approach within a single experiment [53–57]. The use of synthetic *N*-glycopeptides allowed them to optimise analysis conditions that subsequently also improved the software-assisted data analysis [52], which to date remains one of the biggest challenges in glycoproteomics. A similar work based on HCD fragmentation regimes, *HCD.low* and *HCD.step*, was developed by Hoffmann et al. to enable unambiguous identification of the peptide backbone as well as the detailed analysis of the glycan moiety [58]. Based on the unique oxonium ion patterns observed in the product ion MS spectra, they were able to discriminate not only hybrid- and oligo-mannose-type *N*-glycans but also glycan epitope features such as antenna vs bisecting GlcNAc, antenna and core fucosylation. Nevertheless, caution needs to be taken when making these assignments on glycopeptide level as in particular deoxy-sugars such as fucose are prone to gas phase migration when fragmenting protonated ions that can give rise to misleading fragments not representative of the actual glycopeptide [59, 60]. Recently, Sanda et al. described a similar analytical workflow utilising optimised collision energy for quantitative structure and site-specific glycopeptide analysis. Like Hinneburg et al. reported previously [52], they, too, found that glycopeptide spectra acquired at low collision energy settings allowed resolving glycan structure motifs of *N*-glycopeptides, while high collision energy fragmentation was crucial for peptide sequence determination [61].

Alternatively, ion-based fragmentation techniques such as electron transfer dissociation (ETD) and electron capture dissociation (ECD) are non-vibrational approaches to dissociate precursor ions. The glycan portion stays intact during these fragmentation conditions, which almost exclusively fragment peptide bonds at the N-Cα bond to deliver c and z-ions do provide amino acid sequence information complementary to CID fragmentation [62, 63]. Nevertheless, while these work very well on unmodified peptides or peptides with small modifications, we recently demonstrated that for glycopeptides the number and quality of assignable peptide backbone fragments in ETD fragmentation significantly depends on glycan size and the position of the modification within a peptide sequence [64]. Highly charged glycopeptides ($z > 3$) with precursor masses of $m/z < 900$ delivered significantly better-quality product ion spectra ETD spectra [64]. In recent times, the use of a hybrid fragmentation technique that combines ETD and HCD, termed EThcD [65], has found wide applicability for glycopeptide characterisation. EThcD generates a dual product ion series that facilitates extensive peptide backbone fragmentation as well as glycan fragmentation. Recently Čaval and co-workers demonstrated that extending the mass scan range from 2000 *m/z* (standard range) to 4,000 *m/z* significantly increased the confidence in EThcD-based *N*-glycopeptide identification [66]. Notably, a different hybrid fragmentation approach called activated ion electron transfer dissociation (AI-ETD), combines infrared photoactivation with ETD and has shown better performance for proteoform fragmentation (top-down) than HCD and standard ETD [67, 68]. Riley and co-workers demonstrated that superior peptide backbone information is obtained in AI-ETD providing 100% sequence coverage for the *N*-glycopeptide TN*SSFIQGFVDHVKEDCDR when carrying an oligomannose type *N*-glycan [69]. Recently, Riley and co-workers systematically explored the advantages and disadvantages of conventional HCD, stepped HCD, ETD, and EThcD for intact glycopeptide analyses and determined their suitability for both *N*- and *O*-glycoproteomics. HCD and stepped HCD generated similar numbers of identifications for *N*-glycopeptides, although stepped HCD generally provided higher-quality spectra. ETD-based methods, especially EThcD, were found to be indispensable for site-specific analyses of *O*-glycopeptides [70].

## 2.2 *Sialic Acid Linkage Differentiation on Glycopeptides*

### 2.2.1 Ion-Mobility MS Enables Differentiation of Isobaric Compounds

Ion-mobility is a technique where ions are separated in the gas phased based on their charge and gas-phase configuration [72]. As such, it can be perfectly incorporated into mass spectrometers providing novel analytical opportunities that are in particular attractive for the analysis of glycoconjugates. Hinneburg et al. employed synthetic *N*-glycopeptides designed based on the human protein C glycopeptide $^{284}$EVFVHPNYSK$^{293}$ (UniProt entry P04070) that just differed in the linkage of the sialic acid residue (either an α2–6 or α2–3) but were otherwise entirely isobaric. While these compounds did not show any significant IM separation when analysed as intact glycopeptides, this could easily be achieved on the level of oxonium ions that were subjected to IM-MS after an initial CID-fragmentation of the glycopeptide precursor. The *m/z* 657 $B_3$ type oxonium ion fragment (NeuAc-Gal-GlcNAc trisaccharide) showed baseline separated different drift times depending on the NeuAc linkage. The α2–6 fragment exhibiting considerably shorter drift times compared to the α2–3 equivalent (Fig. 4) [73], which allowed an easy differentiation of sialic acid linkage directly on the glycopeptide without any additional sample preparation. In addition, the collision cross sections measured in nitrogen drift gas ($^{TW}CCS_{N2}$) differed significantly for these two trisaccharides: 236 Å$^2$ for the α2–6 linked NeuAc and 246 Å$^2$ for the α2–3 linked NeuAc containing fragments [74]. These values were highly diagnostic for the regiochemistry of the underlying NeuAc linkage and could thus be used to gain site-specific information on important glycan structural features directly from individual glycopeptides in a single experiment. When coupling the system with an LC, this approach enables now site-specific sialic acid linkage determination of glycopeptides in an -omics context as these collision cross sections of oxonium ions are independent of the glycopeptide precursor [75, 76].

Recently Barroso and co-workers evaluated the capacity of traveling wave IM-MS to separate isomeric glycoconjugates on three different levels [77]: as an intact glycoprotein, after digestion into glycopeptides and just the released glycans. They put particular focus on the ability to differentiate different types of sialic acid linkage (i.e. α2–3 and α2–6). In agreement with previous studies, isomer separation was achieved for glycans (without fragmentation) and for glycopeptides (after fragmentation as described by Hinneburg et al. and Guttmann et al.) [75, 76]. Under native MS conditions, no baseline isomer separation of intact glycoproteins was observed. However, the drift time of various glyco-isoforms increased with an increase in the carbohydrate fraction, i.e. complexity and branching of the glycoforms [77]. These advancements in glycopeptide analyses by IM-MS are particularly important for the biotherapeutics/biosimilars produced in CHO cells whose glycosylation machinery is similar to that of humans, except that

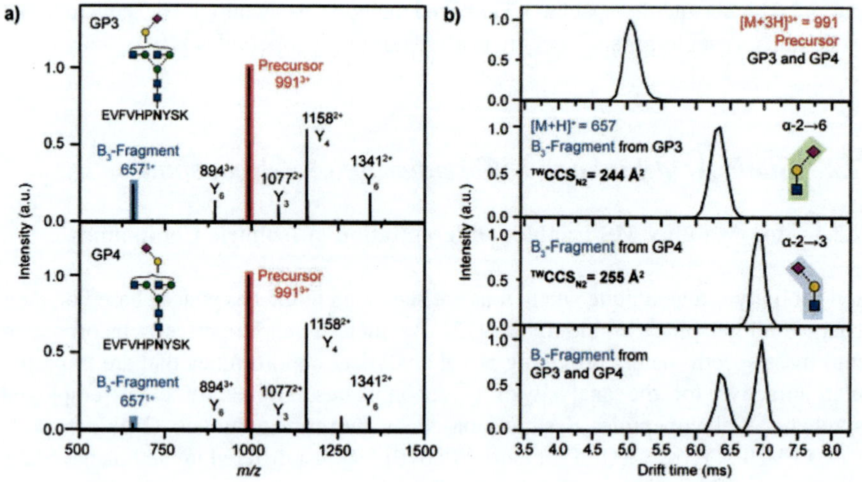


**Fig. 4** Differentiation of *N*-acetylneuraminic acid (NeuAc) linkage isomers using CID fragmentation and subsequent IM-MS analysis. Two isomeric glycopeptides, which either carry α2-6 (GP3) or α2-3 (GP4) linked NeuAc were analysed. (**a**) Both peptides exhibited identical MS/MS spectra, as shown for the triply protonated precursor ion (red). (**b**) When analysed as mixture the intact glycopeptide ions could not be separated by IM-MS (*m/z* 991, red). B3-trisaccharide fragments (*m/z* 657, blue) directly cleaved from the glycopeptide by CID, however, showed characteristic drift times that were dependent on the regiochemistry of the NeuAc linkage. This feature allowed unambiguous identification α2-3 and α2-6 sialic acid linkage directly from the glycopeptide, thus enabling site-specific sialic acid linkage information. Reproduced from Hinneburg et al. [75] with permission of the publisher

they lack a functional ST6, limiting their sialylation capability to the addition of α2–3 linked sialic acid residues [78–80].

### 2.2.2 Sialic Acid Linkage Determination by Multi-stage MS

While tandem MS provides the structural depth sufficient for peptide-sequence determination, linkage specific information on glycan structure features such as sialic acid linkages are more difficult to extract reliably from glycopeptide product ion spectra. Promising progress in this direction has been reported recently by Pett et al. who demonstrated that the oxonium-ion ratios between the signals at *m/z* 204, 274, 292, and 366 at defined can be indicative for the sialic acid linkage when glycopeptides are fragmented [81]. Glycopeptides carrying NeuAc residues in α2–3 linkage exhibited *m/z* 274 and 292 signals with a stronger intensity compared to the *m/z* 204 and 366 oxonium ions, whereas the NeuAc related oxonium ions were much lower when linked α2–6. While this approach might not be ideally suited to quantify α2–3/6 linkage ratios on a specific site as the ion-mobility approach, it offers an easy opportunity to get a picture of the prevalent NeuAc-linkages present on a specific glycopeptide without any additional derivatisation or specific instrumentation.

Recently, Zhu and co-workers developed a different strategy known as GLAMS using glycosyltransferase for unambiguous identification of sialoglycopeptide isomers [82]. Upon enzymatic labelling of glycopeptides using the CgtA enzyme, in HCD C-trap dissociation fragmentation α2–3 sialoglycopeptides generated unique reporter ions with specific *m/z* values that allow differentiation of α2–3 from α2–6 sialoglycopeptide isomers.

### 2.2.3 MALDI-MS

Unlike ESI, sialylated glycans usually undergo decomposition in MALDI due to the presence of a labile carboxylic proton to give rise to focused (in-source fragmentation) and unfocused (post-source fragmentation) ion peaks when measured in reflector-TOF detectors [83]. Thus, several strategies have been developed to stabilise and neutralise sialic acid residues and make them more suitable for MALDI-TOF-MS analyses. This can be achieved by permethylation [84], methyl esterification, matrix [85], or derivatisation with acetohydrazide [86]. In 2009, Harvey and co-workers demonstrated a method for stabilising sialic acids and discriminating α2–3 and α2–6 isomers. Here the glycans were treated with 4-(4,6-dimethoxy-1,3,5-triazin-2-yl)-4-methylmorpholinium chloride in methanol converting α2–6 linked sialic acids to methyl esters (+14 Da) and the α2–3 linked sialic acids formed lactones (−18 Da) [87]. In 2014, Reiding et al. developed a simplified procedure for the derivatisation and discrimination of sialic acids using 1-ethyl-3-(3-(dimethylamino)propyl)carbodiimide and 1-hydroxybenzotriazole as activators in ethanolic solution. In contrast to the previous approach described by Harvey, this reaction converts α2–6 linked sialic acid to dimethylamide (+28 Da) and α2–3 linked sialic acid to a cyclic lactone with the adjacent galactose (−18 Da). This particular approach has successfully been applied to differentiate α2–3 and α2–6 linked sialic acids at both glycan [88] and glycopeptide level [89]. This approach, however, is limited to stabilise and differentiate sialic acids and their linkages while other structure isomers will not be resolved.

### 2.2.4 Capillary Electrophoresis: Electrospray Ionisation–Mass Spectrometry (CE-ESI-MS)

Gahoul and co-workers developed a CE-MS/MS method based on the sheathless CE-ESI-MS (CESI) platform allowing a fast and precise characterisation of a monoclonal antibody digest [90, 91]. This allowed them to achieve 100% sequence coverage for both heavy and light chain in a single analytical experiment including the glycopeptides from 100 fmol of protein digest. In 2017, the Wuhrer group developed a high-resolution separation platform based on capillary electrophoresis–mass spectrometry (CE–MS) for selective differentiation of α2–3 and α2–6-sialylated glycopeptides without any sample pre-treatment [92] (Fig. 5). Unlike other conventional MS techniques, CE enables the baseline separation of sialylated

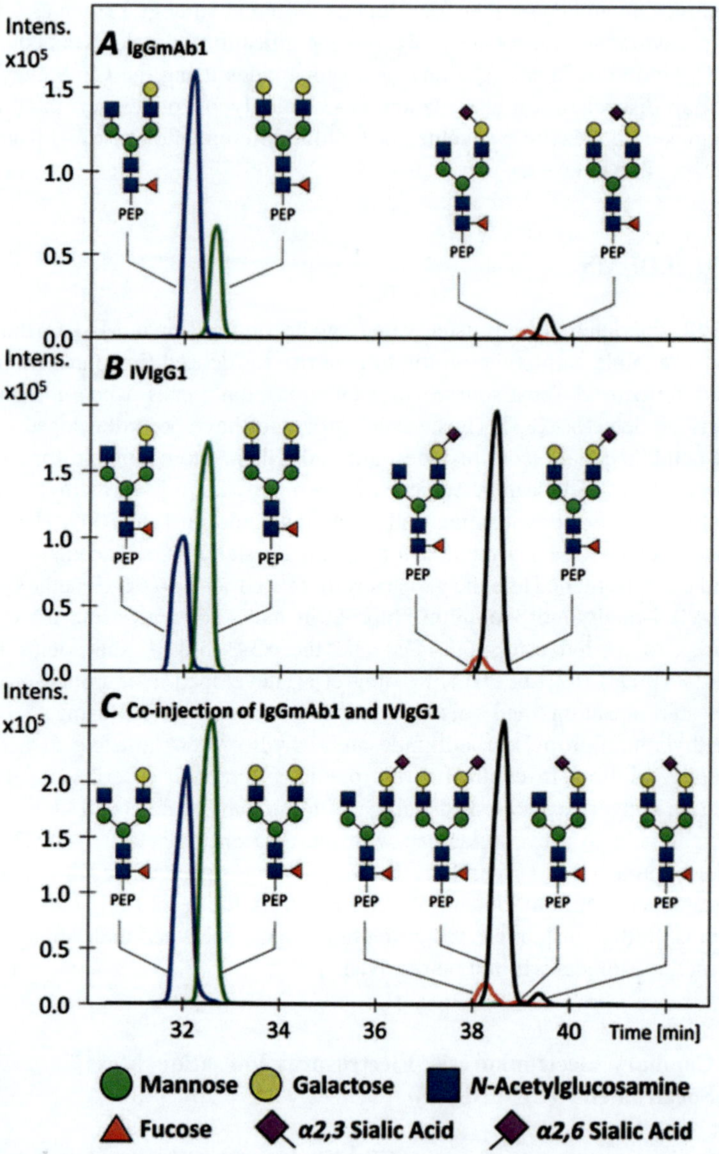


**Fig. 5** Extracted ion electropherograms (EIEs) of IgGmAb1 and IVIgG1 glycopeptides obtained with CE–ESI-MS after targeted alignment. (**a**) EIEs of IgGmAb1 glycopeptides derived from CHO cells, (**b**) EIEs of IVIgG1 retrieved from human plasma, and (**c**) EIEs of a co-injection of IgGmAb1 and IVIgG1. The "PEP" label illustrates the tryptic peptide sequence EEQYNSTYR to which the glycan is attached. Figure taken from Kammeijer et al. [92]with permission from publisher

glycopeptides due to a difference in their electrophoretic mobilities. The method they developed was used for the analysis of tryptic prostate-specific antigen glycopeptides, enabling the identification of 75 PSA glycopeptides, a significantly higher amount compared to the 37 PNGase F released *N*-glycan signals detected by MALDI-TOF-MS.

## 3 Conclusion

The complexity and heterogeneity of glycans present considerable challenges to the biopharmaceutical industry to manufacture biotherapeutics with a reproducible and consistent glycosylation profile. This is highly relevant as increasing evidence confirms a fundamental role for glycosylation of therapeutics that impacts biological activity, physicochemical properties, effector functions, and in vivo bioactivity. With the rapidly increasing use of glycoprotein therapeutics in clinical use, employment of appropriate glycoanalytical tools from early development stage throughout final regulatory approval, and product validation is crucial for the efficient development of both safe and functionally active originator products and biosimilars. Reliable glycoanalytical tools also provide an important basis that leads towards a better understanding of the structure and function of glycosylation in vitro and in vivo, which facilitates the development of the next generation of biotherapeutics with optimized glycoforms and improved therapeutic capabilities.

## References


1. Grassi L, Cabrele C (2019) Susceptibility of protein therapeutics to spontaneous chemical modifications by oxidation, cyclization, and elimination reactions. Amino Acids 51:1409–1431. https://doi.org/10.1007/s00726-019-02787-2
2. Zhong X, Wright JF (2013) Biological insights into therapeutic protein modifications throughout trafficking and their biopharmaceutical applications. Int J Cell Biol 2013:273086. https://doi.org/10.1155/2013/273086
3. Srebalus Barnes CA, Lim A (2007) Applications of mass spectrometry for the structural characterization of recombinant protein pharmaceuticals. Mass Spectrom Rev 26:370–388. https://doi.org/10.1002/mas.20129
4. Rogers RS et al (2017) A view on the importance of “multi-attribute method” for measuring purity of biopharmaceuticals and improving overall control strategy. AAPS J 20:7. https://doi.org/10.1208/s12248-017-0168-3
5. Bui LA et al (2015) Key considerations in the preclinical development of biosimilars. Drug Discov Today 20(Suppl 1):3–15. https://doi.org/10.1016/j.drudis.2015.03.011
6. Mishra V, Thakur S, Patil A, Shukla A (2018) Quality by design (QbD) approaches in current pharmaceutical set-up. Expert Opin Drug Deliv 15:737–758. https://doi.org/10.1080/17425247.2018.1504768
7. Radaev S, Sun P (2002) Recognition of immunoglobulins by Fcgamma receptors. Mol Immunol 38:1073–1083

8. Houde D, Peng Y, Berkowitz SA, Engen JR (2010) Post-translational modifications differentially affect IgG1 conformation and receptor binding. Mol Cell Proteomics 9:1716–1728. https://doi.org/10.1074/mcp.M900540-MCP200
9. Zou G et al (2011) Chemoenzymatic synthesis and Fcgamma receptor binding of homogeneous glycoforms of antibody Fc domain. Presence of a bisecting sugar moiety enhances the affinity of Fc to FcgammaIIIa receptor. J Am Chem Soc 133:18975–18991. https://doi.org/10.1021/ja208390n
10. Mimura Y et al (2001) Role of oligosaccharide residues of IgG1-Fc in Fc gamma RIIb binding. J Biol Chem 276:45539–45547. https://doi.org/10.1074/jbc.M107478200
11. Satoh M, Iida S, Shitara K (2006) Non-fucosylated therapeutic antibodies as next-generation therapeutic antibodies. Expert Opin Biol Ther 6:1161–1173. https://doi.org/10.1517/14712598.6.11.1161
12. Nechansky A, Koller I, Kircheis R (2010) Response to: 'impact of glycosylation on effector functions of therapeutic IgG' (Pharmaceuticals 2010, 3, 146-157). Pharmaceuticals 3:1887–1891. https://doi.org/10.3390/ph3061887
13. Wright A, Morrison SL (1997) Effect of glycosylation on antibody function: implications for genetic engineering. Trends Biotechnol 15:26–32. https://doi.org/10.1016/S0167-7799(96)10062-7
14. Misaizu T et al (1995) Role of antennary structure of N-linked sugar chains in renal handling of recombinant human erythropoietin. Blood 86:4097–4104
15. Mastrangeli R et al (2017) In-vivo biological activity and glycosylation analysis of a biosimilar recombinant human follicle-stimulating hormone product (Bemfola) compared with its reference medicinal product (GONAL-f). PLoS One 12:e0184139. https://doi.org/10.1371/journal.pone.0184139
16. Beck A (2011) Biosimilar, biobetter and next generation therapeutic antibodies. MAbs 3:107–110. https://doi.org/10.4161/mabs.3.2.14785
17. Zhang P et al (2016) Challenges of glycosylation analysis and control: an integrated approach to producing optimal and consistent therapeutic drugs. Drug Discov Today 21:740–765. https://doi.org/10.1016/j.drudis.2016.01.006
18. Kolarich D, Weber A, Turecek PL, Schwarz HP, Altmann F (2006) Comprehensive glycoproteomic analysis of human alpha1-antitrypsin and its charge isoforms. Proteomics 6:3369–3380. https://doi.org/10.1002/pmic.200500751
19. Herndl A et al (2007) Mapping of *Malus domestica* allergens by 2-D electrophoresis and IgE-reactivity. Electrophoresis 28:437–448. https://doi.org/10.1002/elps.200600342
20. Stadlmann J, Pabst M, Kolarich D, Kunert R, Altmann F (2008) Analysis of immunoglobulin glycosylation by LC-ESI-MS of glycopeptides and oligosaccharides. Proteomics 8:2858–2871. https://doi.org/10.1002/pmic.200700968
21. Reiding KR, Bondt A, Franc V, Heck AJR (2018) The benefits of hybrid fragmentation methods for glycoproteomics. TrAC Trends Anal Chem 108:260–268. https://doi.org/10.1016/j.trac.2018.09.007
22. Tian Y, Ruotolo BT (2018) The growing role of structural mass spectrometry in the discovery and development of therapeutic antibodies. Analyst 143:2459–2468. https://doi.org/10.1039/c8an00295a
23. Alagesan K, Khilji SK, Kolarich D (2017) It is all about the solvent: on the importance of the mobile phase for ZIC-HILIC glycopeptide enrichment. Anal Bioanal Chem 409:529–538. https://doi.org/10.1007/s00216-016-0051-6
24. Stavenhagen K et al (2013) Quantitative mapping of glycoprotein micro-heterogeneity and macro-heterogeneity: an evaluation of mass spectrometry signal strengths using synthetic peptides and glycopeptides. J Mass Spectrom 48:627–639. https://doi.org/10.1002/jms.3210
25. Montacir O et al (2018) Physicochemical characterization, glycosylation pattern and biosimilarity assessment of the fusion protein etanercept. Protein J 37:164–179. https://doi.org/10.1007/s10930-018-9757-y

26. Pralow A, Hoffmann M, Nguyen-Khuong T, Rapp E, Reichl U (2017) Improvement of the glycoproteomic toolbox with the discovery of a unique C-terminal cleavage specificity of flavastacin for N-glycosylated asparagine. Sci Rep 7:11419. https://doi.org/10.1038/s41598-017-11668-1
27. Zhang H, Ge Y (2011) Comprehensive analysis of protein modifications by top-down mass spectrometry. Circ Cardiovasc Genet 4:711. https://doi.org/10.1161/CIRCGENETICS.110.957829
28. Timp W, Timp G (2020) Beyond mass spectrometry, the next step in proteomics. Sci Adv 6: eaax8978. https://doi.org/10.1126/sciadv.aax8978
29. Wilm M, Mann M (1996) Analytical properties of the nanoelectrospray ion source. Anal Chem 68:1–8. https://doi.org/10.1021/ac9509519
30. Karas M, Bahr U, Dulcks T (2000) Nano-electrospray ionization mass spectrometry: addressing analytical problems beyond routine. Fresenius J Anal Chem 366:669–676
31. Hui JP, White TC, Thibault P (2002) Identification of glycan structure and glycosylation sites in cellobiohydrolase II and endoglucanases I and II from Trichoderma reesei. Glycobiology 12:837–849
32. Nagy K et al (2004) Electrospray ionization fourier transform ion cyclotron resonance mass spectrometry of human alpha-1-acid glycoprotein. Anal Chem 76:4998–5005. https://doi.org/10.1021/ac040019a
33. Yang Y et al (2016) Hybrid mass spectrometry approaches in glycoprotein analysis and their usage in scoring biosimilarity. Nat Commun 7:13397. https://doi.org/10.1038/ncomms13397
34. Duivelshof BL et al (2019) Glycosylation of biosimilars: recent advances in analytical characterization and clinical implications. Anal Chim Acta 1089:1–18. https://doi.org/10.1016/j.aca.2019.08.044
35. Fornelli L, Ayoub D, Aizikov K, Beck A, Tsybin YO (2014) Middle-down analysis of monoclonal antibodies with electron transfer dissociation orbitrap fourier transform mass spectrometry. Anal Chem 86:3005–3012. https://doi.org/10.1021/ac4036857
36. Tran BQ et al (2016) Comprehensive glycosylation profiling of IgG and IgG-fusion proteins by top-down MS with multiple fragmentation techniques. J Proteome 134:93–101. https://doi.org/10.1016/j.jprot.2015.10.021
37. He L et al (2017) Analysis of monoclonal antibodies in human serum as a model for clinical monoclonal gammopathy by use of 21 tesla FT-ICR top-down and middle-down MS/MS. J Am Soc Mass Spectrom 28:827–838. https://doi.org/10.1007/s13361-017-1602-6
38. Moginger U et al (2016) Cross reactive material 197 glycoconjugate vaccines contain privileged conjugation sites. Sci Rep 6:20488. https://doi.org/10.1038/srep20488
39. Demelbauer UM et al (2004) Characterization of glyco isoforms in plasma-derived human antithrombin by on-line capillary zone electrophoresis-electrospray ionization-quadrupole ion trap-mass spectrometry of the intact glycoproteins. Electrophoresis 25:2026–2032. https://doi.org/10.1002/elps.200305936
40. Balaguer E et al (2006) Glycoform characterization of erythropoietin combining glycan and intact protein analysis by capillary electrophoresis – electrospray – time-of-flight mass spectrometry. Electrophoresis 27:2638–2650. https://doi.org/10.1002/elps.200600075
41. Neususs C, Demelbauer U, Pelzing M (2005) Glycoform characterization of intact erythropoietin by capillary electrophoresis-electrospray-time of flight-mass spectrometry. Electrophoresis 26:1442–1450. https://doi.org/10.1002/elps.200410269
42. Thakur D et al (2009) Profiling the glycoforms of the intact alpha subunit of recombinant human chorionic gonadotropin by high-resolution capillary electrophoresis-mass spectrometry. Anal Chem 81:8900–8907. https://doi.org/10.1021/ac901506p
43. Bagal D, Valliere-Douglass JF, Balland A, Schnier PD (2010) Resolving disulfide structural isoforms of IgG2 monoclonal antibodies by ion mobility mass spectrometry. Anal Chem 82:6751–6755. https://doi.org/10.1021/ac1013139
44. Olivova P, Chen W, Chakraborty AB, Gebler JC (2008) Determination of N-glycosylation sites and site heterogeneity in a monoclonal antibody by electrospray quadrupole ion-mobility time-

of-flight mass spectrometry. Rapid Commun Mass Spectrom 22:29–40. https://doi.org/10.1002/rcm.3330
45. Upton R et al (2019) Hybrid mass spectrometry methods reveal lot-to-lot differences and delineate the effects of glycosylation on the tertiary structure of Herceptin (R). Chem Sci 10:2811–2820. https://doi.org/10.1039/c8sc05029e
46. Beck A et al (2013) Analytical characterization of biosimilar antibodies and fc-fusion proteins. TrAC Trend Anal Chem 48:81–95. https://doi.org/10.1016/j.trac.2013.02.014
47. Zhang H, Cui WD, Gross ML (2014) Mass spectrometry for the biophysical characterization of therapeutic monoclonal antibodies. FEBS Lett 588:308–317. https://doi.org/10.1016/j.febslet.2013.11.027
48. Huang YN, Salinas ND, Chen E, Tolia NH, Gross ML (2017) Native mass spectrometry, ion mobility, and collision-induced unfolding categorize malaria antigen/antibody binding. J Am Soc Mass Spectrom 28:2515–2518. https://doi.org/10.1007/s13361-017-1782-0
49. Huddleston MJ, Bean MF, Carr SA (1993) Collisional fragmentation of glycopeptides by electrospray ionization LC/MS and LC/MS/MS: methods for selective detection of glycopeptides in protein digests. Anal Chem 65:877–884
50. Domon B, Costello CE (1988) A systematic nomenclature for carbohydrate fragmentations in Fab-Ms Ms spectra of glycoconjugates. Glycoconj J 5:397–409. https://doi.org/10.1007/Bf01049915
51. Parker BL et al (2011) Quantitative N-linked glycoproteomics of myocardial ischemia and reperfusion injury reveals early remodeling in the extracellular environment. Mol Cell Proteomics 10:M110 006833. https://doi.org/10.1074/mcp.M110.006833
52. Hinneburg H et al (2016) The art of destruction: optimizing collision energies in quadrupole-time of flight (Q-TOF) instruments for glycopeptide-based glycoproteomics. J Am Soc Mass Spectrom 27:507–519. https://doi.org/10.1007/s13361-015-1308-6
53. Dodds ED (2012) Gas-phase dissociation of glycosylated peptide ions. Mass Spectrom Rev 31:666–682. https://doi.org/10.1002/mas.21344
54. Kolli V, Dodds ED (2014) Energy-resolved collision-induced dissociation pathways of model N-linked glycopeptides: implications for capturing glycan connectivity and peptide sequence in a single experiment. Analyst 139:2144–2153. https://doi.org/10.1039/c3an02342g
55. Jebanathirajah J, Steen H, Roepstorff P (2003) Using optimized collision energies and high resolution, high accuracy fragment ion selection to improve glycopeptide detection by precursor ion scanning. J Am Soc Mass Spectrom 14:777–784. https://doi.org/10.1016/S1044-0305(03)00263-0
56. Vékey K et al (2013) Fragmentation characteristics of glycopeptides. Int J Mass Spectrom 345-347:71–79. https://doi.org/10.1016/j.ijms.2012.08.031
57. Yang H, Yang C, Sun T (2018) Characterization of glycopeptides using a stepped higher-energy C-trap dissociation approach on a hybrid quadrupole orbitrap. Rapid Commun Mass Spectrom 32:1353–1362. https://doi.org/10.1002/rcm.8191
58. Hoffmann M et al (2018) The fine art of destruction: a guide to in-depth glycoproteomic analyses-exploiting the diagnostic potential of fragment ions. Proteomics 18:e1800282. https://doi.org/10.1002/pmic.201800282
59. Rath CB et al (2018) Flagellin glycoproteomics of the periodontitis associated pathogen selenomonas sputigena reveals previously not described O-glycans and rhamnose fragment rearrangement occurring on the glycopeptides. Mol Cell Proteomics 17:721–736. https://doi.org/10.1074/mcp.RA117.000394
60. Wuhrer M, Deelder AM, van der Burgt YE (2011) Mass spectrometric glycan rearrangements. Mass Spectrom Rev 30:664–680. https://doi.org/10.1002/mas.20337
61. Sanda M, Benicky J, Goldman R (2020) Low collision energy fragmentation in structure-specific glycoproteomics analysis. Anal Chem 92:8262–8267. https://doi.org/10.1021/acs.analchem.0c00519

62. Wuhrer M, Catalina MI, Deelder AM, Hokke CH (2007) Glycoproteomics based on tandem mass spectrometry of glycopeptides. J Chromatogr B Analyt Technol Biomed Life Sci 849:115–128. https://doi.org/10.1016/j.jchromb.2006.09.041
63. Alley Jr WR, Mann BF, Novotny MV (2013) High-sensitivity analytical approaches for the structural characterization of glycoproteins. Chem Rev 113:2668–2732. https://doi.org/10.1021/cr3003714
64. Alagesan K, Hinneburg H, Seeberger PH, Silva DV, Kolarich D (2019) Glycan size and attachment site location affect electron transfer dissociation (ETD) fragmentation and automated glycopeptide identification. Glycoconj J 36:487–493. https://doi.org/10.1007/s10719-019-09888-w
65. Frese CK et al (2012) Toward full peptide sequence coverage by dual fragmentation combining electron-transfer and higher-energy collision dissociation tandem mass spectrometry. Anal Chem 84:9668–9673. https://doi.org/10.1021/ac3025366
66. Caval T, Zhu J, Heck AJR (2019) Simply extending the mass range in electron transfer higher energy collisional dissociation increases confidence in N-glycopeptide identification. Anal Chem 91:10401–10406. https://doi.org/10.1021/acs.analchem.9b02125
67. Riley NM, Coon JJ (2018) The role of electron transfer dissociation in modern proteomics. Anal Chem 90:40–64. https://doi.org/10.1021/acs.analchem.7b04810
68. Riley NM, Westphall MS, Coon JJ (2015) Activated ion electron transfer dissociation for improved fragmentation of intact proteins. Anal Chem 87:7109–7116. https://doi.org/10.1021/acs.analchem.5b00881
69. Riley NM, Hebert AS, Westphall MS, Coon JJ (2019) Capturing site-specific heterogeneity with large-scale N-glycoproteome analysis. Nat Commun 10:1311. https://doi.org/10.1038/s41467-019-09222-w
70. Riley NM, Malaker SA, Driessen MD, Bertozzi CR (2020) Optimal dissociation methods differ for N- and O-glycopeptides. J Proteome Res 19:3286–3301. https://doi.org/10.1021/acs.jproteome.0c00218
71. Reiding KR, Bondt A, Franc V, Heck AJR (2018) The benefits of hybrid fragmentation methods for glycoproteomics. TrAC Trend Anal Chem 108:260–268. https://doi.org/10.1016/j.trac.2018.09.007
72. Mucha E et al (2019) In-depth structural analysis of glycans in the gas phase. Chem Sci 10:1272–1284. https://doi.org/10.1039/c8sc05426f
73. Nilsson J (2016) Liquid chromatography-tandem mass spectrometry-based fragmentation analysis of glycopeptides. Glycoconj J 33:261–272. https://doi.org/10.1007/s10719-016-9649-3
74. Hofmann J et al (2014) Estimating collision cross sections of negatively charged N-glycans using traveling wave ion mobility-mass spectrometry. Anal Chem 86:10789–10795. https://doi.org/10.1021/ac5028353
75. Hinneburg H et al (2016) Distinguishing N-acetylneuraminic acid linkage isomers on glycopeptides by ion mobility-mass spectrometry. Chem Commun 52:4381–4384. https://doi.org/10.1039/c6cc01114d
76. Guttman M, Lee KK (2016) Site-specific mapping of sialic acid linkage isomers by ion mobility spectrometry. Anal Chem 88:5212–5217. https://doi.org/10.1021/acs.analchem.6b00265
77. Barroso A et al (2018) Evaluation of ion mobility for the separation of glycoconjugate isomers due to different types of sialic acid linkage, at the intact glycoprotein, glycopeptide and glycan level. J Proteome 173:22–31. https://doi.org/10.1016/j.jprot.2017.11.020
78. Lee EU, Roth J, Paulson JC (1989) Alteration of terminal glycosylation sequences on N-linked oligosaccharides of Chinese hamster ovary cells by expression of beta-galactoside alpha 2,6-sialyltransferase. J Biol Chem 264:13848–13855
79. Jeong YT et al (2008) Enhanced sialylation of recombinant erythropoietin in CHO cells by human glycosyltransferase expression. J Microbiol Biotechnol 18:1945–1952
80. Raymond C et al (2015) Production of alpha2,6-sialylated IgG1 in CHO cells. MAbs 7:571–583. https://doi.org/10.1080/19420862.2015.1029215

81. Pett C et al (2018) Effective assignment of alpha2,3/alpha2,6-sialic acid isomers by LC-MS/MS-based glycoproteomics. Angew Chem 57:9320–9324. https://doi.org/10.1002/anie.201803540
82. Zhu H et al (2020) Identifying sialylation linkages at the glycopeptide level by glycosyltransferase labeling assisted mass spectrometry (GLAMS). Anal Chem 92:6297–6303. https://doi.org/10.1021/acs.analchem.9b05068
83. Harvey DJ (1999) Matrix-assisted laser desorption/ionization mass spectrometry of carbohydrates. Mass Spectrom Rev 18:349–450. https://doi.org/10.1002/(SICI)1098-2787(1999)18:6<349::AID-MAS1>3.0.CO;2-H
84. Ciucanu I, Kerek F (1984) A simple and rapid method for the permethylation of carbohydrates. Carbohydr Res 131:209–217. https://doi.org/10.1016/0008-6215(84)85242-8
85. Selman MH et al (2012) MALDI-TOF-MS analysis of sialylated glycans and glycopeptides using 4-chloro-alpha-cyanocinnamic acid matrix. Proteomics 12:1337–1348. https://doi.org/10.1002/pmic.201100498
86. Toyoda M, Ito H, Matsuno YK, Narimatsu H, Kameyama A (2008) Quantitative derivatization of sialic acids for the detection of sialoglycans by MALDI MS. Anal Chem 80:5211–5218. https://doi.org/10.1021/ac800457a
87. Wheeler SF, Domann P, Harvey DJ (2009) Derivatization of sialic acids for stabilization in matrix-assisted laser desorption/ionization mass spectrometry and concomitant differentiation of alpha(2-->3)- and alpha(2-->6)-isomers. Rapid Commun Mass Spectrom 23:303–312. https://doi.org/10.1002/rcm.3867
88. Reiding KR, Blank D, Kuijper DM, Deelder AM, Wuhrer M (2014) High-throughput profiling of protein N-glycosylation by MALDI-TOF-MS employing linkage-specific sialic acid esterification. Anal Chem 86:5784–5793. https://doi.org/10.1021/ac500335t
89. de Haan N et al (2015) Linkage-specific sialic acid derivatization for MALDI-TOF-MS profiling of IgG glycopeptides. Anal Chem 87:8284–8291. https://doi.org/10.1021/acs.analchem.5b02426
90. Gahoual R et al (2013) Rapid and multi-level characterization of trastuzumab using sheathless capillary electrophoresis-tandem mass spectrometry. MAbs 5:479–490. https://doi.org/10.4161/mabs.23995
91. Moini M (2007) Simplifying CE-MS operation. 2. Interfacing low-flow separation techniques to mass spectrometry using a porous tip. Anal Chem 79:4241–4246. https://doi.org/10.1021/ac0704560
92. Kammeijer GSM et al (2017) Sialic acid linkage differentiation of glycopeptides using capillary electrophoresis – electrospray ionization – mass spectrometry. Sci Rep 7:3733. https://doi.org/10.1038/s41598-017-03838-y

# Glycan Array Technology

**Juana Elizabeth Reyes Martinez, Baptiste Thomas, and Sabine Lahja Flitsch**

## Contents



**Abstract** Glycan (or carbohydrate) arrays have become an essential tool in glycomics, providing fast and high-throughput data on protein-carbohydrate interactions with small amounts of carbohydrate ligands. The general concepts of glycan arrays have been adopted from other microarray technologies such as those used for nucleic acid and proteins. However, carbohydrates have presented their own challenges, in particular in terms of access to glycan probes, linker attachment chemistries and analysis, which will be reviewed in this chapter. As more and more glycan probes have become available through chemical and enzymatic synthesis and robust

The original version of this chapter was revised: Graphical Abstract and Abbreviations included as new additional corrections in this chapter.

J. E. R. Martinez
Divisiön de Ciencias Naturales y Exactas, Departamento de Biología, Universidad de Guanajuato, Guanajuato, Mexico

B. Thomas and S. L. Flitsch (✉)
School of Chemistry and MIB, The University of Manchester, Manchester, UK
e-mail: sabine.flitsch@manchester.ac.uk

linker chemistries have been developed, the applications of glycan arrays have dramatically increased over the past 10 years, which will be illustrated with recent examples.

**Graphical Abstract**

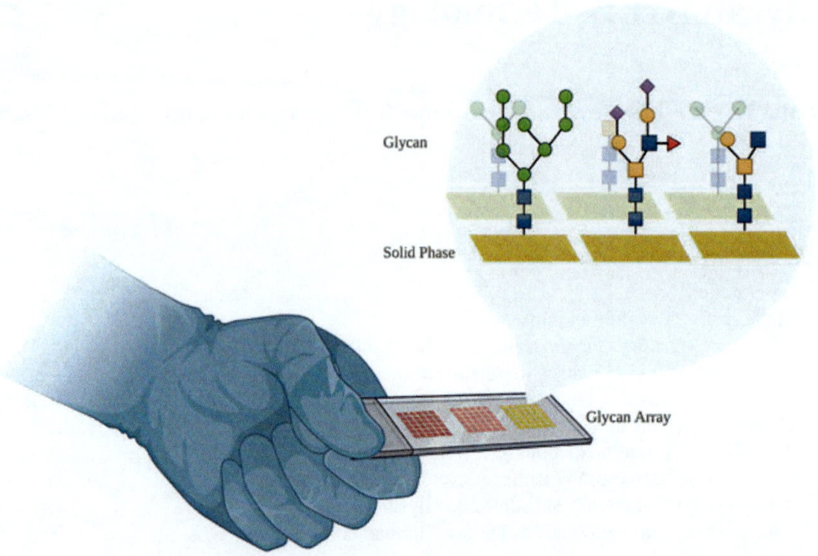


**Keywords** Antibodies, Carbohydrate-binding proteins, Glycan arrays, Glycoenzymes, Lectins, Linkers

## Abbreviations

| | |
|---|---|
| 2-AA | 2-amino-benzoic acid |
| 2-AB | 2-amino-benzamide |
| AEAB | 2-amino-*N*-(2-aminoethyl)-benzamide |
| CAZys | Carbohydrate-active enzymes |
| CFG | Consortium for functional glycomics |
| CSEE | Core synthesis/enzymatic extension |
| DAP | 2,6-diamino-pyridine |
| DDI | DNA-direct immobilisation |
| Fmoc-Cl | 9-fluorenylmethyl chloroformate |
| GBPs | Glycan-binding proteins |
| GPI | Glycosylphosphatidylinositol |
| GTs | Glycosyltransferases |
| HIV | Human immunodeficiency virus |
| hIntL-1 | Human intelectin-1 |

| | |
|---|---|
| JCV | Human JC polyomavirus |
| KDO | 3-deoxy-D-manno-oct-2-ulosonic acid |
| KO | D-glycero-D-talo-oct-2-ulosonic acid |
| LPS | Lipopolysaccharide |
| LSTc | Lactoseries tetrasaccharide c |
| MS | Mass spectrometry |
| NBS | *N*-Bromosuccinimide |
| NGL | Neoglycolipid |
| PMP | 1-phenyl-3-methyl-5-pyrazolone |
| PNGase-F | Peptide-*N*-glycosidase F |
| PNPA | p-nitrophenyl anthranilate |
| SLL | Staphylococcal superantigen-like |
| SV40 | Simian virus 40 |
| TLC | Thin-layer chromatography |
| VCC | Vibrio cholerae cytolysin |

## 1 Introduction

Understanding the interactions between carbohydrates and proteins is very challenging as both binding partners are highly complex and binding can be very weak, often relying on polyvalent interactions to reach binding that is biologically relevant. Because carbohydrate sequences are not directly encoded in the genome, and glycan biosynthesis is controlled by complex networks of proteins, there are fewer genetic tools in glycomics compared to genomics and proteomics. All these issues have driven the establishment of glycan arrays as a prime tool to understand the interactions of carbohydrates with other biomolecules, in particular carbohydrate-binding proteins. The basic principle of glycan arrays is similar to microarrays developed for nucleic acids and proteins, in that the glycan analyte is immobilised on a solid surface using biocompatible linkers either through covalent or non-covalent conjugation (Fig. 1). The surface is then interrogated by the potential binding partner, in most cases a carbohydrate-binding protein, lectin, antibody or enzyme [1–5].

Synthesis and functionalisation of glycans remain a challenge. Many glycans lack functional groups (such as amines, carboxylic acids) that would allow highly orthogonal and selective attachment to solid phase. In many cases (such as animal cell surface glycans), attachment at the reducing end would be predicted to present the glycan in a more accessible way to its partner, given that these carbohydrates are anchored to lipids and proteins via the reducing end. The reducing sugar in glycan chains does of course carry unique functionality, and the sluggish reactivity of free glycosides has been overcome by a range of reactive labelling techniques providing access to glycan probes either from synthetic or from natural sources. Synthetic glycans often carry a stable and reactive functional group which can be installed during chemical or enzymatic synthesis. Both synthetic glycan arrays [1] and natural

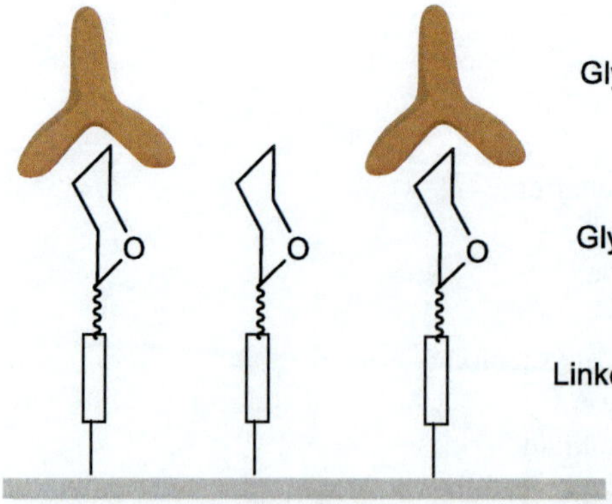


**Fig. 1** General component of glycan arrays: Glycan structures are attached to a solid phase through a linker either covalently or non-covalently. The glycan array is probed for binding of biomolecules, mostly carbohydrate-binding proteins or enzymes

glycan arrays [6] have been developed, with synthetic array generally limited to smaller numbers of different glycans (up to ca 1,000) [4] because of the challenges of carbohydrate synthesis.

## 2 Methods for Glycan Immobilisation

A range of different coupling strategies have been used for glycan immobilisation on array surfaces, and some of the frequently used examples are shown in Fig. 2.

### *2.1 Free Reducing Sugars as Starting Materials*

Direct coupling of free sugars to hydrazide-functionalised glass slides has been reported, but is a sluggish reaction that can give mixtures of products [7]. One of the earliest glycan array platforms reported is the neoglycolipid technology developed by the Feizi group (Fig. 3) [4, 8]. This method uses a two-step protocol, in which free glycans are first reacted with amino-lipid through reductive amination, followed by chromatographic purification of product before printing onto nitrocellulose slides. This platform has been used successfully for interrogation of many glycan-binding proteins from microbial and animal sources. The neoglycolipid technology can be applied to both synthetic and natural glycans, since free glycans are used as starting materials.

Because *Nature* offers such an important source of natural glycans, there has been a great interest in developing glycan conjugation chemistries. Particularly useful has been the labelling of glycans with fluorescent dye tags that can be used both for

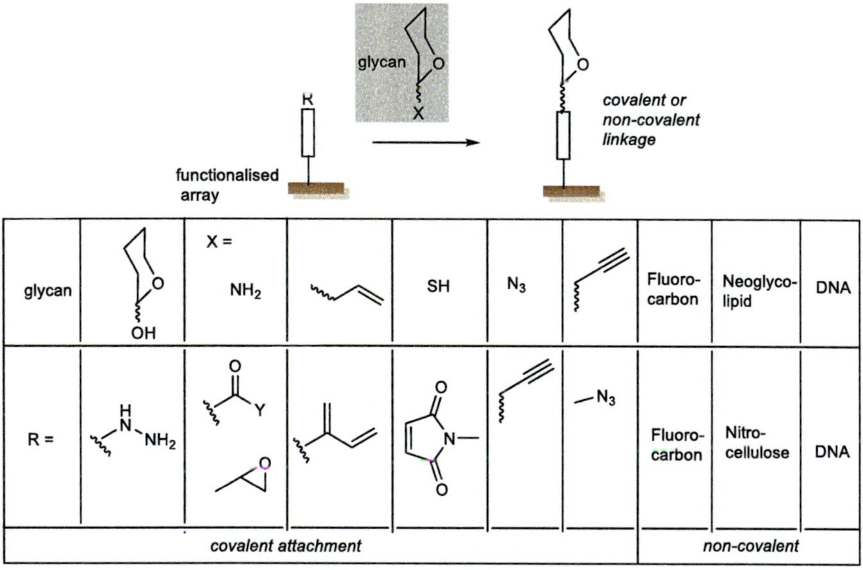


**Fig. 2** Examples of linker chemistry used in glycan arrays for covalent and non-covalent conjugation of glycans to array surfaces. R are those molecules on the array surface, and X in glycan is the chemistry used for their attachment to functionalised surface

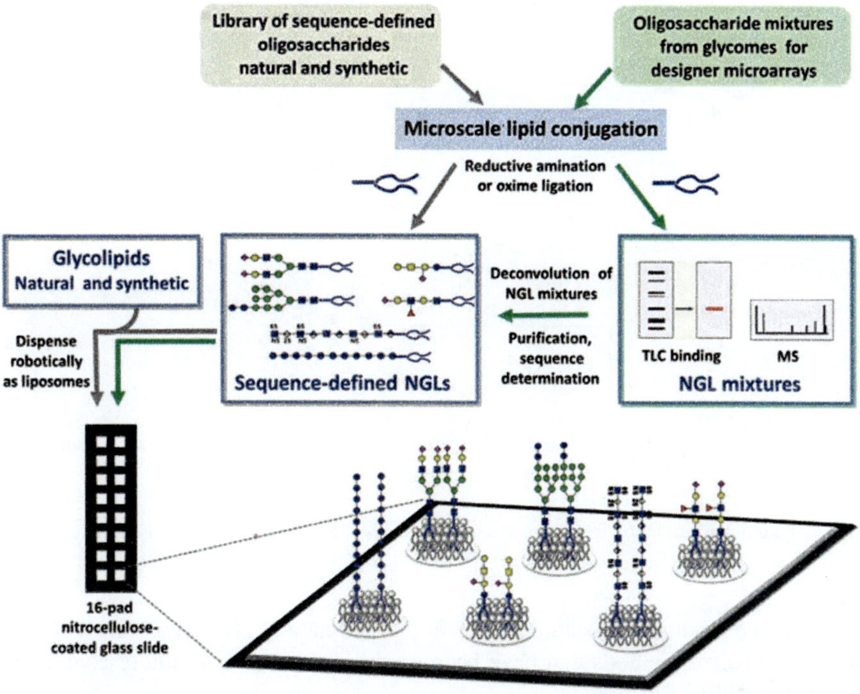


**Fig. 3** Neoglycolipid array technology (adapted with kind permission from Ref. [8])

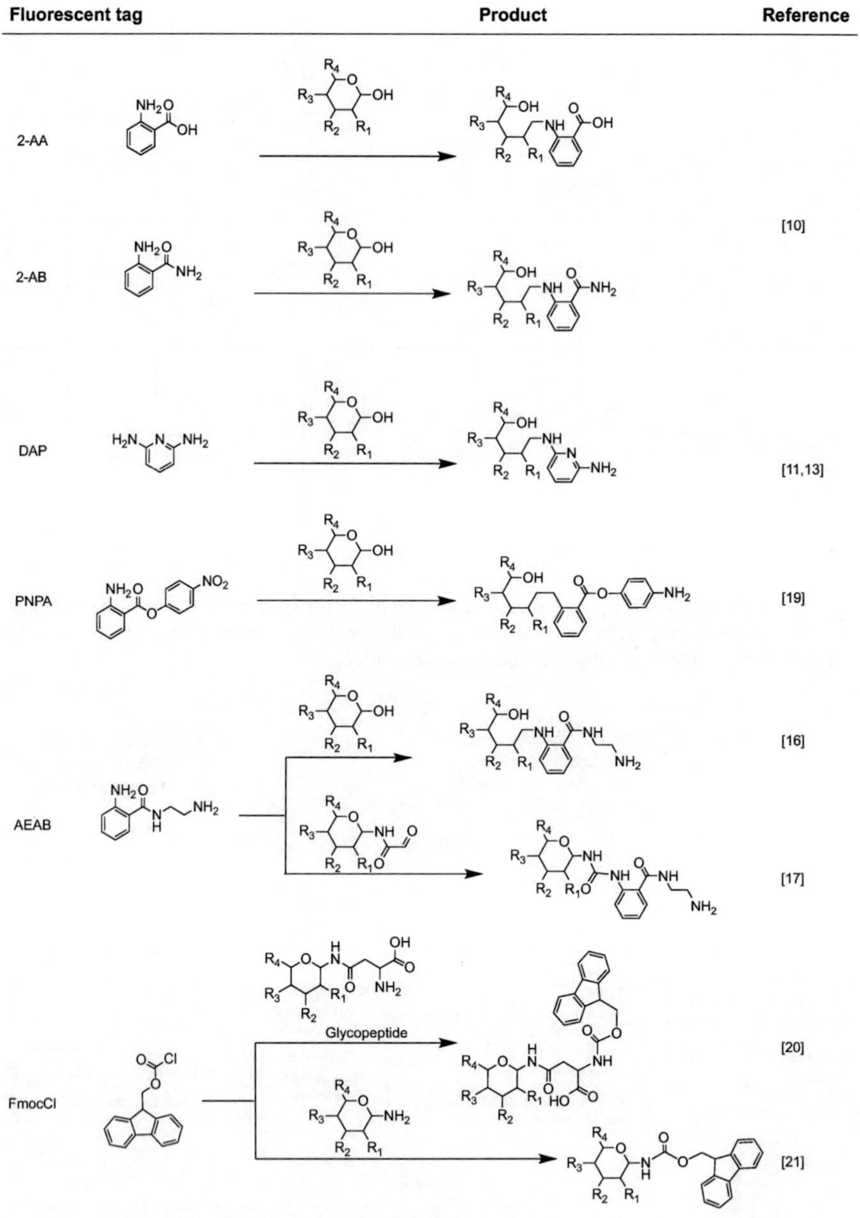


**Fig. 4** Functionalisation of free sugars with fluorescent tags for labelling and for microarray printing

labelling and facilitate separation to obtain purified natural glycans. Figure 4 lists a number of recent methods that have been used to introduce both fluorescent label and more reactive functionality.

Probably the most used strategy to functionalise natural glycans is through reductive amination using the aldehydes at the reducing end of the glycan chain. This strategy was initially explored by using fluorescent probes such as 2-aminobenzamide (2-AB) and 2-aminobenzoic acid (2-AA) [9] and fully explored to immobilise the product on epoxy-activated glass slides via the secondary amine group linking the glycan moiety to the fluorescent tag [10]. Functionalisation with 2,6-diaminopyridine (DAP) to generate fluorescent labelled glycans [11] and oxime formation [12] is also used for glycoarray preparation on NHS- and epoxy-activated glass slides [6, 13, 14]. *N*-Bromosuccinimide (NBS) treatment after the labelling reaction allows to easily eliminate the fluorescent tag and regenerate isolated glycans, which is particularly important if the label interferes with the bioactivity of the glycan [15].

Reductive amination strategies have been further exploited by using 2-amino-*N*-(2-aminoethyl)-benzamide (AEAB) generating fluorescently labelled glycans which have been used for glycoarray printing [16–18] showing good efficiency in solid-phase glycan immobilisation. A bifunctional linker with aryl amine group and a *p*-nitrophenyl ester group (p-phenyl anthranilate, PNPA) was also used in glycoarray preparation with the advantage of increased fluorescent properties upon glycan functionalisation [19]. Fmoc chemistry using 9-fluorenylmethyl chloroformate (Fmoc-Cl) as a cleavable fluorescent tag was used in 2009 by Song and co-workers [17] to label glycans obtained from chicken ovalbumin, bovine fetuin and horseradish peroxidase. The products were used on NHS-activated glass slides to generate a useful glycoarray platform. Fmoc-derivatised *O*-glycans obtained from mucins and carcinoma cells were later shown to have 3.5 times higher sensitivity than those glycans labelled with 2-AA in a glycoarray platform [20, 21]. For labelling purposes, glycans have to be released from the natural sources. A number of enzymes such as pronase, trypsin, peptide-*N*-glycosidase F (PNGase-F) and endoglycosidases have been used as the most common strategies for glycan release [22]. The proteases digest the protein backbone and the endoglycosidases release *N*-linked glycans cleanly before derivatisation [20, 23–25]. The availability of endoglycosidases for *N*-glycans has driven the field forward dramatically. However, whereas *N*-glycan analysis is rapidly becoming a standard tool in glycomics, the availability of endoglycosidases for *O*-glycans and polysaccharides is still limited.

Alternatively chemical strategies for releasing *O*-glycans from natural sources have been investigated, and hydrazinolysis is a common technique for glycan release and can be useful to generate materials for glycan array. Hydrazinolysis has shown to have high efficiency of hydrolysis towards *N*-glycans, and under certain conditions by addition of ethylene diamine tetraacetic acid, this technique has shown good efficiency for *O*-glycan hydrolysis minimising degradation known as "peeling" [26–28]. β-Elimination strategies under basic and mild conditions using dimethylamine or non-reductive *O*-de-glycosylation with ammonia addition have been explored for glycan analysis. In both methods 1-phenyl-3-methyl-5-pyrazolone (PMP) was used for labelling and analysis of released glycans from glycoproteins [29, 30]. The procedure to extract glycans from glycosphingolipids is a multistep process involving organic extraction and subsequent treatment with glycoceramidases [31]. Very recently, an oxidative strategy using NaClO to release free reducing *N*- and *O*-glycans from glycoproteins and glycosphingolipids was shown to be good alternative for glycan analysis and a promising methodology to generate a library for glycan array preparation [32].

### 2.2 Functionalised Glycans for Conjugation

There are a number of advantages in using synthetic over natural glycans for arrays, the first being purity. Given that natural glycans are isolated from complex mixtures, there is always concern about co-purification of a highly bioactive minor contaminant. In terms of functionalisation for linking to solid support, synthetic strategies are highly flexible and allow for a wide range of active groups as illustrated in Fig. 2. These functionalities are generally introduced by chemical synthesis [33]. Although there is a wide range of options, there has been a focus on using aminoethyl and aminopropyl glycosides, which can be attached to activated carboxyl esters, such as NHS or pentafluorophenyl [3, 34, 35]. A number of amino alkyl glycosides are now commercially available.

## 3 Synthesis of Complex Glycans for Glycoarrays

The field of chemical carbohydrate synthesis is still very challenging, and a review of oligosaccharides synthesis is beyond the scope of this article. Given the success of automated synthesis in nucleic acid and peptide chemistry, there have been a number of projects aimed at automated synthesis either on solid support [36, 37] or on soluble tags [38, 39]. Given that carbohydrate synthesis still requires large numbers of steps and usually yield mg quantities of final target material, glycan arrays are particularly attractive in terms of economy of scale, such that thousands of bioassays can be conducted on mgs of material. Chemical synthesis can also generate bioisosteres that have increased stability. A good example is the chemical synthesis of neuraminidase-resistant sialosides for the detection of influenza viruses [40] and applications for glycan-based detection and drug susceptibility of influenza virus [41].

To overcome the challenges in oligosaccharide synthesis, enzymes are increasingly employed as highly selective catalysts for the fast synthesis of natural glycans. These carbohydrate-active enzymes (CAZys) can be used in vitro on their own or in chemoenzymatic strategies which combine chemical and enzymatic methods [42–44]. The advantage of using enzymes is that they can be used in one-pot multiple enzyme strategies, which have achieved excellent yields and selectivity avoiding purification of intermediates [45–48]. Figure 5 is an example of chemoenzymatic routes to *O*-mannosyl glycopeptide synthesis, which can be performed in solution, but also directly on the glycan arrays [49].

Natural glycans can be used as substrates for enzymatic modification either before or after being coupled to a solid support in array format. This strategy has been extensively

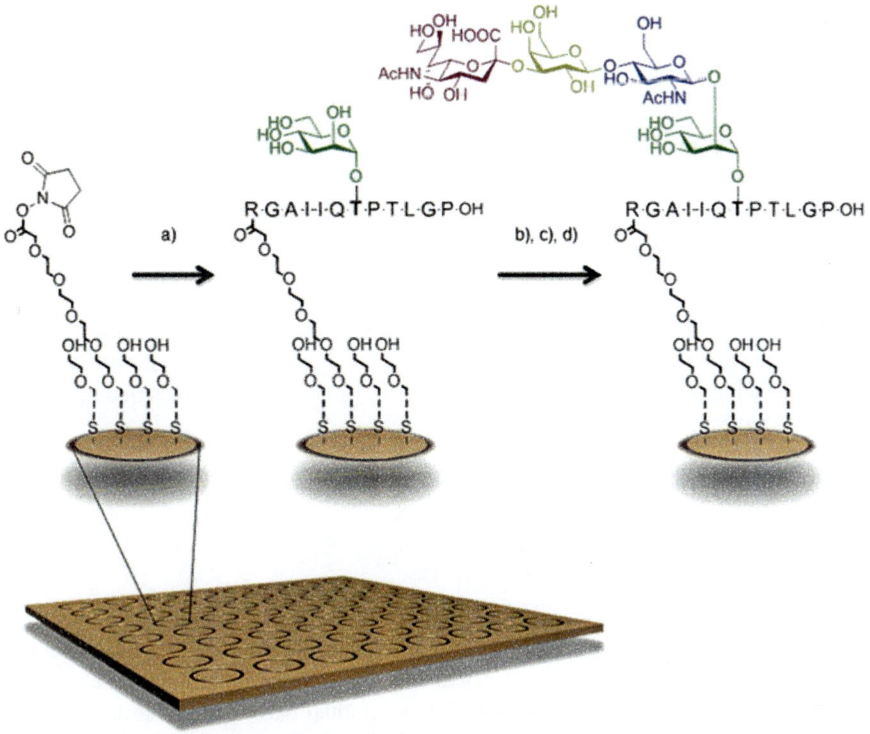


**Fig. 5** Chemoenzymatic synthesis of glycopeptides on glycan arrays [49]

used to increase the complexity of glycan structures. One of the largest and diverse glycan arrays from the Consortium for Functional Glycomics (CFG) has been generated using enzymatic strategies [50]. Enzymatic synthesis of glycans has been widely exploited in glycoarray technology, used to address function-structure of glycans especially terminal sialylated structures [51–54]. Acceptor substrates on array platforms range from glycans, glycopeptides, glycolipids to nucleic acids which have been used as substrates for galactosyltransferases [55] fucosyltransferases [56] and sialyltransferases [52, 57]. Chemoenzymatic strategies have been shown in several studies to generate large-scale production of glycan associated tumour-associated *N*-acetyllactosamine antigens attached on an array platform [58]. The *N*-glycan core pentasaccharide has been selectively extended by glycosyltransferases (GTs) providing unique glycan moieties of asymmetrically branched *N*-glycans printed on an array platform for lectins and influenza virus-hemagglutinin binding studies [59]. More recently, a Core Synthesis/Enzymatic Extension (CSEE) strategy was developed; this strategy is based on 8 *N*-glycan core structures for enzymatic synthesis of 73 *N*-glycans applied to the glycan array technology showing the diversity and applications of GTs to increase the glycan diversity on array platforms [60].

The great potential to generate and increase array diversity has been shown by the generation of a phosphorylated glycan array in which mannose-6-phosphate was attached to the array surface, to study the specificity of mannose-6-phosphate cation-dependent and cation-independent receptors [61]. New routes for enzymatic synthesis of glycans for array technology will grow exponentially thanks to the increasing number of glycoenzymes (catalogued in the Carbohydrate-Active Enzyme database [62]) which can be used for such proposes.

# 4 Glycoarray Applications

Glycoarrays are now widely used for carbohydrate-protein interactions, and a comprehensive coverage of all examples would be beyond the scope of this chapter. In the following, we have selected a number of representative examples that illustrate the breadth of applications.

## *4.1 Glycan-Virus Interactions*

An increased number of glycans, in particular from animal and human sources, are now available for glycan array printing. Array technology has become a powerful tool to understand pathogenicity mechanisms throughout infection process. The specificity of viral capsule proteins towards individual carbohydrate structures has been studied using carbohydrate array technologies [63]. Both viral hemagglutinin and neuraminidase specificities have been extensively interrogated using multiple array technologies [54, 64], and several studies have reported viral specificity from swine or human influenza virus [65–67]. Glycan arrays are also useful to determine drug susceptibility of some influenza virus from clinical isolates [41]. Interestingly, Walther and colleagues found no correlation on binding specificities with infection symptoms and severity of the disease on clinical isolated influenza strains, thus highlighting the need to expand the glycan diversity on array technology specifically to *N*- and *O*-glycans identified in human lung tissues [68].

Binding specificities from other viruses have also been studied. Glycan arrays based on neoglycolipids have been used to identify the specificity for GM1 in simian virus 40 (SV40) [69]. Human JC polyomavirus (JCV) specificity towards the pentasaccharide NeuNAc-α2,6-Gal-β1,4-GlcNAc-β1,3-Gal-β1,4-Glc a sialylated derivate from lactoseries tetrasaccharide c (LSTc) was determined using glycan array technology and viral infectivity studies, confirmed by crystallography and mutational studies, to reveal a close interaction with the terminal sialic acid of the LSTc motif and their importance for the infective process [70]. Those studies are the result of technical advances in glycan printing technology, allowing the screening for drugs with potential binding inhibitor activity. A study by the Seeberger group identified key interactions with the glycan fraction of the glycoprotein gp120 from HIV virus identifying scytovirin as potential binding inhibitor [71]. The binding specificity of rotavirus, one of the main pathogenic agents causing diarrhoea and gastroenteritis in infants, was studied using glycoarray technology. This virus processes a VP8 protein, which showed binding specificity to A-type histo-blood group antigen as confirmed by X-ray crystallography [72, 73].

## *4.2 Glycan-Bacteria Interactions*

Many bacterial pathogens use carbohydrate-protein interactions to facilitate infective processes, and carbohydrate microarrays have been used to determine specific interactions identifying glycan-binding proteins on the membrane of some microorganisms. This technology has been used for identifying glycan recognition patterns of some pathogenic bacteria [74], with the capacity for high-throughput screening for potential new antibiotics. Bacteria express virulence factors that recognise glycans, and the binding specificity of glycan-binding proteins (GBPs) could easily be identified by glycoarray technology: α-Gal-β1,4-Gal glycan terminator showed preferential binding over α-Gal-β1,3-Gal or α-Gal-β1,6-Gal to the lectin A (LacA or PA-IL) from *Pseudomonas aeruginosa.* Binding specificity of lectin B (PA-IIL) from the same organism was fully characterised by using fucosylated DNA (glycoclusters) which then were immobilised into a solid phase by using the well-known DNA-direct immobilisation (DDI) technique arising the DNA-based glycoarray technology, both lectins are involved in host cell adhesion and biofilm formation [75, 76]. On the other hand, the lectin A (BC2L-A) from *Burkholderia cenocepacia* identified by glycan array technology preferential binding towards oligomannose-type oligosaccharides usually present in human glycoproteins [77]. Further studies, in which 377 glycans were printed on a glycan microarray chip, showed a dual recognition to both terminal mannose and fucose oligosaccharides of BC2L-A lectin, suggesting that the lectin forms a dual link between bacteria and epithelial cells facilitating the infective process [78, 79].

Sialylated terminal glycans are usually target molecules for bacterial binding as part of the mechanism used by many pathogens to infect host cells; several studies using glycan arrays have contributed to understand molecular mechanisms of the infection process and have led to the identification of SLL-4, SLL-5 and SLL-11, staphylococcal super antigen-like (SLL) proteins containing highly conserved carbohydrate-binding site also present in SLL-2, SLL-3 and SLL-6. By glycan array technology, sialylated glycans (sialyl Lewis X, sialyl lactosamine) were identified as SLL targets, and subsequent detailed crystallographic studies showed small differences in active site conformation that could explain small affinity preferences [80].

*N*-glycans with the heptasaccharide GlcNAc4Man3 core are present on mammalian cells, and by using glycan arrays, these structures have been identified as targets for *Vibrio cholerae* cytolysin (VCC), a pore-forming toxin secreted by this organism. The studies suggest that the carbohydrate-binding domains on VCC toxin facilitate high affinity targeting mammalian cell membranes contributing to cell lysis at picomolar concentrations [81]. A bacterial lectin with fucose specificity was identified from *Streptomyces rapamycinicus*. The SL2-1 lectin belonging to a new family of fucose-binding proteins shows high specificity towards core α1,6-fucosylated *N*-glycans but not to core α1,3-fucosylated *N*-glycans or other fucosylated oligosaccharides. This lectin has no similarity to other known bacterial fucose-binding proteins, but is showing certain similarity to eukaryotic fucose-binding lectins [82].

Complementary to these interactions, mammalian cell surfaces contain carbohydrate-recognition proteins with the ability to recognise glycans on the surface of many pathogens. Galectins and C-type lectins are the most studied systems in humans. The protector effect of galectins (lectins defined by their binding to β-Gal oligosaccharides) has been addressed by studies on several glycoarrays generated from microbial polysaccharides. Disruption of the membrane integrity and viability loss has been observed on *E. coli* expressing human blood group antigen (ABO-H epitopes), an effect promoted by galectin-4 and galectin-8 expressed in the intestinal tract and by galectin-9 [83–85]. In addition to galectins, interesting defence mechanisms addressing structural understanding of glycan recognition of langerin, a glycan-binding protein expressed by Langerhans cells, revealed their binding specificity to high-mannose *N*-linked oligosaccharides, blood group B antigen and 6-sulfated galactosides all identified in fungi, mycobacteria and virus [86].

So far, the human intelectin-1 (hIntL-1) does not seem to bind to any known human glycan epitope. However, this protein interacts with multiple glycan epitopes found exclusively in bacteria, such as recognition of β-Gal*f*, D-phosphoglycerol-modified glycans, heptoses, D-glycero-D-talo-oct-2-ulosonic acid (KO) and 3-deoxy-D-manno-oct-2-ulosonic acid (KDO), thus promoting protective effects [87].

## *4.3 Glycans as Enzyme Substrates on Glycan Arrays*

The density of carbohydrates on microarrays can have important consequences for lectin binding. Lectins often present multivalent binding sites, and the importance of using a flexible platform in order to perform high- and low-density binding interaction for identification of multivalent binding sites has also been addressed by several studies [48, 88, 89]. Three-dimensional carbohydrate conjugates have been synthesised, mimicking structures present on cell surfaces based on highly branched monodisperse macromolecules known as dendrimers [90–92], glycopolymers [93–95], glycoproteins [96–98], glycolipids and glycolipid-containing liposomes [99–101], DNA glycoclusters [102] and carbohydrate-coated nanoparticles attached to array platform [103].

Glycan array platforms have allowed the analysis of carbohydrate-processing enzymes. One of the major challenges here is to find a reliable readout method to determine enzymes' activity-specificity in a qualitative and quantitative way. This challenging step was initially overcome by using biotinylated glycosyl donors [104] enabling direct detection of incorporated glycan residues on the acceptor molecule attached to the array surface. Indirect detection methods using fluorescently labelled lectins [105] or biotinylated-labelled lectins subsequently detected by fluorescently labelled streptavidin or antibodies have been widely used on several array platforms [3, 14, 82, 106]. However, quantification of enzymatic activity with this method is still challenging. Higher sensitivity was reached by tritium autoradiography allowing more accurate readings which can be used for quantitative on-chip enzymatic assays [107]. More recently, a novel array for exopolysaccharide hydrolysing enzymes, polysaccharide lyases, carbohydrate esterases and lytic polysaccharide mono-oxygenases activity has been reported [108], based on labelled antibodies and lectins for semi-quantitative enzymatic assays, which was used to determine enzymatic activity from biomass samples mainly from fungal organisms. Labelling glycan moieties on the array platform has also been explored with chemoselective labelling glycan residues on the array platform after galactose oxidase or periodate oxidation. Aniline-catalysed oxime ligation has been used for neuraminidase specificity detection; however this method has not been fully exploited and at the moment has very limited glycan substrates [54].

Label-free detection strategies using mass spectrometry analysis to determine transferase and hydrolase activities have been reported for bovine β1,4-galactosyltransferase I activity, 7 ppGalNAcT activity [109, 110], sialyltransferases activity [52, 104, 111], glycosyl hydrolase activity from environmental samples [112] and hydrolase/transferase activity by galactose-processing enzymes [113].

Mass spectrometry can be complemented with fluorescently labelled lectins to provide quantitative readout, demonstrating the ability to screen glycosyltransferase activity in a high-throughput manner [78, 114]. These techniques have led to the discovery of new fucosyltransferases and galactosyl- and *N*-acetylgalactosaminetransferases [115] and have opened the possibilities to test drug inhibitors for potential pharmaceutical applications. Given the very large number of potential CAZy enzymes in genomic databases [62] and the lack of generic high-throughput screens for enzyme activities, glycan arrays are a promising tool to study these glycoenzymes.

## 4.4 *Anti-glycan Antibodies as Disease Biomarkers*

Usually, antibodies are generated as a defence mechanism against pathogens, and it has been suggested that the population of these antibodies change according to the onset of disease, age, exposure to antigens or vaccination. Antibodies circulating in human sera are accessible and very popular for biomarker discovery. Microarray technology has been used to explore anti-glycan antibodies as biomarkers for immune response, infections, autoimmune diseases and cancer [116], and a number of antibodies against glycans and glycopeptides have also been found in human normal sera [117]. Glycans expressed by pathogens have been used as antigens to develop glycan array technology as diagnostic tool. Lipopolysaccharides (LPS) are components of the outer membrane in Gram-negative bacteria. LPS and glycan antigens have been used in array platforms in order to identify antibodies indicative for infection caused by *Burkholderia pseudomallei, Francisella tularensis* and *Bacillus anthracis* [118]. *O*-Antigen-specific antibodies have been found in human sera from salmonellosis patients demonstrating that glycoarray technology as a potential high-throughput technology as diagnostic tool and to follow infection outbreaks [119].

Parasitic infections have been also studied by anti-glycan antibody detection using glycoarrays as diagnostic tools. Using a synthetic GPI array, specific antibodies in malaria-infected patients were detected, with the pentasaccharide Man-α1,2-Man-α1,4-GlcNH2-α1,6-*myo*-inositol-1-PO4 being identified as a minimal epitope for antibody binding [120]. The antigen GalNAc-β1,4-(Fuc-α1,3)-GlcNAc was identified as antigen for antibody production, thus leading to a good target for serodiagnostic tools in patients infected with *Trichinella spiralis* [121]. By using a glycopeptide array, Tn antigen (GalNAc-α-1-Ser/Thr-R)-specific antibodies against *Cryptosporidium parvum* causal agent of gastrointestinal infections and diarrhoea were detected efficiently [122]. Anti-glycan antibody response was identified in the sera of patients with *S. mansoni* infection. When 33 non-mammalian xylosylated and core-fucosylated *N*-glycans were used, clear and immunologically relevant differences between children and adult groups infected with this parasite were observed [123].

A large number of studies have focussed on influenza virus infections, many of which are based in enzymatic assays described earlier in this chapter. Antibody responses against viral infections have been studied on glycoarray platforms to a lesser extent than influenza virus infections, but nevertheless highlighting this technology of great potential for studies of viral infections. Differential antigen binding was identified in antibodies from patients infected with herpes virus 2 and herpes virus 1 or noninfected individuals [124]. A protective response against human immunodeficiency virus (HIV) infection was shown to be driven by antibodies production [71]. By using glycoarray platforms, neutralising effects of those antibodies were observed. The binding specificities to complex *N*-glycans and oligomannose fraction (Man8GlcNAc2 and Man9GlcNAc2) of the envelope protein gp120 have been confirmed as new targets for vaccine development [125–127].

The change in glycosylation of cancer cells has been known for a very long time, and glycan arrays have helped to identify tumour-associated carbohydrate antigens which found exclusively on cancer cells but also altered glycosylation levels due to aberrant glycosylation processes [128]. Using glycoarray platforms, anti-Globo H antibodies have been found in elevated levels in breast cancer patients [129], and antibodies against aberrant *O*-glycopeptides derived from MUC1 glycoprotein have also been detected in sera from breast, ovarian and prostate cancer patients [130]. Extending glycoarrays based on MUC1 to include MUC4 glycopeptides was useful to increase sensitivity and specificity for colorectal cancer detection [131]. Additionally, cancer-associated glycoforms of MUC1 antibodies were detected in a high concentration in early stage and beginning of breast cancer but not in healthy controls. Additionally, anti-CoreMUC1 (GlcNAcβ1,3-GalNAc-MUC1) and STnMUC1 (NeuAcα2,6-GalNAc-MUC1) antibodies were associated with reduced incidence and delay in metastases, revealing their importance as cancer biomarkers and highlighting the importance of antibodies in cancer progression. Despite the promising advances in cancer biomarker detection by glycoarray technology, there is still much to be done. For example, studies using larger number of samples were not able to discriminate between healthy volunteers and patients with ovarian, pancreatic and lung cancer when MUC1 glycoform was used [132]. However, non-mucinous glycans which also have been used in array format including P(1) (Galα-1,4-Galβ-1,4-GalNAcβ) were shown to be useful to determine anti-glycan antibodies in cancer patients with higher sensitivity-specificity compared to the CA125 tumour marker [133]. Different markers such as GM3, tumour-associated Tn antigen Gal-β1,3-GalNAc-α-Ser/Thr and *N*-glycan cryptic antigen have been used on array platform as cancer biomarkers but also as possible targets for vaccine development [116, 134, 135]. Glycan antigens are present not only in cancer cells, bacterial and viral cell walls; they are also in the environment and can act as allergens. Exposure to these external agents can deregulate the immune system causing allergic reactions or autoimmunity [136]. Studies on antibody production in Crohn's disease, multiple sclerosis [137] and systemic sclerosis [138] have demonstrated the potential use of glycoarrays in diagnostics of a wide range of disorders.

## 5 Conclusions

Glycoarray technology has emerged as a key tool in glycosciences, particularly to uncover the extensive networks of carbohydrate-protein interactions in many biological systems, from humans and animals to plants and microorganisms. Arrays can be used for carbohydrate-binding protein and lectin identification, enzyme activity screening and determination of antibody specificity. Several array platforms have been reported and are in broad agreement with each other. Commercial glycoarray platforms are emerging, but there is scope for expansion into more diverse glycan libraries to be displayed on arrays. Many studies have shown that glycan arrays can be used as fast and sensitive discovery tools in fundamental biological studies, but can also identify new biomarkers for disease such as cancer or infection, and have promise as diagnostic tools for the initial detection of disease and as companion diagnostics during treatment.

## References

1. Love KR, Seeberger PH (2002) Carbohydrate arrays as tools for glycomics. Angew Chem Int Ed 41:3583–3586
2. Drickamer K, Taylor ME (2002) Glycan arrays for functional glycomics. Genome Biol 3: Reviews1034
3. Laurent N, Voglmeir J, Flitsch SL (2008) Glycoarrays--tools for determining protein-carbohydrate interactions and glycoenzyme specificity. Chem Commun (Camb):4400–4412
4. Fukui S, Feizi T, Galustian C, Lawson AM, Chai W (2002) Oligosaccharide microarrays for high-throughput detection and specificity assignments of carbohydrate-protein interactions. Nat Biotechnol 20:1011–1017

5. Gray CJ, Weissenborn MJ, Eyers CE, Flitsch SL (2013) Enzymatic reactions on immobilised substrates. Chem Soc Rev 42:6378–6405
6. Song X, Heimburg-Molinaro J, Smith DF, Cummings RD (2015) Glycan microarrays of fluorescently-tagged natural glycans. Glycoconj J 32:465–473
7. Zhi ZL, Laurent N, Powell AK, Karamanska R, Fais M et al (2008) A versatile gold surface approach for fabrication and interrogation of glycoarrays. ChemBioChem 9:1568–1575
8. Palma AS, Feizi T, Childs RA, Chai W, Liu Y (2014) The neoglycolipid (NGL)-based oligosaccharide microarray system poised to decipher the meta-glycome. Curr Opin Chem Biol 18:87–94
9. Bigge JC, Patel TP, Bruce JA, Goulding PN, Charles SM et al (1995) Nonselective and efficient fluorescent labeling of glycans using 2-amino benzamide and anthranilic acid. Anal Biochem 230:229–238
10. de Boer AR, Hokke CH, Deelder AM, Wuhrer M (2007) General microarray technique for immobilization and screening of natural glycans. Anal Chem 79:8107–8113
11. Xia B, Kawar ZS, Ju T, Alvarez RA, Sachdev GP et al (2005) Versatile fluorescent derivatization of glycans for glycomic analysis. Nat Methods 2:845–850
12. Liu Y, Feizi T, Campanero-Rhodes MA, Childs RA, Zhang Y et al (2007) Neoglycolipid probes prepared via oxime ligation for microarray analysis of oligosaccharide-protein interactions. Chem Biol 14:847–859
13. Song XZ, Xia BY, Lasanajak Y, Smith DF, Cummings RD (2008) Quantifiable fluorescent glycan microarrays. Glycoconj J 25:15–25
14. Heimburg-Molinaro J, Song X, Smith DF, Cummings RD (2011) Preparation and analysis of glycan microarrays. Curr Protoc Protein Sci. https://doi.org/10.1002/0471140864.ps1210s64
15. Song X, Johns BA, Ju H, Lasanajak Y, Zhao C et al (2013) Novel cleavage of reductively aminated glycan-tags by N-bromosuccinimide to regenerate free, reducing glycans. ACS Chem Biol 8:2478–2483
16. Song X, Xia B, Stowell SR, Lasanajak Y, Smith DF et al (2009) Novel fluorescent glycan microarray strategy reveals ligands for galectins. Chem Biol 16:36–47
17. Song X, Lasanajak Y, Xia B, Smith DF, Cummings RD (2009) Fluorescent glycosylamides produced by microscale derivatization of free glycans for natural glycan microarrays. ACS Chem Biol 4:741–750
18. Song X, Heimburg-Molinaro J, Smith DF, Cummings RD (2011) Derivatization of free natural glycans for incorporation onto glycan arrays: derivatizing glycans on the microscale for microarray and other applications (ms# CP-10-0194). Curr Protoc Chem Biol 3:53–63
19. Luyai A, Lasanajak Y, Smith DF, Cummings RD, Song XZ (2009) Facile preparation of fluorescent neoglycoproteins using p-nitrophenyl anthranilate as a heterobifunctional linker. Bioconjug Chem 20:1618–1624
20. Song X, Lasanajak Y, Rivera-Marrero C, Luyai A, Willard M et al (2009) Generation of a natural glycan microarray using 9-fluorenylmethyl chloroformate (FmocCl) as a cleavable fluorescent tag. Anal Biochem 395:151–160
21. Yamada K, Hirabayashi J, Kakehi K (2013) Analysis of O-glycans as 9-fluorenylmethyl derivatives and its application to the studies on glycan array. Anal Chem 85:3325–3333
22. O'Neill RA (1996) Enzymatic release of oligosaccharides from glycoproteins for chromatographic and electrophoretic analysis. J Chromatogr A 720:201–215
23. Goetz JA, Novotny MV, Mechref Y (2009) Enzymatic/chemical release of O-glycans allowing MS analysis at high sensitivity. Anal Chem 81:9546–9552
24. Schiel JE, Smith NJ, Phinney KW (2013) Universal proteolysis and MS(n) for N- and O-glycan branching analysis. J Mass Spectrom 48:533–538
25. Song X, Ju H, Zhao C, Lasanajak Y (2014) Novel strategy to release and tag N-glycans for functional glycomics. Bioconjug Chem 25:1881–1887
26. Turyan I, Hronowski X, Sosic Z, Lyubarskaya Y (2014) Comparison of two approaches for quantitative O-linked glycan analysis used in characterization of recombinant proteins. Anal Biochem 446:28–36

27. Patel T, Bruce J, Merry A, Bigge C, Wormald M et al (1993) Use of hydrazine to release in intact and unreduced form both N- and O-linked oligosaccharides from glycoproteins. Biochemistry 32:679–693
28. Kozak RP, Royle L, Gardner RA, Bondt A, Fernandes DL et al (2014) Improved nonreductive O-glycan release by hydrazinolysis with ethylenediaminetetraacetic acid addition. Anal Biochem 453:29–37
29. Zauner G, Koeleman CA, Deelder AM, Wuhrer M (2012) Mass spectrometric O-glycan analysis after combined O-glycan release by beta-elimination and 1-phenyl-3-methyl-5-pyrazolone labeling. Biochim Biophys Acta 1820:1420–1428
30. Wang C, Fan W, Zhang P, Wang Z, Huang L (2011) One-pot nonreductive O-glycan release and labeling with 1-phenyl-3-methyl-5-pyrazolone followed by ESI-MS analysis. Proteomics 11:4229–4242
31. Ito M, Yamagata T (1986) A novel glycosphingolipid-degrading enzyme cleaves the linkage between the oligosaccharide and ceramide of neutral and acidic glycosphingolipids. J Biol Chem 261:14278–14282
32. Song X, Ju H, Lasanajak Y, Kudelka MR, Smith DF et al (2016) Oxidative release of natural glycans for functional glycomics. Nat Methods 13:528–534
33. Weissenborn MJ, Castangia R, Wehner JW, Sardzik R, Lindhorst TK et al (2012) Oxo-ester mediated native chemical ligation on microarrays: an efficient and chemoselective coupling methodology. Chem Commun (Camb) 48:4444–4446
34. Sardzik R, Noble GT, Weissenborn MJ, Martin A, Webb SJ et al (2010) Preparation of aminoethyl glycosides for glycoconjugation. Beilstein J Org Chem 6:699–703
35. Karamanska R, Clarke J, Blixt O, Macrae JI, Zhang JQ et al (2008) Surface plasmon resonance imaging for real-time, label-free analysis of protein interactions with carbohydrate microarrays. Glycoconj J 25:69–74
36. Calin O, Eller S, Seeberger PH (2013) Automated polysaccharide synthesis: assembly of a 30mer mannoside. Angew Chem Int Ed Engl 52:5862–5865
37. Weishaupt M, Eller S, Seeberger PH (2010) Solid phase synthesis of oligosaccharides. Methods Enzymol 478:463–484
38. Tang SL, Pohl NLB (2016) Automated fluorous-assisted solution-phase synthesis of beta-1,2-, 1,3-, and 1,6-mannan oligomers. Carbohydr Res 430:8–15
39. Sittel I, Galan MC (2015) Chemo-enzymatic synthesis of imidazolium-tagged sialyllactosamine probes. Bioorg Med Chem Lett 25:4329–4332
40. He Y, Yang Y, Iyer SS (2016) Neuraminidase resistant sialosides for the detection of influenza viruses. Bioconjug Chem 27:1509–1517
41. Dinh H, Zhang X, Sweeney J, Yang Y, He Y et al (2014) Glycan based detection and drug susceptibility of influenza virus. Anal Chem 86:8238–8244
42. Watt GM, Revers L, Webberley MC, Wilson IBH, Flitsch SL (1997) Efficient enzymatic synthesis of the core trisaccharide of N-glycans with a recombinant beta-mannosyltransferase. Angew Chem Int Ed 36:2354–2356
43. Watt GM, Lowden PAS, Flitsch SL (1997) Enzyme-catalyzed formation of glycosidic linkages. Curr Opin Struct Biol 7:652–660
44. Faber K, Fessner WD, Turner NJ (2015) Glycosyltransferases. Biocatalysis in organic synthesis 1. Georg Thieme Verlag, Stuttgart
45. Hokke CH, Zervosen A, Elling L, Joziasse DH, van den Eijnden DH (1996) One-pot enzymatic synthesis of the Gal alpha 1-->3Gal beta 1-->4GlcNAc sequence with in situ UDP-Gal regeneration. Glycoconj J 13:687–692
46. Wang Z, Zhou L, El-Boubbou K, Ye XS, Huang X (2007) Multi-component one-pot synthesis of the tumor-associated carbohydrate antigen Globo-H based on preactivation of thioglycosyl donors. J Org Chem 72:6409–6420
47. Voglmeir J, Sardzik R, Weissenborn MJ, Flitsch SL (2010) Enzymatic glycosylations on arrays. OMICS 14:437–444

48. Bojarova P, Rosencrantz RR, Elling L, Kren V (2013) Enzymatic glycosylation of multivalent scaffolds. Chem Soc Rev 42:4774–4797
49. Sardzik R, Green AP, Laurent N, Both P, Fontana C et al (2012) Chemoenzymatic synthesis of O-mannosylpeptides in solution and on solid phase. J Am Chem Soc 134:4521–4524
50. Song X, Heimburg-Molinaro J, Cummings RD, Smith DF (2014) Chemistry of natural glycan microarrays. Curr Opin Chem Biol 18:70–77
51. Song X, Yu H, Chen X, Lasanajak Y, Tappert MM et al (2011) A sialylated glycan microarray reveals novel interactions of modified sialic acids with proteins and viruses. J Biol Chem 286:31610–31622
52. Sardzik R, Sharma R, Kaloo S, Voglmeir J, Crocker PR et al (2011) Chemoenzymatic synthesis of sialooligosaccharides on arrays for studies of cell surface adhesion. Chem Commun (Camb) 47:5425–5427
53. Padler-Karavani V, Song X, Yu H, Hurtado-Ziola N, Huang S et al (2012) Cross-comparison of protein recognition of sialic acid diversity on two novel sialoglycan microarrays. J Biol Chem 287:22593–22608
54. McCombs JE, Diaz JP, Luebke KJ, Kohler JJ (2016) Glycan specificity of neuraminidases determined in microarray format. Carbohydr Res 428:31–40
55. Ban L, Mrksich M (2008) On-chip synthesis and label-free assays of oligosaccharide arrays. Angew Chem Int Ed Engl 47:3396–3399
56. Serna S, Yan S, Martin-Lomas M, Wilson IBH, Reichardt NC (2011) Fucosyltransferases as synthetic tools: glycan array based substrate selection and core fucosylation of synthetic N-glycans. J Am Chem Soc 133:16495–16502
57. Serna S, Etxebarria J, Ruiz N, Martin-Lomas M, Reichardt NC (2010) Construction of N-glycan microarrays by using modular synthesis and on-chip nanoscale enzymatic glycosylation. Chem A Eur J 16:13163–13175
58. Vasiliu D, Razi N, Zhang Y, Jacobsen N, Allin K et al (2006) Large-scale chemoenzymatic synthesis of blood group and tumor-associated poly-N-acetyllactosamine antigens. Carbohydr Res 341:1447–1457
59. Wang Z, Chinoy ZS, Ambre SG, Peng W, McBride R et al (2013) A general strategy for the chemoenzymatic synthesis of asymmetrically branched N-glycans. Science 341:379–383
60. Li L, Liu Y, Ma C, Qu J, Calderon AD et al (2015) Efficient chemoenzymatic synthesis of an N-glycan isomer library. Chem Sci 6:5652–5661
61. Song X, Lasanajak Y, Olson LJ, Boonen M, Dahms NM et al (2009) Glycan microarray analysis of P-type lectins reveals distinct phosphomannose glycan recognition. J Biol Chem 284:35201–35214
62. Lombard V, Golaconda Ramulu H, Drula E, Coutinho PM, Henrissat B (2014) The carbohydrate-active enzymes database (CAZy) in 2013. Nucleic Acids Res 42:D490–D495
63. van Breedam W, Pohlmann S, Favoreel HW, de Groot RJ, Nauwynck HJ (2014) Bitter-sweet symphony: glycan-lectin interactions in virus biology. FEMS Microbiol Rev 38:598–632
64. Zhu X, McBride R, Nycholat CM, Yu W, Paulson JC et al (2012) Influenza virus neuraminidases with reduced enzymatic activity that avidly bind sialic Acid receptors. J Virol 86:13371–13383
65. Stevens J, Blixt O, Glaser L, Taubenberger JK, Palese P et al (2006) Glycan microarray analysis of the hemagglutinins from modern and pandemic influenza viruses reveals different receptor specificities. J Mol Biol 355:1143–1155
66. Bradley KC, Jones CA, Tompkins SM, Tripp RA, Russell RJ et al (2011) Comparison of the receptor binding properties of contemporary swine isolates and early human pandemic H1N1 isolates (Novel 2009 H1N1). Virology 413:169–182
67. de Vries RP, Zhu X, McBride R, Rigter A, Hanson A et al (2014) Hemagglutinin receptor specificity and structural analyses of respiratory droplet-transmissible H5N1 viruses. J Virol 88:768–773

68. Walther T, Karamanska R, Chan RW, Chan MC, Jia N et al (2013) Glycomic analysis of human respiratory tract tissues and correlation with influenza virus infection. PLoS Pathog 9: e1003223
69. Campanero-Rhodes MA, Smith A, Chai W, Sonnino S, Mauri L et al (2007) N-glycolyl GM1 ganglioside as a receptor for simian virus 40. J Virol 81:12846–12858
70. Neu U, Maginnis MS, Palma AS, Stroh LJ, Nelson CD et al (2010) Structure-function analysis of the human JC polyomavirus establishes the LSTc pentasaccharide as a functional receptor motif. Cell Host Microbe 8:309–319
71. Adams EW, Ratner DM, Bokesch HR, McMahon JB, O'Keefe BR et al (2004) Oligosaccharide and glycoprotein microarrays as tools in HIV glycobiology; glycan-dependent gp120/ protein interactions. Chem Biol 11:875–881
72. Hu L, Crawford SE, Czako R, Cortes-Penfield NW, Smith DF et al (2012) Cell attachment protein VP8* of a human rotavirus specifically interacts with A-type histo-blood group antigen. Nature 485:256–259
73. Vinson M, Strijbos PJ, Rowles A, Facci L, Moore SE et al (2001) Myelin-associated glycoprotein interacts with ganglioside GT1b. A mechanism for neurite outgrowth inhibition. J Biol Chem 276:20280–20285
74. Mahal LK (2004) Catching bacteria with sugar. Chem Biol 11:1602–1604
75. Blanchard B, Nurisso A, Hollville E, Tetaud C, Wiels J et al (2008) Structural basis of the preferential binding for globo-series glycosphingolipids displayed by Pseudomonas aeruginosa lectin I. J Mol Biol 383:837–853
76. Gerland B, Goudot A, Pourceau G, Meyer A, Dugas V et al (2012) Synthesis of a library of fucosylated glycoclusters and determination of their binding toward Pseudomonas aeruginosa lectin B (PA-IIL) using a DNA-based carbohydrate microarray. Bioconjug Chem 23:1534–1547
77. Lameignere E, Malinovska L, Slavikova M, Duchaud E, Mitchell EP et al (2008) Structural basis for mannose recognition by a lectin from opportunistic bacteria Burkholderia cenocepacia. Biochem J 411:307–318
78. Geissner A, Seeberger PH (2016) Glycan arrays: from basic biochemical research to bioanalytical and biomedical applications. Annu Rev Anal Chem (Palo Alto Calif) 9:223–247
79. Sulak O, Cioci G, Lameignere E, Balloy V, Round A et al (2011) Burkholderia cenocepacia BC2L-C is a super lectin with dual specificity and proinflammatory activity. PLoS Pathog 7: e1002238
80. Hermans SJ, Baker HM, Sequeira RP, Langley RJ, Baker EN et al (2012) Structural and functional properties of staphylococcal superantigen-like protein 4. Infect Immun 80:4004–4013
81. Levan S, De S, Olson R (2013) Vibrio cholerae cytolysin recognizes the heptasaccharide core of complex N-glycans with nanomolar affinity. J Mol Biol 425:944–957
82. Vainauskas S, Duke RM, McFarland J, McClung C, Ruse C et al (2016) Profiling of core fucosylated N-glycans using a novel bacterial lectin that specifically recognizes alpha1,6 fucosylated chitobiose. Sci Rep 6:34195
83. Stowell SR, Arthur CM, Dias-Baruffi M, Rodrigues LC, Gourdine JP et al (2010) Innate immune lectins kill bacteria expressing blood group antigen. Nat Med 16:295–301
84. Knirel YA, Gabius HJ, Blixt O, Rapoport EM, Khasbiullina NR et al (2014) Human tandem-repeat-type galectins bind bacterial non-betaGal polysaccharides. Glycoconj J 31:7–12
85. Stowell SR, Arthur CM, McBride R, Berger O, Razi N et al (2014) Microbial glycan microarrays define key features of host-microbial interactions. Nat Chem Biol 10:470–476
86. Feinberg H, Taylor ME, Razi N, McBride R, Knirel YA et al (2011) Structural basis for langerin recognition of diverse pathogen and mammalian glycans through a single binding site. J Mol Biol 405:1027–1039
87. Wesener DA, Wangkanont K, McBride R, Song X, Kraft MB et al (2015) Recognition of microbial glycans by human intelectin-1. Nat Struct Mol Biol 22:603–610

88. Zhang Y, Li Q, Rodriguez LG, Gildersleeve JC (2010) An array-based method to identify multivalent inhibitors. J Am Chem Soc 132:9653–9662
89. Narla SN, Nie H, Li Y, Sun XL (2015) Multi-dimensional glycan microarrays with glyco-macroligands. Glycoconj J 32:483–495
90. Branderhorst HM, Ruijtenbeek R, Liskamp RM, Pieters RJ (2008) Multivalent carbohydrate recognition on a glycodendrimer-functionalized flow-through chip. ChemBioChem 9:1836–1844
91. Parera Pera N, Branderhorst HM, Kooij R, Maierhofer C, van der Kaaden M et al (2010) Rapid screening of lectins for multivalency effects with a glycodendrimer microarray. ChemBioChem 11:1896–1904
92. Zhou X, Turchi C, Wang D (2009) Carbohydrate cluster microarrays fabricated on three-dimensional dendrimeric platforms for functional glycomics exploration. J Proteome Res 8:5031–5040
93. Godula K, Rabuka D, Nam KT, Bertozzi CR (2009) Synthesis and microcontact printing of dual end-functionalized mucin-like glycopolymers for microarray applications. Angew Chem Int Ed Engl 48:4973–4976
94. Godula K, Bertozzi CR (2010) Synthesis of glycopolymers for microarray applications via ligation of reducing sugars to a poly(acryloyl hydrazide) scaffold. J Am Chem Soc 132:9963–9965
95. Narla SN, Sun XL (2011) Orientated glyco-macroligand formation based on site-specific immobilization of O-cyanate chain-end functionalized glycopolymer. Org Biomol Chem 9:845–850
96. Zhang Y, Gildersleeve JC (2012) General procedure for the synthesis of neoglycoproteins and immobilization on epoxide-modified glass slides. Methods Mol Biol 808:155–165
97. Kerekgyarto M, Fekete A, Szurmai Z, Kerekgyarto J, Takacs L et al (2013) Neoglycoproteins as carbohydrate antigens: synthesis, analysis, and polyclonal antibody response. Electrophoresis 34:2379–2386
98. Godula K, Bertozzi CR (2012) Density variant glycan microarray for evaluating cross-linking of mucin-like glycoconjugates by lectins. J Am Chem Soc 134:15732–15742
99. Ma Y, Zhang H, Gruzdys V, Sun XL (2011) Azide-reactive liposome for chemoselective and biocompatible liposomal surface functionalization and glyco-liposomal microarray fabrication. Langmuir 27:13097–13103
100. Zhang H, Ma Y, Sun XL (2011) Chemically selective liposome surface glyco-functionalization. Methods Mol Biol 751:269–280
101. Shen L, Wang Y, Lin C-I, Liu HW, Guo A et al (2014) Membrane environment can enhance the interaction of glycan binding protein to cell surface glycan receptors. ACS Chem Biol 9:1877–1884
102. Chevolot Y, Laurenceau E, Phaner-Goutorbe M, Monnier V, Souteyrand E et al (2014) DNA directed immobilization glycocluster array: applications and perspectives. Curr Opin Chem Biol 18:46–54
103. Tong Q, Wang X, Wang H, Kubo T, Yan M (2012) Fabrication of glyconanoparticle microarrays. Anal Chem 84:3049–3052
104. Blixt O, Allin K, Bohorov O, Liu X, Andersson-Sand H et al (2008) Glycan microarrays for screening sialyltransferase specificities. Glycoconj J 25:59–68
105. Park S, Shin I (2007) Carbohydrate microarrays for assaying galactosyltransferase activity. Org Lett 9:1675–1678
106. Frederiksen RF, Yoshimura Y, Storgaard BG, Paspaliari DK, Petersen BO et al (2015) A diverse range of bacterial and eukaryotic chitinases hydrolyzes the LacNAc (Galbeta1-4GlcNAc) and LacdiNAc (GalNAcbeta1-4GlcNAc) motifs found on vertebrate and insect cells. J Biol Chem 290:5354–5366
107. Serna S, Hokke CH, Weissenborn M, Flitsch S, Martin-Lomas M et al (2013) Profiling glycosyltransferase activities by tritium imaging of glycan microarrays. ChemBioChem 14:862–869

108. Vidal-Melgosa S, Pedersen HL, Schuckel J, Arnal G, Dumon C et al (2015) A new versatile microarray-based method for high throughput screening of carbohydrate-active enzymes. J Biol Chem 290:9020–9036
109. Gray CJ, Sanchez-Ruiz A, Sardzikova I, Ahmed YA, Miller RL et al (2017) Label-free discovery array platform for the characterization of glycan binding proteins and glycoproteins. Anal Chem 89:4444–4451
110. Both P, Green AP, Gray CJ, Šardzík R, Voglmeir J et al (2013) Discrimination of epimeric glycans and glycopeptides using IM-MS and its potential for carbohydrate sequencing. Nat Chem 6:65
111. Laurent N, Haddoub R, Voglmeir J, Wong SCC, Gaskell SJ et al (2008) SPOT synthesis of peptide arrays on self-assembled monolayers and their evaluation as enzyme substrates. ChemBioChem 9:2592–2596
112. Beloqui A, Sanchez-Ruiz A, Martin-Lomas M, Reichardt NC (2012) A surface-based mass spectrometry method for screening glycosidase specificity in environmental samples. Chem Commun (Camb) 48:1701–1703
113. Sanchez-Ruiz A, Serna S, Ruiz N, Martin-Lomas M, Reichardt NC (2011) MALDI-TOF mass spectrometric analysis of enzyme activity and lectin trapping on an array of N-glycans. Angew Chem Int Ed Engl 50:1801–1804
114. Zhi ZL, Powell AK, Turnbull JE (2006) Fabrication of carbohydrate microarrays on gold surfaces: direct attachment of nonderivatized oligosaccharides to hydrazide-modified self-assembled monolayers. Anal Chem 78:4786–4793
115. Ban L, Pettit N, Li L, Stuparu AD, Cai L et al (2012) Discovery of glycosyltransferases using carbohydrate arrays and mass spectrometry. Nat Chem Biol 8:769–773
116. Muthana SM, Gildersleeve JC (2014) Glycan microarrays: powerful tools for biomarker discovery. Cancer Biomark 14:29–41
117. Huflejt ME, Vuskovic M, Vasiliu D, Xu H, Obukhova P et al (2009) Anti-carbohydrate antibodies of normal sera: findings, surprises and challenges. Mol Immunol 46:3037–3049
118. Parthasarathy N, Saksena R, Kovac P, Deshazer D, Peacock SJ et al (2008) Application of carbohydrate microarray technology for the detection of Burkholderia pseudomallei, Bacillus anthracis and Francisella tularensis antibodies. Carbohydr Res 343:2783–2788
119. Blixt O, Hoffmann J, Svenson S, Norberg T (2008) Pathogen specific carbohydrate antigen microarrays: a chip for detection of Salmonella O-antigen specific antibodies. Glycoconj J 25:27–36
120. Kamena F, Tamborrini M, Liu X, Kwon YU, Thompson F et al (2008) Synthetic GPI array to study antitoxic malaria response. Nat Chem Biol 4:238–240
121. Aranzamendi C, Tefsen B, Jansen M, Chiumiento L, Bruschi F et al (2011) Glycan microarray profiling of parasite infection sera identifies the LDNF glycan as a potential antigen for serodiagnosis of trichinellosis. Exp Parasitol 129:221–226
122. Heimburg-Molinaro J, Priest JW, Live D, Boons GJ, Song X et al (2013) Microarray analysis of the human antibody response to synthetic Cryptosporidium glycopeptides. Int J Parasitol 43:901–907
123. Mickum ML, Prasanphanich NS, Song X, Dorabawila N, Mandalasi M et al (2016) Identification of antigenic glycans from Schistosoma mansoni by using a shotgun egg glycan microarray. Infect Immun 84:1371–1386
124. Clo E, Kracun SK, Nudelman AS, Jensen KJ, Liljeqvist JA et al (2012) Characterization of the viral O-glycopeptidome: a novel tool of relevance for vaccine design and serodiagnosis. J Virol 86:6268–6278
125. Walker LM, Huber M, Doores KJ, Falkowska E, Pejchal R et al (2011) Broad neutralization coverage of HIV by multiple highly potent antibodies. Nature 477:466–470
126. Julien JP, Sok D, Khayat R, Lee JH, Doores KJ et al (2013) Broadly neutralizing antibody PGT121 allosterically modulates CD4 binding via recognition of the HIV-1 gp120 V3 base and multiple surrounding glycans. PLoS Pathog 9:e1003342

127. Mouquet H, Scharf L, Euler Z, Liu Y, Eden C et al (2012) Complex-type N-glycan recognition by potent broadly neutralizing HIV antibodies. Proc Natl Acad Sci U S A 109:E3268–E3277
128. Cazet A, Julien S, Bobowski M, Burchell J, Delannoy P (2010) Tumour-associated carbohydrate antigens in breast cancer. Breast Cancer Res 12:204
129. Wang CC, Huang YL, Ren CT, Lin CW, Hung JT et al (2008) Glycan microarray of Globo H and related structures for quantitative analysis of breast cancer. Proc Natl Acad Sci U S A 105:11661–11666
130. Wandall HH, Blixt O, Tarp MA, Pedersen JW, Bennett EP et al (2010) Cancer biomarkers defined by autoantibody signatures to aberrant O-glycopeptide epitopes. Cancer Res 70:1306–1313
131. Pedersen JW, Blixt O, Bennett EP, Tarp MA, Dar I et al (2011) Seromic profiling of colorectal cancer patients with novel glycopeptide microarray. Int J Cancer 128:1860–1871
132. Burford B, Gentry-Maharaj A, Graham R, Allen D, Pedersen JW et al (2013) Autoantibodies to MUC1 glycopeptides cannot be used as a screening assay for early detection of breast, ovarian, lung or pancreatic cancer. Br J Cancer 108:2045–2055
133. Jacob F, Goldstein DR, Bovin NV, Pochechueva T, Spengler M et al (2012) Serum antiglycan antibody detection of nonmucinous ovarian cancers by using a printed glycan array. Int J Cancer 130:138–146
134. Vuskovic MI, Xu H, Bovin NV, Pass HI, Huflejt ME (2011) Processing and analysis of serum antibody binding signals from Printed Glycan Arrays for diagnostic and prognostic applications. Int J Bioinform Res Appl 7:402–426
135. Lawrie CH, Marafioti T, Hatton CS, Dirnhofer S, Roncador G et al (2006) Cancer-associated carbohydrate identification in Hodgkin's lymphoma by carbohydrate array profiling. Int J Cancer 118:3161–3166
136. Kearney JF, Patel P, Stefanov EK, King RG (2015) Natural antibody repertoires: development and functional role in inhibiting allergic airway disease. Annu Rev Immunol 33:475–504
137. Dotan I, Fishman S, Dgani Y, Schwartz M, Karban A et al (2006) Antibodies against laminaribioside and chitobioside are novel serologic markers in Crohn's disease. Gastroenterology 131:366–378
138. Grader-Beck T, Boin F, von Gunten S, Smith D, Rosen A et al (2011) Antibodies recognising sulfated carbohydrates are prevalent in systemic sclerosis and associated with pulmonary vascular disease. Ann Rheum Dis 70:2218–2224

# Erratum to: Animal Cell Expression Systems

M. Butler and U. Reichl

## Erratum to:
## Chapter "Animal Cell Expression Systems" in:
## M. Butler and Ing. U. Reichl, Adv Biochem Eng Biotechnol,
## DOI: 10.1007/10_2017_31

The "Dr.-Ing" is deleted from the name Prof. Dr.-Ing. U. Reichl in the above mentioned online published chapter.

The updated online version of this chapter can be found at
DOI: 10.1007/10_2017_31

# Correction to: Animal Cell Expression Systems

M. Butler and U. Reichl

## Correction to: Chapter "Animal Cell Expression Systems" in: M. Butler and U. Reichl, Adv Biochem Eng Biotechnol, DOI: 10.1007/10_2017_31

On page 3, as per the author's request the current list of abbreviations is replaced with complete list of abbreviations.

## Abbreviations

| | |
|---|---|
| ADCC | Antibody-dependent cellular cytotoxicity |
| AGE1.CR | Muscovy duck cell line |
| AIDS | Acquired immune deficiency syndrome |
| B-cell | White blood cell of the lymphocyte subtype |
| CAP cells | Primary human amniocyte-derived cell line |
| Cas | Castanospermine |
| $CD4^{+}T$ | T helper cell type |
| CGE | Capillary gel electrophoresis |
| cGMP | Current good manufacturing practice |
| CHO cells | Chinese hamster ovary cell line |
| CMP | Cytidine monophosphate |
| CQA | Crucial quality attribute |
| DO | Dissolved oxygen |
| DOE | Design-of-experiment |

The updated online version of this chapter can be found at
DOI 10.1007/10_2017_31

| | |
|---|---|
| EB66 cells | Cell line derived from duck embryonic stem cells |
| EMA | European Medicines Agency |
| ER | Endoplasmic reticulum |
| F243A | Replacement of phenylalanine 243 by alanine |
| Fc | Fragment crystallizable region |
| FDA | Food and Drug Administration |
| flublok® | Recombinant insect cell-derived influenza vaccine |
| G2 | Digalactosylated oligosaccharide |
| Gal | Galactose |
| GalT | Galactosyl transferase |
| GDP | Guanosine diphosphate |
| GI | Galactosylation index |
| Glc | Glucose |
| GlcNAc | *N*-acetylglucosamine |
| Gln | Glutamine |
| GlycoVis | Software to describe glycosylation pathways |
| GT | Glycotransferase |
| HA | Hemagglutinin |
| HA0 | Uncleaved form of the HA protein |
| HA1 | Hemagglutinin subunit |
| HA2 | Hemagglutinin subunit |
| HEK293 | Human embryonic kidney cells |
| HexNAc | *N*-acetyl hexosamine |
| HILIC | Hydrophilic interaction liquid chromatography |
| HIV | Human immunodeficiency virus |
| IgG | Immunoglobulin G |
| IVA | Influenza A virus |
| Kif | Kifunensine |
| LIF | Laser-induced fluorescence |
| Mab | Monoclonal antibody |
| MALDI | Matrix assisted laser desorption ionization |
| Man | Mannose |
| ManNAc | *N*-acetylmannosamine |
| MRC-5 cells | Diploid human fibroblast cells derived from lung tissue |
| MDCK cells | Madin Darby canine kidney cell line |
| MMR | Measles-mumps-rubella |
| MS | Mass spectrometry |
| MTU′ | Migration time units |
| NA | Neuraminidase |
| NANA | *N*-acetylneuraminic acid |
| NGNA | *N*-glycolylneuraminic acid |
| NS0 | Non-immunoglobulin-secreting murine myeloma cells |
| PAT | Process analytical technology |
| PER.C6 | Human cell line |
| PNGase F | Peptide-*N*-glycosidase-F |

| | |
|---|---|
| PR8/34 | Puerto Rico/8/34 |
| QbD | Quality by design |
| Qp | High specific productivity |
| RFU | Relative fluorescence units |
| RPH | Relative peak height |
| Sabin | Live attenuated oral polio vaccine |
| SARS | Severe acute respiratory syndrome |
| Salk | Inactivated polio vaccine |
| SF$^{+}$ | Serum-free insect cell line |
| STR | Stirred tank reactor |
| Swa | Swainsonine |
| TOF | Time of flight |
| TPH | Total peak height |
| UDP | Uridine diphosphate |
| UMG | Uridine, manganese and galactose |
| Vero cells | African green monkey kidney cell line |
| WHO | World Health Organization |

# Correction to: Enzymatic Synthesis of Glycans and Glycoconjugates

**Thomas Rexer, Dominic Laaf, Johannes Gottschalk, Hannes Frohnmeyer, Erdmann Rapp, and Lothar Elling**

## Correction to: Chapter "Enzymatic Synthesis of Glycans and Glycoconjugates" in: T. Rexer et al., Adv Biochem Eng Biotechnol, https://doi.org/10.1007/10_2020_148

The affiliation of the author Prof. Dr. Erdmann Rapp has been updated in this chapter to:

Max Planck Institute for Dynamics of Complex Technical Systems, Magdeburg, Germany

glyXera GmbH, Magdeburg, Germany

rapp@mpi-magdeburg.mpg.de; e.rapp@glyxera.com

The updated online version of this chapter can be found at
https://doi.org/10.1007/10_2020_148